AF868632

Glycoprotein Methods and Protocols

The Mucins

METHODS IN MOLECULAR BIOLOGY™

John M. Walker, SERIES EDITOR

156. **Antigen Processing and Presentation Protocols,** edited by *Joyce C. Solheim, 2000*
155. **Adipose Tissue Protocols,** edited by *Gérard Ailhaud, 2000*
154. **Connexin Methods and Protocols,** edited by *Roberto Bruzzone and Christian Giaume, 2000*
153. **Neuropeptide Y Protocols**, edited by *Ambikaipakan Balasubramaniam, 2000*
152. **DNA Repair Protocols:** *Prokaryotic Systems*, edited by *Patrick Vaughan, 2000*
151. **Matrix Metalloproteinase Protocols,** edited by *Ian M. Clark, 2000*
150. **Complement Methods and Protocols,** edited by *B. Paul Morgan, 2000*
149. **The ELISA Guidebook**, edited by *John R. Crowther, 2000*
148. **DNA–Protein Interactions:** *Principles and Protocols* **(2nd ed.)**, edited by *Tom Moss, 2000*
147. **Affinity Chromatography:** *Methods and Protocols*, edited by *Pascal Bailon, George K. Ehrlich, Wen-Jian Fung, and Wolfgang Berthold, 2000*
146. **Protein and Peptide Analysis:** *New Mass Spectrometric Applications*, edited by *John R. Chapman, 2000*
145. **Bacterial Toxins:** ***Methods and Protocols***, edited by *Otto Holst, 2000*
144. **Calpain Methods and Protocols**, edited by *John S. Elce, 2000*
143. **Protein Structure Prediction:** *Methods and Protocols*, edited by *David Webster, 2000*
142. **Transforming Growth Factor-Beta Protocols**, edited by *Philip H. Howe, 2000*
141. **Plant Hormone Protocols**, edited by *Jeremy A. Roberts and Gregory A. Tucker, 2000*
140. **Chaperonin Protocols**, edited by *Christine Schneider, 2000*
139. **Extracellular Matrix Protocols**, edited by *Charles Streuli and Michael Grant, 2000*
138. **Chemokine Protocols**, edited by *Amanda E. I. Proudfoot, Timothy N. C. Wells, and Christine Power, 2000*
137. **Developmental Biology Protocols, Volume III**, edited by *Rocky S. Tuan and Cecilia W. Lo, 2000*
136. **Developmental Biology Protocols, Volume II,** edited by *Rocky S. Tuan and Cecilia W. Lo, 2000*
135. **Developmental Biology Protocols, Volume I,** edited by *Rocky S. Tuan and Cecilia W. Lo, 2000*
134. **T Cell Protocols:** *Development and Activation*, edited by *Kelly P. Kearse, 2000*
133. **Gene Targeting Protocols,** edited by *Eric B. Kmiec, 2000*
132. **Bioinformatics Methods and Protocols,** edited by *Stephen Misener and Stephen A. Krawetz, 2000*
131. **Flavoprotein Protocols,** edited by *S. K. Chapman and G. A. Reid, 1999*
130. **Transcription Factor Protocols,** edited by *Martin J. Tymms, 2000*
129. **Integrin Protocols,** edited by *Anthony Howlett, 1999*
128. **NMDA Protocols,** edited by *Min Li, 1999*
127. **Molecular Methods in Developmental Biology:** Xenopus *and Zebrafish*, edited by *Matthew Guille, 1999*
126. **Adrenergic Receptor Protocols,** edited by *Curtis A. Machida, 2000*
125. **Glycoprotein Methods and Protocols:** *The Mucins*, edited by *Anthony P. Corfield, 2000*
124. **Protein Kinase Protocols,** edited by *Alastair D. Reith, 2000*
123. ***In Situ*** **Hybridization Protocols (2nd ed.),** edited by *Ian A. Darby, 2000*
122. **Confocal Microscopy Methods and Protocols,** edited by *Stephen W. Paddock, 1999*
121. **Natural Killer Cell Protocols:** *Cellular and Molecular Methods*, edited by *Kerry S. Campbell and Marco Colonna, 2000*
120. **Eicosanoid Protocols,** edited by *Elias A. Lianos, 1999*
119. **Chromatin Protocols,** edited by *Peter B. Becker, 1999*
118. **RNA–Protein Interaction Protocols,** edited by *Susan R. Haynes, 1999*
117. **Electron Microscopy Methods and Protocols,** edited by *M. A. Nasser Hajibagheri, 1999*
116. **Protein Lipidation Protocols,** edited by *Michael H. Gelb, 1999*
115. **Immunocytochemical Methods and Protocols (2nd ed.),** edited by *Lorette C. Javois, 1999*
114. **Calcium Signaling Protocols,** edited by *David G. Lambert, 1999*
113. **DNA Repair Protocols:** *Eukaryotic Systems*, edited by *Daryl S. Henderson, 1999*
112. **2-D Proteome Analysis Protocols,** edited by *Andrew J. Link, 1999*
111. **Plant Cell Culture Protocols,** edited by *Robert D. Hall, 1999*
110. **Lipoprotein Protocols,** edited by *Jose M. Ordovas, 1998*
109. **Lipase and Phospholipase Protocols,** edited by *Mark H. Doolittle and Karen Reue, 1999*
108. **Free Radical and Antioxidant Protocols,** edited by *Donald Armstrong, 1998*
107. **Cytochrome P450 Protocols,** edited by *Ian R. Phillips and Elizabeth A. Shephard, 1998*
106. **Receptor Binding Techniques,** edited by *Mary Keen, 1999*
105. **Phospholipid Signaling Protocols,** edited by *Ian M. Bird, 1998*
104. **Mycoplasma Protocols,** edited by *Roger J. Miles and Robin A. J. Nicholas, 1998*
103. ***Pichia*** **Protocols,** edited by *David R. Higgins and James M. Cregg, 1998*
102. **Bioluminescence Methods and Protocols,** edited by *Robert A. LaRossa, 1998*
101. **Mycobacteria Protocols,** edited by *Tanya Parish and Neil G. Stoker, 1998*
100. **Nitric Oxide Protocols,** edited by *Michael A. Titheradge, 1998*
99. **Stress Response:** *Methods and Protocols*, edited by *Stephen M. Keyse, 2000*
98. **Forensic DNA Profiling Protocols**, edited by *Patrick J. Lincoln and James M. Thomson, 1998*
97. **Molecular Embryology:** *Methods and Protocols*, edited by *Paul T. Sharpe and Ivor Mason, 1999*
96. **Adhesion Protein Protocols**, edited by *Elisabetta Dejana and Monica Corada, 1999*
95. **DNA Topoisomerases Protocols:** *II. Enzymology and Drugs*, edited by *Mary-Ann Bjornsti and Neil Osheroff, 2000*
94. **DNA Topoisomerases Protocols:** *I. DNA Topology and Enzymes*, edited by *Mary-Ann Bjornsti and Neil Osheroff, 1999*
93. **Protein Phosphatase Protocols**, edited by *John W. Ludlow, 1998*
92. **PCR in Bioanalysis**, edited by *Stephen J. Meltzer, 1998*

METHODS IN MOLECULAR BIOLOGY™

Glycoprotein Methods and Protocols

The Mucins

Edited by

Anthony P. Corfield

Bristol Royal Infirmary, Bristol, UK

Humana Press Totowa, New Jersey

© 2000 Humana Press Inc.
999 Riverview Drive, Suite 208
Totowa, New Jersey 07512

All rights reserved. No part of this book may be reproduced, stored in a retrieval system, or transmitted in any form or by any means, electronic, mechanical, photocopying, microfilming, recording, or otherwise without written permission from the Publisher. Methods in Molecular Biology™ is a trademark of The Humana Press Inc.

All authored papers, comments, opinions, conclusions, or recommendations are those of the author(s), and do not necessarily reflect the views of the publisher.

This publication is printed on acid-free paper. ∞
ANSI Z39.48-1984 (American Standards Institute) Permanence of Paper for Printed Library Materials.

Cover design by Patricia F. Cleary.

For additional copies, pricing for bulk purchases, and/or information about other Humana titles, contact Humana at the above address or at any of the following numbers: Tel: 973-256-1699; Fax: 973-256-8341; E-mail: humana@humanapr.com, or visit our Website at www.humanapress.com

Photocopy Authorization Policy:
Authorization to photocopy items for internal or personal use, or the internal or personal use of specific clients, is granted by Humana Press Inc., provided that the base fee of US $10.00 per copy, plus US $00.25 per page, is paid directly to the Copyright Clearance Center at 222 Rosewood Drive, Danvers, MA 01923. For those organizations that have been granted a photocopy license from the CCC, a separate system of payment has been arranged and is acceptable to Humana Press Inc. The fee code for users of the Transactional Reporting Service is: [0-89603-720-7/00 $10.00 + $00.25].

Printed in the United States of America. 10 9 8 7 6 5 4 3 2 1
Libarary of Congress Cataloging-in-Publication Data

Glycoprotein methods and protocols : the mucins / edited by Anthony P. Corfield.
p. cm. -- (Methods in molecular biology ; v. 125)
Includes bibliographical references and index.
ISBN 0-89603-720-7 (alk. paper)
1. Mucins--Laboratory manuals. I. Corfield, Anthony P.
II. Series: Methods in molecular biology (Totowa, N.J.) ; v. 125.
QP552.M8G59 2000
572'.68--dc21 99-25785
CIP

Preface

The mucins (mucus glycoproteins) have long been a complex corner of glycoprotein biology. While dramatic advances in the separation, structural analysis, biosynthesis, and degradation have marked the progress in general glycoprotein understanding, the mucins have lagged behind. The reasons for this lack of progress have always been clear and are only now being resolved. The mucins are very large molecules; they are difficult to separate from other molecules present in mucosal secretions or membranes; they are often degraded owing to natural protective functions or to isolation methodology and their peptide and oligosaccharide structures are varied and complex. Understanding these molecules has demanded progress in several major areas. Isolation techniques that protect the intact mucins and allow dissociation from other adsorbed but discrete molecules needed to be developed and accepted by all researchers in the field. Improved methods for the study of very large molecules with regard to their aggregation and polymerization were also needed. Structural analysis of the peptide domains and the multitude of oligosaccharide chains was required for smaller sample sizes, for multiple samples, and in shorter time. In view of these problems it is perhaps not surprising that the mucins have remained a dilemma, of obvious biological importance and interest, but very difficult to analyze.

The driving force behind the production of *Glycoprotein Methods and Protocols: The Mucins* has been the accumulation of novel advances in the ability to analyze mucins reliably and the impact of molecular biology and immunology on the general awareness of mucins as important molecules. This volume is overdue as there is no comprehensive compendium of methods for mucin analysis. It is vital to gather together protocols from those groups who have sorted out the fundamental methods in order that others wanting to use these advances have a reference to follow. In this way *Glycoprotein Methods and Protocols: The Mucins* will make a major contribution in eliminating variation between individual labs and enable the mucin field as a whole to make genuine comparative studies. The range of analytical techniques presented here represents the culmination of the recent advances in the mucin field alluded to above. In several cases this is the result of many years' continuous struggle and it is very satisfying to bring together these new methods in one volume.

The initial problems of mucin analysis were related directly to their purification from secretions and tissues. These methods have been refined to include extraction in denaturing solvents, protection with antiproteolytic agents, and combinations of repeated density gradient centrifugation, gel filtration, ion-exchange chromatography, and electrophoresis, especially in agarose gels (Chaps. 1, 2, 7, and 8). Parallel to these developments have been the efforts to detect and quantify mucins in tissues and in extracts during purification (Chaps. 3–6, 29, and 30); this is still a growing area.

Much of the current knowledge of mucin polypeptide structure has been derived from direct peptide analysis and sequencing (Chaps. 10–13). Confirmation of much of these data and considerably more information with regard to molecular organization and tissue-specific expression patterns has been derived from the molecular biological description of mucin genes (Chaps. 24–28). This has led to the identification of mucin domains, variable number tandem repeat sequences, and new proposals for the way in which mucins are assembled and for their tissue-specific function.

In keeping with the high proportion of carbohydrate typically present in mucins, the latest sensitive methods for the total monosaccharide composition and sequence determination of oligosaccharides is covered (Chaps. 14–16). This is often a large undertaking since the number of individual oligosaccharide chains in a purified mucin is often high (i.e., at least 20–50 structures). Further modifications of the oligosaccharide chains are also common, especially sulfation (Chap. 17), and these additions present their own analytical problems.

The biosynthesis of mucins has been studied in a variety of tissue and cell culture systems. The new developments in separation and mucin gene structure have focused the direction of this work on the design of new specific reagents (Chaps. 18–21). In addition, the glycosylation and sulfation reactions and their inhibition have opened new concepts in the approach to mucin carbohydrate biology (Chaps. 22 and 23).

Study of the degradation of mucins has been hampered by the limited availability of suitable mucin-related substrates. This is still an area of development, one that has benefited from the new information appearing on the detailed structure and organization of the mucins. The concept of a whole "mucinase" activity is also addressed in this volume (Chap. 31) and is backed up by a more detailed consideration of the known members of the total mucin degrading activity (Chaps. 32–34).

One of the most exciting and novel aspects of mucin biology to appear in the last few years has been the interaction of mucins with organisms. This volume would not be complete without these novel concepts concerning bacterial interaction in biofilms (Chap. 36) and the general interactions of bacteria with mucins (Chap. 35).

The cellular and humoral responses to mucins (largely MUC1) has proved to be a major item of interest in cancer biology. As a result it is appropriate that representation of this methodology is also part of the volume (Chaps. 37–41).

The compilation of this practical handbook has been made easier by the trouble taken by the authors to fit their protocols to the format. This volume represents a start in the collection of a reliable and comprehensive collection of methods for the mucin researcher.

Anthony P. Corfield

Contents

Preface .... *v*
Contributors .... *xiii*

Part I Purification of Mucin

1 Isolation of Large Gel-Forming Mucins *3*
Julia R. Davies and Ingemar Carlstedt

2 Preparation of Membrane Mucin *15*
Kermit L. Carraway

Part II Detection and Quantitation of Mucin

3 Histologically Based Methods for Detection of Mucin *29*
Michael D. Walsh and Jeremy R. Jass

4 Detection and Quantitation of Mucins Using Chemical, Lectin, and Antibody Methods *45*
Michael A. McGuckin and David J. Thornton

5 The Gastrointestinal Adherent Mucous Gel Barrier *57*
Adrian Allen and Jeffrey P. Pearson

6 Quantitation of Biosynthesis and Secretion of Mucin Using Metabolic Labeling *65*
Jan Dekker, B. Jan-Willem Van Klinken, Hans A. Büller, and Alexandra W. C. Einerhand

Part III Separation, Identification, and Physical Characterization of Mucins

7 Separation and Identification of Mucins and Their Glycoforms *77*
David J. Thornton, Nagma Khan, and John K. Sheehan

8 Heterogeneity and Size Distribution of Gel-Forming Mucins *87*
John K. Sheehan and David J. Thornton

Part IV Rheology of Mucin

9 Rheology of Mucin *99*
Jeffrey P. Pearson, Adrian Allen, and David A. Hutton

Part V Mucin Peptide Analysis

10 Amino Acid Analysis of Mucins *113*
Jun X. Yan and Nicolle H. Packer

11 Identification of Glycosylation Sites in Mucin Peptides by Edman Degradation *121*
Natasha E. Zachara and Andrew A. Gooley

12 Synthetic Peptides for the Analysis and Preparation of Antimucin Antibodies *129*
Andrea Murray, Deirdre A. O'Sullivan, and Michael R. Price

13 Mucin Domains to Explore Disulfide-Dependent Dimer Formation *143*
Sherilyn L. Bell and Janet F. Forstner

Part VI Carbohydrate Structural Analysis of Mucins

14 Monosaccharide Composition of Mucins *159*
Jean-Claude Michalski and Calliope Capon

15 *O*-Linked Oligosaccharide Chain Release and Fractionation *181*
Elizabeth F. Hounsell

16 Structural Analysis of Mucin-Type *O*-Linked Oligosaccharides *191*
André Klein, Gérard Strecker, Geneviève Lamblin, and Philippe Roussel

17 Measurement of Sulfate in Mucins *211*
Mathew J. Harrison and Nicolle H. Packer

Part VII Mucin Biosynthesis

18 Biosynthesis of Mucin Cell and Organ Culture Methods for Biosynthetic Study *219*
Anthony P. Corfield, Neil Myerscough, Alexandra W. C. Einerhand, B. Jan-Willem Van Klinken, Jan Dekker, and Christos Paraskeva

19 Metabolic Labeling Methods for the Preparation and Biosynthetic Study of Mucin *227*
Anthony P. Corfield, Neil Myerscough, B. Jan-Willem Van Klinken, Alexandra W. C. Einerhand, and Jan Dekker

20 Identification of Mucins Using Metabolic Labeling, Immunoprecipitation, and Gel Electrophoresis *239*
B. Jan-Willem Van Klinken, Hans A. Büller, Alexandra W. C. Einerhand, and Jan Dekker

21 Mucin Precursors: *Identification and Analysis of Their Intracellular Processing* *249*
Alexandra W. C. Einerhand, B. Jan-Willem Van Klinken, Hans A. Büller, and Jan Dekker

22 Inhibition of Mucin Glycosylation *261*
Guillemette Huet, Philippe Delannoy, Thécla Lesuffleur, Sylviane Hennebicq, and Pierre Degand

23 *O*-Linked Chain Glycosyltransferases *273*
Inka Brockhausen

PART VIII MUCIN GENE DETECTION

24 Mucin cDNA Cloning *297*
Jean-Pierre Aubert and Nicole Porchet

25 Northern Blot Analysis of Large mRNAs *305*
Nicole Porchet and Jean-Pierre Aubert

26 Southern Blot Analysis of Large DNA Fragments *313*
Nicole Porchet and Jean-Pierre Aubert

27 *In Situ* Hybridization Techniques for Localizing Mucin mRNA *323*
Ilene K. Gipson

28 Detection of Mucin Gene Polymorphism *337*
Lynne E. Vinall, Wendy S. Pratt, and Dallas M. Swallow

PART IX PREPARATION OF ANTIMUCIN ANTIBODIES

29 Polyclonal and Monoclonal Techniques *353*
Francisco X. Real, Carme de Bolós, and Egbert Oosterwijk

30 Monoclonal Antibodies to Mucin VNTR Peptides *369*
Pei Xiang Xing, Vasso Apostolopoulos, Jim Karkaloutsos, and Ian F. C. McKenzie

PART X ENZYMATIC DEGRADATION OF MUCINS

31 Mucinase Activity *385*
Roger M. Stark, Rebecca Wiggins, Elizabeth Walley, Sally J. Hicks, Gulnaz A. Gill, Stephen D. Carrington, and Anthony P. Corfield

32 Proteinase Activity *393*
David A. Hutton, Adrian Allen, and Jeffrey P. Pearson

33 Glycosidase Activity *403*
Anthony P. Corfield and Neil Myerscough

34 Assays for Bacterial Mucin-Desulfating Sulfatases *417*
Anthony M. Roberton, Douglas I. Rosendale, and Damian P. Wright

PART XI MUCIN-BACTERIAL INTERACTIONS

35 Mucin-Bacterial Binding Assays 429
Nancy A. McNamara, Robert A. Sack, and Suzanne M. J. Fleiszig

36 Growth of Mucin Degrading Bacteria in Biofilms 439
George T. Macfarlane and Sandra Macfarlane

PART XII CELLULAR AND HUMORAL RESPONSES TO MUCINS

37 Generation of MUC1 Cytotoxic T-Cells in Mice and Epitope Mapping 455
Vasso Apostolopoulos, Ian F. C. McKenzie, and Geoffrey A. Pietersz

38 Analysis of the Frequency of MHC-Unrestricted MUC1-Specific Cytotoxic T-Cells in Peripheral Blood by Limiting Dilution Assay 463
John R. McKolanis and Olivera J. Finn

39 Expression of MUC1 in Insect Cells Using Recombinant Baculovirus 471
Pawel Ciborowski and Olivera J. Finn

40 Mucin Antigen Presentation Using Dendritic Cells 487
Joy Burchell, Rosalind Graham, and Joyce Taylor-Papadimitriou

41 Detection of Humoral Immune Responses to Mucins 495
Silvia von Mensdorff-Pouilly, Claus Vennegoor, and Joseph Hilgers

Index 501

Contributors

ADRIAN ALLEN • *Department of Physiological Sciences, Medical School, University of Newcastle Upon Tyne, Newcastle Upon Tyne, UK*
VASSO APOSTOLOPOULOS • *Austin Research Institute, Austin and Repatriation Medical Centre, Heidelberg, Victoria, Australia*
JEAN-PIERRE AUBERT • *Biologie et Physiopathologie des Cellules Mucipares, Laboratoire de Recherche Gérard Biserte, Institut National de la Santé et de la Recherche Médicale, Lille, France*
SHERILYN L. BELL • *Division of Structural Biology and Biochemistry, Research Institute, The Hospital for Sick Children; Department of Biochemistry, University of Toronto, Toronto, Ontario, Canada*
CARME DE BOLÓS • *Institut Municipal Investigació Mèdica, Barcelona, Spain*
INKA BROCKHAUSEN • *Division of Rheumatology, Department of Medicine, Queen's University, Kingston, Ontario, Canada*
HANS A. BÜLLER • *Pediatric Gastroenterology and Nutrition, Laboratory Pediatrics, dr Molewaterplein, Rotterdam, The Netherlands*
JOY BURCHELL • *Breast Cancer Biology, Imperial Cancer Research Fund, London, UK*
CALLIOPE CAPON • *Laboratoire de Chimie Biologique, Université des Sciences et Technologies de Lille, Lille, France*
INGEMAR CARLSTEDT • *Section for Molecular Pathogenesis, Department of Cell and Molecular Biology; Mucosal Biology Group, Lund University, Lund, Sweden*
KERMIT L. CARRAWAY • *Department of Cell Biology and Anatomy, University of Miami School of Medicine, Miami, FL*
STEPHEN D. CARRINGTON • *Veterinary Anatomy, Veterinary College, University College, Dublin, Ireland*
PAWEL CIBOROWSKI • *School of Medicine, University of Pittsburgh, Pittsburgh, PA*
ANTHONY P. CORFIELD • *Mucin Research Group, Division of Medicine, Bristol Royal Infirmary, Bristol, UK*
JULIA R. DAVIES • *Section for Molecular Pathogenesis, Department of Cell and Molecular Biology; Mucosal Biology Group, Lund University, Lund, Sweden*

PIERRE DEGAND • *Biologie et Physiopathologie des Cellules Mucipares, Laboratoire de Recherche Gérard Biserte, Institut National de la Santé et de la Recherche Médicale, Lille, France*

JAN DEKKER • *Pediatric Gastroenterology and Nutrition, Laboratory Pediatrics, Rotterdam, The Netherlands*

PHILIPPE DELANNOY • *Laboratoire de Chimie Biologique, Université de Sciences et Technologies de Lille, Lille, France*

ALEXANDRA W. C. EINERHAND • *Pediatric Gastroenterology and Nutrition, Laboratory Pediatrics, dr Molewaterplein, Rotterdam, The Netherlands*

OLIVERA J. FINN • *School of Medicine, University of Pittsburgh, Pittsburgh, PA*

SUZANNE M. J. FLEISZIG • *Morton D. Sarver Laboratory for Cornea and Contact Lens Research, School of Optometry, University of California, Berkeley, CA*

JANET F. FORSTNER • *Division of Structural Biology and Biochemistry, Research Institute, The Hospital for Sick Children; Department of Biochemistry, University of Toronto, Toronto, Ontario, Canada*

GULNAZ A. GILL • *Mucin Research Group, Division of Medicine, Bristol Royal Infirmary, Bristol, UK*

ILENE K. GIPSON • *Schepens Eye Research Institute; Department of Ophthalmology, Harvard Medical School, Boston, MA*

ANDREW A. GOOLEY • *Proteome Systems Ltd., Sydney, New South Wales, Australia*

ROSALIND GRAHAM • *Breast Cancer Biology, Imperial Cancer Research Fund, London, UK*

MATHEW J. HARRISON • *Macquarie University Centre for Analytical Biotechnology, School of Biological Sciences, Macquarie University, Sydney, Australia*

SYLVIANE HENNEBICQ • *Biologie et Physiopathologie des Cellules Mucipares, Laboratoire de Recherche Gérard Biserte, Institut National de la Santé et de la Recherche Médicale, Lille, France*

SALLY J. HICKS • *Department of Anatomy, School of Veterinary Science, Bristol, UK*

JOSEPH HILGERS • *Department of Obstetrics and Gynaecology, Academic Hospital Vrije Universiteit, Amsterdam, The Netherlands*

ELIZABETH F. HOUNSELL • *Department of Biochemistry and Molecular Biology, University College London, London, UK*

GUILLEMETTE HUET • *Biologie et Physiopathologie des Cellules Mucipares, Laboratoire de Recherche Gérard Biserte, Institut National de la Santé et de la Recherche Médicale, Lille, France*

David A. Hutton • *Bioprocessing Ltd., Consett, County Durham, UK*
Jeremy R. Jass • *Department of Pathology, The University of Queensland Medical School, Herston, Brisbane, Queensland, Australia*
Nagma Khan • *Wellcome Trust Centre for Cell Matrix Research, Division of Biochemistry, University of Manchester, Manchester, UK*
Jim Karkaloutsos • *Austin Research Institute, Austin and Repatriation Medical Centre, Heidelberg, Victoria, Australia*
André Klein • *Laboratoire de Chimie Biologique et Unité Mixte de Recherche du CNRS 111, Université des Sciences et Technologies de Lille, Lille, France*
Geneviève Lamblin • *Laboratoire de Chimie Biologique et Unité Mixte de Recherche du CNRS 111, Université des Sciences et Technologies de Lille, Lille, France*
Thécla Lesuffleur • *Unité INSERM 505, Paris, France*
George T. Macfarlane • *MRC Microbiology and Gut Biology Group, Dundee University, Dundee, UK*
Sandra Macfarlane • *MRC Microbiology and Gut Biology Group, Dundee University, Dundee, UK*
Michael A. McGuckin • *Mater Medical Research Institute, Mater Misericordiae Hospitals, South Brisbane, Queensland, Australia*
Ian F. C. McKenzie • *Austin Research Institute, Austin and Repatriation Medical Centre, Heidelberg, Victoria, Australia*
John R. McKolanis • *School of Medicine, University of Pittsburgh, Pittsburgh, PA*
Nancy A. McNamara • *Morton D. Sarver Laboratory for Cornea and Contact Lens Research, School of Optometry, University of California, Berkeley, CA*
Silvia von Mensdorff-Pouilly • *Department of Obstetrics and Gynaecology, Academic Hospital Vrije Universiteit, Amsterdam, The Netherlands*
Jean-Claude Michalski • *Laboratoire de Chimie Biologique, UMR CNRS 8576, Université des Sciences et Technologies de Lille, Lille, France*
Andrea Murray • *Cancer Research Laboratories, School of Pharmaceutical Sciences, University of Nottingham, Nottingham, UK*
Neil Myerscough • *Mucin Research Group, Division of Medicine, Bristol Royal Infirmary, Bristol, UK*
Egbert Oosterwijk • *Department of Urology, Urology Research Laboratory, University Hospital of Nijmegen, Nijmegen, The Netherlands*
Deirdre A. O'Sullivan • *Biochemical Recovery Group, School of Chemical Engineering, University of Birmingham, Birmingham, UK*

NICOLLE H. PACKER • *Proteome Systems Ltd., Sydney, New South Wales, Australia*
JOYCE TAYLOR-PAPADIMITRIOU • *Breast Cancer Biology, Imperial Cancer Research Fund, London, UK*
CHRISTOS PARASKEVA • *CRC Colorectal Tumour Biology Group, Department of Pathology and Microbiology, School of Medical Sciences, University of Bristol, Bristol, UK*
JEFFREY P. PEARSON • *Department of Physiological Sciences, Medical School, University of Newcastle Upon Tyne, Newcastle Upon Tyne, UK*
GEOFFREY A. PIETERSZ • *Austin Research Institute, Austin and Repatriation Medical Centre, Heidelberg, Victoria, Australia*
NICOLE PORCHET • *Biologie et Physiopathologie des Cellules Mucipares, Laboratoire de Recherche Gérard Biserte, Institut National de la Santé et de la Recherche Médicale, Lille, France*
WENDY S. PRATT • *MRC Human Biochemical Genetics Unit, The Galton Laboratory, University College of London, London, UK*
MICHAEL R. PRICE • *Cancer Research Laboratories, School of Pharmaceutical Sciences, University of Nottingham, Nottingham, UK*
FRANCISCO X. REAL • *Institut Municipal d'Investigació Mèdica, Universität Pompeu Fabra, Barcelona, Spain*
ANTHONY M. ROBERTON, • *School of Biological Sciences, The University of Auckland, Auckland, New Zealand*
DOUGLAS I. ROSENDALE • *School of Biological Sciences, The University of Auckland, Auckland, New Zealand*
PHILIPPE ROUSSEL • *Laboratoire de Chimie Biologique et Unité Mixte de Recherche du CNRS 111, Université des Sciences et Technologies de Lille, Lille, France*
ROBERT A. SACK • *State College of Optometry, State University of New York, New York, NY*
JOHN K. SHEEHAN • *Wellcome Trust Centre for Cell Matrix Research, Division of Biochemistry, University of Manchester, Manchester, UK*
ROGER M. STARK • *Mucin Research Group, Division of Medicine, Bristol Royal Infirmary, Bristol, UK*
GÉRARD STRECKER • *Laboratoire de Chimie Biologique et Unité Mixte de Recherche du CNRS 111, Université des Sciences et Technologies de Lille, Lille, France*
DALLAS M. SWALLOW • *MRC Human Biochemical Genetics Unit, The Galton Laboratory, University College of London, London, UK*

DAVID J. THORNTON • *Wellcome Trust Centre for Cell Matrix Research, Division of Biochemistry, University of Manchester, Manchester, UK*

B. JAN-WILLEM VAN KLINKEN • *Pediatric Gastroenterology and Nutrition, Laboratory Pediatrics, dr Molewaterplein, Rotterdam, The Netherlands*

CLAUS VENNEGOOR • *Department of Obstetrics and Gynaecology, Academic Hospital Vrije Universiteit, Amsterdam, The Netherlands*

LYNNE E. VINALL • *MRC Human Biochemical Genetics Unit, The Galton Laboratory, University College of London, London, UK*

ELIZABETH WALLEY • *Department of Anatomy, School of Veterinary Science, Bristol, UK*

MICHAEL D. WALSH • *Department of Pathology, The University of Queensland Medical School, Herston, Brisbane, Queensland, Australia*

REBECCA WIGGINS • *Mucin Research Group, Division of Medicine, Bristol Royal Infirmary, Bristol, UK*

DAMIAN P. WRIGHT • *School of Biological Sciences, The University of Auckland, Auckland, New Zealand*

PEI XIANG XING • *Austin Research Institute, Austin and Repatriation Medical Centre, Heidelberg, Australia*

JUN X. YAN • *Macquarie University Centre for Analytical Biotechnology, School of Biological Sciences, Macquarie University, Sydney, Australia*

NATASHA E. ZACHARA • *Macquarie University Centre for Analytical Biotechnology, School of Biological Sciences, Macquarie University, Sydney, Australia*

I

Purification of Mucin

1

Isolation of Large Gel-Forming Mucins

Julia R. Davies and Ingemar Carlstedt

1. Introduction

The large gel-forming mucins, which form the major macromolecular components of mucous secretions, are members of the mucin "superfamily." Nine mucin genes (*MUC1–MUC4, MUC5AC, MUC5B,* and *MUC6–MUC8*) have been identified (for reviews see ***refs. 1*** and ***2***), with each gene showing expression in several tissues. Only the *MUC1*, *MUC2*, *MUC4*, *MUC5*, and *MUC7* mucins have been sequenced completely ***(3–11)*** although large stretches of *MUC5AC* ***(12–15)*** as well as the C-terminal sequences of *MUC3* ***(16)*** and *MUC6* ***(17)*** are now known.

A characteristic feature of mucins is the presence of one or more domains rich in serine and/or threonine residues that, owing to a high degree of oligosaccharide substitution, are resistant to proteolysis. Mucins comprise cell-associated, usually monomeric species, as well as those that are secreted; the latter can be subdivided into large, gel-forming glycoproteins and smaller, monomeric ones. The gel-forming mucins (M_r = 10–30 million Dalton) are oligomers formed by subunits (monomers) joined via disulfide bonds (for a review see ***ref. 18***), and treatment with reducing agents will release the subunits and cause unfolding of regions stabilized by intramolecular disulfide bonds. Thus, after reduction, we term the monomers *reduced subunits*. Reduced subunits are more sensitive to protease digestion than the intact mucin molecules.

The isolation procedures that we use for the large oligomeric mucins depend on their source. In secretions such as respiratory tract sputum, tracheal lavage fluid, and saliva, the material is centrifuged to separate the gel from the sol phase, allowing the identification of the gel-forming mucins. Repeated extraction of the gel phase solubilizes the "soluble" gel-forming species, leaving the "insoluble" mucin complex in the extraction residue. Mucin subunits may be isolated from the "insoluble" glycoprotein complex following reduction of disulfide bonds. When mucins are isolated from tissue samples, it may be an advantage to "physically" separate histologically defined areas of the tissue such as the surface and the submucosa of an epithelium. For example, material from the surface epithelium may be enriched by gently scraping the surface mucosa, thereby allowing gland material to be obtained from the remaining tissue.

From: *Methods in Molecular Biology, Vol. 125: Glycoprotein Methods and Protocols: The Mucins*
Edited by: A. Corfield © Humana Press Inc., Totowa, NJ

To isolate mucins, the bonds that hold the mucous gel together and those that anchor cell-associated glycoproteins to the plasma membrane must be broken. In our laboratory, high concentrations of guanidinium chloride are used for this purpose, and high-shear extraction procedures are avoided to minimize the risk of mechanical degradation. Protease inhibitors are used to protect the protein core and a thiol blocking agent is added to prevent thiol-disulfide bond exchange. However, breaking intermolecular bonds with highly denaturing solvents will most likely cause unfolding of ordered regions within the mucins, and properties dependent on an intact protein core structure may be lost. Following extraction, mucins are subjected to isopycnic density gradient centrifugation in the presence of guanidinium chloride. This method allows the group separation of large amounts of mucins from nucleic acids and proteins/lipids under dissociative conditions without the problems associated with matrix-based methods such as gel chromatography.

2. Materials

2.1. Extraction of Mucins

2.1.1. Guanidinium Chloride Stock Solution

We use practical grade guanidinium chloride that is treated with activated charcoal and subjected to ultrafiltration before use. We request small samples from several companies and test them for clarity after filtration as well as absorbance at 280 nm. Once we have established a suitable source, we purchase large batches of guanidinium chloride, which considerably reduces the cost. Ultrapure grade guanidinium chloride, which is much more expensive, may be used without prior purification.

1. Dissolve 765 g of guanidinium chloride in 1 L of distilled water, stirring constantly.
2. Add 10 g of activated charcoal and stir overnight.
3. Filter solution through double filter paper to remove the bulk of the charcoal.
4. To remove the remaining charcoal, filter solution through an Amicon PM10 filter (Amicon, Beverley, MA), or equivalent, using an ultrafiltration cell. A Diaflow system is a practical way to increase the filtration capacity.
5. Measure the density of the solution by weighing a known volume in a calibrated pipet, and calculate the molarity of the guanidinium chloride stock solution (*see* **Note 1**). The molarity should be approx 7.5 *M* with this procedure.

2.1.2. Solutions for Mucin Extractions

1. 6 *M* Guanidinium chloride extraction buffer: 6 *M* guanidinium chloride, 5 m*M* EDTA, 10 m*M* sodium phosphate buffer, pH 6.5 (adjusted with NaOH). This solution can be stored at room temperature. Before extraction, cool to 4°C and immediately before use, add *N*-ethyl maleimide (NEM) and diisopropyl phosphofluoridate (DFP) to final concentrations of 5 and 1 m*M*, respectively. DFP is extremely toxic (*see* **Note 2**).

2. Phosphate buffered saline (PBS) containing protease inhibitors: 0.2 *M* sodium chloride, 10 m*M* EDTA, 10 m*M* NEM, 2 m*M* DFP, 10 m*M* sodium phosphate buffer, pH 7.4 (adjusted with NaOH).

3. 6 *M* Guanidinium chloride reduction buffer: 6 *M* guanidinium chloride, 5 m*M* EDTA, 0.1 *M* Tris/HCl buffer, pH 8.0 (adjusted with HCl). This solution can be stored at room temperature.

2.2. Isopycnic Density Gradient Centrifugation

Density gradient centrifugation in our laboratory is carried out using CsCl in a two-step procedure (*see* **Notes 3** and **4**).

1. Small samples of high-quality CsCl are obtained from several companies and tested for clarity in solution, absorbance at 280 nm, and spurious color reactions with the analyses for, e.g., carbohydrate that we use. Once we have established a suitable source, we purchase large batches, which considerably reduces the cost. As with guanidinium chloride, more expensive ultrapure grade may also be used.
2. Beckman Quick Seal polyallomer centrifuge tubes (Beckman Instruments, Palo Alto, CA) or equivalent.
3. 6 *M* Guanidinium chloride extraction buffer, pH 6.5 (*see* **Subheading 2.1.2., step 1**).
4. Sodium phosphate buffer: 10 m*M* sodium phosphate buffer, pH 6.5 (adjusted with NaOH).
5. 0.5 *M* Guanidinium chloride buffer: 0.5 *M* guanidinium chloride, 5 m*M* EDTA, 10 m*M* sodium phosphate buffer, pH 6.5 (adjusted with NaOH).

2.3. Gel Chromatography

2.3.1. 4 M Guanidinium Chloride Buffer

1. Elution buffer: 4 *M* guanidinium chloride, 10 m*M* sodium phosphate buffer, pH 7.0 (can be stored at room temperature).

2.3.2. Gels and Columns

We use either Sepharose CL-2B or Sephacryl S-500HR (Pharmacia Biotech, Uppsala, Sweden) for the separation of mucins, reduced mucin subunits, and proteolytic fragments of mucins. Both "whole" mucins and subunits are usually excluded on Sephacryl S-500, but since Sepharose CL-2B is slightly more porous, mucin subunits are included and can often be separated from whole mucins on this gel. In our experience, whole mucins show a tendency to adhere to Sephacryl gels, which is not seen with Sepharose gels.

2.4. Ion-Exchange High-Performance Liquid Chromatography

Ion-exchange high performance liquid chromatography is carried out in our laboratory using a Mono Q HR 5/5 (Pharmacia Biotech) column and eluants based upon a piperazine buffer system with lithium perchlorate as the elution salt (*see* **Note 5**).

2.4.1. Separation of Reduced Mucin Subunits and Proteolytic Fragments of Mucins (see **Note 6**).

1. Buffer A: 0.1% (w/v) CHAPS in 6 *M* urea, 10 m*M* piperazine/perchlorate buffer, pH 5.0 (adjusted with perchloric acid).
2. Buffer B: 0.1% (w/v) CHAPS in 6 *M* urea, 0.25–0.4 *M* $LiClO_4$, 10 m*M* piperazine/perchlorate buffer, pH 5.0 (adjusted with perchloric acid).
3. Buffer C: 10 m*M* piperazine/perchlorate buffer, pH 5.0 (adjusted with perchloric acid).
4. Buffer D: 0.25–0.4 *M* $LiClO_4$ in 10 m*M* piperazine/perchlorate buffer, pH 5.0 (adjusted with perchloric acid).

3. Methods

3.1. Extraction of Mucins from Mucous Secretions

1. Thaw secretions, if necessary, preferably in the presence of 1 m*M* DFP.
2. Mix the secretions with an equal volume of ice-cold PBS containing protease inhibitors.
3. Centrifuge secretions at 4°C in a high-speed centrifuge (23,000*g* average [av]).
4. Pour off the supernatant, which represents the sol phase.
5. Add 6 *M* guanidinium chloride extraction buffer to the pellet (which represents the gel phase) and stir gently overnight at 4°C. If samples are difficult to disperse, the material can be suspended using two to three strokes in a Dounce homogenizer (Kontes Glass Co., Vineland, NJ) with a loose pestle.
6. Centrifuge secretions at 4°C in a high-speed centrifuge (23,000*g* av).
7. Pour off the supernatant corresponding to the "soluble" gel phase mucins.
8. If necessary, repeat **steps 5–7** another two to three times or as long as mucins are present in the supernatant.
9. Add 6 *M* guanidinium chloride reduction buffer containing 10 m*M* dithiothreitol (DTT) to the extraction residue (equivalent to the "insoluble" gel mucins).
10. Incubate for 5 h at 37°C.
11. Add iodoacetamide to give a 25 m*M* solution, and incubate overnight in the dark at room temperature.
12. Centrifuge secretions at 4°C in a high-speed centrifuge (23,000*g* av).
13. Pour off the supernatant corresponding to the reduced/alkylated "insoluble" mucin complex.

3.2. Extraction of Mucins from Tissue Samples

Tissue pieces are usually supplied to our laboratory frozen at –20°C. If mucins are to be prepared from the surface epithelium and the submucosa separately, begin with **step 1**. If mucins are to be extracted from the whole tissue, begin with **step 4**.

1. Thaw the tissue in the presence of 10 m*M* sodium phosphate buffer, pH 7.0, containing 1 m*M* DFP.
2. Scrape the surface epithelium away from the underlying mucosa with a glass microscope slide.
3. Place the surface epithelial scrapings in ice-cold 6 *M* guanidinium chloride extraction buffer and disperse with a Dounce homogenizer (two to three strokes, loose pestle).
4. Cut the submucosal tissue into small pieces and submerge in liquid nitrogen. Pulverize or grind the tissue (for this purpose we use a Retsch tissue pulverizer, Retsch, Haan, Germany).
5. Mix the powdered tissue with ice-cold 6 *M* guanidinium chloride extraction buffer and disperse with a Dounce homogenizer (two to three strokes, loose pestle).
6. Gently stir samples overnight at 4°C.
7. Centrifuge secretions at 4°C in a high-speed centrifuge (23,000*g* av).
8. Pour off the supernatant corresponding to the "soluble" mucins.
9. Repeat **steps 5–7** three more times, if necessary.
10. Add 6 *M* guanidinium chloride reduction buffer containing 10 m*M* DTT to the extraction residue.
11. Incubate for 5 h at 37°C.
12. Add iodoacetamide to give a 25 m*M* solution and incubate overnight in the dark at room temperature.

13. Centrifuge secretions at 4°C in a high-speed centrifuge (23,000*g* av).
14. Pour off the supernatant corresponding to the reduced/alkylated "insoluble" mucin complex.

3.3. Isopycnic Density Gradient Centrifugation in CsCl/Guanidinium Chloride

3.3.1. Isopycnic Density Gradient Centrifugation in CsCl/4 M Guanidinium Chloride

1. Dialyze samples against 10 vol of 6 *M* guanidinium chloride extraction buffer. The volume of the sample that can be run in each tube is two-thirds of the total volume held by the tube.
2. For practical purposes, the preparation of gradients is carried out by weighing rather than measuring volumes. Check the volume by weighing (the density of 6 *M* guanidinium chloride is 1.144 g/mL; *see* **Note 1**). If the sample volume is less than two-thirds of the total, fill up to the required volume with 6 *M* guanidinium chloride.
3. Weigh the required amount of CsCl to give the correct density into a beaker (*see* **Note 3**).
4. Add the sample to the CsCl and stir gently.
5. The final weight of the sample is calculated from the volume of the tube and the final density of the solution. Add sodium phosphate buffer to give the final weight and stir the sample gently.
6. Measure the density of the sample prior to loading with a syringe and cannula into the tubes. Balance the tubes carefully and seal according to the manufacturer's instructions.
7. Centrifuge the samples. We use a Beckman L-70 Optima centrifuge and either a 50.2Ti rotor (tube capacity 40 mL), with a starting density 1.39 g/mL, or a 70.1Ti rotor (tube capacity 13 mL), with a starting density of 1.40 g/mL. Samples are centrifuged at 36,000 rpm (50.2Ti rotor) or 40,000 rpm (70.1Ti rotor) at 15°C for 72–96 h (*see* **Note 7**). These conditions give gradients of approx 1.25–1.60 g/mL but will vary according to the rotor geometry, starting density, and speed used. Care should be taken to ensure that the starting concentration of CsCl at a given rotor speed and temperature does not exceed that recommended so that CsCl does not precipitate at the bottom of the tubes during the centrifugation run. This information should be available in the manufacturer's rotor handbook.
8. After centrifugation, recover 20–40 fractions from the gradients by piercing the bottom of the tubes and collecting fractions with a fraction collector equipped with a drop counter. Analyze the fractions for density (by weighing a known volume) and absorbance at 280 nm, as well as the appropriate carbohydrate and antibody reactivities.
9. Large amounts of proteins/lipids in the samples may lead to a poor separation between these molecules and mucins. In this case, mucin-containing fractions may be pooled and subjected a second time to density gradient centrifugation in CsCl/4 *M* guanidinium chloride. Sodium dodecyl sulfate-polyacrylamide gel electrophoresis of the mucin-containing fractions may be used to determine whether all proteins have been removed.

3.3.2. Isopycnic Density Gradient Centrifugation in CsCl/0.5 M Guanidinium Chloride

Density Gradient Centrifugation in CsCl/4 *M* guanidinium chloride may be followed by subjecting the mucin-containing fractions to a second density gradient step in CsCl/0.5 *M* guanidinium chloride, which gives a better separation between mucins and DNA (*see* **Note 3**). Some mucins show a tendency to precipitate in the presence of CsCl at low concentrations of guanidinium chloride, and CHAPS is sometimes added to the gradients to counteract this effect.

1. Dialyze samples against 10 vol of 0.5 *M* guanidinium chloride buffer.
2. Measure the volume of the sample by weighing (the density of 0.5 *M* guanidinium chloride is 1.015 g/mL; *see* **Note 3**).
3. Weigh cesium chloride to give the required density into a beaker (*see* **Note 3**).
4. Add the sample (volume must not exceed three-fourths of the total volume held by the tube).
5. If required, add 1% CHAPS solution to give a final concentration of 0.01% (i.e., 1% of the total volume).
6. The concentration of guanidinium chloride in the final volume must be 0.5 *M*, and the volume of the CsCl and CHAPS must therefore be compensated for by the addition of a small volume of 8 *M* guanidinium chloride.
7. The final weight of the sample is calculated from the volume of the tube and the final density of the solution. Add sodium phosphate buffer to give the final weight and stir the sample gently.
8. Measure the density of the sample and load into the tubes with a syringe and cannula. Seal the tubes according to the manufacturer's instructions.
9. Centrifuge the samples at 36,000 rpm (50.2Ti rotor, starting density 1.50 g/mL) or 40,000 rpm (70.1Ti rotor, starting density 1.52 g/mL) at 15°C for 72–96 h (*see* **Note 7**). These conditions give gradients of approx 1.35–1.67 g/mL but will vary according to the rotor geometry, starting density, and speed used. Care should be taken to ensure that the starting concentration of CsCl at a given rotor speed and temperature does not exceed that recommended so that CsCl does not precipitate at the bottom of the tubes during the centrifugation run. This information should be available in the manufacturer's rotor handbook.
10. After centrifugation, recover 20–40 fractions from the gradients by piercing a hole in the bottom of the tubes and collecting fractions with a fraction collector equipped with a drop counter. Analyze the fractions for density (by weighing a known volume) and absorbance at 280 nm, as well as the appropriate carbohydrate and antibody reactivities.

3.4. Gel Chromatography

3.4.1. Sepharose CL-2B

1. Elute columns (100 × 1.6 cm) packed according to the manufacturer's specifications with 4 *M* guanidinium chloride buffer at a rate well below the maximum of 15 mL/(cm^{-2}·h^{-2}).
2. Apply samples, the volume of which should be <5% of the column volume, that have been dialyzed against the running buffer to the column through an injector.
3. Monitor the eluate on-line with an ultraviolet (UV) monitor and collect fractions using a fraction collector and subject to the appropriate carbohydrate and antibody analyses.

3.4.2. Sephacryl S-500

1. Elute columns (50 × 1.6 cm) packed according to the manufacturer's specifications are eluted with 4 *M* guanidinium chloride buffer at a flow well below the maximum rate of 40 mL/(cm^{-2}·h^{-2}). We run S-500HR columns on a system consisting of a 2150 LKB titanium head pump and a Pharmacia V-7 injector (Pharmacia Biotech).

2. Apply samples, the volume of which should be <5% of the column volume, that have been dialyzed against the running buffer to the column through an injector.
3. Monitor the eluate on-line with a UV monitor and collect fractions using a fraction collector and subject to the appropriate carbohydrate and antibody analyses.

3.5. Ion-Exchange Chromatography (see Note 5)

1. Run Mono Q columns on a system comprising a 2150 LKB titanium head pump connected to a 2152 LKB controller and a Pharmacia V-7 injector. All connections are made using Teflon tubing.
2. Equilibrate the column with buffer A or C.
3. Dialyze sample exhaustively or dissolve sample in buffer A or C and apply the sample to the column.
4. Run the column in a linear gradient up to 100% buffer B or D.
5. Monitor the eluate on-line with a UV monitor and collect fractions using a fraction collector and subject to the appropriate carbohydrate and antibody analyses.

3.6. Analysis of Mucins

Methods for the detection and analysis of mucins are dealt with in other chapters in this volume. However, three principally different methods are available: solution assays such as colorimetric assays for hexose and sialic acid; membrane-based methods such as slot-blotting and staining with periodic acid-Schiff reagent; antibodies and lectins or coating methods such as the glycan detection method and enzyme-linked immunosorbent assays (ELISA). All these techniques have advantages and disadvantages. Solution methods often crave larger amounts of material than the other two, but selective loss of components is less of a problem. Membrane-based methods allow relatively large volumes of "dilute" sample to be analyzed, thus increasing the sensitivity, but components that do not adhere to the membrane may be lost and the linear range of the technique may be limited. "Coating methods" such as ELISA are prone to artefacts if samples are concentrated, and care must be taken to ensure that the signals obtained are within the linear range for the technique.

4. Notes

1. The molarity of guanidinium chloride solution can be calculated from the density according to the following formula:

$$M = (\rho - 1.003)/0.02359$$

   where M is the molarity and ρ is the density in grams per milliliter.
2. Inhibitors are added to the 6 *M* guanidinium chloride extraction buffer in order to block the activity of the three major classes of proteolytic enzymes: metalloproteases, serine proteases, and thiol proteases. The action of metalloproteases is inhibited by the addition of EDTA to the buffer. This can be added during the initial preparation since it is stable at room temperature. DFP is a potent inhibitor of serine proteases and esterases, including acetylcholinesterase, and should therefore be handled in a fume cupboard with extreme care! DFP is supplied in 1-g vials with a septum, and prior to dilution, vials should be cooled on ice to reduce the vapor pressure. Under supervision, the septum should be pierced with a needle to equilibrate the pressure, and the DFP should be transferred using a syringe and needle. The contents of the vial should be placed directly into the correct volume of

ice-cold dry propan-1-ol to give a 100 m*M* solution. DFP is unstable in water but can be stored at –20°C in propan-1-ol. After dilution, the vial as well as the needles and syringes used may be rinsed with 1 *M* NaOH to inactivate the DFP. Phenylmethylsulfonylfluoridate (PMSF) can be used at a concentration of 0.1 m*M* in place of DFP. However, we find this a less attractive option owing to its low solubility although it is possible to prepare first a stock solution of PMSF in an organic solvent that is miscible with water. Thiol proteases are inactivated through the addition of NEM. In addition, NEM will also block exchange reactions between free thiol groups and disulfide bonds.

3. Samples in our laboratory are usually subjected to a two-stage isopycnic density gradient procedure (*see* **Note 4**). First, samples are centrifuged in CsCl/4 *M* guanidinium chloride, which gives a good separation of higher buoyant density mucins and nucleic acids from low buoyant density proteins, glycoproteins, and lipids while maintaining a denaturing environment. Thus, proteolytic enzymes can be separated from mucins before the concentration of guanidinium chloride is reduced. The second step of the purification is to pool the partially separated mucins and nucleic acids and subject them to a second density gradient step in CsCl/0.5 *M* guanidinium chloride. These conditions give a good group separation between mucins and nucleic acids. The amount of cesium chloride needed to give a required density in 4 or 0.5 *M* guanidinium chloride can be calculated according to the following formula:

$$x = v\,(1.347\rho - 0.0318M - 1.347)$$

where x is CsCl (grams), v is the final volume, M is the molarity of the guanidinium chloride (4 or 0.5*M*), and ρ is the density (grams per milliliter).

4. In our laboratory, CsCl rather than CsBr or $CsSO_4$, is used as the density gradient–forming salt since gradients are run in the presence of guanidinium chloride and the use of CsBr or $CsSO_4$ in the presence of guanidinium chloride gives rise to mixed cesium salts. **Figure 1** shows a comparison of the separation obtained between mucins and DNA using the two-step approach in CsCl/4 *M* guanidinium chloride followed by CsCl/ 0.5 *M* guanidinium chloride with that given by CsBr or $CsSO_4$ in 10 m*M* sodium phosphate buffer. DNA was mixed with purified cervical mucins and gradients prepared using each of the cesium salts. In CsCl/4 *M* guanidinium chloride, there is poor resolution of mucins from DNA (**Fig. 1A**); however, a reduction in the concentration of guanidinium chloride to 0.5 *M* leads to a baseline separation between mucins and DNA in this salt (**Fig. 1B**). In $CsSO_4$, mucins are also completely separated from DNA (**Fig. 1C**). In CsBr, however, DNA and mucins have a similar buoyant density, and DNA trails into the mucin peak (**Fig. 1D**). These data indicate that CsBr is not the salt of choice for samples containing DNA.

5. Traditionally, we have used lithium perchlorate as the elution salt since it is compatible with our colorimetric assays for carbohydrate based on sulfuric acid (e.g., the anthrone procedure). Alternative salt/buffer systems may give at least as good, if not better, separation depending on the nature of the mucins in question.

6. The optimum concentration of $LiClO_4$ in buffers B and D varies between 0.25 and 0.5 *M* depending on the charge densities of the mucins to be separated, although typically we use a concentration of 0.4 *M*. For buffers A and B, stock solutions of 8 *M* urea are freshly prepared and run through a column containing a mixed anion/cation exchanger (e.g., Elgalite or Amberlite resin). The buffer system A and B containing 6 *M* urea and 0.1% CHAPS gives good separation between different populations of reduced mucin subunits, whereas for the separation of proteolytic fragments, buffers C and D are used.

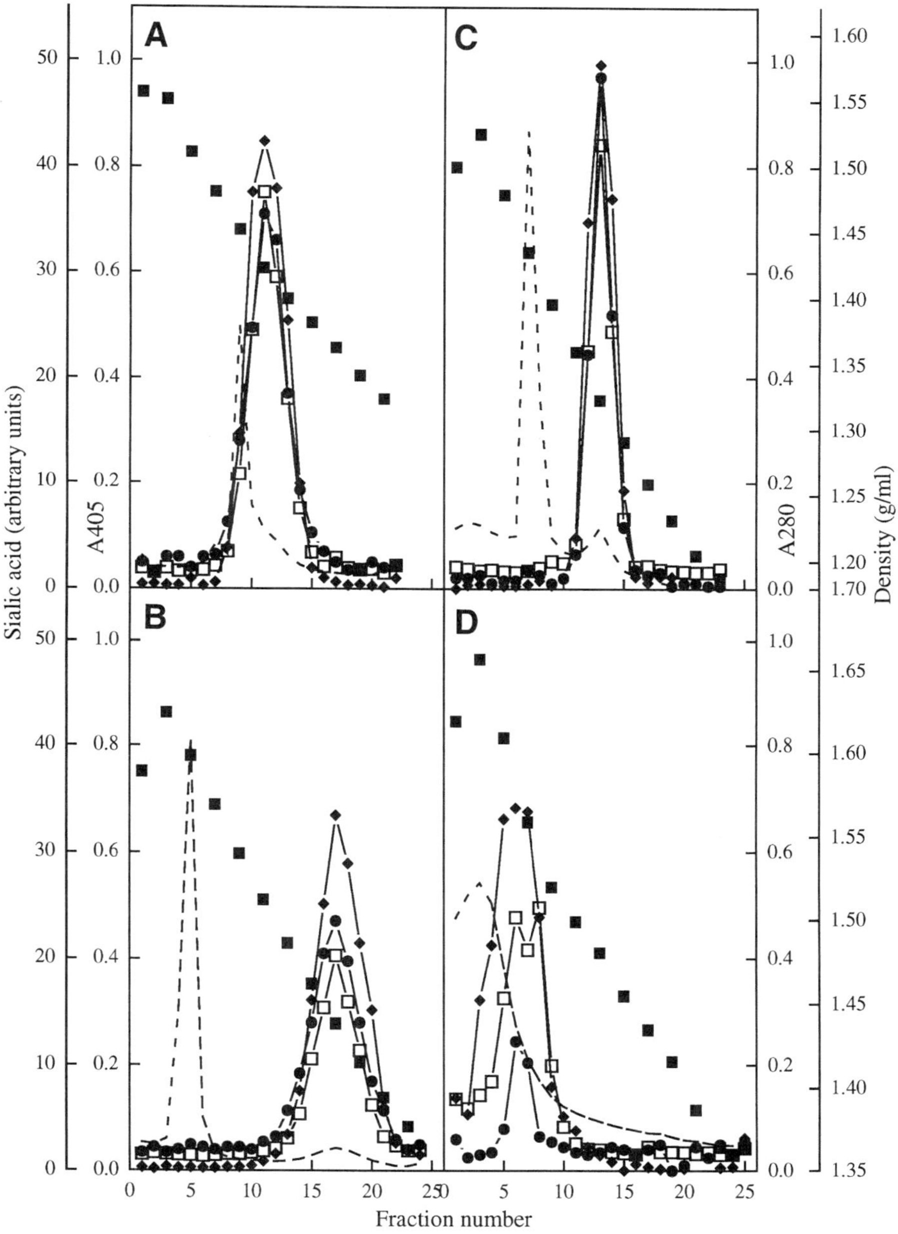

Fig. 1. Density gradient centrifugation of cervical mucins and DNA. Purified cervical mucins were mixed with DNA and subjected to density gradient centrifugation in (**A**) CsCl/4 *M* guanidinium chloride; (**B**) CsCl/0.5 *M* guanidinium chloride; (**C**) $CsSO_4$/10 m*M* sodium phosphate buffer, pH 6.5; and (**D**) CsBr/10 m*M* sodium phosphate buffer, pH 6.5. After centrifugation in a Beckman L70 centrifuge (70.1Ti rotor, 40,000 rpm, 15°C, 65 h, starting density: [**A**] 1.41 g/mL, [**B**] 1.52 g/mL, [**C**] 1.34 g/mL, and [**D**] 1.49 g/mL), fractions were collected from the bottom of the tubes and analyzed for sialic acid (●), carbohydrate (glycan detection method) (□), *MUC5B* antibody reactivity (◆), absorbance at 280 nm (----), and density (■).

7. The rotor type, speed and starting densities rather than the g-force are given for the runs. In our experience, the rotor geometry (the tube angle within the rotor), starting density, and the speed at which the run is conducted are the most important factors determining the gradient formed and the conditions cannot necessarily be reproduced by using the same g-force in another rotor.

Acknowledgments

The authors acknowledge support from the Swedish Medical Research Council (grant no. 7902, 9823), the Medical Faculty of Lund, Centrala Försökdjursnämnden, Swedish Match AB, the Swedish Cancer Fund, Greta och Johan Kocks Stiftelse, Stiftelsen Riksförbundet Cystisk Fibros, Vårdals Stiftelse, and the Smokeless Tobacco Research Council, Inc. (USA).

References

1. Gendler, S. J. and Spicer, A. P. (1995) Epithelial mucin genes. *Annul. Rev. Physiol.* **57,** 607–634.
2. Rose, M. C. and Gendler, S. J. (1997) *Airway Mucus: Basic Mechanisms and Clinical Perspectives* (Rogers, D. F. and Lethem, M. I., eds.), Birkhauser Verlag, Basel, pp. 41–66.
3. Lancaster, C. A., Peat, N., Duhig, T., Wilson, D., Taylor-Papadimitriou, J. and Gendler, S. J. (1990) Structure and expression of the human polymorphic epithelial mucin gene: An expressed VNTR unit. *Biochem. Biophys. Commun.* **173,** 1019–1029.
4. Gum, J. R., Hicks, J. W., Toribara, N. W., Siddiki, B., and Kim, Y. S. (1994) Molecular cloning of human intestinal mucin (MUC 2) cDNA: Identification of the amino acid functions and overall sequence similarity to prepro- von Willebrand factor. *J. Biol. Chem.* **269,** 2440–2446.
5. Moniaux, N., Nollet, S., Porchet, N., Degand, P., Laine, A., and Aubert, J.-P. (1999) Complete sequence of the human mucin *MUC4*: a putative cell membrane-associated mucin. *Biochem. J.* **338,** 325–333.
6. Desseyn, J.-L., Guyonet-Dupérat, V., Porchet, N., Aubert, J.-P., and Laine, A. (1997) Human mucin gene MUC5B, the 10.7-kb large central exon encodes various alternate subdomains resulting in a super-repeat. *J. Biol. Chem.* **272,** 3168–3178.
7. Desseyn, J.-L., Aubert, J.-P., Van Seuningen, I., Porchet, N., and Laine, A. (1997) Genomic organization of the 3' region of the human mucin gene MUC5B. *J. Biol. Chem.* 272, **16,** 873–16,883.
8. Desseyn, J.-L., Buisine, M.-P., Porchet, N., Aubert, J.-P., and Laine, A. (1998) Genomic organization of the human mucin gene *MUC5B. J. Biol. Chem.* **273,** 30,157–30,164.
9. Troxler, R. F., Offner, G. D., Zhang, F., Iontcheva, I., and Oppenheim, F. G. (1995) Molecular cloning of a novel high molecular weight mucin (MG1) from human sublingual gland. *Biochem. Biophys. Res. Commun.* **217,** 1112–1119.
10. Nielsen, P. A., Bennett, E. P., Wandall, H., Therkildsen, M. H., Hannibal, J., and Clausen, H. (1997) Identification of a major human high molecular weight salivary mucin (MG1) as tracheobronchial mucin MUC5B. *Glycobiology* **7,** 413–419.
11. Bobek, L. A., Tsai, H., Biesbrock, A. R., and Levine, M. J. (1993) Molecular cloning, sequence and specificity of expression of the gene encoding the low molecular weight human salivary mucin (MUC 7). *J. Biol. Chem.* **268,** 20,563–20,569.
12. Meerzaman, D., Charles, P., Daskal, E., Polymeropoulos, M. H., Martin, B. M., and Rose, M. C. (1994) Cloning and analysis of cDNA encoding a major airway glycoprotein, human tracheobronchial mucin (MUC 5). *J. Biol. Chem.* **269,** 12,932–12,939.

13. Guyonnet-Duperat, V., Audie, J.-P., Debailleul, V., Laine, A., Buisine, M.-P., Galiegue-Zouitina, S., Pigny, P., Degand, P., Aubert, J.-P., and Porchet, N. (1997) Characterization of the human mucin gene MUC5AC: a consensus cysteine-rich domain for 1lpl5 mucin genes? *Biochem. J.* **305,** 211–219.
14. Lesuffleur, T., Roche, F., Hill, A. S., Lecasa, M., Fox, M., Swallow, D. M., Zweibaum, A., and Real, F. X. (1995) Characterization of a mucin cDNA clone isolated from HT-29 mucus-secreting cells. *J. Biol. Chem.* **270,** 13,665–13,673.
15. Li, D., Gallup, M., Fan, N., Szymkowski, D. E., and Basbaum, C. B. (1998) Cloning of the aminoterminal and 59-flanking region of the human MUC5AC mucin gene and transcriptional up regulation by bacterial exoproducts. *J. Biol. Chem.* **273,** 6812–6820.
16. Gum, J. R. J., Ho, J. J. L., Pratt, W. S., Hicks, J. W., Hill, A. S., Vinall, L. E., Roberton, A. M., Swallow, D. M. and Kim, Y. S. (1997) MUC3 human intestinal mucin. Analysis of gene structure, the carboxyl terminus, and a novel upstream repetitive region. *J. Biol. Chem.* **272,** 26,678– 26,686.
17. Toribara, N. W., Ho, S. B., Gum, E., Gum, J. R. J., Lau, P., and Kim, Y. S. (1997) The carboxy-terminal sequence of the human secreltory mucin, MUC6. *J. Biol. Chem.* **272,** 16,398–16,403.
18. Strous, G. J. and Dekker, J. (1992) Mucin-type glycoproteins. *Crit. Rev. Biochem. Mol. Biol.* **27,** 5792.

2

Preparation of Membrane Mucin

Kermit L. Carraway

1. Introduction

The first task in this chapter is to define the term *membrane mucin.* In a classical sense, the term is an oxymoron, because mucins were defined as the major glycoproteins of mucous secretions. However, the recognition of the importance of mucinous tumor cell surface glycoproteins and their prominence in the early work on the cloning of mucin has led to a shift in usage ***(1,2)***, in which both secreted and membrane components are recognized as mucins. This usage has led to another complication, in which membrane components with highly *O*-glycosylated mucinlike domains are called *mucins* ***(3,4)***. Such mucinous domains are present in many cell surface molecules, most of which have few of the characteristics of other mucins ***(4)***. For the purpose of this chapter I have assumed a simple definition of membrane mucins. They must exhibit two characteristics: (1) they must be strongly bound to the membrane, and (2) they must have a large, highly *O*-glycosylated domain of mucin. Eventually this definition should evolve to require that membrane mucins contain a defined membrane-binding domain, such as a hydrophobic transmembrane sequence, and mucin repeat sequences. However, application of that requirement at present would exclude epiglycanin, the first membrane mucin to be discovered, for which such information is not available, because it has not been cloned. The present definition still restricts the number of membrane mucins to four examples: epiglycanin, MUC1, sialomucin complex ([SMC], also ASGP-1/ASGP-2 and MUC4), and rat MUC3. Another aspect of these mucins needs to be considered. All three, which have been sufficiently characterized, are found in soluble forms as well as membrane forms. SMC is found in goblet cell secretory granules in the intestine and is secreted via a regulated mechanism ***(5)***. Thus, the term *membrane mucin* is somewhat of a misnomer, although it remains the best descriptor until functional descriptive names become feasible.

Epiglycanin was originally discovered through its implication in the allotransplantability of the Ha subline of the TA3 mouse mammary carcinoma ***(6)***. The allotransplantable Ha subline contained much greater amounts of cell surface sialic acid

From: *Methods in Molecular Biology, Vol. 125: Glycoprotein Methods and Protocols: The Mucins*
Edited by: A. Corfield © Humana Press Inc., Totowa, NJ

than did the nonallotransplantable St subline. The sialoglycoconjugate was demonstrated to be a glycoprotein by trypsin treatment of the TA3-Ha cells. The released epiglycanin was purified by a single gel filtration step, eluting in the void volume of a Bio-Gel P-100 column *(7)*. Subsequent analyses of this epiglycanin from metabolically labeled cells by gel filtration on Bio-Gel A-5m showed the presence of two components, which were compositionally similar and appeared to be proteolysis fragments of the same glycoprotein. An extensive series of analyses demonstrated epiglycanin to be a high M_r mucin-type glycoprotein with numerous short oligosaccharides ***(6)***. Proteolysis is frequently used for the identification of cell surface glycoproteins and in some cases can be used for purification. The major advantage is that it eliminates the need for cell lysis or cell fractionation. A major disadvantage is that this method necessarily fragments the protein and results in the loss of the membrane attachment site, a critical feature of membrane proteins. To avoid proteolysis, epiglycanin can also be isolated from ascites fluid *(8)*, taking advantage of the high-level concentration of the glycoprotein released during cell growth. However, since ascites fluid epiglycanin is a soluble protein, it also does not contain a membrane anchor domain. Two issues have further complicated studies of epiglycanin: the lack of a peptide-specific antibody by which epiglycanin could be unequivocally identified, and the lack of sequence data from either cloning or peptide sequencing by which epiglycanin can be compared with other known mucins. Thus, it is not entirely certain that epiglycanin is not a mouse form of one of the known mucins, as SMC is the rat form of MUC4

MUC1 is one of the two mucins that have been described as components of milk membranes ***(9)***. As such, it was first isolated as the major sialoglycoprotein (GP-2) of bovine milk fat globule membranes (MFGMs) ***(10)***. However, the major interest in MUC1 did not arise until antibodies prepared against defatted human cream fraction from milk were shown to exhibit recognition of human neoplasms ***(11)***. A glycoprotein bearing the antigen, originally called epithelial membrane antigen (EMA), was purified from skim milk by sequential chromatographic methods, with the final step being peanut lectin affinity chromatography ***(12)***. Subsequent studies using human milk, breast tumor cells, or their membranes elicited monoclonal antibodies against the same glycoprotein, which has been variously called DF3 antigen, episialin, epitectin, and polymorphic epithelial mucin ***(9)***. The latter designation arose from its identification with the polymorphic mucin from human urine ***(13)***.

With the development of highly specific antibodies and rapid cloning methods, the strategy for purifying and characterizing mucins has changed substantially, as exemplified for MUC1 ***(14)***. The glycoprotein was purified from human skim milk by immunoaffinity chromatography, deglycosylated with hydrogen fluoride, and used to elicit polyclonal antibodies for screening a cDNA library. Sequencing of these clones initially identified the 20 amino acid repeat, which carries most of the *O*-linked oligosaccharides ***(15,16)***. Subsequent cloning and sequencing characterized the sequence of the remainder of the molecule, including a transmembrane domain and a highly conserved cytoplasmic domain ***(17)***. As the first mucin to be cloned and sequenced, the human gene was assigned the designation *MUC1*. The protein is most frequently called MUC1 or MUC1 protein.

Although cloning methods have greatly enhanced our understanding of mucin structures, they provide an incomplete characterization. Two examples help to illustrate this point. First, biochemical and biophysical studies have shown that the carbohydrate and oligosaccharide composition of MUC1 is tissue and differentiation dependent *(18)*. Pancreatic tumor MUC1 contains 80% carbohydrate, in contrast to the 50% observed for breast tumors or milk. Changes in MUC1 oligosaccharides result in altered accessibility of the polypeptide to antibodies and differences in antibody recognition of tumor cells compared to their normal counterparts *(18)*. Second, biosynthesis studies have shown that membrane MUC1 is a heterodimeric glycoprotein, which is cleaved into two subunits during the early part of its transit to the cell surface *(19)*. Although MUC1 is usually described as a membrane mucin, the purification procedure from skim milk undoubtedly isolated a soluble form. Soluble MUC1 is also released from tumor cells and has been used as a diagnostic serum marker *(20)*. Two mechanisms have been proposed for the release—alternative splicing and proteolysis *(2)*; each may be operative in different contexts. An alternative form of MUC1 that is missing the mucin repeats has been described *(21)*. Obviously, it is not a mucin by my definition.

Ascites sialoglycoprotein-1 (ASGP-1) was first recognized by cell surface and metabolic labeling and proteolysis studies as the major glycoprotein on the cell surface of highly metastatic rat ascites 13762 mammary adenocarcinoma cells *(22,23)*. It could be rapidly purified from cell lysates or cell membranes by CsCl density-gradient centrifugation in 4 *M* guanidine hydrochloride, and was recognized as a mucin by its large size and high content of *O*-linked oligosaccharides *(24)*. The membrane association mechanism was identified with a second subunit (ASGP-2) by isolation under nondissociating conditions *(25)*. Biochemical studies indicated that ASGP-2 is heavily *N*-glycosylated and strongly associated with both ASGP-1 and the ascites cell membranes (**Fig. 1**). Although the two subunits are quite different compositionally, biosynthesis studies indicate that they are derived from a common precursor and single gene *(26)*. This result was verified by cloning and sequencing, demonstrating the transmembrane domain in ASGP-2 and multiple repeats of ≈125 amino acids in ASGP-1 *(27,28)*. ASGP-1 overexpression in tumor cells has been demonstrated to reduce their adhesiveness to both cells and the extracellular matrix *(29)*, providing one explanation for its previous implication in metastasis *(30)*. Interestingly, ASGP-2 also contains two epidermal growth factor (EGF) domains, both of which have the consensus residues found in such domains exhibiting the ability to act as ligands for ErbB receptor tyrosine kinases *(27)*. By transfection studies, I and my colleagues have shown that ASGP-2 can form a stable complex with ErbB2, the central member of that family in cellular signaling *(31)*. In addition, we have proposed that ASGP-2 can modulate ErbB signaling through its interaction with ErbB2, although the mechanism remains uncertain *(32)*.

SMC is found in a large group of normal epithelial tissues, including the ependymal epithelium of the brain, lactating mammary gland, trachea, oral cavity, intestine, cornea, and uterus *(33)*. In contrast to the 13762 ascites cells, which contain predominantly membrane SMC, most of these tissues express both membrane and soluble

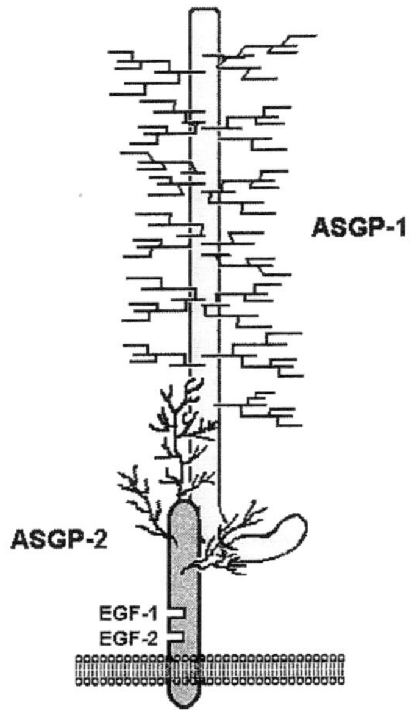

Fig. 1. Model for the structure of SMC. Generically, this model also applies to MUC1, which has both a mucin and a transmembrane subunit synthesized from a precursor encoded by a single gene.

forms *(33)*. An exception is the intestine, in which SMC is predominantly in a soluble form *(5)*. SMC in the intestine is intracellular instead of at the apical surface of the epithelium, where it is found in many other tissues *(33–35)*. These results suggest that SMC provides multiple functions for the protection of the epithelium. Possibilities include acting as a classical secreted mucin, serving as a membrane-blocking (antiadhesive) agent, and modulating cell survival and proliferation in damaged epithelia through its interaction with ErbB2.

Rat Muc3 was recently identified as a potential membrane mucin by cloning and sequencing *(36)*. The original clones were isolated from a cDNA library screened with antibody prepared against deglycosylated rat mucin isolated from rat mucosa by sequential gel filtration and CsCl gradient centrifugation *(37)*. Interestingly, Muc3, like ASGP-2, has two EGF-like domains, one of which is a juxtamembrane domain similar to EGF-2 of ASGP-2 (**Fig. 2**). Furthermore, direct sequence comparisons show that the EGF domains from these different mucins have a substantial number of amino acid identities—eight cysteines rather than the usual six for EGF domains—a similar, but not identical spacing of the cysteine residues, and a similar spacing between the last cysteine and the putative transmembrane domain. Now that the C-terminal peptide sequence is known, it will be interesting to see whether a membrane form can be detected on intestinal cells and isolated in intact form. In contrast to rat Muc3, human

B

EGF2

```
DGVTCVSPCSEGY----CHNGGQCKHLPDGPQCTCATFSIYTSWGERCEHLSVKLGAFF   ASGP-2
NKWYCVTPCSSGYSTSKNCSYGKCQLQRSGPRCLCLSTDTNWYSGENCDWGTQKSLVYG   RMuc3
```

EGF1

```
CACLPGFSGDRCQLQTR---CQNGGQWDGLKCQCPSTFYGSSCDFAV      HMUC3
CVCPNGFSGDRCQNRVPVVDCQNGGTWDGLKCQCTGLFYGPRCEEVM      RMuc3
SEFCQNHSCPVNYCYNHGHCDISGPPDCQPTCTCAPAFTGNRCFLAG      RASGP-2
```

Fig. 2. Comparisons of the C-terminal domains of ASGP-2, rat Muc3, and human MUC3, **(A)** domain organizations; **(B)** sequence comparisons between EGF-2 domains of rat ASGP-2 and rat Muc3, and sequence comparisons between the EGF-1 domains of rat ASGP-2, rat Muc3, and human MUC3. Since the full sequences of the MUCs are not known, position 1 for all sequences has been assigned strictly for comparisons. Likewise, the EGF domain sizes are only estimates. Sequence identities are double underlined. All cysteines are shown in bold.

MUC3 appears not to be a membrane protein. The recently published sequence of human MUC3 *(38)* suggests that it is truncated at the C-terminus compared to the rat analog, missing the juxtamembrane EGF, transmembrane, and cytoplasmic domains. Human and rat MUC3 do have one highly similar EGF-like domain (**Fig. 2**), suggesting that it serves some function in the mucins.

Membrane mucins can be isolated either by classical biochemical techniques or by immunoaffinity methods. The former most appropriately takes advantage of specific attributes of mucins—their large size and high density—to simplify purification. Regardless of which approach is used, I would argue that membrane mucins should be isolated for characterization from membranes or cells. Furthermore, they should be isolated under the least dissociating conditions feasible. Characterization of soluble forms of these molecules provides useful information but yields an incomplete story and may actually hinder progress. For example, isolation of the membrane form of epiglycanin might have yielded unglycosylated peptide that was more amenable to producing peptide-specific, instead of carbohydrate-specific, antibodies. Alternatively, it might have provided peptides for sequencing as a step toward cloning. Similarly, isolation and characterization of intact MUC1 should give information about its transmembrane subunit, which is still lacking and might be useful in understanding MUC1 functions, particularly in normal epithelia. After all, it is the functions of these molecules that are most important, yet still inadequately understood.

2. Materials

1. Ascites cells and tissues: TA3-Ha ascites mouse mammary carcinoma cells grown in strain A mice for isolation of epiglycanin; MAT-B1 or MAT-C1 ascites sublines of 13762 mammary adenocarcinoma cells grown in female Fischer 344 rats for isolation of ASGP-1 or SMC; rat tracheal tissue snap-frozen in N_2.
2. Cultured cells for isolation of MUC1: H.Ep.2 cells were maintained as monolayers or in suspension culture in Eagle's minimum essential medium supplemented with glutamine, sodium pyruvate, nonessential amino acids, and 10% (v/v) heat-inactivated fetal calf serum. At confluence the cells were harvested by versene treatment, centrifuged at 1000g, and washed with phosphate-buffered saline (PBS).
3. Bovine milk or human milk for isolation of MUC1.
4. Gel filtration columns: BioGel P-100, G-200, Sepharose 6B, and Sepharose CL-2B (Amersham Pharmacia Biotech, Piscataway, NJ).
5. Immunoaffinity columns: anti-MUC1 HMFG-1 on protein A-Sepharose, anti-MUC1 Ca1 or bovine γ-globulin coupled to Sepharose 4B with cyanogen bromide ***(39)***, anti-ASGP-2 on Immunopure® Protein A IgG Orientation Kit (Pierce, Rockford, IL).
6. Other chromatography columns: hydroxyapatite column (Bio-Gel HT, BioRad, Hercules, CA), carboxymethyl-Sephadex G-25, peanut lectin (Sigma, St. Louis, MO) immobilized on Sepharose 4B.
7. CsCl gradients: Preformed CsCl/guanidine gradients were prepared by gently layering 2 mL each of 4 *M* guanidine hydrochloride in 10 m*M* Tris-HCl, pH 7.4, containing 1.58, 2.0, 2.37, 2.79, and 3.15 *M* CsCl. CsCl/Triton gradients were formed by successively underlaying five CsCl solutions of 1.3, 1.35, 1.40, 1.45, and 1.5 g/mL densities buffered with PBS containing 0.2% Triton X-100.
8. Collodion bags for vacuum dialysis/concentration.
9. Tosylphenylalanine chloromethyl ketone-trypsin (Sigma) for proteolysis.
10. Sodium dodecyl sulfate polyacrylamide gel electrophoresis (SDS-PAGE) apparatus for electrophoretic analyses; Coomassie blue for staining proteins; periodate-Schiff reagent for staining glycoproteins; immunoblot transfer apparatus and chemiluminescence reagents for detection of antigens.
11. RIPA buffer for immunoprecipitation: 150 mM NaCl, 1% Nonidet P-40, 0.5% deoxycholate, 0.1% SDS, 50 m*M* Tris-HCl, pH 8.0.
12. Protease inhibitors: 0.5 m*M* EDTA, 1 m*M* phenylmethylsulfonylfluoride, 1 kIU/mL aprotinin, 1 m*M* leupeptin, 1 m*M* pepstatin.

3. Methods

3.1. Purification of Epiglycanin by Proteolytic Release (7)

1. Treat washed TA3-Ha cells (1 to 2 × 10^9 cells in 20 mL) in balanced salt solution with 18 μg of TPCK-trypsin at 0–4°C for 20 min with rotation. Repeat each treatment five to seven times, combine the batches, centrifuge at 20,000*g* to remove cellular debris, and lyophilize.

2. Purify by gel filtration on Bio-Gel P-100 at 4°C using 0.05 *M* pyridine acetate (pH 5.3) as the eluent.
3. Assay fractions for epiglycanin with peanut lectin blots.

3.2. Purification of Epiglycanin from Ascites Fluid (8)

1. Collect ascites fluid from mice on d 7 after tumor cell injection and centrifuge sequentially at 80 and 37,000*g* to remove cells and cellular debris, respectively.
2. Treat supernatant with perchloric acid by dropwise addition at 0°C to a concentration of 0.25 *M*. After 20 min remove the precipitate by centrifugation at 10,000*g* and neutralize with 2.5 *M* $KHCO_3$. Remove insoluble K_2ClO_4 by centrifugation at 10,000*g*, and concentrate supernatant by vacuum dialysis in collodion bags against 0.05 *M* pyridine acetate, pH 5.3.
3. Fractionate the concentrated material on Sepharose 4B.

3.3. Preparation of GP2 (Bovine MUC1) from Bovine Milk Fat Globule Membranes (10) (see Note 1)

1. Prepare MFGMs from cream fraction of bovine milk by a freeze-thaw procedure to release the membranes from lipid globules, followed by centrifugation at 40,000*g* for 1 h.
2. Solubilize MFGMs (10 mg protein/mL) in 1% SDS, 10 m*M* phosphate buffer (pH 6.4), and 1% mercaptoethanol and dialyze overnight against 0.1% SDS in the phosphate buffer with 2 m*M* mercaptoethanol.
3. Fractionate the sample on Bio-Gel HT, eluting with a gradient of phosphate buffer.
4. Purify GP2-containing fractions further by gel filtration on Sephadex G-200 or Sepharose 4B.
5. Assay GP2 by SDS-PAGE and periodic acid-Schiff (PAS) staining.

3.4. Biochemical Purification of EMA (MUC1) from Human Milk (12) (see Note 2)

1. Fractionate human skim milk by sequential precipitations with 40 and 80% saturated ammonium sulfate.
2. Dissolve the latter precipitate in distilled water, dialyze against water, bring to 1% Triton X-100, and fractionate on Sepharose 6B.
3. Extract EMA-containing fractions with chloroform-methanol (2:1).
4. Dialyze the aqueous phase against water, concentrate, bring to 0.01 *M* acetate, pH 6.0, and fractionate on carboxymethyl–Sephadex G-25.
5. Purify the eluted fractions containing EMA further by peanut lectin affinity chromatography.

3.5. Immunoaffinity Purification of MUC1 from Human Skim Milk (14) (see Notes 3 and 4)

1. Pass human skim milk through an affinity column of anti-MUC1 MAb HMFG-1 on protein A-Sepharose.
2. Wash the column with PBS and elute with 0.1 *M* glycine (pH 2.5).

3.6. Immunoaffinity Purification of Epitectin (MUC1) from H.Ep.2 Cells (39) (see Notes 4 and 5)

1. Suspend washed cells (approx 10^8) in 20 mL of 10 m*M* Tris-HCl, pH 8.0, and 0.2% sodium deoxycholate with protease inhibitors and break by passage several times through a 19-gage syringe needle.
2. Stir the lysate for 1 h at 4%C and centrifuge at 60,000*g* for 30 min.

3. Heat the supernatant in boiling water for 10 min, recentrifuge to remove denatured proteins, and apply to tandem affinity columns of bovine γ-globulin and anti-MUC1 (Ca1) antibody.
4. Wash the column with 0.5% deoxycholate-1% Triton X-100 and elute with 3 *M* KSCN.

3.7. Isolation of ASGP-1 from Ascites Cell Membranes (40)

1. Suspend washed ascites cells in 10 vol of 10 m*M* Tris HCl, pH 8.0, and keep at 0°C for 2 min prior to centrifugation at 600*g* for 2 min.
2. Suspend the pellet of swollen cells in 10 vol of the same buffer and homogenize by four to five strokes of a Dounce homogenizer with a tight pestle.
3. Bring the suspension immediately to a concentration of 3 m*M* in Mg^{2+} by addition of 30 m*M* $MgCl_2$, and 10 m*M* NaCl.
4. Centrifuge the homogenate successively at 1000*g* for 1 min and at 10,000*g* for 10 min.
5. Collect membrane vesicles by centrifugation at 100,000*g* for 90 min.
6. Suspend vesicles in 4 *M* guanidine hydrochloride in 10 m*M* Tris, pH 8.0, by homogenization.
7. Layer 1.0 mL (≈5 mg of protein) onto the CsCl/4 *M* guanidine hydrochloride gradient and centrifuge in a Beckman 75 Ti rotor at 4°C for 16–24 h at 55,000 rpm (or another rotor at approx. 100,000*g*).
8. Collect fractions and assay for ASGP-1 by SDS-PAGE.
9. Pool fractions containing ASGP-1 and remove CsCl and guanidine hydrochloride by dialysis/concentration with a collodion bag apparatus.

3.8. Purification of SMC and ASGP-2 from Ascites Cell Microvilli or Membranes (41)

1. Extract ascites cell membranes (500 µL, ≈5 mg protein) in 3.2 mL of 0.2% Triton X-100, 5 m*M* glycine, 2 m*M* EDTA, pH 9.5, for 15 min at room temperature.
2. Load the extract onto a two-phase gradient in which the upper phase consists of 1.0 mL of 0.2% Triton X-100/PBS, pH 7.4, in 4% sucrose, and the lower phase consists of 0.5 mL of 2% SDS/30 m*M* imidazole, pH 7.4, in 10% sucrose.
3. Centrifuge the gradient at 100,000*g* for 1 h in an SW50.1 rotor.
4. Dialyze/concentrate the upper layer, which is enriched in detergent-soluble membrane proteins and SMC, and apply to a discontinuous CsCl/Triton density gradient.
5. Centrifuge the gradients in an SW28 rotor at 100,000*g* for at least 40 h at 4°C.
6. Analyze the gradient fractions for protein and SMC by SDS-PAGE.
7. Pool fraction(s) containing SMC free of lower molecular weight protein contaminants from the CsCl density gradient centrifugation step and dialyze/concentrate against 6 *M* guanidine-HCl, 5 m*M* dithiothreitol, 0.2% Triton X-100, 10 m*M* Tris-HCl, pH 8.0 (complex dissociating buffer).
8. Concentrate to 0.5–1.0 mL, layer onto a column (1.0 × 40 cm) of Sepharose CL-2B, and elute with complex dissociating buffer.
9. Collect fractions and analyze for protein and ASGP-2 by SDS-PAGE.

3.9. Immunoaffinity Purification of Membrane and Soluble Forms of SMC from Rat Trachea (34) (see Note 6)

1. Collect rat tracheas from adult Fischer 344 female rats, snap-freeze in liquid N_2, and pulverize with a mortar and pestle.
2. Solubilize powdered tissues in RIPA buffer/proteinase inhibitors and homogenize with a probe sonicator.

3. Centrifuge the lysate at 2000*g* and use the supernatant for affinity purification.
4. Prepare affinity columns with anti-ASGP-2 or anti-C-pep polyclonal antisera (antipeptide Ab elicited with cytoplasmic domain of rat ASGP-2) and goat antimouse IgG1 (heavy chain specific)-agarose or with anti-ASGP-2 MAb 13C4, following the supplied protocols.
5. Fractionate first on the anti-C-pep column, collecting the soluble form in the flowthrough.
6. Elute column with acid (0.1 *M* glycine, pH 2.8) to obtain the membrane form of SMC.
7. Purify the soluble form further on the anti-ASGP-2 column and elute with acid.
8. Add Triton X-100 to both eluents to a final concentration of 0.05%.
9. Neutralize eluents with 1 *M* Tris (pH 9.5), dialyze against water, and analyze by SDS-PAGE and immunoblotting.

4. Notes

1. This method yields the mucin subunit of MUC1, which is the major glycoprotein detected in the bovine MFGM by PAS staining.
2. This procedure will give purified human MUC1, presumably containing both the mucin and transmembrane subunits, although this has not been established.
3. This procedure yields purified MUC1 mucin subunit. Whether part of the transmembrane subunit is included in this soluble material is unclear, because the form of MUC1 in the skim milk fraction has not been characterized.
4. Other anti-MUC1 antibodies can be used but must be tested for reactivity to the source material, e.g., by immunoprecipitation or immunoblotting.
5. This procedure may yield both mucin and transmembrane subunits of MUC1, although this has not been determined. The stringency of the sample preparation, including a heat step in deoxycholate, may cause subunit dissociation.
6. Anti-SMC antibodies are available from K. L. Carraway for appropriate projects.

Acknowledgments

Original research was supported by NIH grants CA52498 and HD35472. The author wishes to thank Dr. Coralie Carraway for providing Figure 1 and Dr. Carraway and Lynn Herbert for critically reading the manuscript.

References

1. Carraway, K. L. and Hull, S. R. (1991) Cell surface mucin-type glycoproteins and mucin-like domains. *Glycobiology* **1,** 131–138.
2. Gendler, S. J. and Spicer, A. P. (1995) Epithelial mucin genes. *Annu. Rev. Physiol.* **57,** 607–634.
3. Shimizu, M. and Shaw, S. (1993) Mucins in the mainstream. *Nature* **366,** 630, 631.
4. van Klinken, B. J.-W., Dekker, J., Buller, H. A., and Einerhand A. W. C. (1995) Mucin gene structure and expression: protection vs. adhesion. *Am. J. Physiol.* **269,** G613–G627.
5. Rossi, E. A., McNeer, R., Price-Schiavi, S. A., Komatsu, M., Van den Brande, J. M. H., Thompson, J. F., Carraway, C. A. C., Fregien, N. L., and Carraway, K. L. (1996) Sialomucin complex, a heterodimeric glycoprotein complex: expression as a soluble, secretable form in lactating mammary gland and colon. *J. Biol. Chem.* **271,** 33,476–33,485.
6. Codington, J. F. and Frim, D. M. (1983) Cell-surface macromolecular and morphological changes related to allotransplantability in the TA3 tumor. *Biomembranes* **11,** 207–258.
7. Codington, J. F., Sanford, B. H., and Jeanloz, R. W. (1972) Glycoprotein coat of the TA3 cell: isolation and partial characterization of a sialic acid containing glycoprotein fraction. *Biochemistry* **11,** 2559–2564.

8. Codington, J. F., Cooper, A.G., Miller, D. K., Slayter, H. S., Brown, M. C., Silber, C., and Jeanloz, R. W. (1979) Isolation and partial characterization of an epiglycanin-like glycoprotein from a new non-strain-specific subline of TA3 murine mammary adenocarcinoma. *J. Natl. Cancer Inst.* **63,** 153–161.
9. Patton, S., Gendler, S. J., and Spicer, A. P. (1995) The epithelial mucin, MUC1, of milk, mammary gland and other tissues. *Biochim. Biophys. Acta* **1241**, 407–424.
10. Snow, L. D., Colton, D. G., and Carraway, K. L. (1977) Purification and properties of the major sialoglycoprotein of the milk fat globule membrane. *Arch. Biochem. Biophys.* **179,** 690–697.
11. Heydeman, E., Steele, K., and Ormerod, M. G. (1979) A new antigen on the epithelial membrane: its immunoperoxidase localisation in normal and neoplastic tissue. *J. Clin. Pathol.* **32,** 35–39.
12. Ormerod, M. G., Steele, K., Westwood, J. H., and Mazzini, M. N. (1983) Epithelial membrane antigen: partial purification, assay and properties. *Br. J. Cancer* **48,** 533–541.
13. Swallow, D. M., Gendler, S., Griffiths, B., Corney, G., Taylor-Papadimitriou, J., and Bramwell, M. E. (1987) The human tumour-associated epithelial mucins are coded by an expressed hypervariable gene locus PUM. *Nature* **328,** 82–84.
14. Gendler, S., Burchell, J., Duhig, T., Lamport, D., White, R., Parker, M., and Taylor-Papadimitriou, J. (1988) Cloning of partial cDNA encoding differentiation and tumor-associated mucin glycoproteins expressed by human mammary epithelium. *Proc. Natl. Acad. Sci. USA* **84,** 6060–6064.
15. Gendler, S., Taylor-Papadimitriou, J., Duhig, T., Bothbard, J., and Burchell, J. (1988) A highly immunogenic region of a human polymorphic epithelial mucin expressed by carcinomas is made up to tandem repeats. *J. Biol. Chem.* **263,** 12,820–12,823.
16. Siddiqui, J., Abe, M., Hayes, D., Shani, E., Yunis, E., and Kufe, D. (1988) Isolation and sequencing of a cDNA coding for the human DF3 breast carcinoma-associated antigen. *Proc. Natl. Acad. Sci. USA* **85,** 2320–2323.
17. Gendler, S. J., Lancaster, C. A., Taylor-Papadimitriou, J., Duhig, T., Peat, N., Burchell, J., Pemberton, L., Lalani, E., and Wilson, D. (1990) Molecular cloning and expression of human tumor-associated polymorphic epithelial mucin. *J. Biol. Chem.* **265**, 15,286–15,293.
18. Gendler, S. J., Spicer, A. P., Lalani, E.-N., Duhig, T., Peat, N., Burchell, J., Pemberton, L., Boshell, M., and Taylor-Papadimitriou, J. (1991) Structure and biology of a carcinoma-associated mucin, MUC1. *Am. Rev. Respir. Dis.* **144,** S42–S47.
19. Ligtenberg, M. J. L., Kruijshaar, L., Buijs, F., van Meijer, M., Litvinov, S. V., and Hilkens, J. (1992) Cell-associated episialin is a complex containing two proteins derived from a common precursor. *J. Biol. Chem.* **267,** 6171–6177.
20. Hayes, D. F., Ohno, T., Abe, M., Sekine, H., and Kufe, D. W. (1985) Detection of elevated plasma DF3 antigen levels in breast cancer patients. *J. Clin. Invest.***75,** 1671–1678.
21. Zrihan-Licht, S., Vos, H. L., Baruch, A., Elroy-Stein, O., Sagiv, D., Keydar, I., Hilkens, J., and Wreschner, D. H. (1994) Characterization and molecular cloning of a novel MUC1 protein, devoid of tandem repeats, expressed in human breast cancer tissue. *Eur. J. Biochem.* **224,** 787–795.
22. Sherblom, A. P., Huggins, J. W., Chesnut, R. W., Buck, R.L., Ownby, C. L., Dermer, G. B., and Carraway, K. L. (1980) Cell surface properties of ascites sublines of the 13762 rat mammary adenocarcinoma: relationship of the major sialoglycoprotein to xenotransplantability. *Exp. Cell Res.* **126,** 417–426.
23. Buck, R. L., Sherblom, A. P., and Carraway, K. L. (1979) Sialoglycoprotein differences between xenotransplantable and nonxenotransplantable ascites sublines of the 13762 rat mammary adenocarcinoma. *Arch. Biochem. Biophys.* **198,** 12–21.

24. Sherblom, A. P., Buck, R. L., and Carraway, K. L. (1980) Purification of the major sialoglycoproteins of 13762 MAT-B1 and MAT-C1 rat ascites mammary adenocarcinoma cells by density gradient centrifugation in cesium chloride and guanidine hydrochloride. *J. Biol. Chem.* **255,** 783–790.
25. Sherblom, A. P. and Carraway, K. L. (1980) A complex of two cell surface glycoproteins from ascites mammary adenocarcinoma cells. *J. Biol. Chem.* **255,** 12,051–12,059.
26. Sheng, Z., Hull, S. R., and Carraway, K. L. (1990) Biosynthesis of the cell surface sialomucin complex of ascites 13762 rat mammary adenocarcinoma cells from a high Mr precursor. *J. Biol. Chem.* **265,** 8505–8510.
27. Sheng, Z., Wu, K., Carraway, K. L., and Fregien, N. (1992) Molecular cloning of the transmembrane component of the 13762 mammary adenocarcinoma sialomucin complex: A new member of the epidermal growth factor superfamily. *J. Biol. Chem.* **267,** 16,341–16,346.
28. Wu, K., Fregien, N., and Carraway, K. L. (1994) Molecular cloning and sequencing of the mucin subunit of a heterodimeric, bifunctional cell surface glycoprotein complex of ascites rat mammary adenocarcinoma cells. *J. Biol. Chem.* **269,** 11,950–11,955.
29. Komatsu, M., Carraway, C. A. C., Fregien, N. L., and Carraway, K. L. (1997) Reversible disruption of cell-matrix and cell-cell interactions by overexpression of sialomucin complex. *J. Biol. Chem.* **272,** 33,245–33,254.
30. Steck, P. A. and Nicolson, G. L. (1983) Cell surface glycoproteins of 13762NF mammary adenocarcinoma clones of differing metastatic potential. *Exp. Cell Res.* **147,** 255–267.
31. Carraway, K. L. III, Rossi, E.A., Huang, D., Price-Schiavi, S. A., Komatsu, M., Carvajal, M. E., Cantley, L. C., Fregien, N., Carraway, C. A. C., and Carraway, K. L. An intramembrane modulator of the ErbB2 receptor tyrosine kinase that potentiates neuregulin signalling. *J. Biol. Chem.* **274,** 5263–5266.
32. Carraway, K. L., Carraway, C. A. C., and Carraway, K. L. III (1997) ErbB-3 and ErbB-4 in the physiology and pathology of the mammary gland. *J. Mammary Gland Biol. Neoplasia* **2,** 187–198.
33. McNeer, R. R., Price-Schiavi, S., Komatsu, M., Fregien, N., Carraway, C. A. C., and Carraway, K .L. (1997) Sialomucin complex in tumors and tissues. *Front. Biosci.* **2,** 449–459.
34. McNeer, R. R., Huang, D., Fregien, N. L., and Carraway, K. L. (1998) Sialomucin complex in the rat respiratory tract: a model for its role in epithelial protection. *Biochem. J.* **330,** 737–744.
35. McNeer, R. R., Carraway, C. A. C., Fregien, N. L., and Carraway, K. L. (1998) Characterization of the expression and steroid hormone control of sialomucin complex in the rat uterus: implications for uterine receptivity. *J. Cell. Physiol.*, **176,** 110–119.
36. Khatri, I. A., Forstner, G. G. and Forstner, J. F. (1997) The carboxyl-terminal sequence of rat intestinal mucin RMuc3 contains a putative transmembrane region and two EGF-like motifs. *Biochim. Biophys. Acta* **1326,** 7–11.
37. Gum, J. R. Jr., Hicks, J. W., Lagace, R. E., Byrd, J. C., Toribara, N. W., Siddiki, B., Fearney, F. J., Lammport, D. T. A., and Kim, Y. S. (1991) Molecular cloning of rat intestinal mucin: lack of conservation between mammalian species. *J. Biol. Chem.* **266,** 22,733–22,738.
38. Gum, J. R. Jr., Ho, J. J. L., Pratt, W.S., Hicks, J. W., Hill, A.S., Vinall, L. E., Roberton, A. M., Swallow, D. M., and Kim, Y. S. (1997) MUC3 human intestinal mucin: analysis of gene structure, the carboxyl terminus, and a novel upstream repetitive region. *J. Biol. Chem.* **272,** 26,678–26,686.
39. Bardeles, R., Bhavanandan, V. P., Wiseman, G., and Bramwell, M. E. (1989) Purification and characterization of the epitectin from human laryngeal carcinoma cells. *J. Biol. Chem.* **264,** 1980–1987.

40. Hull, S. R., Laine, R. A., Kaizu, T., Rodriguez, I., and Carraway, K. L. (1984) Structures of the O-linked oligosaccharides of the major cell surface sialoglycoprotein of MAT-B1 and MAT-C1 ascites sublines of the 13762 rat mammary adenocarcinoma. *J. Biol. Chem.* **259,** 4866–4877.
41. Hull, S. R., Sheng, Z., Vanderpuye, O., David, C., and Carraway, K. L. (1990) Isolation and partial characterization of ascites sialoglycoprotein-2 (ASGP-2) of the cell surface sialomucin complex of 13762 rat mammary adenocarcinoma cells. *Biochem. J.* **265,** 121–129.

II

Detection and Quantitation of Mucin

3

Histologically Based Methods for Detection of Mucin

Michael D. Walsh and Jeremy R. Jass

1. Introduction

Morphologically based studies on mucins allow structural characterization to be linked to specific sites of synthesis and secretion. The histochemical approach to the study of mucin is therefore highly informative. There is a correspondingly large body of literature documenting the tissue distribution of mucins as demonstrated by mucin histochemistry, lectin histochemistry, and immunohistochemistry (and various combinations of these methods). Two principal issues need to be considered in order to maximize the potential value of morphologically based methodologies: (1) nature and limitations of the individual techniques, and (2) interpretation and reporting of mucin staining.

1.1. Nature and Limitations of Mucin-Staining Methods

Mucin histochemistry, lectin, and immunohistochemistry bring their own advantages and disadvantages to the identification and characterization of epithelial mucin. Remember that mucin can be well visualized with hematoxylin; Ehrlich's hematoxylin stains acid mucins (e.g., of salivary glands and intestinal goblet cells) deep blue. The appearance is sufficiently characteristic to allow a mucin-secreting adenocarcinoma to be diagnosed without the use of specific mucin stains.

Methods of tissue fixation influence mucin-staining. Formalin fixation is adequate for most techniques using light microscopy, but fails to preserve the surface mucous gel layer found throughout the gastrointestinal (GI) tract. Alcohol-based fixatives such as Carnoy's are required to demonstrate this structure *(1)*. The duration of fixation and nature of fixative used play significant roles in determining optimal protocols for the demonstration of glycoproteins including mucins. The exact mechanisms of fixation, particularly aldehyde fixation, remain unclear, although it appears that formalin, e.g., blocks protein amido groups and forms methylene bridges between amino acids, which disturb the natural tertiary structure of proteins, rendering epitopes less amenable to antibody binding to varying degrees *(2)*. Since the initial description by Shi et al.

From: *Methods in Molecular Biology, Vol. 125: Glycoprotein Methods and Protocols: The Mucins*
Edited by: A. Corfield © Humana Press Inc., Totowa, NJ

(3) of a technique for microwave treatment of sections to restore antigenicity, a number of "antigen retrieval" or "antigen unmasking" techniques relying on heat to "unfix" tissues have been rapidly incorporated into the routine histochemical repertoire. Previously efforts to reverse fixation alterations in tissue hinged on the use of proteolytic digestion of sections with enzymes such as trypsin and pepsin. In all cases, the success or failure of these techniques must be determined empirically. **Subheading 3.4.** and **3.5.** discuss a by-no-means exhaustive selection of these techniques.

1.1.1. Mucin Histochemistry

The first specific stain to be used for the demonstration of mucin was mucicarmine *(4)*, but this stain has now been largely supplanted by methods based on more strictly histochemical approaches that utilize a specific chemical reaction (organic, enzymic, or immunological) in which staining intensity correlates directly with the amount of substrate. Periodic acid-Schiff (PAS) is the quintessential mucin histochemical technique *(5)*, with much of current practice bound up with the PAS reaction. Periodic acid breaks the C–C bond in 1:2 glycols of monosaccharides, converting the glycol groups into dialdehydes that are not oxidized further but localized with Schiff's reagent. The intensity of the magenta color reaction is directly proportional to the number of reactive glycol structures.

Several modifications of the PAS stain have been described. These relate to the variable structure of sialic acid and specifically to the presence of *O*-acetyl groups at C_4 and/or the $C_{7\text{-}9}$ side chain. *O*-Acetylation means that the 1:2 glycol groups are no longer available for conversion to dialdehydes. For example, colonic sialic acid is heavily *O*-acetylated and relatively PAS nonreactive. *O*-Acetyl groups can be removed by a saponification step. If preexisting dialdehyde reactivity is first blocked (using borohydride), the sequence periodate borohydride/KOH/PAS will demonstrate *O*-acetyl sialic acid *(6)*. This technique was developed further in the form of periodic acid/thionin Schiff/KOH/PAS (PAT/KOH/PAS) *(6)* to allow simultaneous demonstration of both *O*-acetyl (magenta) and non-*O*-acetyl (blue) sialic acid. The interposition of phenylhydrazine (P) (to block neutral sugar reactivity) and borohydride (Bh) (to improve specificity) represented a subsequent improvement *(7)*. These PAS modifications are complex and have not been incorporated into routine diagnostic practice. They are important, nonetheless, because they provide the only reliable means of differentiating sialic acid variants. A simple modification using mild periodic acid at 4°C (mild PAS) has proved particularly useful for the specific identification of non-*O*-acetyl sialic acid *(8)*.

Acid mucins may be demonstrated by means of cationic dyes (electrostatic binding). Alcian blue (AB) was the first of a family of alcian dyes to be introduced by the ICI chemist Haddock (*see* **ref. 9**). Used initially as a mucin stain by Steedman *(10)*, the dye binds to the carboxyl group of sialic acid or sugars with sulfate substitution. The more highly acidic sulfated mucins can be demonstrated selectively by lowering the pH, as first shown by Mowry *(11)*. AB is often used in combination with PAS. Neutral mucins stain magenta whereas acid mucins stain blue. Many acid mucins are PAS as well as AB reactive and therefore give a deep purple with the AB/PAS sequence.

Sulfate can be stained and differentiated from carboxy groups by aldehyde fuchsin or high-iron diamine (HID), either alone or in combination with AB: aldehyde fuchsin/AB ***(12)*** and HID/AB ***(13)***. The HID/AB technique has been used extensively to distinguish "sialomucin" (blue) from "sulfomucin" (brown). However, since HID and AB are in ionic competition, a brown reaction does not indicate the absence of sialic acid nor does a blue reaction indicate the absence of sulfate. Nevertheless, a change from brown to blue (in colorectal cancer mucin as compared to normal goblet cell mucin) will indicate a generalized alteration of the ratio of sialic acid:sulfate in favor of sialic acid. Despite the requirement for care in the interpretation in results, the carcinogenicity of diamine compounds, and a certain fickleness in the technique itself ***(14)***, the HID/AB technique remains the best method for staining acid mucins.

The structural information that can be obtained from classical mucin histochemistry is, of course, limited. Sialic acid features as a peripheral sugar in virtually all acid mucins, and the strength of mucin histochemistry lies in its ability to demonstrate sialic acid and its *O*-acetylated variants. Conversely, we learn nothing of the actual composition of the oligosaccharide chains or the nature of the sugars substituted with sulfate. For this information, we must turn to lectin histochemistry and immunohistochemistry.

1.1.2. Lectin Histochemistry

Lectins are a diverse group of proteins or glycoproteins found primarily in plant seeds, but also in the fleshy parts of some plants and various invertebrates. They bind to sugars comprising the oligosaccharide chains of glycoproteins and glycolipids along cell membranes as well as those of secretory glycoproteins (mucins). They have been used as hemagglutinins and for stimulating lymphocyte transformation and proliferation. Some lectins, such as *Ricinus communis* agglutinin, are highly toxic. Using either direct or indirect visualization techniques ***(15)***, lectins have been utilized extensively in the study of specific sugars in glycoproteins and glycolipids. Lectins are not only relatively specific, but may react only when sugars are expressed within particular structural configurations. For example, *Ulex europaeus* agglutinin (UEA-1) binds to α-fucose when presented as blood group substance H type 2 or Lewis^{y} but not H type 1 or Lewis^{b} ***(16)***. Similarly, *Sambucus nigra* lectin binds to sialic acid in α2,6 linkage (e.g., as STn) but not in α2,3 linkage ***(17)***. *Trichosanthes japanonica* lectin is even more specific, binding to sialic acid in α2,3 linkage to type 2 backbone structures ***(18)***.

Despite the previously discussed examples, lectins are not necessarily as specific in their binding affinities as is suggested in commercial data sheets or the literature. For example, peanut agglutinin (PNA) binds not only to T-antigen (β-d-Gal1-3GalNAc), but also to structures found within the backbone of oligosaccharides (β-d-Gal1-3/4GlcNAc) ***(19)***. Demonstration of PNA binding is not necessarily evidence of T-antigen expression.

Lectins will bind only to peripherally situated sugars within oligosaccharide chains, the most common are sialic acid, fucose, and *N*-acetylgalactosamine (GalNAc). Since sialic acid may be attached to galactose or GalNAc, lectin binding to these sugars may be demonstrated by removing sialic acid. This has been achieved for galactose using PNA and for GalNAc using *Dolichos biflorus* agglutinin (DBA) within normal and diseased colon *(20,21)*. Strikingly different patterns are observed depending on whether sialic acid has been removed or not. However, note that removal of sialic acid is affected by the presence of *O*-acetyl sialic acid. Colonic sialic acid is heavily *O*-acetylated and therefore resistant to neuraminidase digestion. In various pathological conditions of the colon, *O*-acetyl groups are lost and sialic acid becomes sensitive to neuraminidase. Therefore, the lectin-binding pattern with PNA and DBA is influenced by the specific structural characteristics of substituted sialic acid, which, in turn, is influenced by disease states *(20,21)*.

1.1.3. Immunohistochemistry

Whereas mucin histochemical reagents bind to parts of sugars and lectins bind to whole sugars, antibodies recognize specific sequences of sugars forming blood group substances or still larger molecular arrangements. The structure may be exclusively carbohydrate, a combination of carbohydrate and apomucin (*MUC* gene product), or exclusively apomucin when antibodies have been raised against synthetic MUC peptide sequences *(22)*. Carbohydrate structures may include sialic acid or substituted sulfate *(23)*. The antibody is generally highly specific, but sensitivity for individual components may be low. For example, antibodies generated against STn, SLe^x, or SLe^a only identify sialic acid within the relevant structural conformation. Furthermore, even the correct conformation may not be recognized when the structure of sialic acid is subtly modified by the presence of *O*-acetyl substituents *(24,25)*. Therefore, the high specificity of monoclonal antibodies (MAbs), although advantageous, may lead to errors in interpretation. As in the case of lectin histochemistry, MAb reactivity may be modified by the removal of sialic acid *(20)* and neutral sugars *(26)*. The main advantage of MAbs is in their application to the study of specific blood group substances, core structures, and apomucins, bearing in mind that reactivity may be influenced by relatively small chemical changes or modification in carbohydrate linkages.

Immunohistochemistry is prone to many technical errors. Factors influencing staining patterns and their intensity include the duration and type of fixation, section thickness, the use of various antigen retrieval procedures such as trypsin digestion or heat retrieval, as well as the antibody concentrations. (Note that stored paraffin sections may lose their antigenicity.) These variables should be standardized as much as possible, and negative and positive controls should be incorporated into immunohistochemical staining runs. Many of these caveats apply also to both mucin and lectin histochemistry.

1.2. Interpretation and Reporting of Mucin Staining

The interpretation of mucin staining will be incomplete or even misleading if the results are not integrated with microscopic anatomy in sufficient detail or fail to heed variation that may be owing to differences between anatomical regions or genetic factors.

1. Relationship of the distribution of mucin should be linked to specific cell lineages
   a. Columnar cells elaborating trace amounts of mucin, e.g., "absorptive" cells of the GI tract.

   b. Columnar cells elaborating mucin in intermediate amounts, e.g., the duct epithelium lining the pancreatico-biliary system and the anal glands.
   c. Columnar cells elaborating abundant mucin, e.g., gastric foveolar epithelium and endocervical epithelium.
   d. Classical goblet cells, e.g., within intestinal and bronchial epithelium.
   e. Cuboidal cells lining glands, e.g., bronchial, salivary, submucosal esophageal, pyloric, Brunner's, and mucous neck cells of the stomach.
2. Correlation of normal and malignant lineages: Do malignant mucous-secreting cells have normal counterparts and are these found within the tissue of origin or a different tissue (metaplasia)?
3. Precise localization of mucin within cellular and extracellular compartments
   a. Golgi apparatus.
   b. Cytoplasm.
   c. Apical theca (columnar cells).
   d. Goblet cell theca.
   e. Glycocalyx.
   f. Lumina.
   g. Intracytoplasmic lumina.
   h. Interstitial tissues.
4. Regional variation
   a. Blood group substances (A, B, H, Le^b) and terminal fucose are not expressed by goblet cells in the adult distal colon and rectum *(27)*.
   b. Goblet cells of the proximal colon show more DBA lectin binding than those of the distal colon *(28)*.
   c. There is variation among regions of the GI tract.
5. Cellular maturation
   a. The immature cells of the crypt base epithelium in large intestine express small amounts of apical or glycocalyceal mucin: MUC1 carrying a variety of carbohydrate epitopes (Le^x, Le^y, T-antigen). MUC1 disappears from cells that have entered the mid-crypt compartment *(29)*.
   b. Goblet cells of the lower half of small and large intestinal mucosa express more STn than superficial goblet cells *(24)*.
   c. Goblet cells of the upper crypt and surface epithelium of large intestine show more DBA binding than those of the lower crypt *(28)*.
   d. Columnar and goblet cells of the lower crypt epithelium of large intestine express MUC4 whereas MUC3 is more evident in superficial columnar cells.
6. Hereditary and racial factors
   a. Expression of A, B, and H blood group structures *(27)*.
   b. Blood group secretor status *(27)*.
   c. *O*-acetyl transferase status influencing the structure of colorectal sialic acid ***(30,31)***.

Once a particular anatomical site has been selected for study, it is desirable that the results be presented in a standardized manner. The size of the area to be assessed may be predetermined, but this is more likely to be important for deriving proliferative indices, e.g., rather than interpreting mucin stains. It is necessary to grade random fields, yet, at the same time, the selection of particular fields must be valid. For example, the invasive margin of a tumor may be more informative than an *in situ* component or areas of tumor necrosis.

Assessment may be based on the fraction of positive cells, the intensity of staining (0, +, ++, +++), or a combination of both *(21)*. In general, the fraction of positive cells is likely to be more informative, whereas both factors are critical, e.g., in the assessment of estrogen receptors. Nevertheless, tumor heterogeneity may be problematic, and particular approaches may be required to distinguish focal but intense staining and diffuse but weak staining. Grading of staining intensity is notoriously unreliable in the intermediate range *(32)*. Image analysis is laborious and expensive. Furthermore, immunostaining is only stoichiometric (giving a linear relationship between amount of color absorption and amount of antigen) with low staining intensities that would not be used routinely *(33)*.

Cutoff points may be determined by comparison with existing biochemical findings or by pragmatic clinical correlations. The latter could include survival, tumor recurrence, or response to therapy. The cutoff points will be valid if generated by one observer and verified on additional data sets and by other observers.

By combining the various technical approaches to the demonstration of mucins in tissues and heeding the previously enumerated caveats, it is possible to construct meaningful insights into the structure of mucin and the significance of changes that occur in various disease processes.

2. Materials

1. Mayer's Hematoxylin (*see* **Note 1**): Dissolve 1 g of hematoxylin (BDH, Poole, UK) in 1000 mL of distilled water using heat. Add 50 g of aluminium potassium sulfate ($AlK[SO_4]_2 \cdot 12H_2O$) and dissolve using heat. Then add 0.2 g of sodium iodate ($NaIO_3 \cdot H_2O$) followed by 1 g of citric acid and then 50 g of chloral hydrate ($CCl_3 \cdot CH[OH]_2$). Cool and filter before use.
2. Silanized (adhesive) slides: Clean slides using 2% Deconex detergent and then rinse in distilled water. Rinse in acetone for 2–5 min and treat with 2% 3-aminopropyltriethoxysilane (Sigma, St. Louis, MO) in acetone for 5–15 min. Rinse in two changes of acetone and then one change of distilled water for 2–5 min each. Dry slides overnight and store in dustproof container (*see* **Notes 2** and **3**).
3. Phosphate-buffered saline (PBS): 0.1 *M* phosphate buffer with 0.15 *M* NaCl, pH 7.2–7.4.
4. Tris-buffered saline (TBS): 0.1 *M* Tris-HCl, 0.15 *M* NaCl, pH 7.2–7.4.
5. Histochemical solution—Schiff reagent (Barger and DeLamater) *(34)*: Dissolve 1 g of basic fuchsin (BDH) in 400 mL of distilled water using gentle heat if necessary. Add 1 mL of thionyl chloride ($SOCl_2$), stopper the flask, and allow to stand for 12 h. Add 2 g of activated charcoal, shake, and filter. Store in a stoppered, dark bottle at 4°C. (*see* **Notes 4** and **5**).
6. Freshly filtered 1% Alcian Blue 8GX (BDH) in 3% acetic acid (pH 2.5) and 1% Alcian Blue 8GX in 0.1 *N* HCl (pH 1.0).
7. HID: Dissolve 120 mg of *N,N*-dimethyl-*m*-phenylenediamine dihydrochloride (Sigma) and 20 mg of *N,N*-dimethyl-*p*-phenylenediamine dihydrochloride (Sigma) in 50 mL distilled water. Then add 1.4 mL 40% ferric chloride. The solution pH should be between 1.5–1.6.
8. 0.1% Porcine trypsin (Sigma) in PBS with 0.1% $CaCl_2$.
9. 0.05% 3,3'-diaminobenzidine tetrahydrochloride (Sigma) with 0.0001% H_2O_2 in TBS, pH 7.6.

10. Antigen retrieval solutions: 0.001–0.01 *M* citric acid, pH 6.0 (pH 2.5–6.0), 0.5 *M* Tris-HCl, pH 9.5–10.0 ± 3–6 *M* urea, 0.001 *M* EDTA, pH 8.0, commercial antigen retrieval solutions such as Antigen Retrieval *Glyca* Microwave Solution (BioGenex, San Ramon, CA) or Target Retrieval Solution (Dako, Carpinteria, CA).
11. Deglycosylation reagents: 0.1 U/mL of *Clostridium perfringens* neuraminidase type VI (Sigma) in 0.1 *M* sodium acetate buffer, pH 5.5, 0.6 mU of *O*-glycanase (Genzyme, Cambridge, MA) in 100 mL of 0.1 *M* citrate/phosphate buffer, pH 6.0 containing 100 mg/mL bovine serum albumin (BSA) and 0.02% NaN_3.

3. Methods

3.1. Mucin Histochemistry (see Notes 6–8)

3.1.1. PAS Reaction (see **Note 9**)

1. Dewax 3- to 5-μm paraffin sections in xylene and rehydrate through descending graded alcohols to distilled water.
2. Oxidize for 5-15 min in 1% periodic acid.
3. Wash in running tap water for 5 min and then rinse in distilled water.
4. Treat sections with Schiff's reagent for 10–30 min.
5. Wash for 10 min in running tap water.
6. Counterstain with Mayer's hematoxylin for 2 to 3 min.
7. Wash in running tap water for 5–10 min. Then dehydrate sections through graded alcohols, clear in xylene, and mount with DePeX (BDH) or similar.
8. *Results:* Aldehyde groups formed by oxidation of 1,2-glycol groups are stained deep magenta.

3.1.2. AB Techniques

1. Dewax 3- to 5-μm paraffin sections in xylene and rehydrate through descending graded alcohols to distilled water.
2. Stain in AB 8GX solution, pH 2.5 or 1.0 for 30 min.
3. For sections stained in AB pH 2.5, wash thoroughly in water; for sections stained in AB pH 1.0, rinse briefly in 0.1 *N* HCl and then blot dry on fine-grade filter paper (blotting is not necessary for AB pH 2.5 sections).
4. Dehydrate sections through graded alcohols, clear in xylene, and mount with DePeX or similar.
5. *Results:* AB is a water-soluble copper thalocyanin that binds to acidic groups by an unknown mechanism. Predominantly sulfated mucins will stain blue at pH 1.0, whereas at pH 2.5, acidic mucins will also be stained.

3.1.3. Spicer's (HID) Technique **(13)**

1. Dewax 3- to 5-μm paraffin sections in xylene and rehydrate through descending graded alcohols to distilled water.
2. Stain sections for 24 h in freshly prepared diamine solution.
3. Rinse rapidly in distilled water.
4. Dehydrate sections rapidly through graded alcohols, clear in xylene, and mount with DePeX or similar.
5. *Results:* Sulfomucins are stained grey-purple-black whereas nonsulfated mucins remain unstained.

3.1.4. Methylation (35) (see ***Note 10***)

1. Dewax 3- to 5-μm paraffin sections in xylene and rehydrate through descending graded alcohols to distilled water.
2. Treat sections in preheated 1% HCl in methanol at 60°C for 4 h.
3. Rinse in alcohol.
4. Stain using appropriate histochemical technique.

3.1.5. Saponification (35) (see ***Note 10***)

1. Dewax 3- to 5-μm paraffin sections on adhesive slides in xylene and rehydrate through descending graded alcohols to distilled water.
2. Treat sections with 0.5% KOH in 70% ethanol for 30 min.
3. Rinse carefully in 70% ethanol.
4. Wash in slowly running tap water for 10 min.
5. Stain using appropriate histochemical technique.

3.2. Lectin Histochemistry (see ***Notes 11–13***)

3.2.1. Indirect Peroxidase Technique for Ulex europaeus *Agglutinin I (UEA-I)*

1. Affix 3- to 5-μm sections to adhesive slides and dry overnight at 37°C.
2. Dewax sections and rehydrate through descending graded alcohols to PBS.
3. Incubate the sections in 0.1% trypsin in PBS with 0.1% $CaCl_2$ at 37°C for 20 min.
4. Transfer back to PBS and wash thoroughly in three changes for 5 min.
5. Quench endogenous peroxidase activity by incubating the sections in 1% H_2O_2 and 0.1% NaN_3 in PBS for 10 min.
6. Wash sections in three changes of PBS for 5 min each.
7. Transfer the sections to a humidified chamber and incubate with the lectin, UEA-I (Vector, Burlingame, CA), diluted 1:50 to 1:100 in PBS for 30 min.
8. Wash sections in three changes of PBS for 5 min each.
9. Incubate sections in peroxidase-conjugated rabbit anti-UEA PAb (Dako) diluted 1:100 in PBS.
10. Wash sections in three changes of PBS for 5 min each.
11. Develop color with 3,3'-diaminobenzidine (DAB) with H_2O_2 for 3–5 min (*see* **Note 14**).
12. Wash sections in gently running tap water for 5–10 min to remove excess chromogen.
13. Lightly counterstain sections in Mayer's hematoxylin. Then dehydrate through ascending graded alcohols, clear in xylene, and mount using DePeX or similar.

3.2.2. Inhibition Studies to Confirm Lectin Specificity (37) (see ***Notes 15*** *and* ***16***)

1. Dilute the appropriate competing (inhibiting) sugar in PBS to a concentration in the range of 0.2–0.6 m*M*.
2. Add lectin to a final concentration one-fifth that of the inhibiting sugar and incubate for 30 min to 2 h.
3. Proceed with histochemistry protocol as usual.

3.2.3. Enzymatic Deglycosylation to Confirm Lectin Specificity

See **Subheadings 3.6.1.–3.6.4.**

3.3. Mucin Immunohistochemistry (see Notes 17–19): Standard ABC Immunoperoxidase Technique for Paraffin Sections

1. Affix 3- to 4-µm paraffin sections to adhesive slides and dry overnight at 37°C.
2. Dewax in xylene and rehydrate to water through descending graded alcohols.
3. Transfer to PBS, pH 7.4, and wash in three changes for 5 min each.
4. Block endogenous peroxidase activity using 1.0% H_2O_2 and 0.1% NaN_3 in PBS for 10 min.
5. Wash in three changes of PBS for 5 min each.
6. Incubate sections in 4% commercial skim milk powder in PBS for 15 min (*see* **Note 20**).
7. Wash briefly in PBS to remove excess milk solution. Then place sections flat in a humidified chamber and apply 10% normal (nonimmune) goat serum (Zymed, San Francisco, CA) for 20 min.
8. Decant excess serum and then apply primary antibody in PBS or TBS (*see* **Note 21**).
9. Wash in three changes of PBS for 5 min each.
10. Incubate with biotinylated goat antimouse or rabbit immunoglobulins (Jackson ImmunoResearch, West Grove, PA) diluted 1:300–1:500 for 30 min.
11. Wash in three changes of PBS for 5 min each.
12. Incubate sections with streptavidin-horseradish peroxidase (HRP) conjugate (Jackson) diluted 1:250–1:400 for 30 min.
13. Wash in three changes of PBS for 5 min each.
14. Develop color in DAB solution with H_2O_2 for 3–5 min.
15. Wash well in gently running tap water to remove excess chromogen.
16. Lightly counterstain in Mayer's hematoxylin.
17. Dehydrate through ascending graded alcohols, clear in xylene, and then mount with DePeX or similar.

3.4. Antigen Retrieval Techniques: Microwave Heat Retrieval (see Notes 22 and 23)

Antigen retrieval or unmasking steps are typically inserted into immunohistochemistry protocols following tissue rehydration and prior to the commencement of the staining protocol.

1. Place sections in a rack in a covered, heat proof vessel with antigen retrieval solution.
2. Place in microwave oven and set power on "high" and heat the solution so that it boils for 5 min.
3. Transfer sections to fresh antigen retrieval solution and repeat process.
4. Remove from microwave oven and permit sections to cool in antigen retrieval (AR) solution for approx 20–30 min.
5. Transfer to PBS or TBS and proceed with immunohistochemical protocol.

3.5. Proteolytic Digestion Techniques

Similar to heat antigen retrieval, proteolytic digestion to improve tissue antigenicity is routinely performed following tissue rehydration but prior to the commencement of the staining routine.

1. Transfer to distilled water and wash in three changes for 2 min each.
2. Incubate sections in 0.4% pepsin (Sigma) in 0.1 *N* HCl at 37°C for 90 min or 0.1% pronase E (Sigma) in PBS or 0.1% trypsin (Sigma) in PBS, pH 7.4, with 0.1% $CaCl_2$ at 37°C

for 10–30 min. Sections to be digested with pepsin should be rinsed in 0.1 *N* HCl prior to incubation in the enzyme solution.
3. Wash in three changes of PBS for 5 min each.

3.6. Deglycosylation Techniques (see Note 24)

3.6.1. Alkaline Hydrolysis

Ono et al. **(38)** described a β-elimination protocol that uses prolonged incubation of sections in an alkaline alcohol solution to strip sugars.

1. Dewax sections and rehydrate through graded alcohols.
2. Coat sections with collodion (Acros Organics, Geel, Belgium) in ether alcohol for 2 min (*see* **Note 25**).
3. Air-dry for 1 min and then immerse section in 70% ethanol for 3 min.
4. Incubate sections in alkaline ethanol solution for 3–7 d at 4°C.
5. Rinse in three changes of 70% ethanol.
6. Repeat **steps 2–5.**
7. Rinse in distilled water.
8. Proceed with histochemical protocol of choice.

3.6.2. Periodic Acid Deglycosylation (39,40)

1. Dewax sections and rehydrate through descending graded alcohols to PBS.
2. Incubate sections in 1–100 m*M* periodic acid in 0.05 *M* acetate buffer, pH 5.0, for 30 min at room temperature.
3. Neutralize acidic reactive groups by incubating in 1% glycine in distilled water for 30 min.
4. Wash thoroughly in PBS.
5. Proceed with histochemical or immunohistochemical protocol.

3.6.3. Neuraminidase Deglycosylation (41,42) (see Note 26)

1. Dewax sections and rehydrate to PBS.
2. Incubate sections with neuraminidase solution for 2 h at 37ºC.
3. Rinse in three changes of ice-cold distilled water.
4. Transfer sections back to PBS.
5. Proceed with immunohistochemical protocol.

3.6.4 O-Glycanase Deglycosylation (43)

1. Dewax sections and rehydrate to PBS.
2. Incubate sections in *O*-glycanase solution for 18 h at 37ºC.
3. Wash well in PBS and then proceed with (immuno)histochemical protocol.

4. Notes

1. This may also be purchased from many laboratory chemical suppliers premade.
2. Detachment of sections from slides during processing of histochemical procedures is a common complaint, particularly when the sections are subjected to heat antigen retrieval (**Subheading 3.4.**), proteolytic digestion (**Subheading 3.5.**), or deglycosylation (**Subheading 3.6.**). This problem may be largely resolved by using charged or coated slides. In our experience, the best choice of slide treatment is silanization. Another alternative, although inferior to silanization, is precoating slides with 0.1% poly-L-lysine or 1% gelatin.

3. Alternatively, prepared adhesive slides such as Superfrost Plus®/(Menzel-Gläser, Braunschweig, Germany) may be purchased.
4. There are many other methods for preparation of Schiff reagent apart from the Barger and DeLamater's technique. All rely on the principle of decolorizing basic fuchsin or similar dyes by removal of the dye's quinoid structure using sulfurous acid. In the presence of aldehydes, the quinoid structure, and hence dye color, is restored.
5. Thionyl chloride is noxious and should be handled only in a fume hood.
6. Many if not most (immuno)histochemical studies of mucins utilize tissues that have been fixed in neutral buffered formaldehyde and then embedded in paraffin. The reasons are largely the accessibility of archival material from pathology departments, superior morphological tissue preservation over frozen sections, and the ease of handling of paraffin sections over frozen sections. Note, however, that exposure to the solvents and high temperatures required for paraffin embedding may alter epitopes sufficiently to render them suboptimally detectable. Frozen sections have the advantage of being less chemically altered than processed tissue. It is frequently desirable to compare histochemical reactivity with mucin probes, whether histochemical, lectin, or immunohistochemical in nature, in frozen sections with that achieved using paraffin-embedded tissue to ensure that comparable detection in the latter is being achieved.
7. Fixatives are used to inhibit autolytic changes in tissues following removal of tissues or cells, and may be broadly classified as either crosslinking (e.g., aldehydes) or coagulating or precipitating (e.g., alcohol). The most common fixatives are formaldehyde based (e.g., neutral buffered formaldehyde), and there are many lectins and antibodies that will successfully react with formaldehyde-fixed tissues. It has been demonstrated, however, that some mucin antibodies react best with tissues fixed in solutions other than formalin, e.g., methacarn (methanol:chloroform:acetone, 60:30:10) ***(44)***.
8. It is beyond the scope of the current work to discuss the full range of histochemical techniques that have been described for demonstrating carbohydrate groups commonly associated with mucins. Histotechnology reference texts such as Culling's *Handbook of Histopathological and Histochemical Techniques* (including museum techniques) ***(36)*** or Kiernan's *Histological and Histochemical Methods: Theory and Practice* ***(45)*** provide comprehensive sections on useful histochemical stains. We intend here to present only a limited selection of techniques by way of example. Remember that, in many instances, the mechanism(s) underlying some techniques remain unclear, and hence results must be interpreted with caution.
9. Although many histopathology laboratories prefer to make their own Schiff reagent, premade reagent may be purchased from many commercial suppliers.
10. Methylation converts carboxyl groups to methyl esters as well as desulfating glycoproteins. Saponification may reverse the methyl esters formed by methylation, as well as increase the PAS reactivity of some GI tract mucins.
11. A number of protocols are available for quenching endogenous peroxidase activity in tissues, most relying on incubation with H_2O_2 in methanol or PBS. We routinely use H_2O_2 with NaN_3 because this has been demonstrated to have superior quenching capability ***(46)*** without incurring the cost or inconvenience of using glucose oxidase/glucose to generate nascent H_2O_2. The use of phenylhydrazine appears to have fallen out of favor.
12. For laboratories using alkaline phosphatase-labeled antibodies, the inhibitor of endogenous enzymatic activity is 1 m*M* levamisole ***(47)***. One should be aware that certain alkaline phosphatases (AlkPhos) particularly those of GI origin, are poorly inhibited by levamisole.

13. Endothelial cells within the section act as an in-built positive control for UEA-I histochemistry. It is important to consider the requirements for negative and positive controls in histochemistry because there are a number of pitfalls that may produce deceptively promising yet totally spurious results. As a negative control, the simplest option is to stain a serial section, omitting the primary lectin or antibody, with all other steps in the protocol identical to the test sample. For immunohistochemistry, an even better approach is substitution of another antibody of the same donor species (preferably the same isotype as the test antibody) that is nonreactive with the tissues being examined. For example, we use an antibody directed against α-gliadin found in wheat gluten ***(48)*** as a control antibody for murine IgG1 MAbs. Equally important is the question of positive controls; ideally the tissues under examination will have an in-built control as with the UEA-I reactivity in endothelial cells of blood vessels. Whereas at least some positive structures are not guaranteed in every section, sections from positive tissue that been fixed and processed in the same manner as the test samples should be included with each staining run.
14. The chromogen of choice for most immunohistochemistry remains DAB, owing to its fine localization in sections, resistance to solvents, thus permitting permanent mounting, and its relative stability. There are, however, several other alternatives, such as 3-amino-9-ethylcarbazole (AEC), for use with horseradish peroxidase (HRP); 5-bromo-4-chloro-3-indolyl phosphate/nitro blue tetrazolium (BCIP/NBT); Vector Red (Vector); Fast Red, New Fuschin, and Fast Red (BioGenex) with alkaline phosphatase; and NBT for glucose oxidase. Care should be exercised when choosing the chromogen since only some of these, e.g., New Fuchsin, Vector Red, and BCIP/NBT, are solvent resistant, thus permitting permanent mounting. Other chromogens such as AEC must be either mounted using an aqueous mounting medium or "permanentized" using Ultramount (Dako) prior to permanent mounting.
15. Comprehensive discussions of lectins and their inhibiting hapten sugars may be found in **refs. *49*** and ***50***, as well as from suppliers such as Vector.
16. Vector has an excellent range of lectins suitable for histochemical applications, whereas Sigma has a range of inhibiting sugars and also some lectins.
17. The degree of sensitivity of immunohistochemistry (amount of specific signal) is largely governed by the final number of enzymes (HRP or AlkPhos) deposited/bound to tissue antigen. Ranked from least to most sensitive, the options are direct immunohistochemistry, indirect immunohistochemistry, biotin-streptavidin systems (the ABC techniques), and catabolyzed signal amplification using biotinyl tyramide.
18. The presence of biotin and other molecules capable of binding (strept)avidin may elicit apparent positive activity in epithelial tissues when using an ABC-type staining protocol. Generally, the process of paraffin embedding destroys or masks this activity so that it poses little or no problem in immunohistochemistry. In cell preparations and frozen sections, this activity may, however, pose a significant problem. It is easily remedied by preincubation of sections or cell cultures/cytospins with 0.1% avidin in 0.05 *M* Tris-HCl followed by 1% D-biotin in 0.05 *M* Tris-HCl for 10 min each with thorough washing in PBS following each incubation.
19. Several approaches singly or in combination may assist with reduction of superfluous noise to signal ("background") in immunohistochemical preparations:
    a. Reducing the concentrations of antibodies.
    b. Adding a detergent such as Tween-20 (0.05–0.5%) or Triton X-100 (0.1–1%) to the wash buffers.
    c. Increasing the osmolarity of the buffers used for antibody dilution and washing.

   d. Decreasing the concentration of the primary antibody and incubating for longer periods, e.g., overnight at 4°C instead of shorter incubations at room temperature.
   e. Adding protein to the diluent buffer for the primary antibody in the form of 2–5% BSA or up to 15% nonimmune serum from the donor species of the secondary antibody in the buffer.
20. In many laboratories nonspecific antibody binding during immunohistochemistry is inhibited by preincubating sections with milk powder (casein) or nonimmune serum. In instances in which background staining is a problem, it may be beneficial to use the two techniques in tandem, as shown.
21. The length of incubation (from 20–30 min to overnight), the temperature at which the incubation is performed (4°C, room temperature, or 37°C), and the concentration of antibody used (typically primary antibody concentrations for immunohistochemistry are in the range of 0.5–10 µg/mL) must all be determined empirically.
22. In recent years, heat-based antigen retrieval, first described by Shi et al. ***(3)***, has been applied to achieve reactivity in paraffin-embedded tissues, with, in some instances, astounding success. The mechanism by which heat retrieval works remains unclear, but appears to remove crosslinkages between proteins induced by aldehyde. The choice of antigen retrieval solution as well as the duration of exposure of the tissue sections to heating are largely determined by epitope of interest, and should be determined empirically, although 0.01 *M* citric acid, pH 6.0, is the most widely used AR solution. As an alternative to using a microwave oven, antigen retrieval may also be performed in a wet autoclave or pressure cooker ***(51,52)***.
23. One should be aware that heat-based antigen retrieval may restore or induce endogenous biotin or biotin-like activity in tissues, which may produce spurious results if using a biotin-streptavidin detection system. The most commonly affected tissues for observing this artifact are kidney and liver, although, to a lesser extent, strenuous antigen retrieval may also produce this effect in colorectal tissues and thyroid. Any antigen retrieval should be checked by staining with streptavidin/HRP alone to determine whether there is biotin or biotin-like activity in the tissues that should be suppressed prior to the commencement of the immunohistochemistry routine.
24. Deglycosylation of mucin glycoproteins is a useful tool in many immunohistochemical protocols serving to expose peptide epitopes for antibody binding. The immunoreactivity of certain MUC1 antibodies, such as SM-3 and MUSE-11 and the MUC5AC antibody M1, is significantly enhanced following deglycosylation of sections prior to immunostaining. ***(53,54)***. Both fucose and sialic acid have been shown to inhibit the binding of antibodies to MUC1 core peptide epitopes ***(40,55)***.
25. We have found that sections will stay attached to silanized (adhesive) slides during shorter periods of hydrolysis (up to 3 d), obviating the need for using collodion.
26. Desialation of mucins also permits confirmation of the sialylated nature of carbohydrate structures such as sialosyl Tn and sialyl Le^X, as well as confirmation of carbohydrate binding by using a range of techniques, with the absence of staining in degylcosylated sections implying carbohydrate target structures. Like antigen retrieval and enzymatic digestion, these steps are usually incorporated prior to commencement of the histochemistry protocol proper.

References

1. Matsuo, K., Ota, H., Akamatsu, T., Sugiyama, A., and Katsuyama, T. (1997) Histochemistry of the surface mucous gel layer of the human colon. *Gut* **40,** 782–789.

2. Fox, C. H., Johnson, F. B., Whiting, J., and Roller, P. P. (1985) Formaldehyde fixation. *J. Histochem. Cytochem.* **33,** 845–853.
3. Shi, S. R., Key, M. E., and Kalra, K. L. (1991) Antigen retrieval in formalin-fixed, paraffin embedded tissues: an enhancement method for immunohistochemical staining based upon microwave oven heating of tissue sections. *J. Histochem. Cytochem.* **39,** 741–748.
4. Southgate, H. W. (1927) Note on preparing mucicarmine. *J. Pathol. Bacteriol.* **30,** 729.
5. Hotchkiss, R. D. (1948) A microchemical reaction resulting in the staining of polysacchaide structures in fixed tissue preparations. *Arch. Biochem.* **16,** 131–141.
6. Culling, C. F., Reid, P. E., and Dunn, W. L. (1976) A new histochemical method for the identification and visualization of both side chain acylated and nonacylated sialic acids. *J. Histochem. Cytochem.* **24,** 1225–1230.
7. Reid, P. E., Dunn, W. L., Ramey, C. W., Coret, E., Trueman, L., and Clay, M. G. (1984) Histochemical identification of side chain substituted *O*-acetylated sialic acids: the PAT-KOH-Bh-PAS and the PAPT-KOH-Bh-PAS procedures. *Histochem. J.* **16,** 623–639.
8. Veh, R. W., Meesen, D., Kuntz, D., and May, B. (1982) Histochemical demonstration of sidechain substituted sialic acid, in *Colonic Carcinogenesis* (Malt, R. A. and Williamson, R. C. N., eds.), MTP, Lancaster, UK, pp. 355–365.
9. Stevens, A. (1990) *Theory and Practice of Histological Techniques*, Churchill Livingstone, Edinburgh, UK, p. 107.
10. Steedman, H. F. (1950) Alcian 8GS: a new stain for mucin. *Quart. J. Microscop. Sci.* **91,** 477–479.
11. Mowry, R. W. (1958) Alcian blue techniques for the histochemical study of acidic carbohydrates. *J. Histochem. Cytochem.* **6,** 82.
12. Spicer, S. S. and Meyer, D. B. (1960) Histochemical differentiation of acidic mucopolysaccharides by means of combined aldehyde fuchsin-alcian blue staining. *Am. J. Clin. Pathol.* **33**, 453.
13. Spicer, S. S. (1965) Diamine methods for differentiating mucosubstances histochemically. *J. Histochem. Cytochem.* **13,** 211–234.
14. Williams, G. T. (1985) Transitional mucosa of the large intestine. *Histopathology* **9,** 1237–1243.
15. Leathem, A. and Atkins, N. (1983) Lectin binding to formalin-fixed paraffin sections. *J. Clin. Pathol.* **36,** 747–750.
16. Hindsgaul, O., Norberg, T., Le Pendu, J., and Lemieux, R. U. (1982) Synthesis of type 2 human blood-group antigenic determinants: the H, X, and Y haptens and variations of the H type 2 determinant as probes for the combining site of the lectin 1 of Ulex europaeus. *Carbohydr. Res.* **109,** 109–142.
17. Sata, T., Roth, J., Zuber, C., Stamm, B., and Heitz, P. U. (1991) Expression of alpha 2,6-linked sialic acid residues in neoplastic but not in normal human colonic mucosa: a lectin-gold cytochemical study with Sambucus nigra and Maackia amurensis lectins. *Am. J. Pathol.* **139,** 1435–1448.
18. Sakiyama, T., Yamashita, K., Ihida, K., Nishimata, H., Arima, T., and Murata, F. (1995) Trichosanthes Japonica agglutinin I staining of human colonic carcinoma: a comparative study using monoclonal antibody against sialosyl-Tn antigen. *Acta Histochem. Cytochem.* **28,** 155–162.
19. Picard, J. K. and Feizi, T. (1983) Peanut lectin and anti-Ii antibodies reveal structural differences among human gastrointestinal glycoproteins. *Mol. Immunol.* **20,** 1215–1220.
20. Jass, J. R., Allison, L. J., and Edgar, S. G. (1995) Distribution of sialosyl Tn and Tn antigens within normal and malignant colorectal epithelium. *J. Pathol.* **176,** 143–149.

21. Jass, J. R. and Smith, M. (1992) Sialic acid and epithelial differentiation in colorectal polyps and cancer—a morphological, mucin and lectin histochemical study. *Pathology* **24,** 233–242.
22. Xing, P. X., Prenzoska, J., Layton, G. T., Devine, P. L., and McKenzie, I. F. (1992) Second-generation monoclonal antibodies to intestinal MUC2 peptide reactive with colon cancer. *J. Natl. Cancer Inst.* **84,** 699–703.
23. Irimura, T., Wynn, D. M., Hager, L. G., Cleary, K. R., and Ota, D. M. (1991) Human colonic sulfomucin identified by a specific monoclonal antibody. *Cancer Res.* **51,** 5728–5735.
24. Jass, J. R., Allison, L. M., and Edgar, S. (1994) Monoclonal antibody TKH2 to the cancer-associated epitope sialosyl Tn shows cross-reactivity with variants of normal colorectal goblet cell mucin. *Pathology* **26,** 418–422.
25. Ogata, S., Ho, I., Chen, A., Dubois, D., Maklansky, J., Singhal, A., Hahomori, S., Itzkowitz, S. H. (1995) Tumor-associated sialylated antigens are constitutively expressed in normal human colonic mucosa. *Cancer Res.* **55,** 1869–1874.
26. Ajioka, Y., Xing, P. X., Hinoda, Y., and Jass, J. R. (1997) Correlative histochemical study providing evidence for the dual nature of human colorectal cancer mucin. *Histochemical J.* **29,** 143–152.
27. Yuan, M., Itzkowitz, S. H., Palekar, A., Shamsuddin, A. M., Phelps, P. C., Trump, B. F., Kim, Y. S. (1985) Distribution of blood group antigens A, B, H, Lewis a, and Lewis b in human normal, fetal, and malignant colonic tissue. *Cancer Res.* **45,** 4499–4511.
28. Jass, J. R., Allison, L. J., Stewart, S. M., and Lane, M. R. (1994) Dilochus biflorus agglutinin binding in hereditary bowel cancer. *Pathology* **26,** 110–114.
29. Ajioka, Y., Allison, L. J., and Jass, J. R. (1996) Significance of MUC1 and MUC2 mucin expression in colorectal cancer. *J. Clin. Pathol.* **49,** 560–564.
30. Sugihara, K. and Jass, J. R. (1986) Colorectal goblet cell sialomucin heterogeneity: its relation to malignant disease. *J. Clin. Pathol.* **39,** 1088–1095.
31. Fuller, C. E., Davies, R. P., Williams, G. T., and Williams, E. D. (1990) Crypt restricted heterogeneity of goblet cell mucus glycoprotein in histologically normal human colonic mucosa: a potential marker of somatic mutation. *Br. J. Cancer* **61,** 382–384.
32. Van Diest, P. J., Van Dam, P., Henzen-Logmans, S. C., Berns, E., van der Burg, M. E. L., Green, J., Veragte, I. et al. (1997) A scoring system for immunohistochemical staining: consensus report of the task force for basic research of the EORTC-GCCG. *J. Clin. Pathol.* **50,** 801–804.
33. Fritz, P., Wu, X., Tuczek, H., Multhaupt, H., and Schwarzmann, P. (1995) Quantitation in immunohistochemistry: a research method or a diagnostic tool in surgical pathology? *Pathologica* **87,** 300–309.
34. Barger, J. D. and DeLamater, E. D. (1948) The use of thionyl chloride in the preparation of Schiff's reagent. *Science* **108,** 121–122.
35. Spicer, S. S. (1960) A correlative study of the histochemical properties of rodent acid mucopoly saccharides. *J. Histochem. Cytochem.* **8,** 18.
36. Culling, C. F. A. (1974) *Handbook of Histopathological and Histochemical Techniques*, Butterworths, London.
37. Colton, C. A., Abel, C., Patchett, J., Keri, J., and Yao, J. (1992) Lectin staining of cultured CNS microglia. *J. Histochem. Cytochem.* **40,** 505–512.
38. Ono, K., Katsuyama, T., and Hotchi, M. (1983) Histochemical application of mild alkaline hydrolysis for selective elimination of *O*-glycosidically linked glycoproteins. *Stain Technol.* **58,** 309–312.
39. Bara, J., Decaens, C., Loridon-Rosa, B., and Oriol, R. (1992) Immunohistological char-

acterization of mucin epitopes by pre-treatment of gastrointestinal sections with periodic acid. *J. Immunol. Methods* **149,** 105–113.

40. Bara, J., Imberty, A., Perez, S., Imai, K., Yachi, A., and Oriol, R. (1993) A fucose residue can mask the MUC-1 epitopes in normal and cancerous gastric mucosae. *Int. J. Cancer* **54,** 607–613.
41. Ohno, J., Ohshima, Y., Arakaki, Z., Yokoyama, S., and Utsumi, N. (1993) Immunohistochemical detection of sialyl Le(x) antigen on mucosal Langerhans cells of human oral mucosa following neuraminidase pretreatment. *Biotech. Histochem.* **68,** 284–289.
42. Reis, C. A., David, L., Nielsen, P. A., Clausen, H., Mirgorodskaya, K., Roepstorff, P., and Sobrinho-Simoes, M. (1997) Immunohistochemical study of MUC5AC expression in human gastric carcinomas using a novel monoclonal antibody. *Int. J. Cancer* **74,** 112–121.
43. Akashi, K., Hinoda, Y., Itoh, F., Adachi, M., Endo, T., and Imai, K. (1997) A novel gastric-cancer-associated mucin antigen defined by a monoclonal antibody A3D4. *Int. J. Cancer* **73,** 795–801.
44. Bartek, J., Durban, E. M., Hallowes, R. C., and Taylor-Papadimitriou, J. (1985) A subclass of luminal epithelial cells in the human mammary gland, defined by antibodies to cytokeratins. *J. Cell Sci.* **75,** 17–33.
45. Kiernan, J. A. (1990) *Histological and Histochemical Methods: Theory and Practice,* Pergamon, Oxford, UK.
46. Andrew, S. M. and Jasani, B. (1987) An improved method for the inhibition of endogenous peroxidase non-deleterious to lymphocyte surface markers: application to immunoperoxidase studies on eosinophil-rich tissue preparations. *Histochem. J.* **19,** 426–430.
47. Ponder, B. A. and Wilkinson, M. M. (1981) Inhibition of endogenous tissue alkaline phosphatase with the use of alkaline phosphatase conjugates in immunohistochemistry. *J. Histochem. Cytochem.* **29,** 981–984.
48. Hill, A. S. and Skerritt, J. H. (1989) Monoclonal antibody-based two-site enzyme-immunoassays for wheat gluten proteins. 1. Kinetic characteristics and comparison with other ELISA formats. *Food Agric. Immunol.* **1,** 147–160.
49. Rhodes, J. M. and Milton, J. D. (1986) *Lectin Methods and Protocols,* Humana, Totowa, NJ.
50. Liener, I. E., Sharon, N., and Goldstein, I. J. (1998) *The Lectins: Properties, Functions, and Applications in Biology and Medicine,* Academic, Orlando, FL.
51. Bankfalvi, A., Navabi, H., Bier, B., Bocker, W., Jasani, B., and Schmid, K. W. (1994) Wet autoclave pretreatment for antigen retrieval in diagnostic immunohistochemistry. *J. Pathol.* **174,** 223–228.
52. Miller, R. T. and Estran, C. (1995) Heat induced epitope retrieval with a pressure cooker: suggestions for optimal use. *Appl. Immunohistochem.* **3,** 190–193.
53. Cao, Y., Blohm, D., Ghadimi, B. M., Stosiek, P., Xing, P. X., and Karsten, U. (1997) Mucins (MUC1 and MUC3) of gastrointestinal and breast epithelia reveal different and heterogeneous tumor associated aberrations in glycosylation. *J. Histochem. Cytochem.* **45,** 1547–1557.
54. Bara, J., Chastre, E., Mahiou, J., Singh, R. L., Forguelafitte, M. E., Hollande, E., and Godeau, F. (1998) Gastric M1 mucin, an early oncofetal marker of colon carcinogenesis, is encoded by the MUC5AC gene. *Int. J. Cancer* **75,** 767–773.
55. Ho, J. J. L., Cheng, S., and Kim, Y. S. (1995) Access to peptide regions of a surface mucin (MUC1) is reduced by sialic acids. *Biochem. Biophys. Res. Commun.* **210,** 866–873.

4

Detection and Quantitation of Mucins Using Chemical, Lectin, and Antibody Methods

Michael A. McGuckin and David J. Thornton

1. Introduction

Detection and quantitation of mucins can be important in both the research and clinical settings. Applications may range from detection of potentially novel mucins present during purification from mucus, to quantitation of specific mucin core proteins or carbohydrate moieties present in clinical samples. This chapter discusses procedures and limitations of several different strategies available to detect and quantify these glycoproteins from biological samples, with a view to providing guidelines from which to select the best applicable techniques. Example protocols are then provided to give a starting point for development of a technique. Refer to Chapter 3 for detection of mucins in histological preparations *(1)*; note, however, that many of the principles for selection of detection tools discussed herein are applicable to histological detection.

Because of the extreme size and extent of glycosylation of mucins, coupled with the fact that many secreted mucins are capable of forming gels, these glycoproteins can be quite difficult to work with biochemically. It is therefore extremely important before attempting to detect mucins that the researcher has a good understanding of the behavior of these molecules in solution, particularly with regard to their potential lack of solubility in standard physiological buffers. Because of these properties, standard preparative methods for secreted mucins involve extraction in chaotropic agents (usually 6 *M* guanidinium chloride) and purification in CsCl density gradients in either the presence or absence of 4 *M* guanidinium chloride. Therefore, methods often have to be applicable to assay in the presence of high concentrations of these agents. Failure to adhere to these considerations may result in embarrassing false-negative results. Readers are advised to refer to Chapters 1 and 2 *(2,3)* of this volume for the preparation of secreted and membrane-associated mucins, respectively, and to Chapter 7 *(4)* for a discussion of methods for mucin separation.

From: *Methods in Molecular Biology, Vol. 125: Glycoprotein Methods and Protocols: The Mucins*
Edited by: A. Corfield © Humana Press Inc., Totowa, NJ

Selection of a technique to detect mucins should be influenced by several factors, including knowledge of the core protein sequence of and/or carbohydrate structures present on the mucin(s) to be measured, nature of the sample (buffer, presence of potential interfering substances), specificity of the data required, availability of specific detection tools, degree of quantitation required, and the number of samples to be processed. Owing to the high *O*-linked carbohydrate content of mucins (as much as 90% of the total weight), many assays are targeted toward this portion of the molecule. Although these tend to be useful general methods for detecting mucins, they are not good tools for distinguishing between specific mucin (*MUC*) gene products; this is even true of carbohydrate-specific monoclonal antibodies (MAbs), which can show crossreactivity between mucins. However, mucin-specific probes are available; these are commonly antibodies raised against peptide sequences from within the different mucin polypeptides. Although these are more specific detection tools, note that the different *MUC* gene products can share regions of homology and therefore cross-reactivity ***(5)***. Many of the early MUC-specific probes were generated against sequences underlying the highly glycosylated tandem repeat regions of the molecules and, although effective against the protein precursors, were of little use for mature mucins. Nevertheless, chemical and/or enzymatic deglycosylation techniques can be used to increase the effectiveness of these probes for mature mucins ***(6-8)***. Detection of mucin core proteins produced by cultured cells can often be enhanced by culture in the presence of competitive inhibitors of *O*-glycosylation, such as benzyl 2-acetamido-2-deoxy-α-D-galactopyranoside, without adversely affecting cells ***(9)***. With the elucidation of more sequence data for mucins, it has become possible to target probes at less glycosylated portions of the molecules; however, a drawback of these probes is that their epitopes tend to be cryptic and need reduction to be exposed.

In summary, the extent of mucin glycosylation influences both carbohydrate- and peptide-specific techniques and must be considered in choosing or developing detection strategies. Regardless of the technique selected as most appropriate, it is recommended, where possible, to verify mucin detection using an additional technique of differing principle, particularly when quantitation is important. Note that a feature that can be an advantage for one application could be a disadvantage for another application. For example, use of specific peptide-reactive antibodies for known mucin core proteins may be the method of choice when specifically quantitating these mucins, but would not be suitable for detection of the total population of mucin in heterogeneous mixtures during purification from mucus because mucins that are yet to be characterized will be excluded from the determination. Detection with antibodies or lectins reactive with commonly expressed carbohydrate groups or simple detection with periodic acid-Schiff (PAS) is more appropriate for the latter application.

Detection of mucins in solution using chemical techniques relies on reactions involving mucin carbohydrate groups and is probably most useful for rapid semiquantitative determination of mucin recovery during purification steps. The main disadvantages of these techniques are interference from nonglycoproteins (lipids, pigments), a lack of specificity (nonmucin glycoproteins can react, no carbohydrate or core protein specificity), and lower sensitivity than slot-blot and immunoassay methods. Chapter 1 *(2)* discusses these assays and they are mentioned here only for completeness. A number of general carbohydrate assays have been used for the detection of mucins, and two of the more popular are the anthrone assay (as both a manual and an automated procedure) ***(10)*** and the PAS reaction ***(11)***. In addition, manual and automated assays have been developed using periodate oxidation and detection with the resorcinol reagent for the determination of sialic acid, which is quite often a constituent of mucin oligosaccharides ***(12)***. A fluorometric assay utilizing alkaline β-elimination and derivitization with 2-cyanoacetamide has been described but is subject to significant interference by CsCl ***(13)***. Part VI of this volume describes more elaborate carbohydrate-specific analytical techniques. Although the determination of A_{280} should not be used for estimating concentrations of mucin owing to very low content of aromatic amino acids, it can be useful for assessing removal of contaminating nonmucin proteins during purification procedures.

2. Materials

1. Immunoassays are most conveniently performed in 96-well plates using 50- to 100-μL incubation volumes; plates with a range of protein-binding properties are commercially available.
2. Immunoassay buffers: CB = 0.1 *M* carbonate buffer, pH 9.6; phosphate-buffered saline (PBS) = 0.05 *M* phosphate, 0.9% (w/v) NaCl, 0.02% KCl, pH 7.2; Tris-buffered saline (TBS) = 0.01 *M* Tris-HCl, 0.9% (w/v) NaCl, pH 7.5.
3. Blocking solutions for enzyme-linked immunosorbent assay (ELISA) and immunoblotting: 10% (w/v) skim milk powder, 1–5% bovine serum albumin, 1–5% casein, or 10% (v/v) serum (of a different species type to detection antibodies) in PBS; nonionic detergents: 0.05% (v/v) Nonidet P-40 or Tween-20.
4. Enzyme substrates: 2,2'-azino-bis(3-ethylbenzathiazaline 6-sulfonic acid) (ABTS) (1 mg/mL, A_{405nm}), *O*-phenyldiamine (OPD) (1 mg/mL, A_{492nm}), or tetramethylbenzidine (TMB) (0.01 mg/mL, A_{450nm}) in Na acetate with 0.01% H_2O_2 (pH 6.0); *p*-nitrophenyl phosphate (PNPP) (1 mg/mL, A_{405nm}) in 10 m*M* diethanolamine with 0.5 m*M* $MgCl_2$ (pH 9.5).

3. Methods

3.1. Immunoassay in Solution— ELISA and Radioisotope Assays (see Note 1)

*3.1.1. Detection of Mucins in Solution Using Double-Determinant Immunoassays (see **Notes 2** and **3**)*

3.1.1.1. Coating the Capture Antibody

1. Antibodies need to be purified to optimize coating; the concentration should be optimized for each antibody, buffer (CB or PBS) and plate type (range 0.1–2 μg/well).
2. Incubate overnight at room temperature.
3. Wash three times for 1 min each in PBS (if using alkaline phosphatase avoid phosphate buffers, e.g., use TBS).

3.1.1.2. Blocking

1. Block nonspecific binding on coated plates with protein-blocking solution and/or nonionic detergent. Block for 1–24 h at room temperature or 4°C.

2. Wash three times for 1 min each in PBS. Blocked plates can be used immediately; stored in PBS for several days at 4°C; or dried thoroughly, vacuum sealed in a bag with silica gel, and stored at 4°C (storage time can be more than 6 mo; addition of 5% [w/v] sucrose to the blocking buffer can substantially increase the shelf life of dried plates).

3.1.1.3. Sample Incubation (*see* **Note 4**)

1. Incubate in humidified environment for 1–24 h at 4–37°C.
2. Wash as per **Note 4**.

3.1.1.4. Detection Antibody Incubation

1. The required concentration of the detection antibody will need to be determined for each application (usual range 0.1–10 µg/mL). Use buffers as above (do not use Na azide if the antibody is horseradish peroxidase [HRP] conjugated), with an incubation time of 1–24 h at 4–37°C. Wash as per **Note 4**.

3.1.1.5. Secondary Labeled Antibody

1. This step is only required if detection antibody is not labeled. Optimization and conditions are as in **Subheading 3.1.1.4.** Wash as per **Note 4**.

3.1.1.6. Detection

1. For enzyme assays, the choice of substrate and buffer depends on the enzyme: ABTS, OPD, or TMB for HRP; PNPP for alkaline phosphatase (AP). Incubate at room temperature or 37°C for 20–60 min. Reactions can be stopped with an equal volume of 2.5% (w/v) NaF or 1 *M* H_2SO_4 (HRP) or 0.1 *M* EDTA (AP), and plates are read at the appropriate wavelength.
2. For radioisotope detection, gel-forming scintillant should be added to the wells after **Subheading 3.1.1.5., step 1** and the radioactivity determined using a microplate isotope counter.

3.1.1.7. Quantitation

1. Quantitation is best achieved using a standard curve fitted using an appropriate line of best fit; programs are available to interface with microplate readers and isotope counters that store data and compute standard curves.

3.1.2. Detection of Mucins in Solution Using Antibody Capture Competitive Binding Immunoassays (see ***Notes 5–7***)

3.1.2.1. Assay Optimization

1. Serially dilute the mucin down one or more 96-well plates and incubate overnight at room temperature; leave one column with buffer only to control for nonspecific binding (*see* **Subheading 3.1.1.** for plates and coating buffers).
2. Wash three times for 1 min each in PBS.
3. Block plate as in **Subheading 3.1.1.2., steps 1** and **2**.
4. Repeat wash.
5. Prepare antibody at 10 µg/mL in selected assay buffer (*see* **Note 4**) and serially dilute across the plates.
6. Incubate for 1–24 h at room temperature.
7. Wash as in **Subheading 3.1.1.1., step 2** and detect as in **Subheadings 3.1.1.5.** and **3.1.1.6.**

3.1.2.2. Assay

1. Select a dilution of antibody and antigen that gives an absorbance of about 1.5 (or about 75% of maximal radioactivity for isotope detection) and uses the least amount of coating mucin or peptide. Coat and block plates as above; coated plates can be dried and stored in vacuum-sealed bags for at least several weeks at 4°C. Prepare duplicate or triplicate samples and standards (serial dilution of mucin in sample buffer) in assay buffer containing the detection antibody at the final dilution. The sample/antibody mix can be preincubated (1–24 h at 4–37°C) prior to transfer to the mucin-coated plate. Incubate, wash, and detect as in **Subheading 3.1.2.1., step 7**.

3.1.2.3. Quantitation

1. Absorbance values, or radioactivity, are normally expressed as a percentage of the noninhibited (sample blank) controls and appropriate standard curves fitted as in **Subheading 3.1.1.7.**

3.2. Dot-Slot and Western Blotting

*3.2.1. Preparation of Dot-Slot Blots for Detection of Mucins (*see ***Note 8)***

3.2.1.1. Application of Samples

1. Samples can be either applied directly to membranes (*see* **Note 9**) in volumes of 0.5–2.5 µL or added using a commercially available vacuum manifold device (these are preferable owing to more even sample distribution, greater sample volume, and superior washing). For quantitation and comparison across blots, a standard in the same buffer as samples should be titrated for use as a standard curve, and samples should be included on all blots to determine interassay variation. Equivalent amounts of a nonmucin protein should also be titrated to act as a measure of nonspecific binding.

3.2.1.2. Washing and Storage

1. Wash the wells (for manifold devices) and then the entire membrane in three changes of PBS or TBS. Either proceed directly to PAS or detection using antibodies or lectins (*see* **Subheadings 3.2.2.** and **3.2.3.**) or store the membrane sealed in a bag in buffer at 4°C or dry thoroughly and store sealed at –20°C.

*3.2.2. Detection of Mucins on Membranes Using PAS (*see ***Notes 10–12)***

1. Wash the dot-slot or Western blots in three changes of water (1 mL/cm^2) and transfer to a freshly prepared solution of 1% (v/v) periodic acid in 3% (v/v) acetic acid (1 mL/cm^2) for 30 min at room temperature.
2. Rinse twice (2 min, 1 mL/cm^2) in freshly prepared 0.1% (w/v) sodium metabisulfite in 1 m*M* HCl. Transfer to Schiff reagent (commercially available) for 15 min (0.5 mL/cm^2). PAS-reactive glycoproteins will stain a pinkish red. Wash three times for 2 min each in sodium metabisulfite and dry the membrane in a warm airstream.

*3.2.3. Detection of Mucins on Membranes Using Antibodies or Lectins (*see ***Notes 10,12*** *and* ***13)***

3.2.3.1. Blocking

1. Membranes need to be blocked with protein and/or nonionic detergents (*see* **Subheading 3.1.1.**). The optimal blocking protein and buffer, wash buffers and antibody buffers (to prevent nonspecific binding) will vary with different antibodies and lectins but can readily

be optimized using 1×1 cm^2 pieces of membrane incubated within 24-well-plate wells taken through the entire staining detection procedure. In open trays with agitation, incubations and washes should use at least 0.25 mL/cm^2 of membrane. Block for 1–24 h at room temperature or 4°C. Wash three times for 1 min each in TBS.

3.2.3.2. Incubation with Antibody or Lectin

1. The concentration of antibody or lectin needs to be optimized to give the best signal-to-background ratio. As a guideline, optimal antibody and lectin concentrations usually will be in the range of 0.1–10 µg/mL.
2. Antibodies can be used either unconjugated (to be followed with a conjugated antibody against the primary antibody species and class) or conjugated directly with an enzyme (e.g., HRP, AP), or a ligand for a secondary enzyme conjugate (e.g., biotin, digoxigenin), or be radioactively labeled (e.g., ^{125}I).
3. Lectins will need to be conjugated usually with biotin through either amino or carbohydrate groups; biotinylated lectins are commercially available.
4. It is recommended that replicate blots be probed with the same species/isotype irrelevant antibody to control for nonspecific binding. Incubation buffers need to contain protein (50% of blocking concentration) and/or nonionic detergent.
5. Incubate for 1–24 h at 4–37°C with agitation. To save valuable reagents, this step can be performed in sealed bags with 0.125 mL/cm^2 of membrane.

3.2.3.3. Washing

1. Thorough washing with agitation is critical in immunoblotting; a good starting point is three times for 2 min each in TBS, three times for 2 min each in TBS plus 0.05% (v/v) Tween-20, three times for 2 min each in TBS. Less stringent washing may suffice for some antibodies. If nonspecific binding is a problem, try washing in 1% (v/v).
2. Nonidet P-40, 0.05% (w/v) sodium deoxycholate, and 0.1% (w/v) sodium dodecyl sulfate (SDS) in TBS or increase the NaCl concentration until nonspecific binding is reduced and specific binding retained.

3.2.3.4. Secondary Antibodies

1. The concentration of secondary antibody (or streptavidin-peroxidase for biotin) will also need to be optimized to give the best signal-to-background ratio. Affinity-purified antibodies with crossreactivity with other species' antibodies deleted are best.
2. A recommended dilution for blotting (usually in the range of 1/500 to 1/20,000) is often provided with commercial conjugates. Incubation details are as for primary antibody.
3. Wash as in **Subheading 3.2.3.3.**

3.2.3.5. Detection

1. Detection of bound enzyme conjugates can be achieved using insoluble chromogens that leave a colored precipitate on the blot (e.g., 3,3'-diaminobenzidine, 4-chloronapthol mix for HRP ***[14]***; 5-bromo-4-chloro-3'-indolyphosphate toluidine salt, nitro blue tetrazolium chloride mix for AP). Alternatively, chemiluminescent substrates (e.g., ECL [Amersham, Little Chalfont, UK] for HRP) can be utilized, which have the major advantages of high sensitivity, allowing for several different exposures to be recorded on X-ray film, and compatibility with stripping and reprobing blots. However, beware of possible nonspecific results with chemiluminescent substrates on membranes distorted by vacuum manifold devices.
2. Detection of ^{125}I-labeled antibodies is achieved by direct autoradiography with X-ray film.

3.2.3.6. Quantitation

1. Densitometry can be used to quantitate results provided that the samples have not been overloaded (exceeded the membrane binding capacity or detection system capacity). Titration of samples may aid in quantitation.
2. Interassay control samples will need to be included on each membrane (gel for Western blotting) if quantitation between membranes, and particularly between different electrophoresis/transfer/immunodetection runs, is required.

3.2.3.7. Stripping

1. Antibody-probed chemiluminescent-detected blots can be stripped after thorough washing in TBS by incubation at 50°C for 30 min in 2% (w/v) SDS and 100 m*M* 2-mercaptoethanol in 62.5 m*M* Tris, pH 6.8.
2. Thoroughly washed stripped blots can be stored in TBS at 4°C before reblocking and probing.
3. Up to four probings are often possible and, although sensitivity will gradually diminish with each cycle, signal-to-noise ratio often increases concurrently allowing longer development times.

4. Notes

1. Immunoassay techniques rely on reactions between antigens (in this case mucins) and antibodies or lectins; the protocols refer to antibodies but lectins are interchangeable. Details of preparation and characterization of mucin peptide and carbohydrate-specific antibodies can be found in Part IX of this volume. Immunoassays are suitable for quantitative and sensitive detection of mucins in large numbers of samples. A variety of different techniques can be devised with both antigen and antibody being free in solution or with either being fixed to a solid phase such as a tube, bead, or 96-well plate. All the variations require that the antigen, the antibody, or a secondary antibody be labeled with an enzyme, a ligand (e.g., biotin, digoxigenin) for a labeled secondary conjugate, or a radioisotope. Optimization of conditions for these assays is required, and comprehensive texts concerning the theory and practical aspects of immunoassays are available ***(15)***. Double-determinant assays are especially useful for detection, quantitation, and characterization of mucins owing to their large size and the multivalent nature of many mucin peptide and carbohydrate epitopes. For example, a core protein epitope–specific antibody can be used for capture and then several antibodies reactive with different carbohydrate epitopes can be used for detection to both quantitate and characterize a particular mucin. However, it is extremely important that the previous warnings regarding the influence of mucin glycosylation on antibody reactivity be heeded (*see* Introduction). For example, almost all the commercially available MUC1 assays utilize a double-determinant enzyme or radioisotope format. We have shown that assays using antibodies with similar tandem repeat domain epitopes can have vastly different capture and detection characteristics. For example, the cancer-associated serum antigen (CASA) assay detects very high levels of a glycoform of MUC1 present in the serum of pregnant women that is not detected by the CA15.3 assay ***(16)***. Similarly, these two assays show differing specificities for different glycoforms of MUC1 produced by breast and ovarian cancers ***(16,17)***. **Subheading 3.1.1.** describes a double-determinant format and **Subheading 3.1.2.** a competitive binding assay using antibody capture.

2. In this protocol, purified mucin-reactive antibodies or lectins are coated onto microtiter plates and used to capture mucins in biological samples. Detection antibodies or lectins are then introduced to react with the captured mucins. If the detection antibody or lectin is not labeled, secondary enzyme (horseradish peroxidase [HRP] or alkaline phosphatase [AP]) or radioisotope-labeled antibody is used to quantify the amount of captured mucin.
3. The capture antibody/lectin is critical because it determines which mucin molecules will be available for detection by the detection antibody/lectin. Choice is governed by knowledge of the mucin to be measured and availability of specific antibodies. If capture antibodies are to be detected by secondary antibody conjugates, capture and detection antibodies need to be of differing species or isotypes. Although it is possible to use combinations of both capture and detection antibodies with different specificities, interpretation of binding is problematic and performing distinct assays, although more time-consuming, will be more informative.
4. Samples need to be added in a buffer compatible with antibody-antigen reactions. Avoid high concentrations of chaotropic agents, SDS, and reducing reagents (because immunoassays are sensitive, this can often easily be achieved through dilution); interference by specific factors can be tested easily by progressive addition. Some biological fluids can be assayed neat but often cause interference problems. Addition of protein (50% of blocking concentration) and/or 0.05% nonionic detergent should be trialed. Serial dilution of antigen in antigen-free assay fluid needs to be performed to validate the assay; this should also be the form of the standard curve included on each plate along with a sample buffer blank. Each sample should be assayed at least in duplicate. Multiple aliquots of several samples at different levels of the standard curve should be prepared for inclusion on each plate as a measure of interassay variation. Thorough washing is important (e.g., three times for 1 min each in PBS-0.05% Tween-20; three times for 1 min each in PBS). More or less stringent washing may be needed for some antigens/antibodies; if nonspecific binding is a problem, try different detergents and gradually increase the NaCl concentration of the wash buffer. Enzymatic or chemical deglycosylation can be used before, during, or following antigen capture; however, it must be ensured that the techniques are compatible with maintenance of the antibody-antigen reaction. For example, treatment of serum with neuraminidase (0.1 U/mL in 0.05 *M* acetate, 1 m*M* $CaCl_2$, 154 m*M* NaCl, pH 5.5) for 1 h at 37°C prior to the sample incubation resulted in substantially increased signal in a MUC1 immunoassay ***(18)***.
5. In this assay, nonlabeled semipurified mucin or synthetic mucin peptides or carbohydrates are coated to microtiter plates and are used to capture specific antimucin antibodies or lectins. Samples are introduced to this reaction, and those containing the epitopes recognized by the antibody or lectin will compete for antibody binding to the solid-phase antigen. The amount of bound antibody or lectin is then determined using a secondary enzyme or radioisotope-labeled antibody.
6. The concentration of coating mucin antigen and detecting antibody needs to be determined using a checkerboard serial dilution. Higher binding will be achieved if the mucin is purified but crude preparations can work; synthetic peptides and fusion proteins work well in these assays. Selection of the starting dilution for the mucin is somewhat empirical; however, the protein-binding capacity of the plate wells should not be exceeded.
7. A standard curve, sample blank, and appropriate interassay control samples should be included on each plate. These inhibition assays can be more sensitive than double-deter-

minant assays but can also be subject to greater interassay variation unless there is rigid consistency in technique.

8. Blotting techniques rely on binding of mucins onto a membrane filter support and subsequent detection using either chemical, lectin, or antibody detection. Dot and slot blotting is suitable for semiquantitative detection of mucins in reasonably large numbers of samples. The advantages over solution assays include increased sensitivity owing to the potential for concentration of sample on the membrane and reduced problems with interfering substances, which can be filtered through the membrane. The main disadvantages of direct blotting compared with Western blotting are the potential for false-positive results owing to nonspecific antibody binding and the lack of separation and data regarding the M_r of the reactive proteins. False-negative results also occur if sample protein concentrations are very high. However, direct blotting is more amenable to inclusion of standards than Western blotting (owing to restrictions on the number of lanes per gel). Therefore, dot-slot blotting is often the method of choice, especially for monitoring mucins during purification. However, it is highly recommended that representative samples be subjected to electrophoretic separation and Western blotting (see below) to confirm the specificity of dot or slot blot results by demonstrating that the reactivity is restricted to proteins of an expected M_r. The choice of chemical, lectin, polyclonal antibodies or MAbs will differ with the application and the availability of reagents. **Subheading 3.2.1.** outlines the procedures for preparing the membranes, and **Subheading 3.2.2.** describes the use of the PAS reagent to detect mucins immobilized on membranes. Note that other classical histological reagents (e.g., alcian blue and high-iron diamine) have also been used to probe mucins immobilized on membranes ***(19)***. **Subheading 3.2.3.** describes detection of specific mucin epitopes using antibodies or lectins (these are also applicable to Western blotting).
9. Nitrocellulose is the most commonly used membrane although both polyvinylidene fluoride (PVDF) and nylon (inferior protein binding) can be used. PVDF is less brittle than nitrocellulose and is therefore more likely to survive several rounds of stripping and reprobing and is also resistant to chemical deglycosylation with trifluoromethanesulfonic acid ***(6)***. The total protein added should not exceed the protein-binding capability of the membrane, and samples should be titrated if relative quantitation is required. Some mucins, and in particular mucin glycopeptides (fragments of mucin prepared by extensive proteolysis), may not bind well to nitrocellulose, and the addition of poly-L-lysine (100 µg/mL) or a lectin (e.g., wheat germagglatinin) to the membrane prior to application of the samples can increase retention ***(19,20)***.
10. Western blotting refers to detection of proteins first separated by gel electrophoresis and then transferred to membrane filter supports for subsequent detection using either chemical, lectin, or antibody detection. This technique is suitable for specific, sensitive, semiquantitative detection of mucins in moderate numbers of samples. The main advantages are the potential separation of different mucins and the provision of data regarding M_r. The most frequent mistake in published data on mucin Western blotting of mucin concerns not the immunodetection but the electrophoretic separation. Polyacrylamide gels of a percentage that will not allow migration of high M_r mucins even into the stacking gels are often used. Even at the lower limit of polyacrylamide gel formation (3%) many known mucins are too large to penetrate the gel. Agarose gel electrophoresis is often required to achieve separation of these large mucins ***(6)***; Chapter 7 describes appropriate electrophoretic techniques ***(4)***. Mucins can be detected by chemical methods within acrylamide

or agarose gels *(21,22)*; however, transfer to membranes is necessary for antibody or lectin detection. Transfer of mucins from polyacrylamide or agarose gels can be achieved by electrophoretic elution (wet or semidry), vacuum, or capillary transfer (PAS staining of gels can be used to evaluate the transfer *[22]*).

11. This protocol describes the detection of mucins on membranes following dot-slot blotting or transfer following electrophoretic separation. Mucin carbohydrate groups are reacted with periodic acid and then detected using the Schiff reagent; for more details *see* **ref.** ***19***.
12. PAS-stained dot-slots or bands on Western blots can be readily quantitated using densitometry equipment, and the content of mucin can be determined relative to standards included on the same blot.
13. This protocol describes the detection of mucins on membranes following dot-slot blotting or transfer following electrophoretic separation. Blots are incubated with antibodies or lectins reactive with the mucins, which, in turn, are detected with secondary antibodies/ligands labeled with enzymes or isotopes. Chemical and/or enzymatic deglycosylation can be performed before starting these detection procedures *(6)*.

References

1. Walsh, M. D., Jass, J. R. (2000) Histologically-based methods for detection of mucin, in *Glycoprotein Methods and Protocols: The Mucins* (Corfield, T., ed.), Humana, Totowa, NJ.
2. Davies, R., Carlstedt, I. (2000) Isolation of large gel-forming mucins, in *Glycoprotein Methods and Protocols: The Mucins* (Corfield, T., ed.), Humana, Totowa, NJ.
3. Carraway, K. L. (2000) Preparation of membrane mucin, in *Glycoprotein Methods and Protocols: The Mucins* (Corfield, T. ed.), Humana, Totowa, NJ.
4. Nagma, K., Thornton, D. J., Khan, N., and Sheehan, J. K. (2000) Separation and identification of mucins and their glycoforms, in *Glycoprotein Methods and Protocols: The Mucins* (Corfield, T., ed.), Humana, Totowa, NJ.
5. Kim, Y. S., Gum, J., and Brockhausen, I. (1996) Mucin glycoproteins in neoplasia. *Glycoconj. J.* **13,** 693–707 (review).
6. Thornton, D. J., Howard, M., Devine, P. L., and Sheehan, J. K. (1995) Methods for separation and deglycosylation of mucin subunits. *Anal. Biochem.* **227,** 162–167.
7. Gerken, T. A., Gupta, R., and Jentoft, N. (1992) A novel approach for chemically deglycosylating *O*-linked glycoproteins: the deglycosylation of submaxillary and respiratory mucins. *Biochemistry* **31,** 639–648.
8. Raju, T. S. and Davidson, E. A. (1994) New approach towards deglycosylation of sialoglycoproteins and mucins. *Biochem. Mol. Biol. Int.* **34,** 943–954.
9. Huang, J., Byrd, J. C., Yoon, W. H., and Kim, Y. S. (1992) Effect of benzyl-alpha-GalNAc, an inhibitor of mucin glycosylation, on cancer-associated antigens in human colon cancer cells. *Oncol. Res.* **4,** 507–515.
10. Heinegard, D. (1973) Automated procedures for the determination of protein, hexose and uronic acid in column effluents. *Chemica Scr.* **4**, 199–201.
11. Mantle, M. and Allen, A. (1978) A colorimetric assay for glycoprotein based on the periodic acid/Schiff stain. *Biochem. Soc. Trans.* **6,** 607–609.
12. Lohmander, L. S., De Luca, S., Nilsson, B., Hascall, V. C., Caputo, C. B., Kimura, H., and Heinegard, D. (1980) Oligosaccharides on proteoglycans from the swarm rat chondrosarcoma. *J. Biol. Chem.* **255,** 6084–6091.
13. Crowther, R. S. and Wetmore, R. F. (1987) Fluorometric assay of *O*-linked glycoproteins by reaction with 2-cyanoacetamide. *Anal. Biochem.* **163,** 170–174.

14. Young, P. R. (1989) Enhancement of immunoblot staining using a mixed chromogenic substrate. *J. Immunol. Methods* **121,** 295, 296.
15. Diamandis, E. P. and Christopoulos, T. K. (1996) *Immunoassay*, Academic, San Diego.
16. Devine, P. L., McGuckin, M. A., Quin, R. J., and Ward, B. G. (1994) Serum markers CASA and CA 15-3 in ovarian cancer—all MUC1 assays are not the same. *Tumor Biol.* **15,** 337–344.
17. Devine, P. L., Duroux, M.A., Quin, R.J., McGuckin, M.A., Joy, G.J., and Ward, B.G. (1995) CA15-3, CASA, MSA, and TPS as diagnostic serum markers in breast cancer. *Breast Cancer Res. Treat.* **34,** 245–251.
18. McGuckin, M. A., Devine, P. L., Ramm, L. E., and Ward, B. G. (1994) Factors affecting the measurement of tumor-associated MUC1 mucins in serum. *Tumor Biol.* **15,** 33–44.
19. Thornton, D. J., Holmes, D. F., Sheehan, J. K., and Carlstedt, I. (1989) Quantitation of mucus glycoproteins blotted onto nitrocellulose membranes. *Anal. Biochem.* **182,** 160–164.
20. Ayre, D., Hutton, D. A., and Pearson, J. P. (1994) The use of wheat germ agglutinin to improve binding of heterogeneous mucin species to nitrocellulose membranes. *Anal. Biochem.* **219,** 373–375.
21. Jay, G. D., Culp, D. J., and Jahnke, M. R. (1990) Silver staining of extensively glycosylated proteins on sodium dodecyl sulfate-polyacrylamide gels: enhancement by carbohydrate–binding dyes. *Anal. Biochem.* **185,** 324-330.
22. Thornton, D. J., Carlstedt, I., and Sheehan, J. K. (1996) Identification of glycoproteins on nitrocellulose membranes and gels. *Mol. Biotechnol.* **5,** 171–176.

5

The Gastrointestinal Adherent Mucous Gel Barrier

Adrian Allen and Jeffrey P. Pearson

1. Introduction

Three phases of production of mucin can be identified in the gastrointestinal (GI) tract on the basis of their location in vivo: the stored, presecreted, intracellular mucin; the gel phase adherent to the epithelial surfaces; and the viscous, mobile mucin, which is largely in soluble form and mixes with the luminal contents. The layer of adherent mucous gel that lines the epithelial surfaces throughout the gut from the stomach to the colon marks the interface between the mucosal epithelium and the fluid luminal environment, which is teeming with nutrients, bacteria, destructive hydrolases, foreign compounds and so on. The adherent mucous gel thus provides a protective barrier and a stable unstirred layer with its own microenvironment, between the mucosal surface and the lumen *(1,2)*. In the mouth, where salivary mucins are secreted, and the esophagus, there is no discernible adherent mucous gel layer.

A practical definition of the adherent mucous gel layer is that secretion of mucus which remains attached to the mucosal surface after washing away the luminal contents. This mucous gel layer is readily visible on unfixed, transverse sections of mucosa as a translucent layer of variable thickness (5–200 μm), between the dense mucosal surface and the bathing solution *(3)*. Mucous gel scraped from the mucosal surface has the rheological characteristics of a true viscoelastic gel, although it has the ability to slowly flow (30–120 min) and reform when sectioned *(4,5)*. The concentration of mucin in such gels scraped from the mucosal surface is high, e.g., ranging from 50 mg/mL in gastric mucus to 20 mg/mL in colonic mucus *(6)*. The extent of the mucus adherent to the mucosal surface in vivo is determined by a balance between the rate of secretion by the underlying epithelium and the rate of erosion by mechanical shear associated with the digestive processes and digestion by hydrolases, particularly proteases *(1,2)*. For full studies of mucous secretion in vivo, it is essential to quantitate both the adherent mucous gel and mucin in the luminal solution since the latter can arise by degradation of the former as well as by secretion, and changes in these two phases do not necessarily parallel each other. Despite being the primary mucous bar-

From: *Methods in Molecular Biology, Vol. 125: Glycoprotein Methods and Protocols: The Mucins*
Edited by: A. Corfield © Humana Press Inc., Totowa, NJ

rier and the major secretion of mucin of the gut, the adherent mucous gel is frequently ignored because of the practical difficulties in observing and quantitating it.

Quantitative measurement of the amount of mucin in the adherent mucous gel secretion in vivo can only be approximate. Although the mucous gel can be separated from the mobile mucus by washing the mucosal surfaces, separation of the gel layer from the considerable intracellular mucin stores cannot be done satisfactorily. The mucous gel can be removed from the surface by gentle scraping, but inevitably either not all the adherent gel is removed or the scrapings contain substantial numbers of epithelial cells with stored intracellular mucin *(7)*. Researchers have attempted to estimate the amount of the adherent mucous gel by utilizing the binding by the dye alcian blue (AB) to the negatively charged mucin macromolecules *(8)*. However, the amount of negative charge on mucins varies considerably according to the location in the gut, secretory status, and disease. Also other macromolecules present in the mucus, e.g., negatively charged nucleic acids and cell surface components, will bind the dye, and it has yet to be shown that the dye penetrates the mucous layer uniformly. Results from this method are therefore difficult to interpret and this method is not considered further here.

The best approach to quantitating changes in the adherent mucous gel layer is by measuring its thickness. These measurements have the important advantage that continuity and thickness of the layer can be directly related to functional efficacy *(1,2)*. Routine histological methods for fixing and staining of mucosal sections, although readily demonstrating periodic acid-Schiff (PAS)/AB positive intracellular mucin stores, result in little or no adherent mucous gel being visible at the mucosal surfaces. This is because the prolonged use of fixatives and organic solvents together with conventional paraffin embedding causes denaturation, dehydration, and loss of the surface mucous gel layer *(9–11)*. Similarly, standard preparation procedures for electron microscopy distort the mucous layer to give condensed strands or fenestrated patches, although a continuous mucous layer can be observed by scanning electron microscopy if it is first stabilized by antibodies *(12)*. The paucity or absence of a mucous layer on biological sections has led to controversy in the past as to whether an adherent mucous layer existed over the gastric mucosal surface *(13,14)*. However, a variety of different methods (discussed subsequently) have now demonstrated unequivocally that a thick (100–200 μm), continuous, adherent mucous layer exists over the mammalian gastric mucosal surface. A protective mucoid cap can be seen at the surface of damaged mucosa undergoing re-epithelialization, following preparation of sections by standard histological procedures *(1,13,15)*. This mucoid cap has been confused with the adherent mucus that covers the normal undamaged mucosa; however, it is a quite different structure consisting primarily of a fibrin gel and necrotic cells with some mucin staining *(15)*.

Original methods for measuring adherent mucous thickness *in situ* were based on (1) differences in refractive index between the mucous layer and the mucosal surface, measured by a slit lamp and pachymeter *(16)*; (2) the dimensions of the surface pH gradient measured using microelectrodes *(17,18)*; and (3) observation of transverse sections of unfixed mucosa *(3)*. The first two of these methods also measure the unstirred aqueous layer beyond the adherent mucous gel surface, the extent of which depends on the hydration of exposed mucosa and the degree of stirring of the solution above the mucosa, respectively. Consequently, the dimensions of mucous thickness measured by the slit lamp and pachymeter or pH gradient methods can be substantially greater than the actual values for the gel layer itself ***(19)***. Since the mid-1980s, observation of the translucent mucous layer on thick (1.6 mm) sections of unfixed mucosa (detailed subsequently) has been a simple method for measuring mucous thickness, although histological methodology has now reached the stage where preservation of full adherent mucous thickness can be maintained during processing ***(10)***.

An elegant method for observing mucus and measuring its thickness has been developed using in vivo microscopy in the anesthetized rat animal model ***(20–22)***. The GI mucosa is exteriorized, with its blood and nerve supply intact, and the mucosa is placed over an illuminated lucite cone. A lucite chamber is fitted over the mucosa to expose approx 0.8 cm^2 of the surface and is filled with physiological saline. The mucous layer covering the mucosa can be observed microscopically from above, particularly if the luminal surface of the mucus is enhanced visually by the addition of carbon particles to the bathing solution. The thickness of the mucous layer is best measured by recording the distance travelled by a glass microprobe moved by a micromanipulator at a fixed angle through the mucous layer. A continuous mucous coat of mean thickness 200–300 μm is seen in rat stomach and duodenum ***(21,22)***, and a mean thickness of 800 μm is seen in the colon ***(23)***. Strong suction will remove about one-third of the mucous layer in the stomach and nine-tenths of that in the colon to leave a firm adherent gel, whereas in the duodenum all the mucus can be sucked off. It is not yet clear whether the mucus that can be sucked off in vivo is a continuum of the gel at a lower concentration or a separate secretion. This method, which for experimental practicalities can be applied only to the rat (and possibly other experimental animal models), is particularly useful for validating other in vitro methods for measuring mucous thickness, for studying mucous secretion in vivo, and for studying the secretion of acid from the gastric glands through the covering mucous layer ***(20)***. Full details of this method, which requires careful physiological experimentation to ensure full maintenance of in vivo functions during the experimental period, can be found in **refs.** ***20–22***.

Various groups have adapted histological procedures to observe adherent mucus on fixed mucosal sections. Modifications have included the use of cryostat sections ***(24)*** and fixation in Carnoy's solution ***(11,25)***. However, the adherent mucous layer after such procedures is very thin (typically about 25%) compared with that seen on unfixed mucosal sections and even less than that observed in vivo. A particularly interesting method developed by Ota and Katsuyama ***(11,25)*** shows alternating layers within both human gastric and colonic adherent mucous layers following dual staining with galactose-oxidase-cold thionine Schiff and paradoxical concanavalin A. However, this method, which suggests some form of substructure within the mucous gel layer, uses Carnoy's fixative, as well as clearing in xylene with paraffin embedding, and results in considerable shrinkage of the mucous gel to give a thickness less than one-fifth that seen on unfixed sections. A recent method (detailed subsequently), using cryostat sections, does give thickness values in the rat comparable with those measured for the firm adherent mucous gel remaining after suction in vivo *(10)*. In this method, cryostat sections receive no prefixation or extensive dehydration steps characteristically used in conventional staining techniques. Instead, the tissue is given a brief ethanol pretreatment followed by a prolonged staining with PAS/AB and a gentle postfixation in paraformaldehyde vapor, and finally is mounted in a water-soluble gelatin gel. Values for mucous thickness over the rat gastric mucosa by this method give a mean of 147 μm, which approximates that of the adherent gel remaining after suction in vivo. It would appear, therefore, that this method preserves intact the thickness of the stable adherent mucous gel layer. By contrast, mucous thickness by this method is twice that seen using conventional PAS/AB staining techniques on cryostat sections of mucosa and 50% greater than that for unfixed rat gastric mucosal sections.

2. Materials

2.1. Mucous Thickness on Unfixed Sections

1. NaCl solution: 150 mmol/L.
2. Two parallel, sharp razor blades separated by spacers 1.6 mm apart such that one side of each of the two razor blades is left free for sectioning and held together by bolts tightened by hand.
3. Millipore filter paper to act as a backing for the mucosa during sectioning and subsequent manipulation of the section.

2.2. Histological Staining and Fixation of Cryostat Sections of Mucosa

1. NaCl solution: 150 mmol/L.
2. Acetic acid solution: 3% (v/v).
3. Alcian Blue 8 GX (Sigma, Poole, Dorset, UK) 1% by weight in 3% acetic acid (pH 2.5).
4. Periodic acid (aqueous): 1%.
5. Schiff's reagent (fuchsin-sulfite reagent) readily obtained, made up in solution (Sigma).
6. Paraformaldehyde crystals placed in a desiccator for final postfixing of sections at 37°C.
7. Gelatin for water-soluble mounting prepared by dissolving 10 g of gelatin in 60 mL of distilled water mixed with 250 mg of phenol in 70 mL of glycerol.
8. Poly-L-lysine-coated slides: slides are immersed in 0.01% (w/v) poly-L-lysine for 5 min and dried at room temperature.

3. Methods

3.1. Observation and Measurement of Mucous Thickness on Unfixed Mucosal Sections (see Notes 1–4)

This is a rapid and simple method for observing the adherent mucous layer and measuring its thickness on unfixed sections of mucosa *(3)*. This method has been used to measure mucous thickness on resected human mucosa both in the stomach *(26)* and in the colon *(27)* or in experimental animal models, particularly rat *(3,9,28,29)*. This method relies on the adherent gel being sufficiently firm that it is relatively undistorted when the mucosa is sectioned by parallel razor blades 1.6 mm apart.

1. It is essential that the mucosa and sections are immersed in isotonic saline throughout the procedure except during sectioning of the mucosa in order to maintain the gel in its fully hydrated state.

2. Gently wash the luminal mucosal surface with isotonic saline (150 mmol/L of NaCl) to remove food particles and so on as well as nonadhering, soluble mucus.
3. Remove outer muscle layers from serosal side of mucosa by blunt dissection.
4. Mount the mucosa, luminal surface up, on a Millipore filter on a corkboard and cut sections with a pair of parallel, sharp razor blades separated by a distance of 1.6 mm. The razor blades are held together with spacers (1.6 mm) between them by bolts tightened by hand. It is often useful to loosen partially the bolts in order to remove the newly cut mucosal section from between the razor blades.
5. Position sections transversely on a microscope slide. Check the positioning of the sections with a low-powered (e.g., ×3.5) stereoscopic microscope. Care should be taken not to distort or stretch the sections during manipulation and positioning.
6. The surface mucous layer can be observed under an inverse microscope (e.g., ×200) using either light- or dark-field illumination or phase contrast. The adherent mucous layer appears as a translucent layer between the dense mucosa and the clear bathing solution. The thickness of the mucous layer is measured using an eyepiece graticule.
7. Individual mucous thickness values should be measured on at least three to six sections per mucosal specimen. Measurements of gel thickness should be taken at regular intervals along each section; A good distance is 500 μm between each reading (about 6–10 readings per section, depending on its length).
8. Mucous thickness values on unfixed mucosal sections usually show a nonparametric distribution with more readings at the lower end of the range. Therefore, results are best treated for statistical significance by methods that compare nonparametric distributions such as the Mann-Whitney U-test. Values for thickness should therefore be expressed in terms of medians and upper and lower quartiles.
9. When this technique is used for the first time on a new tissue, it is useful to validate the method by checking that the thickness of the adherent mucous layer on the sections does not change significantly over a period of 45 min and that the same mean values for mucous thickness are obtained for individual sections when read from either side (sections turned over 180° and remounted).

3.2. Histological Fixation and Staining of Adherent Mucous Gel on Cryostat Sections of Mucosa (see Notes 5–9)

Jordan et al.'s ***(10)*** method preserves the mucous layer on cryostat sections of resected mucosa or biopsies without apparently causing shrinkage and retaining its original thickness *in situ*. The cryostat sections undergo no prefixation or long ethanol dehydration steps characteristic of previous methods. Instead, the tissue is given a brief ethanol pretreatment followed by a prolonged aqueous staining, and a gentle postfixation in paraformaldehyde vapor (37°C for 45 min), and is finally mounted in a water-soluble gelatin gel.

3.2.1. Preparation of Sections

Mucosal segments are snap-frozen in liquid nitrogen and stored at –20°C. For smaller gut segments, e.g., rat, a good method is to leave the stomach and intestine intact, ligate one end, fill with 0.9% NaCl, and ligate the other end before snap-freezing. This ensures minimum disturbance of the mucous layers and prevents contact of the adjacent luminal walls of the tissue. Biopsies, e.g., from human endoscopy (e.g., 2 mm^2) need support, which can be provided by wrapping them in a 1–cm^2 sandwich of a suitable piece of tissue. Pig's liver is satisfactory for this procedure. The sandwiched biopsies are immediately snap-frozen and stored at –20°C. Serial cryostat sections, 6–20 μm thick, are cut from representative portions of the tissue and mounted on poly-L-lysine–coated slides and stained. Sections on slides can be conveniently stored at –20°C before staining.

3.2.2. Staining Procedure

Sections are stained by the following sequence of steps:

1.	Defrost:	20 min
2.	Pretreatment in 100% ethanol:	10 min.
3.	Rinse in running tap water:	10 min.
4.	Rinse in 3% (v/v) acetic acid:	2 min.
5.	Stain in 1% Alcian Blue 8GX in 3% acetic acid (pH 2.5):	2.5 h.
6.	Rinse in 3% acetic acid and then rinse in running tap water:	2 min.
7.	Oxidize in 1% periodic acid (aqueous) at room temperature:	10 min.
8.	Wash in running tap water:	5 min.
9.	Immerse in Schiff's reagent:	15 min.
10.	Rinse in running tap water:	5 min.
11.	Rinse in 0.5% sodium metabisulfite, repeat three times:	1 min/rinse.
12.	Rinse in running tap water:	5 min.
13.	Postfix in paraformaldehyde vapor, at 37°C:	45 min.
14.	Mount in gelatin.	

4. Notes

1. Removal of the muscle layers first makes sectioning of unfixed mucosa easier. However, the mucosa can be sectioned with the muscle layer intact, and previous studies have shown that doing so does not affect the thickness of the adherent mucous layer.
2. Various stains, e.g., toluidine blue and PAS have been used to improve visualization of the mucous layer on unfixed sections. However, it is essential that the solutions used are isotonic and do not contain any organic solvent (e.g., alcohol), which will cause dehydration and distortion of the mucous gel layer.
3. The unfixed section method is most satisfactory for gastric mucosa where there is a thick layer of firm adherent gel. Although this method has been used successfully to look at small intestinal and colonic mucous gel, it is less satisfactory for these tissues since the mucous gel is not so firm and the mucosa is more elastic than in the stomach.
4. The unfixed section method has been applied to human biopsy specimens from the gastric mucosa. However, it is very difficult to obtain satisfactory sections from biopsies, and hence, consistent results for mucous thickness. It is necessary to obtain resected mucosa to use this method for human studies.
5. The distribution of mucous thickness readings on histologically stained and fixed sections is usually nonparametric. Therefore, thickness values obtained by this method should be expressed as medians with upper and lower quartiles and comparisons among groups made by methods that compare nonparametric distributions, such as the Mann-Whitney U-test.
6. The thinner the cryostat sections are cut (down to 6 μm) for histological processing the better, because this provides clearer details of cell structure. It may be necessary, in some cases, to use thicker sections (up to 20 μm) in order to obtain a good mucous gel layer.

Where detailed mucosal structure is required, parallel histological methodology can be applied to adjacent mucosal sections from that on which mucous thickness is measured. Note, however, that on such sections, the adherent mucous gel layer will be greatly reduced and discontinuous.

7. Careful handling of the cryostat sections on the slide during fixing and staining is important because excessive washing and so on can remove mucus from the section.
8. Full validation of the staining and fixing methodology for unfixed sections and comparison of results by this method with those of previous methods used for staining the mucous gel can be found in Jordan et al. ***(10)***. An example of the use of this method for studies in humans is given in Newton et al. ***(30)***.
9. This histological and fixation method can be adapted to staining the adherent mucous gel with biotinylated antibodies and lectins coupled to avidin-horseradish peroxidase using conventional methodology in place of **steps 4–11.**

References

1. Allen, A. (1989) Gastrointestinal mucus, in *Handbook of Physiology, The Gastrointestinal Physiology. Salivary, Gastric and Hepatobiliary Secretions,* vol. III (Forte, J. G., ed.), American Physiology Society, Bethesda, MD, pp. 359–382.
2. Allen, A., Flemstrom, G., Garner, A., and Kivilaakso, E. (1993) Gastroduodenal mucosal protection. *Physiol. Rev.* **73(4)**, 823–857.
3. Kerss, S., Allen, A., and Garner, A. (1982) A simple method for measuring thickness of the mucosal gel layer adherent to rat, frog and human gastric mucosa: influence of feeding prostaglandin, N-acetyl-cysteine and other agents. *Clin. Sci.* **63,** 187–195.
4. Bell, A. E., Allen, A., Morris, E. R., and Ross-Murphy, S. B. (1984) Functional interactions of gastric mucus glycoprotein. *Int. J. Biol. Macromol.* **6**, 309–315.
5. Bell, A. E., Sellers, L. A., Allen, A., Cunliffe, W. J., Morris, E. R., and Ross-Murphy, S. B. (1985) Properties of gastric and duodenal mucus: effect of proteolysis, disulfide reduction, bile, acid, ethanol and hypertonicity on mucus gel structure. *Gastroenterology* **88**, 269–280.
6. Sellers, L. A., Allen, A., Morris, E., and Ross-Murphy S. B. (1988) Mucus glycoprotein gels: role of glycoprotein polymeric structure and carbohydrate side-chains in gel-formation. *Carbohydrate Res.* **178**, 93–110.
7. Sellers, L. A., Allen, A., Morris, E., and Ross-Murphy, S. B. (1991) The rheology of pig small intestinal and colonic mucus: weakening of gel structure by non-mucin components. *Biochim. Biophys. Acta* **1115**, 174–179.
8. Bolton, W. P., Palmer, D., and Cohen, M. M. (1978) Stimulation of mucus and nonparietal cell secretion by the E_2 prostaglandins. *Digest. Dis. Sci.* **23,** 359–364.
9. McQueen, S., Allen, A., and Garner, A. (1984) Measurement of gastric and duodenal mucus gel thickness, in *Mechanisms of Mucosal Protection in the Upper Gastrointestinal Tract* (Allen, A., Flemsrom, G., Garner, A., Silen, W., and Turnberg, L. A., eds.), Raven, New York, pp. 215–221.
10. Jordan, N., Newton, J., Pearson, J. P., and Allen, A. (1998) A novel method for the visualization of the *in situ* mucus layer in rat and man. *Clin. Sci.* **95**, 97–106.
11. Matsuo, K., Ota, H., Akamatsu, T., Sugiyama, A., and Katsuyama, T. (1997) Histochemistry of the surface mucous gel layer of the human colon. *Gut* **40,** 782-789.
12. Bollard, J. E., Vanderwee, M. A., Smith, G. W., Tasman-Jones, C., Gavin, J. B., and Lee, S. P. (1986) Preservation of mucus in situ in rat colon. *Dig. Dis. Sci.* **31**, 1338–1344.

13. Morris, G. P., Harding, R. K., and Wallace, J. L. (1984) A functional model for extracellular gastric mucus in the rat. *Virchows Arch. [Cell Pathol.]* **46**, 239–251.
14. Wallace, J. L. (1989) Gastric resistance to acid: is the "mucus-bicarbonate barrier" functionally redundant? *Am. J. Physiol. (Gastrointestin. Liver Physiol. 19)* **256**, G31–G38.
15. Sellers, L., Allen, A., and Bennett, M. K. (1987) Formation of fibrin based gelatinous coat over repairing rat gastric epithelium after acute ethanol damage: interaction with adherent mucus. *Gut* **28**, 835–843.
16. Bickel, M. and Kauffman, G. L. (1981) Gastric gel mucus and effect of distension, 16,16-dimethyl PGE2. *Gastroenterol*ogy **80,** 770–775.
17. Ross, I. N., Bahari, H. M. M., and Turnberg, L. A. (1981) The pH gradient across mucus adherent to rat fundic mucosa in vivo and the effect of potential damaging agents. *Gastroenterology* **81**, 713–718.
18. Williams, S. E. and Turnberg, L. A. (1981) Demonstration of a pH gradient across mucus adherent to rabbit gastric mucosa: evidence for a mucus-bicarbonate barrier. *Gut* **22**, 94–96.
19. Allen, A., Hutton, D. A., McQueen, S., and Garner, A. (1983) Dimensions of gastroduodenal surface pH gradients exceed those on adherent mucus gel layers. *Gastroenterology* **85,** 463–466.
20. Holm, L. and Flemström, G. (1990) Microscopy of acid transport at the gastric surface in vivo. *J. Int. Med.* **228 Suppl. 732**, 91–95.
21. Schade, C., Flemstrom, G., and Holm, L. (1994) Hydrogen ion concentration in the mucus layer on top of acid-stimulated and inhibited rat gastric mucosa. *Gastroenterol*ogy **107**, 180–188.
22. Sababi, M., Nilsson, E., and Holm, L. (1995) Mucus and alkali secretion in the rat duodenum: Effect of indomethacin, Nω-nitro-L-arginine and luminal acid. *Gastroenterology* **109**(5), 1526–1534.
23. Strugala, V., Jordan, N., Pearson, J. P., and Allen, A. (1998) The colonic mucus barrier and changes in inflammatory bowel disease. *Gastroenterology* **114 (suppl S),** A1093.
24. Sakata, T. and v. Engelhardt, W. (1981) Luminal mucin in the large intestine of mice, rats and guinea pigs. *Cell Tissue Res.* **219,** 629–635.
25. Ota, H. and Katsuyama, T. (1992) Alternating laminated array of two types of mucin in the human gastric surface mucous layer. *Histochem. J.* **24**, 86–92.
26. Allen, A., Cunliffe, W. F., Pearson J. P., and Venerables, C. W. (1990) The adherent gastric mucus gel barrier in man and changes in peptic ulceration. *J. Int. Med.* **228**, 83–90.
27. Pullan, R. D., Thomas, G. A. O., Rhodes, M., Newcombe, R. G., Williams, G. T., Allen, A., and Rhodes, J. (1994) Thickness of adherent mucus gel on colonic mucosa in humans and its relevance to colitis. *Gut* **35,** 353–359.
28. Sandzen, B., Blom, H., and Dahlgren, S. (1988) Gastric mucus gel layer thickness measured by direct light microscopy. *Scand. J. Gastroenterol.* **23,** 1160–1164.
29. Sellers, L. A., Carroll, N. J. H., and Allen, A. (1986) Misoprostol-induced increases in adherent gastric mucus thickness and luminal mucus output. *Dig. Dis. Sci.* **31,** 91S–95S.
30. Newton, J. L., Jordan, N., Oliver, L., Strugala, V., Pearson, J., James, O. F. W., and Allen, A. (1998) *Helicobacter pylori in vivo* structural changes in the adherent gastric mucus gel layer but barrier thickness is not compromised. *Gut* **43,** 470–475.

6

Quantitation of Biosynthesis and Secretion of Mucin Using Metabolic Labeling

Jan Dekker, B. Jan-Willem Van Klinken, Hans A. Büller, and Alexandra W. C. Einerhand

1. Introduction

Most epithelial mucins are secretory glycoproteins. The mucin-producing cells are characterized by large intracellular stores of these very large and complex glycoproteins *(1,2)*. These secretory mucins form mucous layers on the apical side of the cells, protecting the vulnerable epithelium, while allowing selective interactions with the apical environments, which is typically the lumen of an organ that is continuous with the outer world. Secretion from mucin-producing cells is regulated. Normally, mucins are constitutively secreted in relatively low amounts, which are sufficient under normal conditions to sustain the thickness of the mucous layer. On acute threats, the accumulated mucins may be secreted in bulk amounts to provide mucus as an effective, yet temporary, means of epithelial protection *(1,2)*. Both types of secretion require synthesis of mucin: constitutive secretion demands a continuous low level of biosynthesis, whereas stimulated secretion requires massive synthesis to replenish the diminished resources.

In particular pathological conditions, mucous production seems either to fall dramatically or to rise excessively, but systematic measurements of the actual changes in mucin production at the various levels of regulation are most often not conducted. For instance, in the chronic inflammatory bowel disease ulcerative colitis, it was debated for many years whether mucous production actually dropped during the inflammation *(3)*. Only recently have researchers been able to show that MUC2 is the predominant mucin in normal colon as well as in colon affected by ulcerative colitis, and that MUC2 production is actually decreased during active inflammation in ulcerative colitis *(4–6)*.

Many researchers in the mucin field may therefore wish to quantify mucin synthesis and secretion in health and disease, in order to determine the sequence and the regulation of events. Only in this way will investigators be able fully to appreciate mucin functions and hope to find ways to interfere in the production and secretion of

From: *Methods in Molecular Biology, Vol. 125: Glycoprotein Methods and Protocols: The Mucins*
Edited by: A. Corfield © Humana Press Inc., Totowa, NJ

mucin. All secretory mucins display specific expression patterns in organs and cell types, implying specific functions of each mucin. Logically, it is essential to be able to measure mucin synthesis in a specific manner, i.e., to determine the level of gene-specific expression of mucin. This can be done at the mucin mRNA level, as described extensively in Chapter 25. However, the presence of mRNA is only indirect proof of the biosynthesis of the encoded mucin. This point was proven by to our knowledge, the only study that actually correlated the levels of mRNA, synthesis of mucin polypeptide, and mature mucin. This study, which formed the basis for this chapter, showed that human colonic MUC2 mRNA, in normal individuals and in patients with ulcerative colitis, did not correlate with MUC2 protein synthesis, but that the MUC2 protein synthesis correlated highly with the total amount of MUC2 present in the tissue ***(6)***. Therefore, this study implies that MUC2 synthesis in the colon is primarily regulated at the posttranscriptional level. Thus, researchers should be cautious to draw conclusions about the amounts of mucins produced by cells or tissues, based on mucinous mRNA levels alone.

Quantitation of production of mucin through metabolic labeling by [^{35}S]amino acids at the polypeptide level has the advantage that it is a vital parameter: it is a measure of the actual capability of cells or tissue to produce mucins ***(1,4–11)***. More important, a distinction can be made with the preexisting, stored mucin in the mucin-producing cells. Metabolic labeling during short periods (up to 60 min) will not add significant mucin to the vast reservoir of stored mucins. Therefore, researchers will be able to distinguish within one experiment the movements of two fundamentally different pools of mucins: the preexisting, unlabeled bulk of the stored mucins; and a small but quite recognizable amount of freshly synthesized mucin ***(1,6–8)***. An extra dimension can be added by labeling mucins at the last step of their synthesis, through metabolic labeling with [^{35}S]sulfate. Thus, another defined pool of mucin molecules can be distinguished and studied—the just synthesized but not yet stored mature mucin ***(1,7,8)***. A mucin precursor is defined as a mucin polypeptide, present in the rough endoplasmic reticulum, containing *N*-glycosylation but no *O*-glycosylation (which occurs only after arrival in the Golgi apparatus) ***(1,8)***. Mature mucins are defined as the end product of the biosynthetic processes.

To ensure meaningful measurement of the biosynthesis of mucin, one must be able to identify unequivocally the mucin precursor of interest. In practice this is done by immunochemical techniques. However, immunoprecipitation of metabolically labeled mucins is seldom quantitative, owing to the large amount of stored yet unlabeled mucin that competes with the labeled mucin for the antibody (*see also* Chapter 20). Fortunately, it appears that the most prevalent mucin precursors in any particular organ or cell line can be distinguished in the homogenates of the tissue or cells in which they are produced, by virtue of two general properties: mucin precursors are (1) extremely large and (2) quite abundant in the tissues of cell lines in which they are produced. Each precursor of the MUC-type mucins appears to display a unique molecular mass on sodium dodecyl sulfate-polyacrylamide gel electrophoresis (SDS-PAGE), and once identified in the homogenates, mucin precursors can be quantified from SDS-PAGE analysis (*see* Chapters 20 and 21 for the identification of the individual mucin precursors).

2. Materials

1. Bicinchoninic acid (BCA) kit to assay protein concentration (Pierce, Rockford, IL).
2. Bovine serum albumin (BSA) solution in homogenization buffer to provide calibration for the protein assay.
3. 5% (w/v) trichloroacetic acid (TCA).
4. 3% stacking/4% running gels for SDS-PAGE, electrophoresis apparatus (mini Protean II system; Bio-Rad, Richmond CA), and Laemmli electrophoresis buffers.
5. PhosphorImager (Molecular Dynamics, Sunnyvale, CA), with ImageQuant software (or equivalent apparatus) to quantify the amount of radioactivity in an identified band on SDS-PAGE.
6. Poly- or monoclonal antibodies (IgG-type) against nonglycosylated regions of the mucin-polypeptide of interest (*see* Chapter 20 for specific antibodies to recognize the major MUC-type mucins).
7. ^{125}I-labeled protein A, specific activity 30 mCi/mg, supplied as solution of 100 μCi/mL (1.1 GBq/mg; 3700 kBq/mL) (Amersham, Little Chalfont, Buckinghamshire, UK).
8. Dot-blot apparatus, vacuum operated (e.g., Bio-Dot, Bio-Rad).
9. Trizol RNA isolation solution (Gibco/BRL, Gaithersburg, MD).
10. Agarose gel electrophoresis apparatus, and 0.8% (w/v) agarose gels.
11. Radiolabeled, homologous cDNA or cRNA probe to quantify the mucin mRNA of interest (*see* Chapters 24 to 27).
12. Radiolabeled, homologous cDNA or cRNA probe to quantify an appropriate control mRNA in each cell sample, i.e., β-actin or glyceraldehydephosphate dehydrogenase.
13. Carbogen-gas container (95% O_2/5% CO_2) and pressure-reduction valve.
14. Tissue homogenizer (Glass/Teflon, Potter/Elvehjem homogenizer).
15. Sterile media for metabolic labeling: Eagle's minimal essential medium (EMEM), described in detail in Chapter 19 **Subheading 2**.
    a. EMEM without methionine and cysteine.
    b. EMEM without sulfate.
    c. Standard EMEM.
16. Radiolabeled compounds (Amersham), described in detail in Chapter 19, Subheading 2.
    a. Pro-Mix^{TM}, containing a mixture of [^{35}S]methionine/[^{35}S]cysteine.
    b. [^{35}S]sulfate.
17. Water bath, 37°C.
18. Whatman 3MM filter paper.
19. Molecular weight marker: nonreduced monomeric and dimeric rat gastric mucin precursors, molecular mass 300 and 600 kDa, respectively (8) (*see* Chapters 20 and 21).
20. Homogenization buffer (pH 7.5, 0°C): 50 m*M* Tris-HCl, 5 m*M* EDTA, 1% Triton X-100 (BDH, Poole, UK), 10 m*M* iodacetamide, 100 μg/mL soybean trypsin inhibitor, 10 μg/mL pepstatin A, aprotinin 1% (v/v) form commercial solution, 1 mM phenylmethylsulfonylfluoride (PMSF), 10 μg/mL leupeptin. All chemicals are from Sigma. PMSF is unstable in water; add just before use, from 100 m*M* stock solution in 2-propanol.
21. Blotto (wash buffer for Western-type dot-blots): 50 m*M* Tris (pH 7.8), 2 m*M* $CaCl_2$, 0.05% (v/v) Nonidet P-40 (BDH), 0.01% antifoam A (Sigma), 5% (w/v) nonfat milk powder (Nutricia, Zoetermeer, The Netherlands).
22. Nitrocellulose paper (Nitran, Schleicher and Schuell, Dassell, Germany).
23. Saran Wrap (Dow Chemicals, Karlsruhe, Germany).
24. Scintillation counter, scintillation vials, and scintillation fluid (Ultima Gold, Packard, Meriden, CT).
25. 50% ethanol/50% diethyl ether (v/v).

3. Methods

3.1. General Assays for Quantitation of Mucin (see Notes 1–3)

Each of these six detailed assays are used to quantify a particular aspect of the biosynthesis of mucin. These assays are essential in the extensive protocol for quantitation of the biosynthesis and secretion of mucin, which is discussed in **Subheadings 3.2.** and **3.3.**

1. Measure protein concentrations according to the BCA protein assay and calculate the protein concentration of each homogenate, with the help of the BSA calibration solutions, in micrograms/milliliter.
2. Measure the incorporation of radiolabel into (glyco-)proteins by TCA precipitation. Spot 5 µL of the homogenate on a pencil-marked location on 3MM Whatman filter paper and air-dry. Immerse the filter paper in ice-cold 5% TCA for at least 10 min. Transfer paper to 5% TCA at 100°C for 10 min. Wash the paper two times for 5 min each in 5% TCA at room temperature. Rinse once in 50% ethanol/50% diethyl ether and air-dry. Quantify the amount of radiolabel incorporated in glycoproteins in the homogenate by liquid scintillation counting as counts per minute/milliliter of homogenate.
3. Identify and quantify the mucin precursor band in the homogenate after separation on reducing 4% SDS-PAGE using the PhosphorImager (*see* Chapters 20 and 21). Calculate the amount of mucin precursor as arbitrary units (au)/milliliter of homogenate.
4. Identify and quantify the mature mucin band in the homogenate and medium after separation on reducing 4% SDS-PAGE using the PhosphorImager (*see* **Note 4**). Calculate the amount of mature mucin as au/milliliter of homogenate or as au/milliliter of medium.
5. Quantify the total amount of the mucin of interest in the homogenate and medium by Western-type dot-blot procedure. Spot aliquots of the homogenates or media on nitrocellulose paper, using the dot-blot apparatus and air-dry for 5 min. Perform all ensuing procedures in Blotto as follows:
   a. Incubate in Blotto for 30 min, and incubate with antibody directed against nonglycosylated peptide epitopes of the mucin of interest for 90 min.
   b. Wash two times for 15 min each.
   c. Incubate with 0.5 µCi of (185 kBq) ^{125}I-labeled protein A for 60 min.
   d. Wash twice for 5 min each in Blotto, and then wash once for 5 min in phosphate-buffered saline.
   e. Dry filter briefly using Whatman 3MM paper, and cover the filter in Saran Wrap.
   f. Place two sheets of Whatman 3MM paper between the filter and the PhosphorImager screen to exclude the ^{35}S radiation, owing to the endogenous radiolabeled compounds in the homogenate, from reaching the screen.
   g. Quantify the ^{125}I label per dot using the PhosphorImager as au/milliliter of homogenate or au/milliliter of medium (*see* **Note 5**).
6. Isolate RNA by the Trizol method, and quantify RNA at A_{260nm}/milliliter. Judge the intactness of the RNA, by analysis of the 18S and 28S rRNA bands on 0.8% agarose electrophoresis. The mucin mRNA of interest is quantified, using a specific homologous cDNA or an antisense cRNA probe, by dot-blotting, using the dot-blot apparatus. The specific signal is quantified using the PhosphorImager as au/A_{260nm}.

3.2. Quantitation of Biosynthesis of Mucin (see Notes 4 and 6–9).

Four mucin-producing cell samples are used: they can be biopsies, tissue explants, or cell line cultures. Metabolic pulse/chase labeling of biopsies and tissue explants is conducted individually submerged in the appropriate medium in small tubes, as described in Chapters 18 and 19.

3.2.1. Cell or Tissue Sample No. 1: Quantitation of Mucin Precursor Synthesis

1. Cell or tissue sample no. 1 is pulse labeled with [^{35}S]methionine/cysteine for 15–60 min, as described in Chapter 19. Homogenize the sample in homogenization buffer on ice, isolate the supernatant by 5 min centrifugation at 12,000g, and measure the following parameters:
   a. Protein concentration (mg/mL).
   b. Protein synthesis, i.e., [^{35}S]amino acids–labeled, TCA-precipitable proteins (cpm/mL).
   c. [^{35}S]Amino acids–labeled mucin precursor-band on 4% SDS-PAGE (au/mL).
   d. Total concentration of mucin by dot-blotting (au/mL).

3.2.2. Cell or Tissue Sample No. 2: Quantitation of Synthesis of Mature Mucin

1. Cell or tissue sample no. 2 is pulse labeled with [^{35}S]sulfate for 30–60 min, as described in Chapter 19. Homogenize the cell sample in homogenization buffer on ice, isolate the supernatant by 5 min centrifugation at 12,000*g*, and measure the following components:
   e. Protein concentration (mg/mL).
   f. Sulfate incorporation as [^{35}S]sulfate-labeled, TCA-precipitable proteins (cpm/mL) (*see* **Note 10**).
   g. Mature [^{35}S]sulfate-labeled, mucin band on 4% SDS-PAGE (au/mL).
   h. Total concentration of mucin by dot-blotting (au/mL).

3.2.3. Cell or Tissue Sample No. 3: Quantitation of Secretion of Mature Mucin

1. Cell or tissue sample no. 3 is pulse labeled with [^{35}S]sulfate, and then chase incubated in the absence of radioactive sulfate for 4–6 h. Isolate medium from the chase incubation, homogenize tissue in homogenization buffer, and isolate the supernatant by 5 min centrifugation at 12,000*g*. Mix the medium with an equal amount of homogenization buffer. Measure the following components:
   i. Protein concentration of tissue homogenate (mg/mL).
   j. Total sulfate incorporation as [^{35}S]sulfate-labeled, TCA-precipitable proteins in tissue homogenate (j1) and medium (j2) (cpm/mL) (**Note 10**).
   k. Mature [^{35}S]sulfate-labeled mucin band in tissue homogenate (k1) and medium (k2) on 4% SDS-PAGE (au/mL).
   m. Total concentration of mucin in tissue homogenate (m1) and medium (m2) by dot-blotting (au/mL).

3.2.4. Cell or Tissue Sample No. 4: Quantitation of Mucin mRNA

1. Cell or tissue sample no. 4 is homogenized in Trizol solution, and the RNA is isolated according to the manufacturer. Measure the following components:

n. Total RNA (A_{260nm}/mL).
p. Specific mucin mRNA by Northern blot, ideally using a radiolabeled cDNA/cRNA probe corresponding to nonrepetitive mucin sequences. Measure mucin mRNA signal relative to the signal of the mRNA of a "housekeeping" protein, i.e., β-actin or glyceraldehydephosphate dehydrogenase, as a measure of sample size (no dimension).

3.3. Calculation of Biosynthesis of Mucin and Level of Regulation (see Notes 1–3, 7, and 11)

1. The sample size (i.e., mucin-producing tissue or cells) is measured as the protein concentration of the homogenates: a, e, and i.
2. The specific mRNA concentration is calculated relative to the total amount of RNA per cell sample, p.
3. The total protein synthesis in the sample (measure of tissue or cell viability), r (**Note 12**), is calculated as follows:

$$r = b/a \text{ (cpm/mg)}$$

4. The specific mucin precursor synthesis relative to the total protein synthesis within the explant, s (**Note 13**), is calculated as follows:

$$s = c/b \text{ (au/cpm)}$$

5. The synthesis of mature mucin can be calculated in two ways: t and u. The synthesis of mature mucin can be calculated relative to the size of the cell sample as indicated by the protein concentration, t:

$$t = g/e \text{ (au/mg)}$$

The synthesis of mature mucin can be related to the total protein synthesis, u. Value t can only be calculated from the duplo sample (i.e., $[^{35}S]$amino acids–labeled, "type no. 1" sample), and has thus to be corrected for different sizes of the samples (i.e., the protein concentrations, values a and e):

$$u = g/b \times e/a \text{ (au/cpm)}$$

6. Secretion of mature mucin can be calculated in two ways: v and w (**Note 14**). The percentage of secretion of total mucin within the duration of the chase incubation (usually 4–6 h) v is calculated as follows:

$$v = m2/(m1 + m2) \times 100 \text{ (\%)}$$

The percentage of secretion of newly synthesized, $[^{35}S]$sulfate-labeled mature mucin within duration of chase incubation (usually 4–6 h) w is calculated as follows:

$$w = k2/(k1 + k2) \times 100 \text{ (\%)}$$

7. The extent of sulfation of the newly synthesized mucin is determined using data from the duplo samples no. 1 and 2: the ratio between the amount of $[^{35}S]$sulfate-labeled mature mucin (determined from sample no. 2; value g) and the amount of $[^{35}S]$amino acids–labeled mucin precursor (determined from sample no. 1; value c): *x*. This calculation then needs to be corrected for the exact sizes of the samples no. 1 and 2 (values a and e):

$$x = g/c \times a/e \text{ (no dimension)}$$

8. The total amount of mucin can be calculated per cell sample, z. This can be done for each of the samples no. 1 and 2, and should yield similar values:

$$z = d/a \text{ (sample no. 1, au/mg) or } z = h/e \text{ (sample no. 2, au/mg)}$$

9. The level of regulation of mucin expression can be determined by calculating the correlation between (1) the mucin mRNA level and the mucin precursor synthesis level, and (2) the mucin precursor synthesis level and the total mucin levels (**Note 11**).

4. Notes

1. All calculations, apart from the mRNA quantitation, are performed in such a way that the values are expressed per milliliter of homogenate or per milliliter of medium.
2. The PhosphorImager is a sensitive apparatus to measure radioactivity in flat materials (gels or filter paper) by autoradiography. It calculates the amount of radioactivity in a band on gel or on a dot on filter paper, taking into account the area of the band/dot and the intensity of the signal. The data are generated in arbitrary units (au).
3. The dimensions of the various calculated values have limited significance, largely because of the use of arbitrary units for each of the measurements of radioactivity, owing to the use of the PhosphorImager.
4. The identification of mucin precursors and mature mucins using polypeptide-specific antisera and SDS-PAGE is elaborated in Chapters 20 and 21, and in several references *(1,4–11)*. The precursor of each mucin is identified by its molecular mass, which is established on each separate gel by the use of molecular mass markers such as the nonreduced rat gastric mucin precursors (**Subheading 2., item 19**). It is strongly advised to run a sample of the specifically immunoprecipitated mucin precursor of interest on the same gel, to be able to identify unequivocally the mucin precursor prior to quantitation by the PhosphorImager.
5. Spot different aliquots of the homogenates on the same sheet of nitrocellulose, using appropriate dilutions of the samples in homogenization buffer, and test whether these diluted samples give the proper linear response in the dot-blot assay. Often around 1 µg of protein per spot is sufficient.
6. Ideally, four cell samples (i.e., tissue explants, biopsies, or cell cultures) are needed to perform all assays. However, the actual size of each cell sample is not important for any of these assays. All measures of the expression of mucin are relative to parameters within the cell sample or its homogenate. But, note that the use of tissue or cell samples of similar sizes makes the analysis a lot simpler, because the values to be measured will fall within the same, linear range of the assays.
7. Each parameter, which can be assessed from each sample, is symbolized by a letter, which is used in the equations under **Subheading 3.3.** to calculate the mucin biosynthesis at the various levels.
8. The length of the metabolic pulse labeling and the concentration of the radioactive label must be optimized for each particular cell line or tissue sample. These parameters have been determined for a wide variety of gastrointestinal tissues and cell lines (e.g., *see* **refs. *4–10***), and are described in detail in Chapter 19.
9. The procedures in this chapter only describe the measurements of the mucin synthesis at each level. Experiments on these mucin-producing cell or tissue samples can be performed prior to or during the metabolic pulse labeling of the mucin-producing cell samples.
10. In most cases [^{35}S]sulfate is primarily incorporated into mucins. Therefore, the values obtained by TCA precipitation of [^{35}S]sulfate-labeled macromolecules (i.e., values f, j1,

and j2) have limited meaning. In fact, these figures might serve as control values for the mature mucin values determined by SDS-PAGE and quantification of the [^{35}S]sulfate-labeled mature mucin (i.e., values g and k).

11. According to this method of calculation, the level of regulation was determined for the colonic synthesis of human MUC2. The decrease in colonic MUC2 synthesis in active ulcerative colitis could be attributed to regulation at the posttranscriptional level, presumably affecting the translational efficiency. There appeared to be no correlation between MUC2 mRNA and MUC2 precursor synthesis, but a very high correlation between MUC2 precursor synthesis and the total levels of MUC2 in each explant ***(6)***.
12. The total protein synthesis of the cell or tissue samples (r) is a value that has a central place in the interpretation of the results. This is a very good indicator of the condition of the tissue (viability), because protein synthesis is an extremely energy-dependent process: decrease in energy supply will be noticed immediately in a drop in the value q. This can be taken one step further, if the patterns of the radiolabeled proteins in each homogenate are analyzed on a higher percentage SDS-PAGE (e.g., 10% polyacrylamide). Such analysis will give an impression of the intensity of the labeling and, more importantly, the intensity of the types of proteins labeled. A change in protein pattern indicates that the biosynthesis of mucin is measured against a "background" of a different mixture of proteins. Therefore, it is essential to compare the [^{35}S]amino acid–labeled protein bands on SDS-PAGE from all samples to identify gross alterations in protein synthesis during the experiment.
13. The specific mucin precursor synthesis (value s) is expressed relative to the total protein synthesis within the explant. Because all calculations are performed per milliliter of homogenate, the amount of protein in the homogenate is of no consequence of this relative value. Therefore, the specific mucin precursor synthesis can be calculated from the amount of mucin precursor (value c, expressed as au [arbitrary units], which is a direct measure of radioactivity) and the total amount of radioactivity incorporated into proteins (value b, in counts per minute [cpm]). This calculation results in value s, expressed in au/cpm, which is not dependent on the protein concentration.
14. Similar experiments can be performed using [^{35}S]amino acids in pulse/chase experiments. The main advantage of this approach is that secretion can be quantified irrespective of the extent of sulfation of the mature mucin (i.e., value χ [no dimension]). However, the measurements are complicated by the fact that two pools of intracellular ^{35}S-labeled mucin molecules exist: precursor and mature mucin. Thus, three bands must be quantified per sample: the precursor, the intracellular mature mucin, and the mature mucin in the medium. The total radioactivity within these three bands is taken as 100% value. The secretion is then calculated as mature mucin in the medium divided by the total mucin, which (after multiplication by 100) gives the percentage of secretion of [^{35}S]amino acids–labeled mucins.

References

1. Strous, G. J. and Dekker, J. (1992) Mucin-type glycoproteins. *Crit. Rev. Biochem. Mol. Biol.* **27,** 57–92.
2. Van Klinken, B. J. W., Dekker, J., Büller, H. A., and Einerhand, A. W. C. (1995) Mucin gene structure and expression updated: protection versus adhesion. *Am. J. Physiol.* **269,** G613–G627.
3. Tytgat, K. M., Dekker, J. and Büller, H. A. (1993) Mucins in inflammatory bowel disease. *Eur. J. Gastroenterol. Hepatol.* **5,** 119–127.

4. Tytgat, K. M. A. J., Büller, H. A., Opdam, F. J. M., Kim, Y. S., Einerhand, A. W. C., and Dekker, J. (1994) Biosynthesis of human colonic mucin: Muc2 is the most prominent secretory mucin. *Gastroenterology* **107,** 1352–1363.
5. Tytgat, K. M. A. J., Opdam, F. J. M., Einerhand, A. W. C., Büller, H. A., and Dekker, J. (1996) MUC2 is the prominent colonic mucin expressed in ulcerative colitis. *Gut* **38,** 554–563.
6. Tytgat, K. M. A. J., Van der Wal, J. W. G., Büller, H. A., Einerhand, A. W. C., and Dekker, J. (1996) Quantitative analysis of MUC2 synthesis in ulcerative colitis. *Biochem. Biophys. Res. Commun.* **224,** 397–405.
7. Dekker, J., Van Beurden-Lamers, W. M. O., and Strous, G. J. (1989) Biosynthesis of gastric mucus glycoprotein of the rat. *J. Biol. Chem.* **264,** 10,431–10,437.
8. Dekker, J. and Strous, G. J. (1990) Covalent oligomerization of rat gastric mucin occurs in the rough endoplasmic reticulum, is N-glycosylation dependent and precedes O-glycosylation. *J. Biol. Chem.* **265,** 18,116–18,122.
9. Tytgat, K. M. A. J., Klomp, L. W. J., Bovelander, F. J., Opdam, F. J. M., Van der Wurff, A., Einerhand, A. W. C., Büller, H. A., Strous, G. J., and Dekker, J. (1995) Preparation of anti-mucin polypeptide antisera to study mucin biosynthesis. *Anal. Biochem.* **226,** 331–341.
10. Van Klinken, B. J. W., Tytgat, K. M. A. J., Büller, H. A., Einerhand, A. W. C., and Dekker, J. (1995) Biosynthesis of intestinal mucins: MUC1, MUC2, MUC3, and more. *Biochem. Soc. Trans.* **23,** 814–818.
11. Van Klinken, B. J. W., Einerhand, A. W. C., Büller, H. A., and Dekker, J. (1998) Strategic biochemical analysis of mucins. *Anal. Biochem.* **265,** 103–116.

III

Separation, Identification, and Physical Characterization of Mucins

7

Separation and Identification of Mucins and Their Glycoforms

David J. Thornton, Nagma Khan, and John K. Sheehan

1. Introduction

This chapter describes a strategy for the separation and identification of the mucins present in mucous secretions or from cell culture, focusing primarily on those mucins involved in gel formation. At present, the mucins MUC2, MUC5AC, MUC5B, and MUC6 are known to be gel-forming molecules *(1–4)*. These mucins share common features in that they are oligomeric in nature and consist of a variable number of monomers (subunits) linked in an end-to-end fashion via the agency of disulfide bonds. In addition, their polypeptides comprise regions of dense glycosylation interspersed with "naked" cysteine-rich domains *(4–7)*.

Histological and biochemical investigations suggested that mucous-producing tissues and their secretions contained a complex mixture of mucin-type glycoproteins. However, until recently and with the advent of the new mucin-specific probes arising from cDNA cloning studies, this theory was not definitively proven. *In situ* hybridization and Northern blotting studies have shown that more than one gel-forming MUC gene product can be expressed in a single mucus-producing epithelia, i.e., MUC5AC and MUC5B in the respiratory tract and MUC5AC and MUC6 in the stomach *(4,8)*. Subsequent biochemical studies on human airway mucus have shown that these two mucin genes are not only expressed but that their glycosylated products are present in respiratory tract secretions *(2,3)*. A further more recent insight into the complex nature of the mucin component of mucous has been the demonstration that a mucin gene product from a single epithelium can have a different oligosaccharide decoration and thus exist in what are termed *glycoforms*. For example, the MUC5B mucin in the respiratory tract can exist in two distinctly charged states *(3)*. Thus, these studies demonstrate the need to have techniques available to dissect these complex mixtures to ascertain mucin type, amount, and glycoform. Such investigations may lead to the identification of novel members of this growing family of molecules.

Owing to the extreme size and polydispersity (M_r = 5–50 × 10^6) of the gel-forming mucins in particular, there are few separation techniques available to use with these

From: *Methods in Molecular Biology, Vol. 125: Glycoprotein Methods and Protocols: The Mucins*
Edited by: A. Corfield © Humana Press Inc., Totowa, NJ

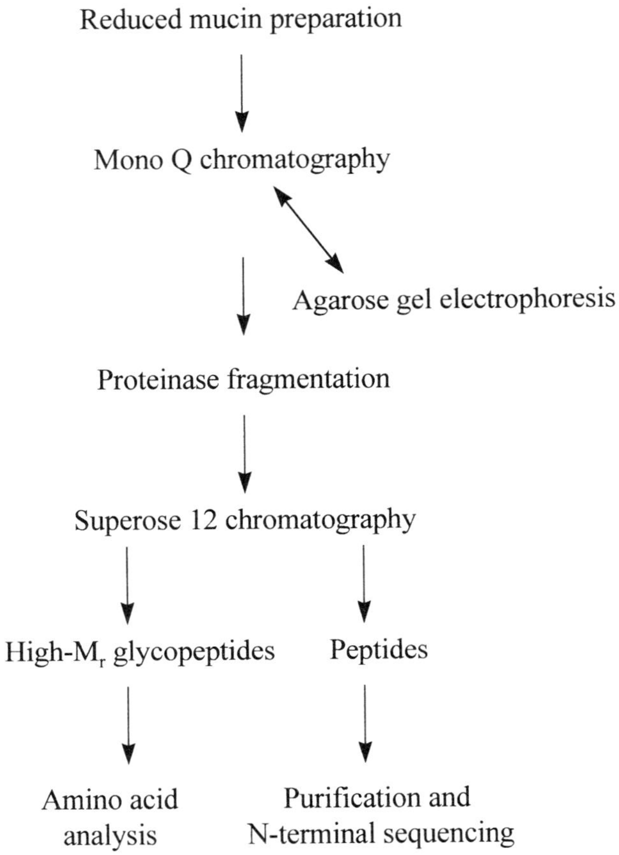

Fig. 1. Outline of the protocol for separation and identification of mucins.

molecules. However, the smaller and more homogeneous reduced mucin monomers generated after reduction (M_r = 2–3 × 10^6) are amenable to conventional biochemical separation methods. We describe a separation protocol based on anion exchange chromatography to fractionate the different species of mucin coupled with agarose gel electrophoresis as a method to measure their homogeneity. Identification of mucin polypeptides is achieved by use of MUC-specific antisera and fragmentation followed by amino acid compositional analysis and peptide purification and sequencing. Mucin glycoforms are detected using carbohydrate-specific probes, i.e., lectins or monoclonal antibodies. **Figure 1** summarizes the overall procedure.

2. Materials

2.1. Extraction and Purification

See Chapter 1 for details.

2.2. Preparation of Reduced Mucin Subunits

1. Reduction buffer: 6 *M* guanidinium chloride, 0.1 *M* Tris-HCl, 5 m*M* EDTA, pH 8.0.
2. Dithiothreitol (DTT).
3. Iodoacetamide.
4. Buffer A: 6 *M* urea, 10 m*M* piperazine pH 5.0, containing 0.02% (w/v) CHAPS.
5. PD-10 (Amersham Pharmacia Biotech, St. Albans, UK) or equivalent desalting column.

2.3. Separation of Mucin

2.3.1. Anion-Exchange Chromatography

1. Mono Q HR5/5 column (Pharmacia).
2. Buffer A (*see* **Subheading 2.2.**).
3. 0.4 *M* lithium perchlorate.

2.3.2. Agarose Gel Electrophoresis

1. Horizontal electrophoresis apparatus, e.g., Bio-Rad DNA subcell Model 96 (25 × 15 cm gels) and the Bio-Rad subcell (15 × 15 cm gels).
2. Agarose (ultrapure).
3. Electrophoresis buffer: 40 m*M* Tris-acetate, 1 m*M* EDTA, pH 8.0, containing 0.1% (w/v) sodium dodecyl sulfate.
4. Loading buffer: electrophoresis buffer containing 30% (v/v) glycerol and 0.002% (w/v) bromophenol blue.
5. Transfer buffer: 0.6 *M* NaCl, 60 m*M* sodium citrate.

2.4. Identification of Mucin

2.4.1. Tryptic Digestion

1. Digestion buffer: 0.1 *M* ammonium hydrogen carbonate, pH 8.0.
2. Modified trypsin or other proteinase.

2.4.2. Trypsic Peptide Analysis

1. Digestion buffer: 0.1 *M* ammonium hydrogen carbonate, pH 8.0.
2. Superose 12 (Pharmacia) or equivalent column.
3. µRPC C2/C18 PC 3.2/3 column (Pharmacia) or other C2/C18 column.
4. 0.1% (v/v) trifluoroacetic acid (TFA).
5. Acetonitrile.

2.4.3. Amino Acid Analysis

1. 6 *M* HCl.
2. Redrying buffer: ethanol:triethylamine:water (2:2:1 [v/v/v]).
3. Coupling buffer: ethanol:triethylamine:water:phenylisothiocyanate (PITC) (70:10:19:1 [v/v/v/v]).
4. 3µ ODS2 column (4.6 × 150 mm) (Phase Separations, Clwyd, UK).
5. Buffer A: 12 m*M* sodium phosphate, pH 6.4.
6. Buffer B: 24 m*M* sodium phosphate, pH 6.4, containing 60% (v/v) acetonitrile.

3. Methods

3.1. Extraction and Purification

For a detailed description of extraction and purification, *see* Chapter 1. In brief, mucins are extracted from samples at 4°C with 6 *M* guanidinium chloride containing proteinase inhibitors and are purified by isopycnic centrifugation in CsCl density gradient first in 4 *M* guanidinium chloride (removal of proteins) and then in 0.2 *M* guanidinium chloride (removal of nucleic acids). Finally, the preparation of mucin is dialyzed into 4 *M* guanidinium chloride for storage.

3.2. Preparation of Reduced Mucin Subunits

Mucin subunits (monomers) are prepared by reduction and alkylation of purified mucins as follows:

1. Transfer (by dialysis or dilution) the mucins into reduction buffer.
2. Reduce the mucins by the addition of DTT to a final concentration of 10 m*M* for 5 h at 37°C.
3. Alkylate the free-thiol groups generated by reduction with the addition of iodoacetamide to a final concentration of 25 m*M*. This step can be performed overnight in the dark at room temperature.
4. Transfer the reduced mucin subunits into buffer A by chromatography on a Pharmacia PD-10 column (follow manufacturer's instructions) or by dialysis.

3.3. Separation of Mucin

Separation of reduced mucin subunits is achieved by using anion-exchange chromatography on a Mono Q HR 5/5 column (Fig. 2A). Assessment of the effectiveness of the separation is monitored with a periodic acid-Schiff (PAS) assay and with a variety of lectins and mucin- or carbohydrate-specific antibodies (for a discussion of the relative merits and drawbacks of these analytical tools, *see* Chapter 4). The homogeneity of the fractions is monitored by agarose gel electrophoresis and Western blots of the gels can be probed with lectins and antibodies. Using this methodology, we have shown that with pooling and rechromatography it is possible to prepare mucin subunit samples enriched in specific MUC gene products and also to isolate different glycosylated forms (glycoforms) of a single MUC gene product (*see*, e.g., **refs.** ***2*** and ***3***).

An example of the data obtained, using this methodology with a salivary mucin reduced subunit preparation, is shown in **Fig. 2**. The range of charge density of the molecules (**Fig. 2a**) is typical of that observed for the reduced subunits prepared from the gel-forming mucins isolated from other epithelial secretions (respiratory and cervical) and from intestinal cell lines (HT-29 and PC/AA). In this example the electrophoretic mobility of the reduced subunits (**Fig. 2b**) is dependent on their charge density since the molecular weight and size (radius of gyration) of the molecules (determined by light scattering) across the charge distribution are essentially the same ***(8a)***. Analysis of Western blots with MUC-specific antisera suggest that the major mucin in this sample is the MUC5B gene product ***(8a)***. Thus, the Mono-Q column appears to be separating differently charged forms of this mucin. The continuum in the electro-

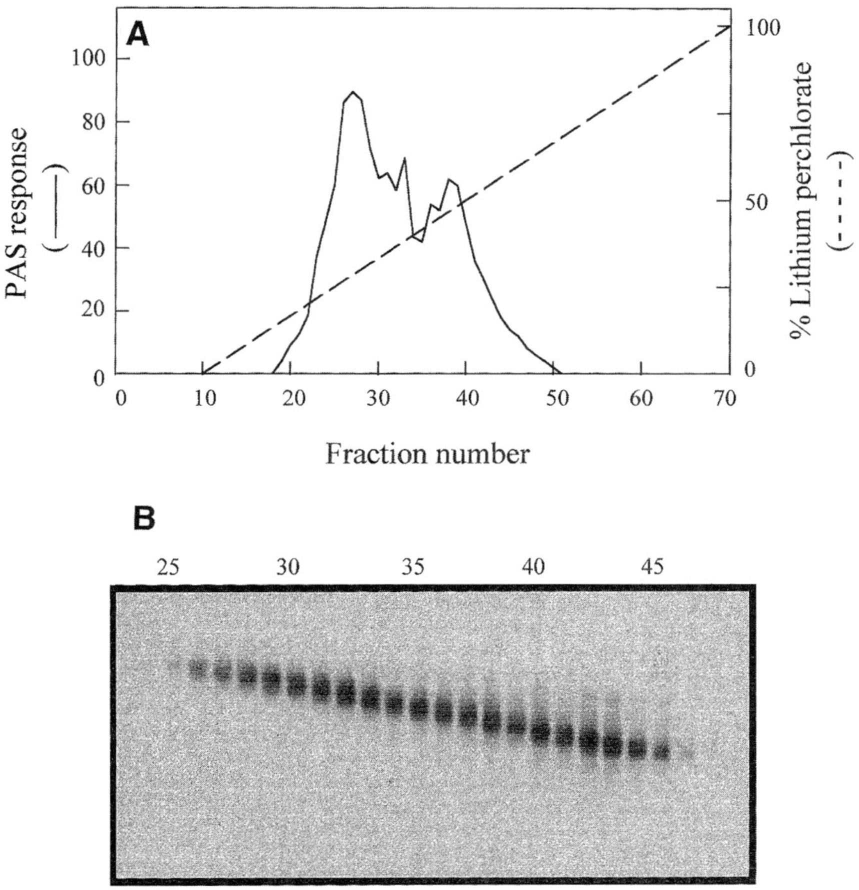

Figure 2. Separation of reduced mucin subunits: (A) Mucins were isolated from saliva, reduced and alkylated, and chromatographed on a Mono-Q HR 5/5 column. The diagram shows the distribution of the reduced subunits as monitored with the PAS-reagent. (B) Aliquots from each fraction across the charge distribution were subjected to 1% (w/v) agarose gel electrophoresis. A Western blot of the gel was probed with a polyclonal antiserum raised against reduced mucin subunits.

phoretic mobility observed with the salivary mucin sample is not normally seen in respiratory mucin subunit preparations. In respiratory samples there are typically two gel-forming mucins, namely MUC5AC and MUC5B. The MUC5B mucin appears by Mono-Q chromatography to be in two differently charged states, and like the salivary mucin subunits, the electrophoretic mobility of the MUC5B subunits is consistent with their charge density. However, the MUC5AC mucin-reduced subunits are of lower charge density than the most highly charged MUC5B molecules, but they migrate farther on a 1% (w/v) agarose gel ***(2)***.

3.3.1. Anion-Exchange Chromatography

The flow rate for anion-exchange is 0.5 mL/min throughout the chromatography, and typically 0.5-mL fractions are collected.

1. Apply the sample (up to 10 mg) to the Mono Q column in buffer A and wash for 10 min after application.
2. Elute the sample with a linear gradient from 0 to 0.4 *M* lithium perchlorate in buffer A over a period of 60 min.
3. Analyze samples with A_{280nm} measurements, an assay for carbohydrate (e.g., PAS reagent) and for lectin and antibody reactivity (*see* **ref. 9** for detailed procedures).

3.3.2. Agarose Gel Electrophoresis

Electrophoresis is performed in 1% (w/v) agarose gels using a standard horizontal gel electrophoresis apparatus.

3.3.2.1. Sample Preparation

1. Dilute or dialyze reduced mucin subunits from the Mono Q separation into electrophoresis buffer (*see* **Note 1**).
2. Add 1/10 of a volume of 30% (v/v) glycerol in electrophoresis buffer containing 0.002% (w/v) bromophenol blue.

3.3.2.2. Gel Preparation

We typically perform electrophoresis in 15 × 15 cm or 25 × 15 cm gels of approx 3 to 4 mm thickness.

1. Dissolve the agarose in electrophoresis buffer in a microwave; for a small gel, use 1.6 g of agarose and 160 mL of buffer, and for a large gel, use 2.8 g of agarose and 280 mL of buffer.
2. Leave to cool before pouring (hand hot) and insert well-forming comb (*see* **Note 2**).
3. Leave to set for at least 1 h prior to use.

3.3.2.3. Electrophoresis

1. Electrophorese sample in electrophoresis buffer for 16 h at 30 V at room temperature. Ensure that the buffer is at least 0.5 cm above the gel surface and always use the same buffer volume.

3.3.2.4. Western Blotting

After electrophoresis, transfer the gel to nitrocellulose or poly(vinylidine difluoride) (PVDF) (*see* **Note 3**) as follows:

1. Wash the gel in transfer buffer for 5 min.
2. Transfer subunits to nitrocellulose or PVDF membrane by vacuum blotting in transfer buffer at a suction pressure of 45 mBar for 1.5 h. We use a Pharmacia Vacu-Gene XL for this procedure.
3. Probe the membrane for lectin and antibody reactivity (*see* **ref. 9** for detailed procedures).

3.4. Identification of Mucin

The easiest route to identification of mucin is by the use of MUC-specific antisera. However, because of sequence similarities among already known mucins, these antisera may not provide an unequivocal answer. Therefore, we also use N-terminal sequencing of proteolytically derived peptides and amino acid compositional analysis of mucin glycopeptides for a more definitive identification. These procedures will also provide a route to obtain information on novel mucins. The starting point for these analyses is a proteolytic fragmentation of reduced mucin subunits. For example, trypsin digestion of reduced mucin subunits liberates high M_r mucin glycopeptides, which correspond to the heavily *O*-glycosylated tandem repeat regions of the molecule, and lower M_r peptides and glycosylated peptides, which arise from the "naked" cysteine-rich regions of the molecule.

3.4.1. Trypsin Digestion

1. Dissolve lyophilized reduced mucin subunits in digestion buffer.
2. Add modified trypsin (**Note 4**) to the subunits in a weight ratio of approx- 1:1000 (*see* **Note 5**) and leave the digestion overnight at 37°C.
3. Separate digestion products into high M_r glycopeptides and tryptic peptides by chromatography on a Superose 12 column (or equivalent gel filtration medium) eluted with digestion buffer (**Note 6**).
4. Lyophilize tryptic fragments on a freeze-drier.

*3.4.2. Tryptic Peptide Analysis (*see ***Note 7****)*

Tryptic peptides can be fractionated by reverse phase chromatography and individual peptides purified and their primary sequence determined by N-terminal sequencing.

3.4.2.1. Peptide Purification and N-Terminal Sequencing

1. Solubilize lyophilized tryptic peptides in 0.1% TFA.
2. Chromatograph on a C2/C18 reverse phase column (*see* **Note 8**).
3. To increase the chances of obtaining unambiguous sequence data, an assessment of peptide homogeneity is advisable. We analyze aliquots of the peaks by using matrix-assisted laser desorption ionization time-of-flight mass spectrometry (MALDI-TOF MS), and purify peptides to homogeneity by rechromatography on the C2/C18 column utilizing shallower gradients centered on their elution point.
4. Dry down peptides by vacuum centrifugation and determine the primary sequence of peptides by automated N-terminal sequencing.
5. Use sequence to search sequence databases.

*3.4.3. Amino Acid Analysis (*see ***Note 9****)*

3.4.3.1. Acid Hydrolysis

1. Dissolve the samples (1 µg to 2 mg) in 500 µL 6 *M* AristaR grade HCl (BDH, Poole, Dorset, UK) (*see* **Note 10**) and transfer to glass hydrolysis tubes (*see* **Note 11**).
2. Flood the tubes with argon gas, seal, and hydrolyze at 110°C for 24 h.

3.4.3.2. Derivatization

1. Transfer the hydrolyzed samples to Eppendorf tubes.
2. Remove all the acid by evaporation under vacuum in a centrifugal evaporator.
3. Add 50 µL of redrying buffer to each sample and dry, under vacuum, for 30 min.
4. Add 50 µL of freshly prepared coupling buffer to each sample and vortex to mix well.
5. Leave at room temperature for 30 min.
6. Remove excess PITC by centrifugal evaporation under vacuum for 1 h.
7. The derivatized samples can be stored at –20°C.

3.4.3.3. Reverse Phase Chromatography

1. Solubilize the derivatized amino acids in buffer A.
2. Chromatograph at 38.6°C on a 3µ ODS2 column (*see* **Note 12**).
3. Monitor column eluent at 254 nm, and determine amino acids in sample by comparison with standard amino acid mixture.

4. Notes

1. Urea can be tolerated up to a concentration of at least 6 *M*, but salts (particularly guanidinium chloride) should be avoided.
2. The comb size (width relative to thickness) is important for the quality of the data. Typically we use combs that are 1.5 mm thick and 1 cm wide. There is a compromise between band broadness and amount of sample to be loaded. Larger amounts of sample tend to yield poorer quality data; i.e., the bands become more smeared.
3. Molecules transferred to PVDF can be treated with trifluoromethanesulfonic acid to remove *O*-linked glycans. This is often essential if using antisera that are directed against core protein epitopes "masked" by oligosaccharides ***(10)***.
4. Treat trypsin by reductive alkylation to modify arginine and lysine residues to prevent autolysis and thus remove the problem of peptides arising from the enzyme.
5. The amount of enzyme added was calculated assuming that protein constitutes 20% of the total mass of the mucin subunit and that approx 50% of this protein is in "naked" regions that are accessible to the proteinase.
6. Chromatography separates the high M_r mucin glycopeptides (void volume) from the lower M_r peptides that elute near or in the total volume; for an example, *see* **ref. *3***.
7. As we purify more mucins, it might become possible by using MALDI-TOF MS to obtain a peptide fingerprint associated with each gene product. We have already performed such an analysis for the MUC5B mucin purified from respiratory, cervical, and salivary secretions, as well as from respiratory cells in culture, and have identified a pattern of four major peptides (masses 1036, 1132, 1688, and 1980 Daltons; some of these peptides contain cysteine residues, which are alkylated in our analyses) that are characteristic of this molecule ***(3)***. These peptides arise from cysteine-rich regions that are repeated several times in the MUC5B polypeptide ***(7)***.
8. We have used a µRPC C2/C18 PC 3.2/3 column eluted at a flow rate of 240 µL/min with 0.1% (v/v) TFA (5 min) followed by a linear gradient of 0–50% (v/v) acetonitrile in 0.1% (v/v) TFA (30 min) using the Pharmacia SMART system.
9. Because of unique sequences of the tandem repeat regions of mucins identified so far, an amino acid composition on the mucin glycopeptides might be a simple way to distinguish between mucins. For example, the MUC2 mucin would be expected to have a much higher threonine content than any of the other identified mucins.

10. Samples must be salt free; otherwise the derivatization is hampered.
11. To decrease the risk of contamination, all glassware should be prewashed in concentrated chromic acid, rinsed in high-quality water, and dried before use.
12. The column is eluted at a flow rate of 0.75 mL/min with buffer A (2.66 min) followed by a linear gradient of 0–50% buffer B (42.34 min). Over the next minute, the concentration of buffer B is brought to 100%.

References

1. Sheehan, J. K., Thornton, D. J., Howard, M., Carlstedt, I., Corfield, A. P., and Paraskeva, C. (1996) Biosynthesis of the MUC2 mucin: evidence for a slow assembly of fully glycosylated units. *Biochem. J.* **315**, 1055–1060.
2. Thornton, D. J., Carlstedt, I., Howard, M., Devine, P. L., Price, M. R., and Sheehan, J. K. (1996) Respiratory mucins: identification of core proteins and glycoforms. *Biochem. J.* **316**, 967–975.
3. Thornton, D. J., Howard, M., Khan N., and Sheehan, J. K. (1997) Identification of two glycoforms of the MUC5B mucin in human respiratory mucus: evidence for a cysteine-rich sequence repeated within the molecule. *J. Biol. Chem.* **272**, 9561–9566.
4. Toribara, N. W., Ho, S. B., Gum, E., Gum, J. R., Lau, P. and Kim, Y. S. (1997) The carboxyl-terminal sequence of the human secretory mucin, MUC6: analysis of the primary amino acid sequence. *J. Biol. Chem.* **272**, 16,398–16,403.
5. Gum, J. R., Hicks, J. W., Toribara, N. W., Rothe, E. M., Lagace, R. E. and Kim, Y. S. (1992) The human MUC2 intestinal mucin has cysteine-rich subdomains located both upstream and downstream of its central repetitive region. *J. Biol. Chem.* **267**, 21,375–21,383.
6. Meerzaman, D., Charles, P., Daskal, E., Polymeropoulos, M. H., Martin, B. M,. and Rose, M. C. (1994) Cloning and analysis of cDNA encoding a major airway glycoprotein, human tracheobronchial mucin (MUC5). *J. Biol. Chem.* **269**, 12,932–12,939.
7. Desseyn, J. L., Guyonnet-Duperat, V., Porchet, N., Aubert, J. P. and Laine, A. (1997) Human mucin gene MUC5B, the 10.7-kb large central exon encodes various alternate subdomains resulting in a super-repeat: structural evidence for a 11p15.5 gene family. *J. Biol. Chem.* **272**, 3168–3178.
8. Audie, J. P., Janin, A., Porchet, N., Copin, M. C., Gosselin, B., and Aubert, J. P. (1993) Expression of human mucin genes in respiratory, digestive, and reproductive tracts ascertained by in situ hybridization. *J. Histochem. Cytochem.* **41**, 1479–1485.

8a. Thornton, D. J., Khan, N., Howard, M., Veerman, E., Packer, N. H. and Sheehan, J. K. (1999) Salivary mucin MG1 is comprised almost entirely of a differently glycosylated populations of the MUC5B gene product. *Glycobiology* **3,** 293–302.

9. Thornton, D. J., Carlstedt, I., and Sheehan, J. K. (1996) Identification of glycoproteins on nitrocellulose membranes and gels. *Mol. Biotechnol.* 5, 171–176.
10. Thornton, D. J., Howard, M., Devine, P. L. and Sheehan, J. K. (1995) Methods for separation and deglycosylation of mucin subunits. *Anal. Biochem.* **227**, 162–167.

8

Heterogeneity and Size Distribution of Gel-Forming Mucins

John K. Sheehan and David J. Thornton

1. Introduction

The rheological properties of human mucus are dominated by the physical properties of large secreted *O*-linked glycoproteins often referred to as gel-forming mucins. These molecules share the ability to assemble into long oligomeric structures via the agency of disulfide bonds. There is evidence that at least four mucins—MUC2, MUC5AC, MUC5B, and MUC6 *(1–4)*—are gel-forming mucins, and it is possible that there are others. In isolating a mixture of mucins from mucus, there is considerable scope for heterogeneity in their mass and length owing to the presence of different gene products that may themselves show polymorphism, different glycoforms of the various gene products (*see* Chapter 7), and variable numbers of subunits contributing to the final polymer. It is possible that heterogeneity may be an important biological property of gel-forming mucins and a key comparative characteristic when studying the change of mucous properties through the course of disease.

This chapter describes only methods for assessing the heterogeneity of mucin populations with regard to mass and size utilizing rate zonal centrifugation and electron microscopy. Other methods for the absolute determination of molecular weight using light scattering and analytical centrifugation are not described here because they require expensive, specialized equipment and a detailed knowledge of their theoretical basis.

2. Materials

2.1. Extraction and Purification

See Chapter 1 for details.

2.2. Rate-Zonal Centrifugation

1. 6 *M* guanidinium chloride (GuHCl).
2. 8 *M* GuHCl.
3. Peristaltic pump.
4. Swing-out rotors and tubes (*see* **Note 1**).
5. Gradient maker (*see* **Note 2**).

From: *Methods in Molecular Biology, Vol. 125: Glycoprotein Methods and Protocols: The Mucins*
Edited by: A. Corfield © Humana Press Inc., Totowa, NJ

6. Magnetic stirrer.
7. Thin glass capillaries (approx 10 cm).
8. Hamilton syringe (100–500 µL).

2.3. Electron Microscopy

2.3.1. Spreading Experiments

1. Spreading agent: 100 µg/mL of benzyldimethylalkylammonium chloride (BAC).
2. Spreading solution: 10 m*M* magnesium acetate (*see* **Note 3**).
3. Staining solution: ethanolic uranyl acetate. This is prepared by making a saturated solution of uranyl acetate in 0.1 *M* HCl and then spinning in a benchtop centrifuge for 5 min. Take 50 µL of the supernatant and make to 1 mL with ethanol.
4. Mica squares (2 × 2 cm).
5. Forceps.
6. Grids (400–600 mesh).
7. Platinum wire (0.2 mm diameter).
8. Rubber O-rings (2 cm diameter).
9. Teflon trough or small plastic Petri dish (approx vol 10 mL).
10. 95% (v/v) ethanol.
11. Liquid nitrogen.
12. 0.1 *M* acetic acid.
13. Carbon rods.
14. Vacuum-coating unit.

2.3.2 Replicas

1. Reagents as for **Subheading 2.3.1.**

3. Methods

3.1. Extraction and Purification

For a detailed description, *see* Chapter 1.

3.2. Rate-Zonal Centrifugation

Because of their extreme size, most gel chromatographic media are not useful for studying size distribution of intact mucins. A useful alternative, however, is rate-zonal centrifugation, which separates molecules not on hydrodynamic volume alone but on their mass-to-volume ratio as well. The basis of rate-zonal centrifugation for this purpose is long established and, in its simplest form, is the overlaying of a small volume of sample onto preformed gradient of a supporting medium of increasing density. The sample is centrifuged and a separation of different species in the mixture is effected on the basis of their sedimentation rates. The role of the preformed gradient is to provide a stable, supporting medium resistant to convective disturbances that are caused by temperature and mechanical instabilities in centrifuges.

We first demonstrated the value of rate-zonal centrifugation for the characterization of cervical mucin heterogeneity and polydispersity *(5)* and have since used this method to analyze the gel-forming mucins from a range of different epithelia and cell cultures *(6,7)*. We typically use 6–8 *M* guanidinium chloride (*see* **Note 4**) as the gradient support medium, and mucins have been shown to have (for certain rotor geometries) isokinetic sedimentation behavior in this system (*see* **Note 5**).

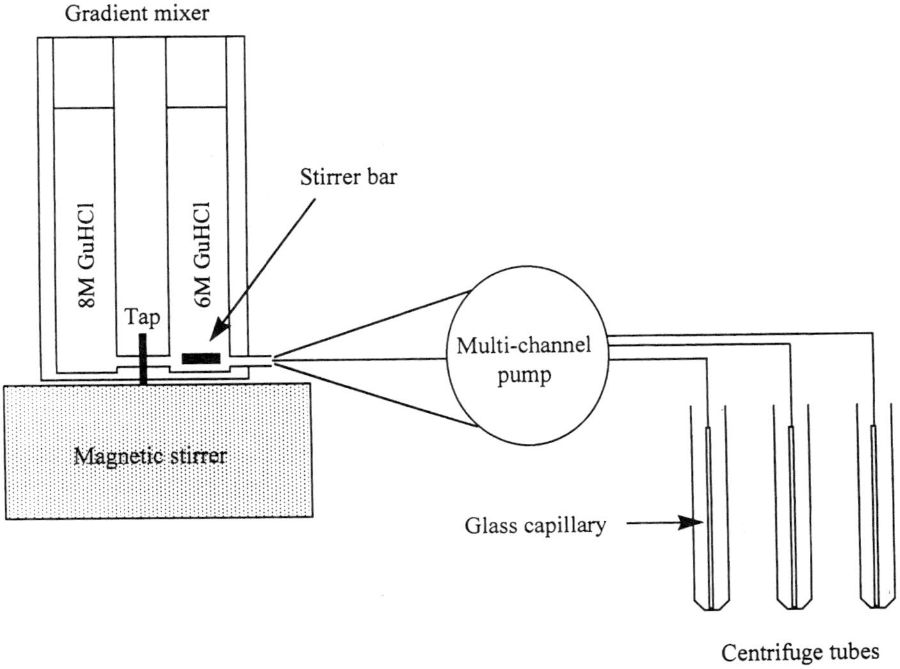

Fig. 1. Schematic diagram of gradient-forming apparatus.

We next describe an experiment that involves the simultaneous preparation of 3 × 12 mL (in 13-mL tubes) 6–8 *M* GuHCl gradients (**Fig. 1**), their subsequent centrifugation in a Beckman (Fullerton, CA) SW 40Ti swing-out rotor, and finally unloading and analysis of the tube contents.

1. Measure out 18 mL of 6 *M* GuHCl into one chamber of the gradient maker and an equal volume of 8 *M* GuHCl into the other chamber (*see* **Note 2**).
2. Mix contents of 6 *M* GuHCl chamber with a magnetic stirrer and open tap.
3. Pump contents of gradient maker into the bottom of three centrifuge tubes (in this case 13-mL tubes) at a flow rate of approx 1.3 mL/min/tube (*see* **Note 6**).
4. Weigh tubes to check that they contain equal volumes.
5. Carefully apply sample (up to 500 µL) with a Hamilton syringe to the top of the gradient (*see* **Note 7**).
6. Centrifuge at 40,000 rpm (Beckman SW 40 Ti swing-out rotor) for desired time at 15°C (in our case 2.5 h) (*see* **Note 8**).
7. Unload tubes from the top with a pipet (for a gradient of this volume we usually take 0.5-mL fractions).
8. Analyze fractions with a general carbohydrate assay (e.g., Periodic acid-Schiff) and for lectin and antibody reactivity (*see* **ref.** ***8*** for detailed procedures). The GuHCl concentration can be determined by measuring the refractive index of each fraction (*see* **Note 9**).
9. **Figure 2** shows an example of the data obtained from a respiratory mucin preparation.

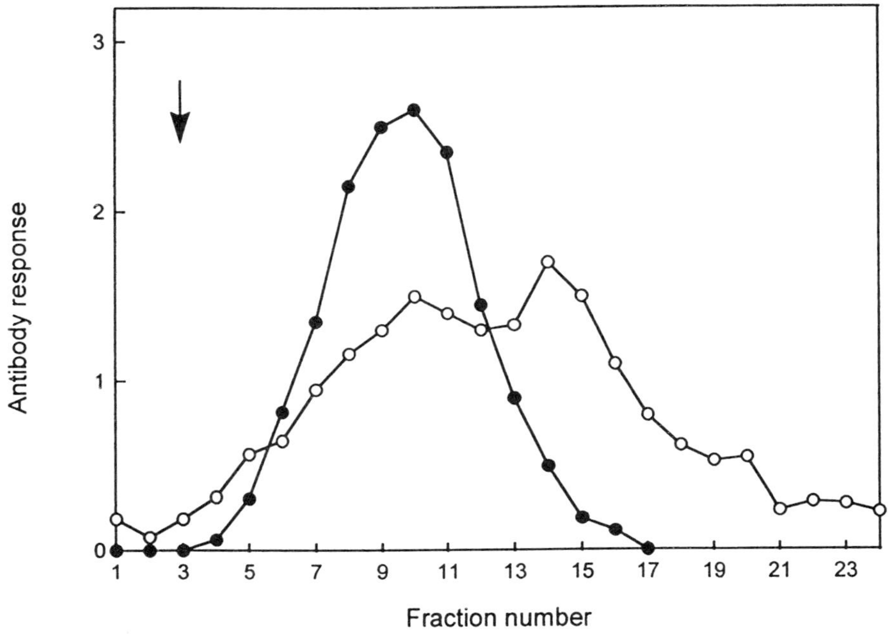

Fig. 2. Rate-zonal centrifugation of respiratory mucous extract. A respiratory mucous extract in 4 *M* GuHCl was applied to a 6–8 *M* GuHCl gradient (12 mL) and centrifuged at 202,000 *g* average (40,000 rpm) for 2.5 h at 15°C in a Beckman SW40 Ti swing-out rotor. The gradient was emptied from the top into 0.5-mL fractions, and these were analyzed for reactivity with antisera for the MUC5AC (•) and MUC5B (○) mucins. The MUC5B mucin is more polydisperse than the MUC5AC mucin and has molecules of an apparent higher molecular size. The arrow denotes the position of sedimentation of the reduced MUC5AC and MUC5B mucins centrifuged under these conditions. The reduced subunits of these mucins can be separated by centrifugation for a longer time (approx 7 to 8 h).

3.3. Electron Microscopy

Electron microscopy has been used to study the size, shape, and structure of both the intact mucins and their subunits ***(9)***. In addition, it can be used to identify the presence of specific epitopes or structural domains ***(10)***. Two methods for preparing mucins for electron microscopic analysis are described: monolayer spreading (adapted directly from the study of DNA) and replica shadowing (commonly used for high-resolution imaging of all types of biomolecules). Rigorous purification of the mucins prior to electron microscopy is essential since lipids and globular proteins interfere with spreading experiments and DNA could be mistaken for the mucins.

3.3.1. Spreading Experiments

Three steps underlie the application of the monolayer method as applied to mucins: the preparation of thin, strong, carbon support films on grids; the deposition of the

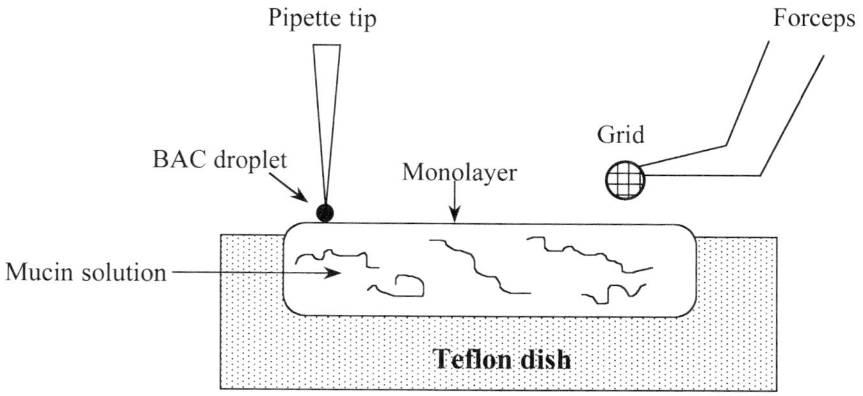

Fig. 3. Schematic diagram of spreading method. A solution of mucins (typically about 10 mL) at a concentration of 0.01–0.1 µg/mL in any aqueous solvent is poured into a Teflon trough. A drop (1 µL) of a solution of BAC (100 µg/mL) is touched to the surface and the solution is left for 5–15 min. In this time the mucins diffuse to the surface and become entrapped in the BAC monolayer. A carbon-coated electron microscope grid (400–600 mesh) is touched to the surface and thereafter washed in 95% ethanol, dried, and rotary shadowed. This can also be performed in a microversion by adding the BAC to the mucin in solution and transferring 40-µL drops to a Teflon surface. Within minutes the BAC forms a monolayer on the solution surface, where the mucin molecules become trapped. The surface film is touched to the carbon-coated grid as described previously.

mucins on these grids; and the addition of contrast, including positive staining and/or metal shadowing.

3.3.1.1. Preparation of Carbon-Coated Grids

1. Prepare a thin carbon film (2–5 nm) by the indirect evaporation of carbon onto 2-cm^2 blocks of freshly cleaved mica (**Fig. 3**).
2. Leave the mica in a water-saturated environment for approx 1 h.
3. Place 15–25 EM grids at the bottom of a water-filled Petri dish on a wire mesh.
4. Remove the film from the mica by floating it off on the surface of the water-filled dish.
5. Gently lower a rubber O-ring onto the floating carbon film. This allows the intact film to be steered on the water surface over the grids.
6. Lower the water level by gentle suction to allow the carbon film to be deposited on the grids.
7. Dry the grids in an oven at 60°C for 2 h.

3.3.1.2. Deposition of Mucins on Grids

The spreading method was originally developed by Kleinschmidt ***(10a)*** using cytochrome C as the spreading agent and was subsequently developed by others ***(11)***. We describe here a modified method first reported for the improved imaging and analysis of DNA ***(12)***. The basis of the method is the creation of a monolayer in which the long filamentous molecules are gently entrapped and thereafter can be removed onto the surface of a grid. We use the spreading agent BAC in a diffusion-/adsorption-based geometry (**Fig. 4**), which requires only small amounts of sample.

1. Pour a solution of mucins in any detergent-free aqueous solvent (typically about 10 mL), at a concentration of 0.01–0.1 μg/mL for intact molecules and 0.1–1.0 μg/mL for reduced subunits, into a Teflon trough or small Petri dish (*see* **Note 10**).
2. Touch a drop (1 μL) of the spreading agent to the surface and leave the solution for 5–15 min. In this time the mucins diffuse to the surface and become entrapped in the BAC monolayer.
3. Touch the carbon surface of a carbon-coated electron microscope grid to the monolayer.
4. If the grid is to be positively stained, *see* **Subheading 3.3.1.3.**; if not, go to the next step.
5. Wash grid in 95% ethanol, remove excess solution on filter paper, and air-dry.

3.3.1.3. Staining and Shadowing

1. Dip grid in staining solution for a few seconds.
2. Wash in 95% ethanol and air-dry.
3. For generating higher contrast, the molecules may also be rotary shadowed with heavy metals such as platinum or tungsten in a standard vacuum apparatus (*see* **Note 11**).

3.3.2. Replicas

This method is also commonly called *rotary shadowing;* however, the shadowing is not the essential principle of the method. There are many variants current in different laboratories and we use a modification described by Mould et al. ***(13)*** because it minimizes the fragmentation of very large molecules that can take place using the more common drop nebulization method.

1. Put a drop of solution in any detergent-free aqueous solvent (20 μL) containing mucins at concentrations from 1 to 0.01 μg/mL (*see* **Subheading 3.3.1.2., step 1**) on the surface of one-half of a freshly cleaved piece of mica.
2. Rejoin the surfaces and leave together for a few minutes.
3. Place the mica sandwich in a beaker of 0.2 *M* ammonium acetate.
4. Separate the two sheets under the solution and leave in the solution for 1–10 min.
5. Remove the two mica sheets and plunge into liquid nitrogen.
6. Place sheets face up on a copper block previously cooled in liquid nitrogen.
7. Place the block in an evaporation unit and pump down until all the frozen condensed water is lost from the block, essentially freeze-drying the molecules on the mica.
8. When dry, rotary shadow the mica with platinum (*see* **Note 11**).
9. Evaporate a thin layer of carbon (approx 10 nm) onto the platinum-shadowed mica surface.
10. Store the mica overnight in a desiccator containing 0.1 *M* acetic acid.
11. Remove the carbon replica the next day onto a clean water surface and transfer onto grids as described in **Subheading 3.3.1.1**.

4. Notes

1. These experiments require high-speed swing-out rotors that can achieve at least 100,000*g*. Rotors are available in a variety of different sizes and should be chosen according to sample volume and concentration.

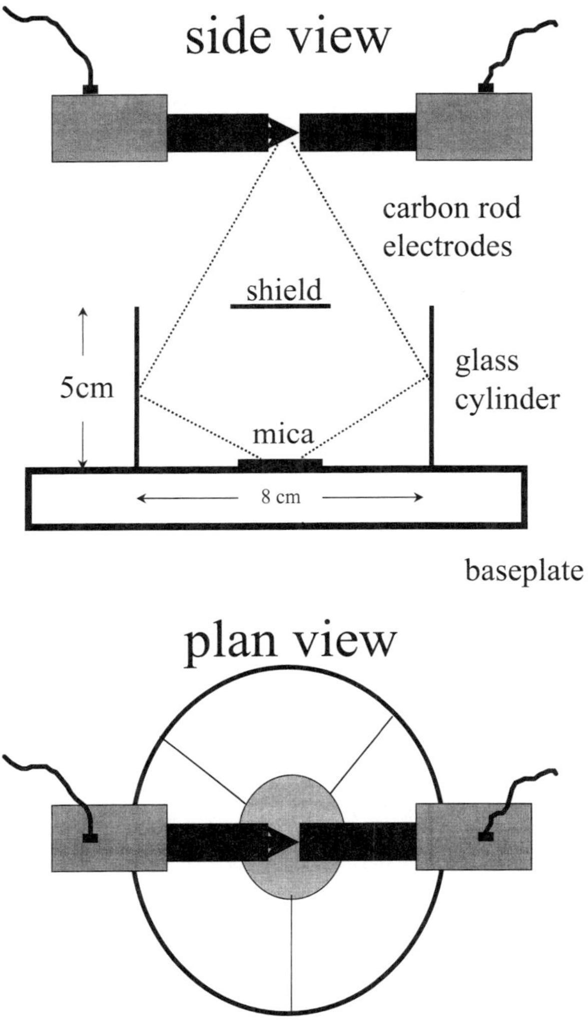

Fig. 4. Indirect carbon evaporation. This figure describes the kind of apparatus used to achieve strong carbon films suitable for coating grids. The precise geometry of the system is not important, but the principle is that the mica is shielded from direct evaporated carbon, which should arrive at the mica surface after reflection from a second surface. This reflecting surface removes large particles of carbon and gives a homogeneous particle distribution that yields films of uniform thickness and high strength. We use a glass cylinder 8 cm in diameter and 5 cm high that is placed on the base plate of the coating unit. A shield approx 2 cm in diameter is suspended by thin wire over the center of the cylinder, and the mica is placed on the base plate directly below the shield. The evaporation electrode is placed above the shield at a height (typically 5 to 6 cm) that would give good line of sight to the inside of the glass cylinder but no direct line to the mica. Evaporation is performed long enough to give a faint tan coloration on a piece of filter paper placed under the edge of the mica. These conditions will have to be sought by experimentation with the available coating unit.

Table 1
Physical Parameters for Guanidinium Chloride Solutions[a]

Concentration (*M*)	ρ (g/mL)	η_{rel}
6.00	1.145	1.620
6.25	1.150	1.685
6.50	1.156	1.770
6.75	1.162	1.845
7.00	1.168	1.925
7.25	1.174	2.030
7.50	1.180	2.125
7.75	1.186	2.230
8.00	1.192	2.400

[a]The values for ρ and $\eta_{rel.}$ are from **ref. *14*.**

2. Linear gradient makers, suitable for making gradients of different volumes, can be purchased from a variety of manufacturers. For the 3 × 12 mL gradients described, we use a 50-mL gradient maker.
3. We have performed this procedure in a wide variety of solutions, including 6 *M* guanidinium chloride, and find it very tolerant of high salt (*see* **Note 10**).
4. Guanidinium chloride (4–6 *M*) is widely used as a solvent to extract and dissolve many mucous gels. It not only prevents interactions among molecules but also prevents proteolysis. Thus, rate-zonal centrifugation in GuHCl can be performed on crude or partially purified mucous extracts.
5. If we assume that the physical size and shape of the molecule are unchanged by the supporting medium and that the rotor speed is constant during the experiment, then the change in the sedimentation rate of the molecule through the gradient is dictated by the following equation:

$$(1-v\rho)r/\eta_{rel}$$

   where v is the partial specific volume (milliliters/gram) and has a value of 0.67 mL/g over a wide range of solvent conditions; ρ is the solution density (grams/milliliter); r is the distance from the center of rotation (centimeters); and η_{rel} is the relative viscosity of the supporting medium at the appropriate value of r. This equation predicts that the molecules will have a tendency to speed up as they move away from the center of the rotor but be slowed down by increasing solvent viscosity and solution density. For the molecules to be separated according to their difference in mass alone, this equation should be approximately constant. Such gradients are generally called *isokinetic*. Different rotors vary primarily in the term r (distance from the center of rotation of the meniscus and bottom of the tube), and this information is available from the rotor data sheet. Using this information together with the data in **Table 1**, isokinetic gradients of guanidinium chloride can be designed for different rotors.
6. We use a multichannel pump, but a single-channel pump can be used and the solvent stream split after the pump or gradients are made one at a time.
7. Sample must be in a solvent with lower density than that of 6 *M* GuHCl (1.145 g/mL).

8. For the Beckman SW 40Ti rotor, we typically centrifuge for 2.5 h to disperse intact mucins across the gradient and 6–8 h for the reduced subunits.
9. This is not usually performed unless there are doubts about the stability of the gradient. The relationship between refractive index and molar concentration of GuHCl is given by the following equation:

$$M = (60.4396 \times \text{refractive index}) - 80.5495.$$

10. If the solution on which the molecules are spread has a high concentration of salt or other reagents, the grids may be washed by floating them on water for 1 h.
11. In our setup the platinum (8 cm of 0.2 mm diameter) is wound onto a tungsten wire 10 cm in length and 0.7 mm in diameter placed between electrodes. The wire is 10 cm from the rotating table, on which the grids are placed, and 5–8° degrees above its plane. The table rotation is adjusted to two to three revolutions per second, and the current in the tungsten wire is gently increased until the platinum is completely evaporated or until the tungsten wire breaks.

References

1. Sheehan, J. K., Thornton, D. J., Howard, M., Carlstedt, I., Corfield, A. P., and Paraskeva, C. (1996) Biosynthesis of the MUC2 mucin: evidence for a slow assembly of fully glycosylated units. *Biochem. J.* **315,** 1055–1060.
2. Thornton, D. J., Carlstedt, I., Howard, M., Devine, P. L., Price, M. R., and Sheehan, J. K. (1996) Respiratory mucins: identification of core proteins and glycoforms. *Biochem. J.* **316,** 967–975.
3. Thornton, D. J., Howard, M., Khan, N., and Sheehan, J. K. (1997) Identification of two glycoforms of the MUC5B mucin in human respiratory mucus: evidence for a cysteine-rich sequence repeated within the molecule. *J. Biol. Chem.* **272,** 9561–9566.
4. Toribara, N. W., Ho, S. B., Gum, E., Gum, J. R., Lau, P., and Kim, Y. S. (1997) The carboxyl-terminal sequence of the human secretory mucin, MUC6: analysis of the primary amino acid sequence. *J. Biol. Chem.* **272,** 16,398–16,403.
5. Sheehan, J. K. and Carlstedt, I. (1987) Size heterogeneity of human cervical mucus glycoproteins. *Biochem. J.* **245,** 757–762.
6. Thornton, D. J., Davies, J. R., Kraayenbrink, M., Richardson, P. S., Sheehan, J. K., and Carlstedt, I. (1990) Mucus glycoproteins from normal human tracheobronchial secretion. *Biochem. J.* **265,** 179–186.
7. Sheehan, J. K., Thornton, D. J., Howard, M., Carlstedt, I., Corfield, A. P., and Paraskeva, C. (1996) Biosynthesis of the MUC2 mucin: evidence for a slow assembly of fully glycosylated units. *Biochem. J.* **315,** 1055–1060.
8. Thornton, D. J., Carlstedt, I., and Sheehan, J. K. (1996) Identification of glycoproteins on nitrocellulose membranes and gels. *Mol.. Biotechnol.* **5,** 171–176.
9. Sheehan, J. K., Oates, K., and Carlstedt, I. C. (1986) Electron microscopy of cervical, gastric and bronchial mucus glycoproteins *Biochem. J.* **239,** 147–153.
10. Sheehan, J. K. and Carlstedt, I. (1990) Electron microscopy of cervical mucus glycoproteins and fragments therefrom. *Biochem. J.* **265,** 169–178.

10a. Kleinschmidt, A. K. and Zahn, R. K. (1959) Monolayer techniques in electron microscopy of nucleic acid molecules. *Meth. Enzymol.* **X11B,** 361–377.

11. Lang D. and Mitani, M. (1970) Simplified quantitative electron microscopy of biopolymers. *Biopolymers* **9,** 373–379.

12. Koller, T., Harford, A. G., Lee, Y. K., and Beer, M. (1969) New methods for the preparation of nucleic acid molecules for electron microscopy. *Micron.* **1**, 110–118.
13. Mould, A. P., Holmes. D. F., Kadler, K. E., and Chapman J. A. (1985) Mica sandwhich technique for preparing macromolecules for rotary shadowing. *J. Ultrastruct. Res.* **91,** 66–76.
14. Kawahara K and Tanford C. (1966) Viscosity and density of aqueous solutions of urea and guanidine hydrochloride. *J. Biol. Chem.* **241,** 3228–3232.

IV

Rheology of Mucin

9

Rheology of Mucin

Jeffrey P. Pearson, Adrian Allen, and David A. Hutton

1. Introduction

Mucins are secreted from epithelial cells of the gastrointestinal (GI), genitourinary, and respiratory tracts to form mucous gels protecting mucosal surfaces from damaging effects and agents *(1,2)*. Mucous gels are not pure mucin but contain other components secreted into the gel, e.g., IgA and protein, lipid, and nucleic acid from shed epithelial cells, and particularly, in the lower GI tract from bacteria. When measuring the rheological properties of mucous/mucin gels, it is essential to consider the effect of these nonmucin components. Mucin alone is the gel-forming component of mucus; other components can inhibit gel formation and strength. Thus, cellular debris has been shown to reduce markedly gel strength and even inhibit gel formation *(3,4)*.

Mucous gels are unique in biology and can be described as viscoelastic and plastic, in that they will deform under load and return to their original shape when the load is removed, albeit in a damped way. These gels can, however, also be deformed in a permanent manner under excess load. They have the inherent properties of flow over long timescales (10–120 min) and will reanneal if sectioned *(4,5)*. Much of the available evidence *(6,7)* points to the interactions maintaining the gel as being noncovalent molecular interactions of the carbohydrate side chains of the mucin molecules. Other gel-forming interactions have been proposed, in particular lectin-like interactions between the exposed mucin protein core and the carbohydrate chains of adjacent mucins *(8)*, but there is, at present, little evidence to support such interactions.

Almost all mucins are very high molecular weight glycoproteins consisting of highly glycosylated subunits joined together by disulfide bridges. This polymeric structure, rather than size of mucin *per se,* is essential for gel formation *(4,7)*. Subunits will not form a gel at the concentrations that mucin is found in native mucous gels, i.e., 50, 30, and 20 mg/mL in gastric, small intestinal, and colonic mucus, respectively, whereas the purified polymeric mucin will *(3,6)*. There have been many approaches to measuring the rheological properties of mucins, including solution viscosity mea-

From: *Methods in Molecular Biology, Vol. 125: Glycoprotein Methods and Protocols: The Mucins*
Edited by: A. Corfield © Humana Press Inc., Totowa, NJ

surements *(9,10)*, analysis of native and reconstituted gels using creep compliance methods *(11)*, spinability *(12)*, magnetic rheogoniometry *(13)*, nuclear magnetic resonance (NMR) *(14)*, light scattering *(15)*, and cone and plate mechanical spectroscopy *(5)*.

Viscosity measurements of mucin solutions, as well as giving information on size and shape of mucins, can give information about the interactions among and within mucin molecules, particularly the interactions preceding gel formation. The viscosity of purified mucin increases asymptotically with increasing concentration of mucin until the solution forms a gel. The key point with pig gastric mucin is about 20 mg/mL, at which the hydrated, expanded mucin molecules fill the whole solution *(16)*. Therefore, at higher concentrations, the hydration spheres start to overlap, allowing gel-forming interactions between the mucin molecules. Solution viscosity measurements can also be useful in screening potential muco-adhesives, because if the compound will not interact with mucins in solution it is unlikely to do so with the intact gel layer in vivo. Polymeric structure investigations can also be carried out in solution, with large falls in viscosity occurring between polymer reduced subunit; and proteolytically derived subunit; for example, human middle ear mucin has intrinsic viscosities of 0.210 ± 0.028, 0.113 ± 0.024, and 0.018 ± 0.010 mL/mg, respectively *(17)*, and pig colonic mucin has intrinsic viscosities of 0.240, 0.100, and 0.02 mL/mg, respectively *(18)*. However, the important form of mucin in vivo is a gel, so it is critical to investigate the interactions involved in forming and maintaining mucous gels.

Many of the methods used to study mucous gels have practical and theoretical problems such as the following:

1. Measurements being made at a single shear rate, which is only suitable for situations in which viscoelasticity is independent of shear rate.
2. Use of continuous shear methods, which can destroy mucous gel structure.
3. Measurement not being made in the viscoelastic region.
4. Sample size and condition—for example, mucous samples are often heterogeneous in consistency, so if only a few microliters are tested, a true picture of the whole gel may not be obtained.
5. Contaminated samples; for example, sputum, the source of many studies on airway mucus, is contaminated with cellular material and saliva.

In addition, storage of samples can be important because the properties can be altered by incorrect freezing, degradative enzyme activity, or dehydration.

Spinability is the simplest although probably the least accurate method for measuring viscoelasticity, and involves drawing out a thread of mucus and recording the length achieved before it breaks. This parameter only gives information about the tensile strength of the gel. It has, however, been correlated to a biological function—the ability of cilia to move mucous blankets using frog palates, giving a relative transport rate *(12)*.

NMR has been used to assess gel strength *(14)*. This method relies on measuring the line width of a water proton signal to give a measure of viscosity; the more viscous the system the broader the line. In theory, this is an elegant system. The problems are that for an accurate measurement one needs a broad line and a homogeneous magnetic field, situations unlikely to be achieved in a mucous gel. The line width is the average of signals (derived from a time constant) from different sites and can therefore be significantly skewed by a small number of very short or very long signals.

Dynamic laser light scattering ***(15)*** has also been used to measure rheological properties of mucins and information about *G*' (storage or elastic modulus) and *G*" (loss or viscous modulus) can be derived. However, major drawbacks with this technique are that it is more applicable to homogeneous solutions than gels and that it only probes at one frequency. The behavior of gels varies with frequency; i.e., concentrated solutions of disordered polymers cannot be resolved from a gel at high frequency. In addition, if *G*" becomes large, the resonances will be wide and shallow, making the data from light scattering unresolvable.

The two most reliable methods to assess physical behavior of mucous gels are dynamic and transient testing. In both methods, a stress is applied to the gel and the resulting strain or rate of strain is measured. In essence, energy is put into the gel system and the amount stored is the elastic component and the amount of energy dissipated is the viscous component. The parameters that represent these components are *G*' and *G*", respectively.

Creep testing, a transient method, has been used extensively to measure the rheological behavior of mucous secretions and reconstituted mucin gels ***(19)***; generally, cone and plate air turbine viscometers have been used. A creep compliance curve is obtained by applying a sudden, constant, low stress to the gel. It is particularly important that the applied stress be set at a very small value so that compliance is a function of only time and not magnitude of stress. Creep-relaxation curves for mucus show a characteristic (instantaneous response) sharp increase in resultant strain from which J_o, the instantaneous shear compliance, is obtained, and this gives information relating to elastic stretching of the bonds within the gel. The rate of change of strain then slows, and in this region viscoelastic behavior predominates. The plot then tends toward a linear steady-state region where molecular chains are flowing over one another. $Je°$, the steady-state compliance, which is a measure of elastic deformation during steady flow and η_0, the residual viscosity, can be derived from this linear region. The data obtained can be transformed to give elastic moduli. The problem with this type of testing is that, in some mucous gels, flow behavior is limited, and, consequently, the time taken to reach a linear part of the curve can be quite long making this method impractical.

In dynamic testing the gel is subjected to a sinusoidally oscillating strain or stress and the resulting stress or strain in the gel is detected. If the gel were entirely elastic, stress and strain would be in phase; if purely viscous, stress and strain would be 90° out of phase. Any lag between the input and output signals is the phase angle Δ. The main advantage of this method is the ability to make measurements over a wide frequency range, i.e., 0.002–20 Hz (gel behavior is frequency dependent). Basically two methods of dynamic testing have been used. The first technique, magnetic rheogoniometry ***(13,20,21)***, is based on the photoelectric analysis of the motion of a small steel ball in the mucous sample. The ball is driven by the application of sinusoidally varying magnetic fields. *G*', *G*", and the dynamic viscosity can be calculated from

$$G' = (F_0/6\Pi r X_0) \cos \Delta + (2/9)\, \rho_s r^2 \omega^2$$

$$G'' = (F_0/6\Pi r X_0) \sin \Delta$$

$$\text{Dynamic viscosity } \eta' = G''/\omega$$

where X_0 = displacement of the ball, Δ = phase lag, F_0 = oscillatory driving force, r = radius of the ball, ρ_s = density of the ball, and ω = angular frequency. The advantage of this method is that small samples of the gel can be tested, i.e., a few microliters, but it also has some serious disadvantages. First, the strain is not fixed at all frequencies but is modulated by the test gel. Second, analysis of small samples of a gel are prone to inaccuracies if the gel is heterogeneous. Third, optical clarity of the sample can present difficulties with photoelectric analysis.

The second method, conventional cone and plate mechanical spectroscopy ***(3–6)***, is at present probably the most reliable and informative. In this system, either the cone or the plate is driven by a sinusoidal signal, and the response is recorded from the nonactive component. This method is superior to magnetic rheogoniometry in that strain is fixed at all frequencies, larger samples (1–3 mL) will average out areas of heterogeneity, and the gel does not need to be optically clear.

The methods described here are for solution viscosity using a cup-and-bob low-shear viscometer and mechanical spectroscopy using a cone and plate system. Appropriate references are given in the preceding text for various other systems developed for studying the rheology of mucin.

2. Materials (*see* Notes 1–4)

2.1. Chemicals

1. Phenylmethylsulfonylfluoride (PMSF) (Sigma, Poole, UK).
2. Propan-2-ol (Lab-Scan, Dublin, Ireland).
3. Iodoacetamide (Sigma).
4. α-Amino hexanoic acid (Aldrich, Gillingham, UK).
5. Benzamidine hydrochloride (Sigma).
6. Disodium EDTA (BDH, Poole, UK).
7. *N*-Ethyl maleimide (NEM) (Sigma).
8. Cesium chloride (Fisher, Loughborough, UK).
9. Periodic acid (BDH).
10. Sodium metabisulfite (BDH).
11. Schiff's reagent (Sigma).
12. Papain-digested pig gastric mucin (Sigma).
13. Protein assay solution (Bio-Rad, Munich, Germany).

2.2. Equipment

1. Couette rotating-cup viscometer (Contraves 30, Contraves, Zurich, Switzerland).
2. Rheogoniometer (Bohlin, Cirencester, Gloucestershire, UK; Rheometrics, Springfield, NJ).

2.3. Native Mucous Gel

Much of the previous work applying mechanical spectroscopy to mucous gels has been carried out using GI mucous gels, which can be obtained in relatively large amounts.

1. Collect fresh tissue from animal or human sources, i.e., immediately after removal at slaughter and operation. Transport to the laboratory on ice.
2. Expose the mucosal surface by cutting open the gut, and gently wash the gel with tap water to remove food debris and so on.
3. Gently scrape off the gel from the mucosal surface with a microscope slide, and deposit the gel into a beaker cooled on ice.
4. The gel should be stored frozen at –20°C until required. It is essential that it be equilibrated at the measuring temperature for 2 h after thawing. This allows the molecular interactions within the gel distorted by freezing to reform *(5)*.

2.4. Reconstituted Mucin Gel (see Notes 5–8)

1. Collect the gel as in **Subheading 2.3.** but scrape into an inhibitor cocktail containing PMSF (dissolve the PMSF in a small volume of propan-2-ol and add to the solution containing the other inhibitors immediately before the cocktail is used), 1 m*M* iodacetamide, 100 m*M* α-amino hexanoic acid, 5 m*M* benzamidine hydrochloride, 10 m*M* disodium EDTA, and 10 m*M* NEM in 0.067 *M*, pH 6.7, sodium hydrogen phosphate buffer *(17)*, to give a final concentration of 1:10 (wt/wt) gel:buffer and then homogenize for a maximum of 1 min at three-quarters speed in a Waring laboratory blender.
2. Centrifuge the resulting homogenate (relative centrifugal force [RCF] av = 8000*g*) for 1 h at 4°C to pellet any cell debris or insoluble material.
3. Dissolve solid cesium chloride in the supernatant to give a starting density of 1.42 g/mL. Then subject this solution to equilibrium density gradient centrifugation (RCF av = 120,000*g*) for 24–36 h at 4°C in a fixed angle (20°) rotor, 8 × 35 mL.
4. After centrifugation, divide each tube into eight equal fractions and carefully remove the solution from the top of the tube using a Pasteur pipet to give eight fractions each with an average volume of 6.8–7.7 mL. Determine the density of the fractions by weighing 1 mL.
5. Exhaustively dialyze each fraction against distilled water to remove the cesium chloride. Measure glycoprotein concentration using the periodic acid-Schiff's (PAS) solution assay *(22)*. Add 0.2 mL of freshly made 0.2% PAS (20 µL of periodic acid in 10 mL of 7% acetic acid) to 1 mL of a 20X dilution of the dialyzed fractions. Add 1 g of sodium metabisulfite to 6 mL of commercial Schiff's reagent (Sigma) just prior to use. Incubate mucin/periodic acid solution for 1 h at 37°C. Then add 0.2 mL of Schiff's reagent to mucin/periodic acid solution and incubate for 30 min at room temperature. Read the absorbance at 555 nm, using 0–200 µg of papain-digested pig gastric mucin (Sigma) as a standard.
6. Measure the levels of protein and nucleic acid at optical density (OD) 280 and 260 nm, respectively. Protein can also be measured by the assay based on Coomassie blue (Bio-Rad protein assay, Bio-Rad).
7. Pool the mucin-rich fractions (PAS rich), density range 1.42–1.52 g/mL, freeze-dry, and store over desiccant at –20°C until required for viscosity measurements.
8. Alternatively, mix the mucin-rich pool from the CsCl gradient 1:1 (v/v) with 300 m*M* NaCl, and 0.004% NaN_3 to give an isotonic solution. Then concentrate the solution at 4°C until gel formation takes place. Concentration can be achieved both by vacuum dialysis or by ultrafiltration with a membrane cutoff of <100,000 mol wt.

2.5. Homogenized, Diluted Native Mucus for Viscosity Studies

Homogenization and dilution of native mucus allows viscosity studies on a preparation containing the total mucin and nonmucin components of the gel.

1. Mucous gel scraped from the mucosal surface can be homogenized three-quarters speed in a Waring laboratory blender at 1:1 up to 1:10 gel:buffer (wt/wt), for a maximum of 1 min. The buffer given in **Subheading 2.4.** can be used to prevent any breakdown of the mucins in the homogenate during 37°C incubation experiments; however, remember that if reduction experiments are being carried out, the iodacetamide and NEM should be omitted.

3. Methods

3.1. Couette Rotating-Cup Viscometer Measurements (see Notes 7 and 9–11)

1. Use variable shear viscometer, e.g., a Contraves low shear 30 (Contraves A.G.). Place approx 2 mL of the mucin-containing solution into the temperature-controlled cup. (The sample must be equilibrated for ≃10 min at the measurement temperature before measurements are taken. Evaporation is avoided by covering the sample with parafilm between measurements, because this will cause viscosity to increase). Lower the bob into the solution ensuring that no bubbles are trapped under it.
2. Shear the sample through the shear range twice (controlled by an electronic speed programmer), and then take a reading on the third run. The signal from the viscometer is converted into chart recorder output on a PC or pen recorder, giving a plot of percentage deflection (amount of torque transferred from the rotating cup via the viscous solution to the bob) on the *y*-axis and shear rate (speed the cup turns) on the *x*-axis.
3. Repeat **step 2** using a sample of the same buffer in which the sample of mucin is solubilized.
4. In experiments in which the effects of agents are being studied on solutions of mucin, a control sample of mucin without the agent must be included and incubated and measured over the same time period and measurement interval as the experimental sample.

3.2. Interpretation

The viscosity of solutions of mucin can be determined as follows.

1. The gradients of the deflection vs shear plots are directly proportional to viscosity; therefore, the relative viscosity (η_{rel}) can be determined from

$$\eta_{rel} = \text{gradient of solution of mucin/gradient of solvent}$$

Note that since both gradients are in the same units, η_{rel} is dimensionless.

2. The specific viscosity (η_{sp}) can be calculated from the relative viscosity:

$$\eta_{sp} = \eta_{rel} - 1$$

3. Reduced specific viscosity (η_{red}) can be calculated from the specific viscosity divided by the concentration of mucin:

$$\eta_{red} = \eta_{sp}/C$$

If the concentration is in milligrams/milliliter, then η_{red} is milliliters/milligram.

4. The intrinsic viscosity [η] can be calculated if the viscosities of serial dilutions by weight of a solution of mucin of known concentration are measured. Plot either η_{red} vs concentration of mucin (Huggins plot) or ln η_{rel}/concentration of mucin vs concentration of

mucin (Kraemer plot). The intercept on the viscosity axis is the intrinsic viscosity; if a mucin has an [η] of 0.2 mL/mg, then 1 mg occupies 0.2 mL in solution.

5. Solution viscosity data can also be used to examine interactions between mucin:mucin molecules and mucin:polymer interactions (an example of a polymer interacting with mucin is the polyacrylate carbopol). If mucin interacts with a polymer, mixtures of mucin/polymer would be expected to show synergism in which the specific viscosity of mucin/polymer increases in a concentration-dependent manner over and above the sums of the specific viscosity of the individual components.
6. The concentration at which intermolecular interactions occur in polymer solutions can be determined by plots of log η_{sp} against log $C(\eta)$ ($C[\eta]$ is the coil overlap parameter where C is concentration of solute). On such plots, a sharp change in gradient indicates significant intermolecular interactions. Further information about interactions can be obtained if log η_{sp} is plotted against log shear rate (γ) at coil overlaps above that at which interaction has been shown to occur, i.e., above the point of the sharp change in gradient. For example, hyaluronic acid, a purely entangled system with some short-lived positive interactions, will give a plot where log η_{sp} is essentially independent of shear (at low shear rates) *(23)*. Hence, a plot for mucin/polymer showing shear dependence would suggest longer-term positive interactions, i.e., more than only entanglement *(24)*.

3.3. Measurement of Dynamic Rheological Properties of Mucous/Mucin Gel (see Note 12)

1. Use rheogoniometers such as Bohlin or Rheometrics.
2. Place the mucin/mucous gel sample (approx 3 g wet wt) between the plates, which are closed up to a set gap of 0.150 mm (for a Bohlin rheogoniometer). Carefully trim away the excess sample and cover the exposed edges with a low-viscosity oil. Surround the sample by a Perspex cover to prevent evaporation, and control the temperature. Remember to allow reequilibration of frozen samples. Then test the sample by applying electronically controlled sinusoidal deformations.
3. The first thing that must be determined is the linear viscoelastic region, i.e., where the moduli G' and G'' remain constant over a range of shear stress. This is described as an amplitude sweep and is carried out at a frequency of 1 Hz at 25°C. For pig gastric mucus, shear stress over the range of 0.1–20 Pa gives constant values for G'. At higher shear stress values above 20 Pa, G' and G'' both collapse in value; this is the nonviscoelastic region where the gel is being disrupted.
4. Once the viscoelastic region has been determined, a frequency sweep can be performed at 25°C between 0.002–30 Hz with the rheogoniometer operating in auto stress mode. The initial stress is set in the middle of the viscoelastic region with the target strain also set in this region.

3.4. Interpretation of Data

1. The following parameters can be obtained:

$$\text{Phase angle } \Delta \text{ (degrees)} = \text{lag between input and output signals}$$

$$G' \text{ (Pa) storage modulus} = \text{(stress/strain)} \times \cos \text{(phase angle)}$$

$$G'' \text{ (Pa) loss modulus} = \text{(stress/strain)} \times \sin \text{(phase angle)}$$

$$G^* \text{ (Pa) complex modulus} = \sqrt{(G')^2 + (G'')^2}$$

$$\eta^* \text{ Pascal seconds (Pa·s) complex viscosity} = G^*/\text{frequency}$$

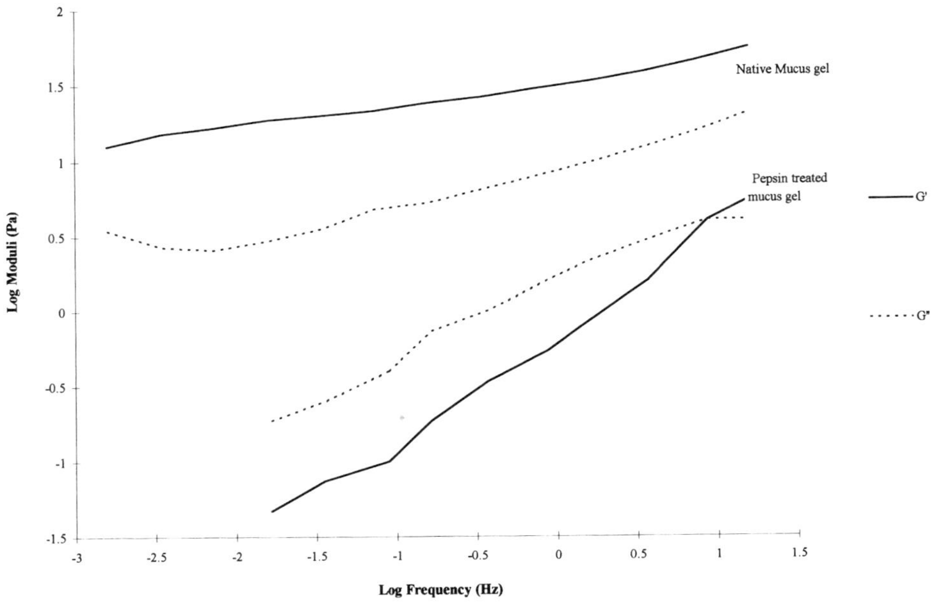

Fig. 1. Dynamic oscillatory measurements carried out using a Rheometrics mechanical spectrometer (Rheometrics). Plots of storage modulus (——) and loss modulus (- - -). Plots show native pig gastric mucous gel, taken directly from the mucosal surface, glycoprotein concentration 55 mg/mL. Native pig gastric mucous gel incubated for 10 h at 37°C with 1 mg/mL of porcine pepsin, pH 2.2.

2. What do these parameters tell us about gel systems? *G*' represents elasticity, *G*" represents viscosity, and *G** is a measure of rigidity or stiffness. Complex viscosity η^* can be related to static viscosity through the Cox-Merz rule ***(25)***. The tangent of the phase angle (tan Δ) gives a measure of the quality of the gel (approximately equivalent to gel strength). It is derived from *G*"/*G*'. Therefore, the smaller the tan Δ the more elastic the system, and the larger the tan Δ the more fluid and viscous the system. In a native gel system, e.g., pig gastric mucus, the preceding parameters vary as described (*see* **Fig. 1**).
   a. *G*' is always above *G*" over the frequency range used (0.002–30 Hz).
   b. Native mucous gels show a characteristic dip in *G*" at lower frequency values (0.008 Hz), giving a minimum value for *G*".
   c. Absolute values of *G'* depend on the content of polymeric mucin in the gel.
   d. Reconstituted mucin gels (as long as the mucin is mainly polymeric) behave as native mucous gel, evidence that mucin is the gel-forming component of mucus.
   e. If the gel structure is destroyed by disulfide bond–cleaving agents or proteolytic enzymes, the frequency sweep data are altered dramatically. At low frequencies, *G*" is above *G*', indicating that the gel is nonelastic and has become a viscous liquid. At high frequencies, it still appears as a gel with *G*' above *G*", consequently, measurements at a single frequency can give anomalous results.
3. The best way to assess gel strength is to use tan Δ, but it must be remembered that tan Δ is frequency dependent; hence, any comparison among gels must be done at the same fre-

quency. A clear demonstration of how the percentage of polymeric mucin in the gel affects gel strength can be seen from plots of tan Δ vs percentage polymer *(6,26)*. Native gastric mucous gels have tan Δ values of ≈0.2 and contain 70–80% polymeric mucin. Gastric mucous gel from gastric ulcer patients have a tan Δ of about 1 and a content of polymeric mucin of 30–35%. Since tan Δ = G''/G', one can see that as tan Δ approaches 1 the gel is becoming a viscous liquid.

4. Notes

1. When scraping the gel from the mucosa, scrape gently to avoid extensive contamination of the gel with cells. The rheological properties of mucous gel can be greatly altered by the presence of mucosal cellular material *(3)*.
2. Mucous gel should be stored in aliquots because serial freezing and thawing can damage the gel structure.
3. In experiments studying the effects of agents on gel structure, mucus from the same individual should be used for the control and the experimental samples. If this is not possible, mucus from many individuals should be pooled and stirred gently to form as homogeneous a preparation as possible.
4. Frozen gels behave the same as fresh gels if they are allowed to thaw and reequilibrate for 2 h at room temperature before rheological measurements.
5. The gel can be solubilized only by homogenization or high ionic strength solvents, i.e., 6 *M* GuHCl, and not by reduction or proteolysis, which will destroy the polymeric structure essential for gel formation. However, note that the use of GuHCl may denature the gel-forming interactions.
6. If the mucous gel is obtained from a site such as the colon or the small intestine, where the gel is extensively contaminated with nonmucin components, a gel filtration step (Sepharose 2B or 4B pooling the excluded volume as the fraction of mucin) or a second CsCl density gradient may be necessary to obtain an essentially pure preparation of mucin.
7. No studies on mucin gel structure should be attempted with commercial preparation of mucin, e.g., pig gastric mucin from Sigma. This preparation is essentially degraded mucin and will not form a gel but behaves as a viscous sol with G'' higher than G' even at concentrations as high as 150 mg/mL *(27)*.
8. Equilibrium density centrifugation in CsCl is best carried out in angle rotors, rather than vertical rotors, because broader gradients are achieved with angle rotors.
9. The viscosity of mucin is dependent on ionic strength. Therefore, all viscosity and rheological studies must be carried out in salt concentrations of 100–150 m*M*; otherwise, lack of charge shielding will cause expansion of the mucin molecules to nonphysiological sizes.
10. In all viscosity studies with solutions of mucin, a control sample must be set up in order to discount any changes in viscosity arising from the activity of components in the solution of mucous/mucin, e.g., traces of proteinases, and ionic effects.
11. At high concentrations, some solutions of mucin show shear thinning when measured in a cup-and-bob viscometer (i.e., percentage deflection vs speed is nonlinear). In such cases, the gradient of the viscosity curve is taken as close to the origin (zero shear) as possible.
12. In mechanical rheological setups, both parallel plate and cone and plate geometry have been used. Parallel plate geometry is useful for nonhomogeneous samples, but remember that the shear on the mucus will vary across the plate. By contrast, cone and plate geometry gives a constant shear rate across the face of the cone, and any edge effect is significant only if the sample is small in volume, i.e., below 250 μL.

References

1. Allen, A. and Pearson, J. P. (1993) Mucus glycoproteins of the normal gastrointestinal tract. *Eur. J. Gastroenterol. Hepatol.* **5(4),** 193–199.
2. Forstner, J. F. and Forstner, G. G. (1994) Gastrointestinal mucus, in *Physiology of Gastrointestinal Tract* (Johnson, L. R., ed). Raven, New York, pp. 1255–1283.
3. Sellers, L. A., Allen, A., Morris, E. R., and Ross-Murphy, S. B. (1991) The rheology of pig small intestinal and colonic mucus: weakening of gel structure by non-mucin components. *Biochim. Biophys. Acta* **1115,** 174–179.
4. Bell, A., Sellers, L. A., Allen, A., Cunliffe, W. J., Morris, E. R., and Ross-Murphy, S. B. (1985) Properties of gastric and duodenal mucus: effect of proteolysis, disulfide reduction, bile, acid, ethanol and hypertonicity on mucus gel structure. *Gastroenterology* **88,** 269–280.
5. Bell, A. E., Allen, A., Morris, E. R., and Ross-Murphy, S. B. (1984) Functional interactions of gastric mucus glycoprotein. Int. *J. Biol./Macromol.* **6,** 309–315.
6. Sellers, L. A., Allen, A., Morris, E. R. and Ross-Murphy, S. B. (1988) Mucus glycoprotein gels. Role of glycoprotein polymeric structure and carbohydrate side-chains in gel formation. *Carbohydr. Res.* **178,** 93–110.
7. McCullagh, C. M., Gupta, R., Jamieson, A. M., and Blackwell, J. (1996) Gelation of fractionated canine submaxillary mucin in a chaotropic solvent. *Int. J. Biol. Macromol.* **18,** 247–253.
8. Silberberg, A. (1989). Mucus glycoprotein, its biophysical and gel-forming properties, in *Mucus and Related Topics* (Chantler, E. and Ratcliffe, N. A., eds.), Company of Biologists, Cambridge, UK. pp. 43–63.
9. Snary, D., Allen, A., and Pain, R. H. (1971) The structure of pig gastric mucus conformational transitions induced by salt. *Eur. J. Biochem.* 24, 183–189.
10. Moretti, M., Gorrini, M., Donnetta, A. M., Baldini, E. V., Venturini, R., Marchioni, C. F., Iadarola, P., and Luisetti, M. (1994) Effects of MR889, a cyclic thiol neutrophil elastase inhibitor, on rheologic properties of mucus. *Life Sci.* **54(7),** 463–469.
11. Marriott, C and Irons, L. I. (1974) A study of the rheology and biochemistry of saliva and tracheobronchial mucus gels. *Biorheology* **11,** 119–128.
12. Puchelle, E., Zahm, J. M., and Duvivier, C. (1983). Spinability of bronchial mucus relationship with viscoelasticity and mucus transport properties. *Biorheology* **20,** 239–249.
13. Litt, M. and Khan, M. A. (1976). Mucus rheology: relation to structure and function. *Biorheology* **13,** 37–48.
14. Odeblad, E. (1974) Measurement of sputum viscosity with NMR-method. *Scand. J. Respir. Dis.* **90(Suppl.),** 37–39.
15. Steiner, C. A., Litt, M., and Nossal, R. (1984). Studies on rheologically active biological macromolecules by quasielastic light scattering. *Biorheology* **21(Suppl. 1),** 335–344.
16. Allen, A. Pain, R.H., and Robson, T. (1976) Model for the structure of gastric mucus gel. *Nature* **264,** 88–89.
17. FitzGerald, J. E., Green, G. G. R., Stafford, F. W., Birchall, J. P., and Pearson, J. P. (1987) Characterisation of human middle ear mucus glycoprotein in chronic secretory otitis media. *Clin. Chim. Acta* **169,** 281–298.
18. Fogg, F .J. J., Hutton, D. A., Jumel, K., Pearson, J. P., Harding, S. E., and Allen, A. (1996) Characterisation of pig colonic mucins. *Biochem. J.* **316,** 937–942.
19. Martin, G. P. Loveday, B. E. and Marriott, C. (1990) The effect of bromhexine hydrochloride on the viscoelastic properties of mucus from the mini pig. *Eur. Respir. J.* **3,** 392–396.
20. Bell, A. E., James, S. L. and Marriott, C. (1984) Selected methods for the evaluation of the

effects of drugs on mucus viscoelasticity and mucociliary clearance. *Biorheology,* **21(Suppl. 1),** 127–130.
21. Silveria, P. S. P., Böhm, G. M., Yang, H. M., Wen, C. L., Guimarèes, E. T., Parada, M. A. C., King, M., and Saldiva, P. H. N. (1992) Computer-assisted rheological evaluation of microsamples of mucus: *Computer Methods and Programs in Biomedicine* **39,** 51–60.
22. Mantle, M. and Allen, A. (1987) A colorimetric assay for glycoproteins based on the Periodic acid/Schiff stain. *Biochem. Soc. Trans.* **6,** 607–609.
23. Welsh, E. J., Rees, D. A., Morris, E. R., and Madden, J. K. (1980) Competitive inhibition evidence for specific intermolecular interactions in hyaluronate solutions. *J. Mol. Biol.* **138,** 375–382.
24. Foster, S. N. E., Pearson, J. P., Hutton, D. A., Allen, A., and Dettmar, P. W. (1994) Interaction of polyacylates with porcine pepsin and the gastric mucus barrier: a mechanism for mucosal protection. *Clin. Sci.* **87,** 719–726.
25. Cox, W. P. and Merz, E. H. (1958) Correlation of dynamic and steady flow viscosities. *J. Polymer Sci.* **28,** 619–622.
26. Allen, A., Newton, J., Oliver, L., Jordan, N., Strugala, V., Pearson, J. P., and Dettmar, P. W. (1997) Mucus and H. Pylori. *J. Physiol. Pharmacol* **48(3),** 297–305.
27. Madsen, F., Eberth, K., and Smart, D. (1996) A rheological evaluation of various mucus gels for use in in-vitro mucoadhesion testing. *Pharmaceu. Sci.* **2,** 563–566.

V

Mucin Peptide Analysis

10

Amino Acid Analysis of Mucins

Jun X. Yan and Nicolle H. Packer

1. Introduction

Amino acid analysis is a commonly used technique that provides quantitative estimation of the amounts of proteins/amino acids present in a sample and/or qualitative information on the amino acid composition of a protein. For protein analysis, the technique essentially involves acid hydrolysis of amino acid peptide bonds within the protein; chemical derivatization of hydrolysate (amino acids) of the protein; and high-performance liquid chromatography (HPLC) separation, detection, and analysis of those derivatized amino acids.

The commercially available amino acid analyzers (e.g., Waters Pico-Tag system [Waters Corp., Milford, MA]; GBC AminoMate system [GBC Scientific, Dandenong, Victoria, Australia]) have made amino acid analysis more practical and feasible in routine protein analysis laboratories. The sensitivity of the analysis has been dramatically increased to low picomole levels of proteins, including those low molecular weight (10–20 kDa) ones (low amount of total amino acids analyzed) *(1)*.

In this chapter, we describe a 9-fluorenylmethyl oxycarbonyl chloride (FMOC)-based precolumn derivatization amino acid analysis that has been extensively validated *(1,2)*. Although the detailed protocols on the use of the automated GBC AminoMate (GBC Scientific) amino acid analyzer have been described elsewhere *(3)*, here, we emphasize the procedures that are used in a manual operation. Thus, this technique can be easily adapted in any laboratory where an HPLC system with a fluorescent detector and gradient controller is available.

Acid hydrolysis is the first and most important step to release the amino acids from the proteins, and it must be carefully controlled in the analysis of mucins. The acid hydrolysis described here recovers 16 amino acids (asparagine and glutamine are deamidated to their corresponding acids, whereas tryptophan and cysteine are destroyed). During the acid hydrolysis, the carbohydrate side chains on the mucins are degraded. Because of the high carbohydrate content of mucins (up to 90% of the dry weight), the sugars can be caramelized and further charcoaled, and the acid hydrolysis results in a black residue. This residue appears to precipitate protein/amino acids, interferes with

From: *Methods in Molecular Biology, Vol. 125: Glycoprotein Methods and Protocols: The Mucins*
Edited by: A. Corfield © Humana Press Inc., Totowa, NJ

chromatography, and leads to a significantly lower recovery of all the amino acids, especially serine and threonine. Strong acid, high temperature, and short time acid hydrolysis (12 *N* HCl at 155°C for 1 h) is generally used in our laboratory for less glycosylated proteins ***(1)***. We have found that weaker acid, lower temperature, and longer time (ensuring the completion of the hydrolysis) acid hydrolysis (6 *N* HCl at 105°C for 24 h) is a *must* to obtain reproducible quantitative analysis of mucin glycoproteins. The hydrolysate of mucin is then derivatized with Fmoc on the α-amino groups under alkaline conditions. The derivatized amino acids are separated by a C18 reversed-phase column and detected by fluorescence. Our routine analyses have shown that the method described here is reliable and useful for both quantitative and qualitative mucin glycoprotein amino acid analysis.

2. Materials

2.1. Apparatus

2.1.1. Hydrolysis Equipment

1. Hydrolysis vessel: the design can be viewed on the World Wide Web at http://www.bio.mq.edu.au/APAF (*see* **Note 1**).
2. Vacuum pump: Savant Speedvac (Savant Industries, Farmingdale, NY) (or equivalent) with a vacuum gauge.
3. Two-way line connection to an argon line and a vacuum pump.
4. Autosampler glass vials made from inert Chromacol Gold™ grade glass (Chromacol, cat. no. 02-MTVWG, Herts, UK).

2.1.2. HPLC System (see **Note 2**)

1. LC-pump with ternary gradient controller.
2. Fluorescence detector: excitation, λ = 270 nm; detection, λ = 316 nm.
3. Sample injector.
4. Degasser: alternatively, buffers can be degassed by continuous flow of helium.
5. Hypersil C18 reversed-phase column: 150 × 4.6 mm inner diameter 5 µm (Keystone, Bellefonte, PA).
6. In-line filter (2 µm) (Upchurch, cat. no. 100-10).

2.2. Chemicals

2.2.1. Hydrolysis and Derivatization Reagents

1. Hydrochloric acid: constant boiling temperature 6 *N* (Pierce, cat. no. 24309).
2. Ultrapure phenol (ICN, cat. no. 800672) kept at 4°C.
3. Borate buffer: 250 m*M* boric acid (analytical reagent grade [AR] grade) in water, adjusted to pH 8.5 with NaOH. Buffer may be kept up to 1 mo at 4°C.
4. Fmoc reagent: 4 mg/mL fluorenylmethyl chloroformate (Sigma, cat. no. F0378) in acetonitrile (HPLC grade). Reagent may be stored up to 1 wk at 4°C.
5. Cleavage reagent: 680 µL of 0.85 *M* NaOH (AR grade), 150 µL of 0.5 *M* hydroxylamine hydrochloride (Sigma, cat. no. H2391), and 20 µL of 2-methylthio-ethanol (Sigma, cat. no. M9268). Stock solution can be kept for 1 mo at 4°C. Working cleavage reagent must be made freshly prior to each use.
6. Quenching reagent: 2 mL of acetic acid (AR grade) and 8 mL of acetonitrile.

2.2.2. Chromatography Reagents

1. Amino acid standards H: protein hydrolysate standard in 0.1 *N* HCl, containing a solution of 17 amino acids (Sigma, cat. no. A9781).
2. L-hydroxyproline used as internal standard (Sigma, cat. no. H1637).
3. Phosphate buffer (2 *M*): 2 *M* anhydrous ammonium monohydrogen phosphate (AR grade) solution, adjusted to pH 6.5 with 2 *M* anhydrous dihydrogen phosphate (AR grade) solution. Buffer may be kept up to 6 mo at 4°C.
4. Mobile phase A (30 m*M* ammonium phosphate [pH 6.5]): Into a 1000 mL volumetric flask, add 15 mL of 2 *M* phosphate buffer and dilute to volume with mobile phase B.
5. Mobile phase B: 15% (v/v) methanol in water.
6. Mobile phase C: 90% (v/v) acetonitrile in water.

3. Methods

3.1. Sample Preparation and Hydrolysis

1. Add an aliquot of a solution of a sample of mucin into an autosampler glass vial and dry under vacuum (*see* **Note 3**).
2. Place the vials, 400 µL of HCl (6 *N*), and a crystal of phenol into the bottom of the hydrolysis vessel.
3. Assemble the vessel tightly. Connect the vessel to a two-way argon and vacuum line.
4. Evacuate the vessel to 3 torr, and flush with argon. Repeat this step twice, and seal the vessel after the third evacuation step (*see* **Note 4**).
5. Place the vessel in a 105°C oven for 24 h.
6. Remove the vessel from the oven and open the vacuum tap immediately within a fume hood (*see* **Note 4**).
7. Place the vials into a vacuum centrifuge for 10 min to evaporate excess HCl.

3.2. Derivatization (see Note 5)

1. Dissolve the mucin hydrolysate in 10 µL of 250 m*M* borate buffer, pH 8.5.
2. Add 10 µL of Fmoc reagent, mix, and then wait 1 min.
3. Add 10 µL of cleavage reagent, mix, and then wait 4 min.
4. Add 10 µL of quenching reagent mix.

3.3. Chromatography (see Note 6)

1. Set up the fluorescent detector with excitation $\lambda = 270$ nm and detection $\lambda = 316$ nm.
2. The mobile phase is a ternary solution system using the gradient shown in **Table 1**.
3. The stationary phase is a 5 µm Hypersil C18 column that is temperature controlled at 38°C. Flow rate is 1.0 mL/min. Allow the column to equilibrate with two gradient runs before the sample injection.
4. Inject an aliquot (≥5 µL) of each derivatized sample (*see* **Note 6**). Run time is 35 min. Allow at least 2 min for pump and column equilibration with the initial composition of the mobile phase.
5. **Figures 1** and **2** show typical chromatograms of the separation of amino acid standards and hydrolyzed bovine conjunctival mucin (10–20 µg).
6. After the run, the peaks can be integrated and the picomoles of each mucin amino acid calculated by comparison with the peak areas given by the amino acid standard. Since the total number of amino acids (or the molecular mass) in a mucin glycoprotein is unknown (or difficult to determine), the actual number of picomoles of mucin protein cannot be determined by this analysis. In practice, the amount of mucin can usually be referred to as

Table 1
Gradient for the Baseline Separation of 16 Fmoc–Derivatized Amino Acids Within 35 min

Time (min)	% Mobile phase A (30 m*M* ammonium phosphate, pH 6.5)	% Mobile phase B (15% [v/v] methanol/water)	% Mobile phase C (90% [v/v] acetonitrile/water)
0	17	68	15
1	17	68	15
32	10.8	43.2	46
32.05	0	0	100
34	0	0	100
34.05	17	68	15
35	17	68	15

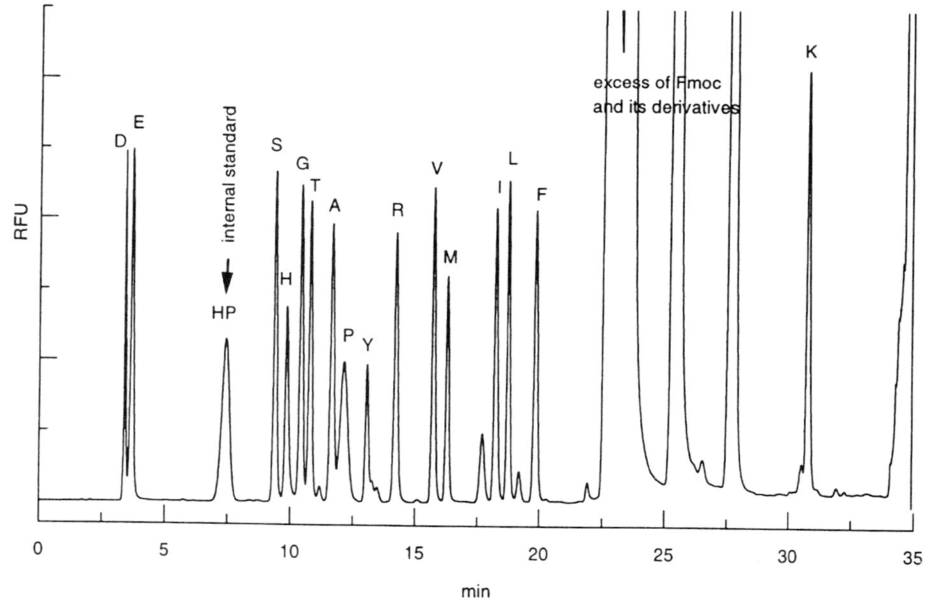

Fig. 1. Typical chromatogram of separation of 16 standard Fmoc-amino acids. RFU = relative fluorescence units; Amino acid code: D = aspartic acid, E = glutamic acid, HP = hydroxy proline, S = serine, H = histidine, G = glycine, T = threonine, A = alanine, P = proline, Y = tyrosine, R = arginine, V = valine, M = methionine, I = isoleucine, L = leucine, F = phenylalanine, K = lysine.

the amount or ratio of certain amino acids, so a relative comparison in quantitative analysis, e.g., carbohydrate analysis, can be done. The amino acid composition of the mucin (%pmol) = pmol of mucin amino acid/total pmol of total amino acids x 100% (*see* **Table 2** for an example) (*see* **Note 7**).

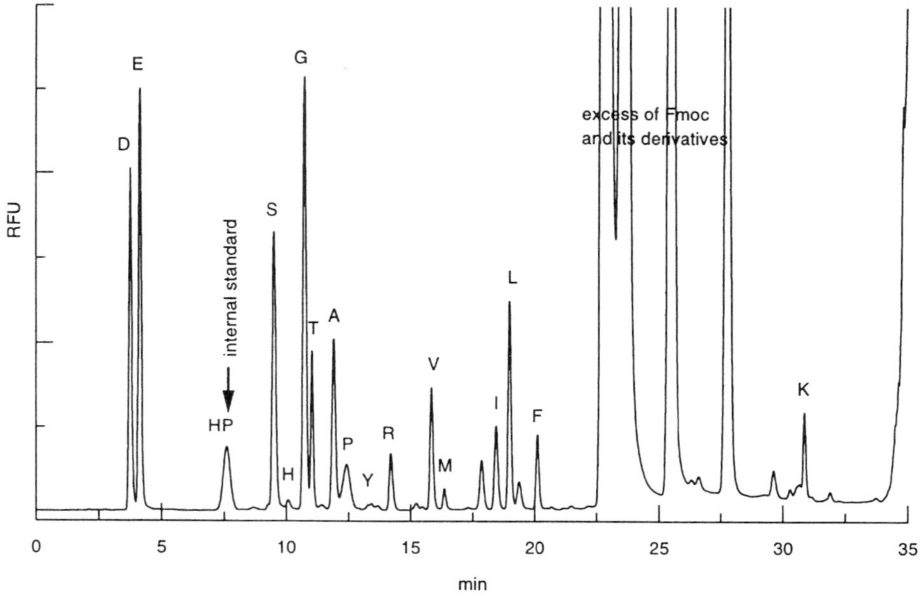

Fig. 2. Typical chromatogram of separation of amino acids from bovine conjunctival mucin hydrolysate. RFU = relative fluorescence units. Amino acid code as defined in **Fig. 1.**

4. Notes

1. The hydrolysis vessel can be made by any glassblower. It has to withstand high temperature in the presence of strong acid vapor, and hold positive and negative gas pressure during each hydrolysis run.
2. The detection of amino acids can be achieved by any HPLC system. However, if Fmoc is used as the derivatization reagent, a fluorescence detector is essential. The HPLC must consist of a ternary gradient controller in order to run the gradient program necessary for the baseline separation of 16 amino acids. We have investigated different types of C18 reversed-phase column and have found that the Hypersil C18 column manufactured by Keystone provides the best separation and resolution. We use the GBC automated AminoMate HPLC system controlled by GBC WinChrom Windows software that provides an automatic integration of the chromatogram and assignment of peaks *(3)*.
3. Sample preparation is an important step in amino acid analysis. Ideally, samples should be dried and free of salts, amines, and detergents. The sample of mucin must be desalted (e.g., by using a size-exclusion desalting column), or the mucin can be dot-spotted onto polyvinylidene difluoride (PVDF) membrane, or gel separated by electrophoresis and electroblotted onto PVDF. Note that mucins are often difficult to bind to PVDF. Wilkins et al. *(3)* have described a detailed protocol for PVDF-bound protein samples.
4. During the evacuation and argon flush of the hydrolysis vessel, the bottom of the vessel will become cold and the acid will boil under vacuum. These are the signs that the vessel is sealed tightly. Following hydrolysis, the vessel should be opened as soon as it is removed from the oven. Acid condensation inside the vessel is an indication of a complete hydrolysis. Handling the vessel requires heat-resistant gloves, and safety goggles must be worn at all times.

Table 2
Example of Quantitative Calculation of Mucin Amino Acid Composition

Amino acid	Standard (area)[a]	Bovine conjunctival mucin (area)	pmol[b]	% Composition[c]
Asx	258,795,760	404,279,488	391	9.6
Glx	215,720,688	532,183,264	617	15.1
Hydroxy-Pro	249,767,456	243,187,696	243	0.0
Ser	209,927,392	426,117,952	507	12.4
His	143,969,008	20,205,276	35	0.9
Gly	213,371,584	607,125,312	711	17.4
Thr	203,866,048	179,105,712	220	5.4
Ala	213,553,008	271,715,648	318	7.8
Pro	216,618,864	149,426,064	172	4.2
Tyr	91,930,856	7,537,291	20	0.5
Arg	200,286,320	84,959,272	106	2.6
Val	203,002,480	169,595,904	209	5.1
Met	144,096,656	29,927,380	52	1.3
Ile	202,357,184	132,131,184	163	4.0
Leu	201,993,024	270,782,624	335	8.2
Phe	206,174,464	104,213,960	126	3.1
Lys	292,284,352	109,133,576	93	2.3
Total			4077	100

[a]The on-column amount of each standard amino acid is generally 125 pmol. For example, for a 10 μL injection of 50 μ*M* stock solution, on-column pmol of standard amino acid = 50 μ*M* × 10 μL/4, where 4 is the dilution factor in derivatization.

[b]The pmol of mucin amino acid = peak area of mucin amino acid/peak area of standard amino acid × 25 pmol × 2, where 2 is the dilution factor from the half of the sample injected.

[c]%pmol = pmol mucin amino acid/total mucin amino acids × 100%, where total mucin amino acids is the sum of mucin amino acids (except cysteine and tryptophan; asparagine and glutamine are recovered as their acid form: aspartic acid and glutamic acid, respectively).

5. The mucin hydrolysate can be analyzed for its amino acid composition by any other chemical derivatization procedure and HPLC separation, e.g., phenylisothiocyanate derivatization with ultraviolet detection. If an autosampler is available, the derivatization procedure can be automated. Nevertheless, manual derivatization is adequate when thorough mixing after the addition of each reagent is practiced.
6. The HPLC separation is highly reproducible, requiring no modification of the gradient or conditions from day to day, and only minor maintenance throughout the life of the column, which is usually capable of 800 injections. Wilkins et al. *(3)* describe detailed troubleshooting of the chromatography.
7. Note that this amino acid analysis does not take into account the amount of cysteine and tryptophan. We are currently developing a method for quantitating cysteine by amino acid analysis.

Acknowledgments

The authors acknowledge support for their research on amino acid analysis through grants from the Australia Research Council, Macquarie University Research Grants, National Health and Medical Research Council, and GBC Scientific Equipment (Dandenong, Victoria, Australia). Jun Yan acknowledges financial support through an MUCAB scholarship. The authors also thank Malcom Ball for donating mucin samples and Dr. Andrew Gooley and Prof. Keith Williams for their support.

References

1. Yan, J. X., Wilkins, M .R., Ou, K., Gooley, A. A., Williams, K. L., Sanchez, J.-C., Golaz, O., Pasquali, C., and Hochstrasser, D.F. (1996) Large-scale amino-acid analysis for proteome studies. *J. Chromatogr.* **736,** 291–302.
2. Ou, K., Wilkins, M.R., Yan, J.X., Gooley, A.A., Fung, Y., Sheumack, D., and Williams, K. L. (1996) Improved high-performance liquid chromatography of amino acids derivatized with 9-fluorenylmethyl chloroformate. *J. Chromatogr.* **723,** 219–225.
3. Wilkins, M. R., Yan, J. X., and Gooley, A. A. (1999) 2-DE spots amino acid analysis with 9-fluorenylmethyl chloroformate, *2-D Protein Gel Electrophoresis Protocols* (Link, A. J., ed.), Humana, Totowa, NJ, 445–460.

11

Identification of Glycosylation Sites in Mucin Peptides by Edman Degradation

Natasha E. Zachara and Andrew A. Gooley

1. Introduction

Although it is possible to determine and characterize the total carbohydrate profile after release of the mucin oligosaccharides (usually by β-elimination), it is challenging to assign the sites of glycosylation (macroheterogeneity) and the carbohydrate heterogeneity at a given glycosylation site (microheterogeneity). Typically, the characterization of macro- and microheterogeneity has been dependent on the isolation of small peptides with only one glycosylation site. However, this is not possible with high molecular weight, heavily glycosylated domains such as those found in mucins.

The two methods for determining the sites of glycosylation in proteins are mass spectrometry and Edman sequencing. Multiple sites of glycosylation cannot easily be detected using mass spectrometry; one strategy involves the β-elimination of the carbohydrate, which results in the conversion of serine to dehydro-alanine and threonine to α-amino butyric acid. These amino acids have unique masses and can be used to map glycosylation sites *(1,2)*. However, the macro- and microheterogeneity of the carbohydrates can not be determined unless the peptide has just one glycosylation site.

In 1950, Edman sequencing was introduced as a repetitive degradation of proteins with phenylisothiocyanate *(3)*. In the mid-1960s, the process was automated *(4)*, resulting in a machine where the N-terminal amino acid is derivatized, cleaved, and transferred to a separate reaction vessel, in which it undergoes a conversion to a phenylthiohydantoin (PTH)-amino acid (**Fig. 1**). It is the PTH-amino that is separated by reversed-phase chromatography and detected.

Although the modern pulsed liquid sequenator is pmol sensitive, glycosylated amino acids are only recovered in low yield. Samples are sequenced (**Fig. 1**) on glass-fiber supports or membranes such as polyvinylidene difluoride (PVDF). Following cleavage, the released amino acid is transferred from the reaction cartridge to the conversion flask by nonpolar solvents such as ethyl acetate or chlorobutane. The more

From: *Methods in Molecular Biology, Vol. 125: Glycoprotein Methods and Protocols: The Mucins*
Edited by: A. Corfield © Humana Press Inc., Totowa, NJ

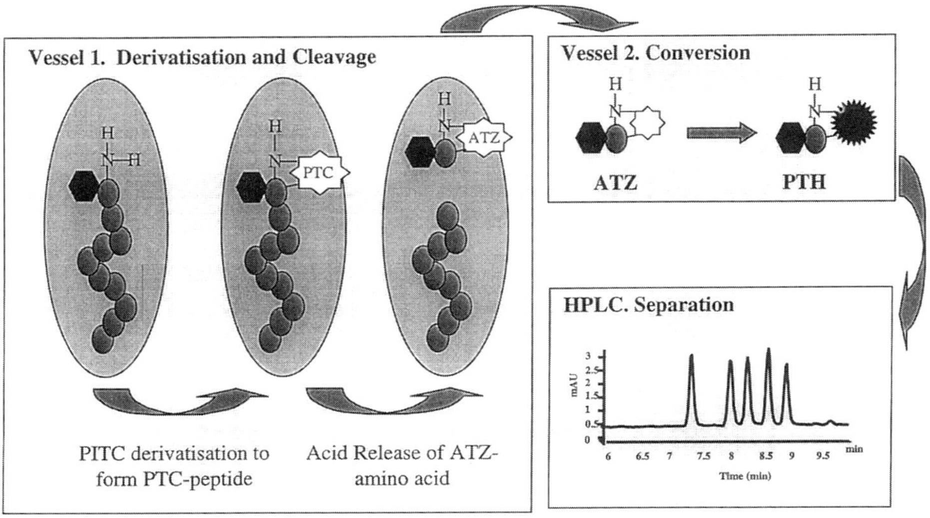

Fig. 1. Edman degradation. Schematic version of Edman degradation in the modern sequenator. (PITC, phenylisothiocyanate).

polar glycosylated amino acids are not soluble and remain in the support matrix (*see* **Fig. 1**, vessel 1) and a blank cycle is observed in the sequencer.

An alternative approach is to use solid-phase sequencing to extract glycosylated amino acids *(5)*. Peptides and proteins are covalently attached to solid supports *(6,7)*, allowing the delivery of polar reagents such as trifluoroacetic acid (TFA) and methanol, which facilitates the transfer of glycosylated amino acids. More importantly, it has been shown that amino acids modified by different carbohydrates have different retention times on high-performance liquid chromatography (HPLC) *(8)*. Thus, for the first time, both macro- and microheterogeneity can be determined for a glycopeptide or glycoprotein in a single experiment *(5)*. This technique has allowed mucin (and mucin-like) glycopeptides and trifluoromethane sulfonic acid (TFMSA)-treated mucin tandem repeats to be sequenced through the heavily glycosylated regions and the macroheterogeneity to be assigned and quantitated *(9–11)*.

Simple modifications to most modern sequenators facilitate the identification of glycosylated amino acids *(9,11,12)*. Although it is possible to identify amino acids substituted with a monosaccharide using standard programs, the glycosylated amino acids are not recovered quantitatively. Modifications of the transfer solvents to more polar reagents, such as TFA or methanol, increase the recovery of glycosylated amino acids modified by monosaccharides and larger oligosaccharides (degree of polymerization [DP]=19). Elsewhere we have published alternative HPLC methods for the separation of PTH-glycosylated amino acids on a modified Beckman LF3600 sequenator *(12)*. These methods were developed for two reasons: (1) to provide a distinct elution window for the PTH-glycosylated amino acids, and (2) to be of low enough salt concentration for the direct infusion of the PTH-glycosylated amino acid into an electrospray ionisation mass spectrometer. However, the chromatography described in **ref.** *12* is not commercially available and must be made in-house. The following examples from the Hewlett Packard G1000A and the Procise™ from PE Biosystems show HPLC separations (**Fig. 2**) of PTH-glycosylated amino acids achieved with commercial kits.

2. Materials

2.1. Chemicals

1. Human Glycophorin A, blood type NN (Sigma, cat. no. G-9266).
2. Bovine κ-Casein glycopeptide (Sigma, cat. no. C-2728).
3. Sequencing grade TFA.
4. HPLC grade solvents, Milli Q water.

2.2. Reagents for Adsorption onto Hyperbond

1. Hyperbond membranes were from Beckman.

2.3. Protein Coupling to Arylamine Membranes (see Note 1)

1. Reagents are available in kit form from Beckman and PerSeptive Biosystems (MA; *see* **Note 2**).
2. Milli Q water and analytical grade methanol are required for wash steps.

2.4. Edman Degradation

1. Reagents are obtained in the form of quality-assured sequencing kits, from the manufacturers of the sequenators.

2.5. Apparatus

1. Heating blocks at 80°C and 50°C.
2. Speed-Vac™ vacuum centrifuge (Savant Instruments, NY)
3. Beckman LF3600 protein sequenator, or Hewlett Packard G1000A protein sequencer, or PE Biosystem's Procise™.

3. Methods

3.1. Sample Preparation

Edman sequencing is a molar-dependent chemistry, and it is necessary to determine how many moles of glycopeptide are to be sequenced (*see* **Note 3**). Samples for Edman degradation (*see* **Note 4**) must be salt free and not contain sialic acid (*see* **Note 5**). Samples can be electroblotted (*see* **Note 6**) or adsorbed onto Hyperbond or PVDF (*see* **Note 7**), or covalently coupled to an arylamine-derivatized solid support (*see* **Note 8**). For qualitative analysis of glycosylation sites, as little as 10 pmol of material is required, whereas for quantification of the macro- and microheterogeneity, more material is required, depending on the length of the peptide and the variation in the structures present (*see* **Note 9**).

3.1.1. Desialylation of Glycopeptides

1. Place glycopeptide (5–50 pmol/μL) in salt-free buffer (up to 20% organic modifier [v/v]) in a screw-capped polypropylene Eppendorf tube.

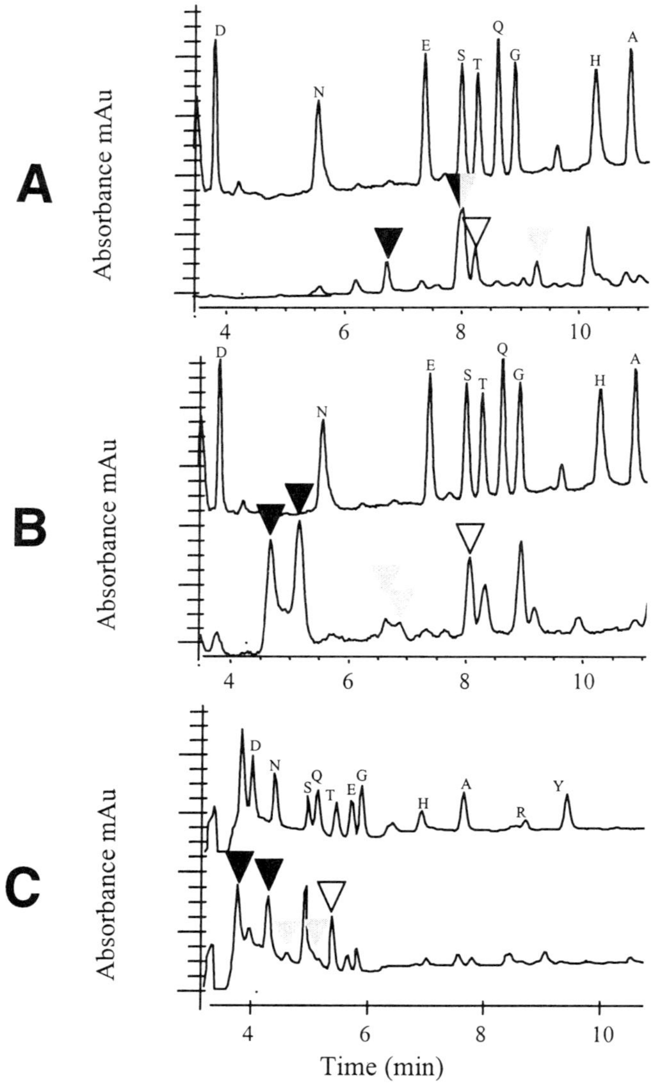

Fig. 2. Glycosylated PTH-amino acid profiles. **(A)** The glycosylated amino acid window with the elution profiles of PTH-threonine-GalNAc-Gal (black arrowheads) and PTH-threonine-GalNAc (gray arrowheads) from Hewlett Packard G1000A protein sequencer. Standard amino acids are shown for comparison (threonine, open arrowhead). These amino acids were separated using the manufacturer's suggested solvent system and program according to the Program 3.1 chemistry. **(B)** The glycosylated amino acid window with the elution profiles of PTH-serine-GalNAc-Gal (black arrowheads) and PTH-serine-GalNAc (gray arrowheads) from Hewlett Packard G1000A protein sequencer. Standard amino acids are shown for comparison (serine, open arrowhead). These amino acids were separated using the manufacturer's suggested solvent system and program according to the 3.1 chemistry. **(C)** The glycosylated amino acid window with elution profiles of PTH-threonine-GalNAc-Gal (black arrowheads) and PTH-threonine-GalNAc (gray arrowheads) from the Applied Biosystems Procise. Standard amino acids are shown for comparison (threonine, open arrowhead). These amino acids were separated using the manufacturer's suggested solvent system.

2. Add an equal volume of 0.2 *M* TFA.
3. Incubate samples at 80°C for 30 min.
4. Remove TFA by lyophilization in a vacuum centrifuge (*see* **Note 10**).
5. Resuspend desialylated glycopeptide in 10–20% (v/v) acetonitrile.

3.1.2. Electroblotting

Glycopeptides can be desialylated before or after electrophoresis. Electroblotted glycopeptides are desialylated as in **Subheading 3.1.1.** with the following precautions:

1. Prior to addition of TFA, place membranes in the reaction vessel and wet with 10 µL of methanol.
2. Add enough 0.1 *M* (not 0.2 *M* as in **Subheading 3.1.1.**) TFA to cover the membrane.
3. Following incubation at 80°C, remove the membrane from the TFA, wash in Milli Q water (three times) and dry.

3.1.3. Adsorption of Glycopeptides onto Hyperbond or PVDF

1. Place a piece of Hyperbond or PVDF (a round membrane is preferable, diameter ~0.5 cm; *see* **Note 11**) in a lid of an Eppendorf tube and place on a hot plate at 50°C.
2. Wet with 10 µL of methanol.
3. When the excess methanol has evaporated, but before the membrane has dried, apply the sample in 10-µL aliquots (*see* **Note 12**).
4. When all of the sample is applied dry the membrane.

3.1.4. Covalent Attachment of Glycopeptides

Glycopeptides are covalently attached to arylamine-derivatized membranes via the activation of peptide carboxyl groups using water-soluble EDC (*see* **Note 13**). Covalent coupling of the peptide to arylamine-derivatized membrane is as described in the kit user's guide.

1. Place an arylamine-derivatized membrane in an Eppendorf tube lid at 50°C.
2. Apply samples in 10-µL aliquots, allowing the membrane to come to near dryness between each aliquot (*see* **Note 14**).
3. Dry the membrane after all of the sample has been applied.
4. Mix approx 1 mg of EDC in 50 µL of reagent attachment buffer (*see* **Note 15**) and carefully pipet 10–50 µL onto the arylamine-derivatized membrane.
5. Incubate the sample at 4°C for 30 min (*see* **Note 16**).
6. Wash the membrane alternately in 1 mL of methanol and 1 mL of Milli Q water three times and dry.

3.2. Edman Sequencing

3.2.1. Glycopeptides Electroblotted or Adsorbed onto Hyperbond or PVDF

Glycopeptides adsorbed onto PVDF or Hyperbond can be sequenced with conventional Edman degradation (Program 3.1 PVDF on the Hewlett Packard G1000A or the PVDF routine for the blot cartridge on the Procise, or a modified program 40 recommended by Beckman for PVDF on the Beckman LF series). While amino acids modified by monosaccharides will be identified, their recovery (as well as that of the larger oligosaccharides) is not quantitative.

3.2.2. Glycopeptides Covalently Attached to Membranes

Covalently attached glycopeptides can be sequenced using a modified Hewlett Packard G1000A routine 3.1 PVDF that includes methanol rather than ethyl acetate for the extraction of the cleaved amino acid from the cartridge (**Fig. 1**, vessel 1; *see* **Note 17**) to the converter (**Fig. 1**, vessel 2; *see* **Note 18**).

4. Notes

1. The arylamine-coated membranes are stable at –20°C; the *N*-ethyl-*N'*-dimethylamino-propylcarbodiimide (EDC reagent is stable at 4°C in a dry environment, and the buffer is stable at 4°C.
2. Arylamine-derivatized membranes, Sequelon AA™, were originally produced by Milligen Biosearch, a division of Millipore. Although Millipore no longer manufactures these membranes, both PerSeptive Biosystems and Beckman supply this product.
3. Quantitation of glycopeptides is quite difficult, and we recommend that 10% of the sample be first sequenced using the standard program prior to glycosylation site mapping.
4. Mucins must be digested extensively by proteases and the glycopeptide purified, desialylated (*see* **Note 5**), and desalted before sequencing. For heavily glycosylated peptides substituted with large oligosaccharides it may be necessary to treat with TFMSA *(13)* or glycosidases before sequencing.
5. There are two principle reasons for desialylating glycopeptides:
   a. If the peptide is to be covalently bound to an arylamine-derivatized support, the sialic acid must be removed to prevent the formation of an amide bond between the carboxyl group of the sialic acid and the amine of the support. Although the alternative of using diisothiocyanate-derivatized membranes is possible, we have found them of little practical use, since the peptides must contain lysine, preferably at the C-terminus, and this residue is uncommon in mucins.
   b. For glycopeptides adsorbed onto Hyperbond or PVDF supports, we have observed poor recovery of sialylated glycosylated amino acids. They elute as a series of multiple peaks and, in many cases, coelute with standard amino acids (unpublished results).
6. Electroblotting is not a usual method for the preparation of mucin glycopeptides since they are large in apparent molecular weight and are rarely separated by sodium dodecyl sulfate-polyacrylamide gel electrophoresis (SDS-PAGE). However, if glycopeptides can be separated by SDS-PAGE, the glycopeptides can be electroblotted onto hydrophobic membranes such as Hyperbond or PVDF and stained with amido black.
7. Hyperbond performs better in this application than PVDF.
8. Covalent coupling gives a quantitative recovery of amino acids modified by more than one carbohydrate residue. Other methods only identify an amino acid modified by one carbohydrate residue.
9. Because mucin glycopeptides are usually of very high molecular weight, an initial yield of 10 pmol in the sequenator represents approx 100 µg of a 200-kDa glycopeptide. Quantitating the carbohydrate component (*see* **Notes 3** and **8**) and the length of sequence obtained is dependent on the number and distribution of proline residues. Proline cleaves with reduced efficiency in Edman sequencing (**Fig. 1**, vessel 1), and a significant lag is introduced when several prolines occur in a cluster and more than one signal is detected in each cycle.
10. Freeze-drying mucins is not recommended because the solid is often poorly soluble. The sample should not be taken to complete dryness.

11. Round pieces of PVDF or Hyperbond have better surface tension and allow larger aliquots to be loaded. Alternatively, a strip of PVDF or Hyperbond can be used, approx 2 × 10 mm. Care must be taken when loading to maintain surface tension on the strip.
12. Membranes should not be allowed to dry between application of sample aliquots.
13. Although small peptides couple well, the high molecular weight of mucin glycopeptides reduces the efficiency of coupling to ~5%.
14. Amine buffers such as Tris interfere with coupling to arylamine and should be avoided.
15. The EDC reagent should be mixed with the reaction buffer just prior to coupling.
16. Incubation at 4°C increases the initial yield *(7)*.
17. To incorporate a methanol extraction into the routine, methanol is placed in an unused bottle position, and solvent delivery from this bottle is substituted for ethyl acetate during the extraction procedure.
18. Poor yields of aspartic and glutamic acid will be observed with arylamine-coupled peptides or proteins because of the coupling of the χ- and δ-carboxyl groups to the membrane.

Acknowledgments

NEZ was supported by an Australian Postgraduate Award. AAG acknowledges the support of the National Health and Medical Research Council. The authors would like to acknowledge the support and useful comments of Dr. Nicolle Packer and Prof. Keith Williams (Macquarie University).

References

1. Rademaker, G. J., Haverkamp, J., and Thomas-Oates, J. (1993) Determination of glycosylation sites in O-linked glycopeptides: a sensitive mass spectrometric protocol. *Organ. Mass Spectrom.* **28,** 1536–1541.
2. Gries, K. D., Hayes, B. K., Comer, F. I., Kirk, M., Barnes, S., Lowary, T. L., and Hart, G. W. (1996) Selective detection and site-analysis of O-GlcNAc-modified glycopeptides by β-elimination and tandem electrospray mass spectrometry. *Anal. Biochem.* **234,** 38–49.
3. Edman P. (1950) Method for the determination of the amino acid sequence in peptides. *Acta. Chem. Scand.* **4,** 283–293.
4. Edman, P and Begg, G., (1967) A protein sequencer. *Eur. J. Biochem.* **1,** 80–91.
5. Gooley, A. A., Classon, B. J., Marschalek, R., and Williams, K. L. (1991) Glycosylation sites identified by detection of glycosylated amino acids released from Edman degradation: the identification of Xaa-Pro-Xaa-Xaa as a motif for Thr-O-glycosylation. *Biochem. Biophys. Res. Commun.* **178,** 1194–1201.
6. Laursen, R. A. (1971) Solid-phase Edman degradation: an automated peptide sequencer. *Eur. J. Biochem.* **20,** 89–102.
7. Laursen, R. A., Lee, T. T., Dixon, J. D., and Liang S.-P. (1991) Extending the performance of the solid-phase sequenator, in *Methods in Protein Sequence Analysis* (Jornvall, H., Hoog, J.-O., and Gustavsson, A.-M., eds.), Birkhauser Verlag, Switzerland, pp. 47–54.
8. Gooley, A. A. and Williams, K. L. (1997) How to find, identify and quantitate the sugars on proteins. *Nature* **358,** 557–559.
9. Muller, S., Goletz, S., Packer, N., Gooley, A., Lawson, A. M., and Hanisch, F. G. (1997) Localisation of O-glycosylation sites on glycopeptide fragments from lactation-associated MUC1: all putative sites within the tandem repeat are glycosylation targets *in vivo*. *J. Biol. Chem.* **272,** 24,780–24,793.
10. Zachara, N. E., Packer, N. H., Temple, M. D., Slade, M. B., Jardine, D. R., Karuso, P., Moss, C. J., Mabbutt, B. C., Curmi, P. M. G., Williams, K. L., and Gooley, A. A. (1996)

Recombinant prespore-specific antigen from *Dictyostelium discoideum* is a β-sheet glycoprotein with a spacer peptide modified by *O*-linked *N*-acetylglucosamine. *Eur. J. Biochem.* **238,** 511–518.
11. Gerken, T. A., Owens, C. L., and Pasumarthy, M. (1997) Determination of the site-specific O-glycosylation pattern of the porcine submaxillary mucin tandem repeat glycoprotein. *J. Biol. Chem.* **272,** 9709–9719.
12. Pisano, A., Jardine, D. R. Packer, N. H., Farnsworth, V., Carson, W., Cartier, P., Redmond, J. W., Williams, K. L., and Gooley, A. A. (1996) Identifying sites of glycosylation in proteins, in *Techniques in Glycobiology* (Townsend, R. and Hotchkiss, A., eds.), Marcel Dekker, New York, pp. 299–320.
13. Gerken, T. A., Gupta, R., and Jentoft, N. (1992) A novel approach for chemically deglycosylating O-linked glycoproteins: the deglycosylation of submaxillary and respiratory mucins. *Biochemistry* **31,** 639–648.

12

Synthetic Peptides for the Analysis and Preparation of Antimucin Antibodies

Andrea Murray, Deirdre A. O'Sullivan, and Michael R. Price

1. Introduction

Since the mid-1980s, the family of high molecular weight glycoproteins known as mucins have evoked considerable interest among those in the field of cancer research. Mucins, which are constituents of mucus, have a lubricating and protective function in normal epithelial tissue *(1)*. However, expression of mucin by the cancer cell is often highly disorganized and upregulated, sometimes to the extent that mucin can be detected in the circulation of the cancer patient. These changes in expression of mucin observed in neoplasia have led to the exploitation of some members of the mucin family as circulating tumor markers *(2,3)* or targets for diagnostic imaging *(4–6)* and therapy of cancer.

The first mucin to have its primary amino acid sequence determined, MUC1, is also the most extensively studied. This molecule is highly immunogenic, and a considerable number of anti-MUC1 monoclonal antibodies (mAbs) and fragments have been produced by various methods. Some of these have found applications for radioimmunoscintigraphy and targeted therapy of cancer, and others have been used to detect circulating MUC1. Although such studies have yielded promising results, their present application is somewhat restricted. In this age of genetic and protein engineering, we have, at our disposal, the technology to design antibodies with ideal characteristics of size, affinity, and specificity for any desired application. However, before considering such ambitions, we must first gain an understanding of the molecular interactions between epitope and paratope when an antibody binds to its antigen. It is essential that key residues involved in the interaction are identified so that a model of how the interaction takes place on a three-dimensional level can be constructed. This identification will enhance our ability to design antibodies with the correct characteristics for our chosen application.

From: *Methods in Molecular Biology, Vol. 125: Glycoprotein Methods and Protocols: The Mucins*
Edited by: A. Corfield © Humana Press Inc., Totowa, NJ

1.1. Immunoassays

Both enzyme-linked immunosorbant assays (ELISAs) and radioimmunoassays have been used in various formats to test antibody binding to synthetic peptides. The indirect ELISA has the advantages of being easy to perform, having no requirement for radioactive tracers, and producing results that are simple to interpret. The disadvantage of the indirect ELISA is that the procedure requires that the antigen, in this case a synthetic peptide, be immobilized on to the surface of a microtiter plate well. Classically this would be achieved by dispensing a solution of antigen into the wells of a microtiter plate to allow adsorption, leaving the plate coated with antigen. However, short synthetic peptides adsorbed on to plates in this way provide unpredictable and inconsistent results. This problem may be owing to the fact that the orientation of the peptide on the plate cannot be controlled or simply that short peptides do not adhere well to polystyrene plates. Several methods of peptide modification have been utilized to overcome these problems. One such procedure involves preparing branched-chain polypeptides in which MUC1 immunodominant peptides ware conjugated to a polylysine backbone *(7)*. These polylysine conjugates provide very potent MUC1-related antigens for the interrogation of antibody specificity; however, the methodology for their preparation is beyond the scope of this chapter. By far the most widely used method for modifying short peptides so that they can be used as antigens in indirect ELISA procedures is to conjugate the peptides to a large carrier protein such as bovine serum albumin (BSA) (*see* **Subheading 3.1.** and **Notes 1–3**).

1.2. Tethered Peptide Libraries for Exploring Antibody Specificity

The peptide synthesis techniques developed by Geysen and colleagues *(8)* represent a significant development in the study of epitopes defined by antibodies reactive with antigens of known primary structure. Unlike most other methods of simultaneous peptide synthesis, this technique allows the concurrent synthesis of hundreds to thousands of peptides so that libraries can be produced and simultaneously used as targets for antibody binding. The peptides are synthesized on derivatized polyethylene or polypropylene gears that are held on stems (**Fig. 1**) arranged in a microtitre plate format so that a simple ELISA procedure can be used to measure antibody binding. Peptides are tethered via the carboxyl terminus.

Several different strategies have been described for peptide sequence design that all provide different information on epitope structure and the fine specificity of an antibody-peptide interaction. The Pepscan approach has been the most widely used and involves the synthesis of a set of overlapping peptides that span the length of the antigenic sequence (**Subheading 3.2.**). In a short peptide sequence, such as that of the MUC1 variable number of tandem repeat (VNTR), each peptide may overlap the next by all but one amino acid, giving rise to a set of 21 heptapeptides that spans the VNTR sequence (**Fig. 2**). For larger proteins, it is more appropriate to produce longer sequences that overlap each other by less residues, thereby spanning the length of the antigenic sequence with a feasible number of peptides (*see* **Note 4**). In the Pepscan approach, peptides are assayed for antibody-binding capacity by ELISA (**Subheading 3.3.**), and residues that are common to all the antibody-binding pins represent the mini-

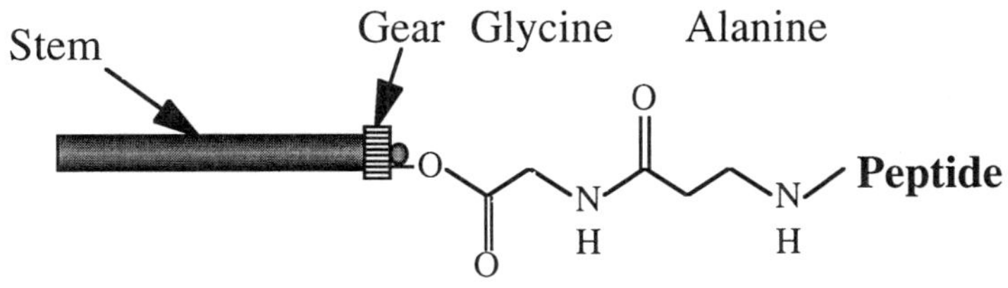

Fig. 1. The Multipin Peptide Synthesis System contains detachable polyethylene gears that fit on to the end of stems. The stems are held in a block in an 8 x 12 microtiter plate format. The surface of the gear is derivatized to give a solvent-compatible polymer matrix on which the peptides are coupled during synthesis. The matrix also provides a two amino acid spacer group.

MUC1 VNTR Amino Acid Sequence

1 2 3 4 5 6 7 8 9 10 11 12 13 14 15 16 17 18 19 20 | 1 2 3 4 5 6 7

P P A H G V T S A P D T R P A P G S T A P P A H G V T

Heptamer Number	Heptamer Sequence	Typical Assay Result
1	P P A H G V T	-
2	P A H G V T S	-
3	A H G V T S A	-
4	H G V T S A P	-
5	G V T S A P D	-
6	V T S A P D T	-
7	T S A [P D T R]	+
8	S A [P D T R] P	+
9	A [P D T R] P A	+
10	[P D T R] P A P	+
11	D T R P A P G	-
12	T R P A P G S	-
13	R P A P G S T	-
14	P A P G S T A	-
15	A P G S T A P	-
16	P G S T A P P	-
17	G S T A P P A	-
18	S T A P P A H	-
19	T A P P A H G	-
20	A P P A H G V	-
21	P P A H G V T	-

Fig. 2. Schematic representation of the overlapping peptides corresponding to the MUC1 VNTR sequence synthesized according to the Pepscan approach to epitope mapping. Antibodies are allowed to react with each peptide, and those containing the epitope or minimum binding unit produce positive results. In this example, the epitope can be deduced as consisting of the amino acids that are common to all positive pins (7–10). Hence, the epitope is **PDTR**.

mum binding unit or epitope for that antibody (**Fig. 2**). Having identified the epitope defined by an antibody using Pepscan, it may be useful to prepare a number of analogs of that sequence in order to investigate the role of individual amino acids in the epitope and to identify critical contact residues. Such peptide design stategies include ommission analysis, alanine substitution and replacement net (RNET) analysis (*see* **Notes 5–9**).

Libraries of peptides on pins can be obtained that comprise 400 different dipeptides prepared with all possible combinations of the 20 natural amino acids. This approach provides qualitative information on antibody specificity and permits identification of significant features of an epitope that may contribute to antibody recognition and binding (*see* **Notes 10** and **11**).

1.3. Purification of Antibodies Using Peptide Affinity Chromatography

The identification of a linear peptide epitope within a protein sequence facilitates the design of peptide affinity matrices that can be used to purify antibodies from biological feedstocks. Such an epitope affinity matrix has been produced by covalently linking a synthetic peptide corresponding to the MUC1-immunodominant domain to cyanogen bromide-activated Sepharose (Pharmacia, Uppsala, Sweden) ***(9)***. The resulting matrix was remarkably efficient for the purification of a range of anti-MUC1 mAbs from biological feedstocks containing high levels of contaminating proteins such as ascitic fluid and hybridoma supernatant (*see* **Note 12**).

Epitope affinity chromatography matrices have an advantage over other affinity adsorbents in that the antibody is bound to the matrix specifically via the paratope. Thus, eluted antibody is fully immunoreactive and of only the desired specificity. Sepharose-peptide conjugates are simple to prepare and affinity chromatography is more robust than other conventional chromatographic techniques in terms of column packing and operation (see **Subheading 3.4.**, **Notes 13–15**, and **Fig. 3**).

1.4. General Comments

The techniques described for the analysis of antimucin antibodies using synthetic peptides can provide a great deal of information on epitope topography and structure. The identification of critical binding residues within an epitope can provide clues to the forces and residues involved in the antibody-antigen interaction. However, bear in mind that the use of linear synthetic peptides can only provide a one-dimensional solution to what is essentially a three-dimensional problem. Further structural studies such as X-ray crystallography, nuclear magnetic resonance spectroscopy, and computational molecular modeling are essential if the knowledge gained is to be confirmed and translated into a useful model on which to base antibody design strategies.

The structural information provided by studies such as those previously described may be of use in peptide vaccine design. However, the analyses performed so far have been mainly concerned with the interaction of murine antibodies, and it may be naive to assume that the human immune system will process mucin-related antigens in the same way. Preliminary epitope-mapping studies on human serum would suggest that the immune response to MUC1 may differ considerably from that observed in the mouse ***(10)***.

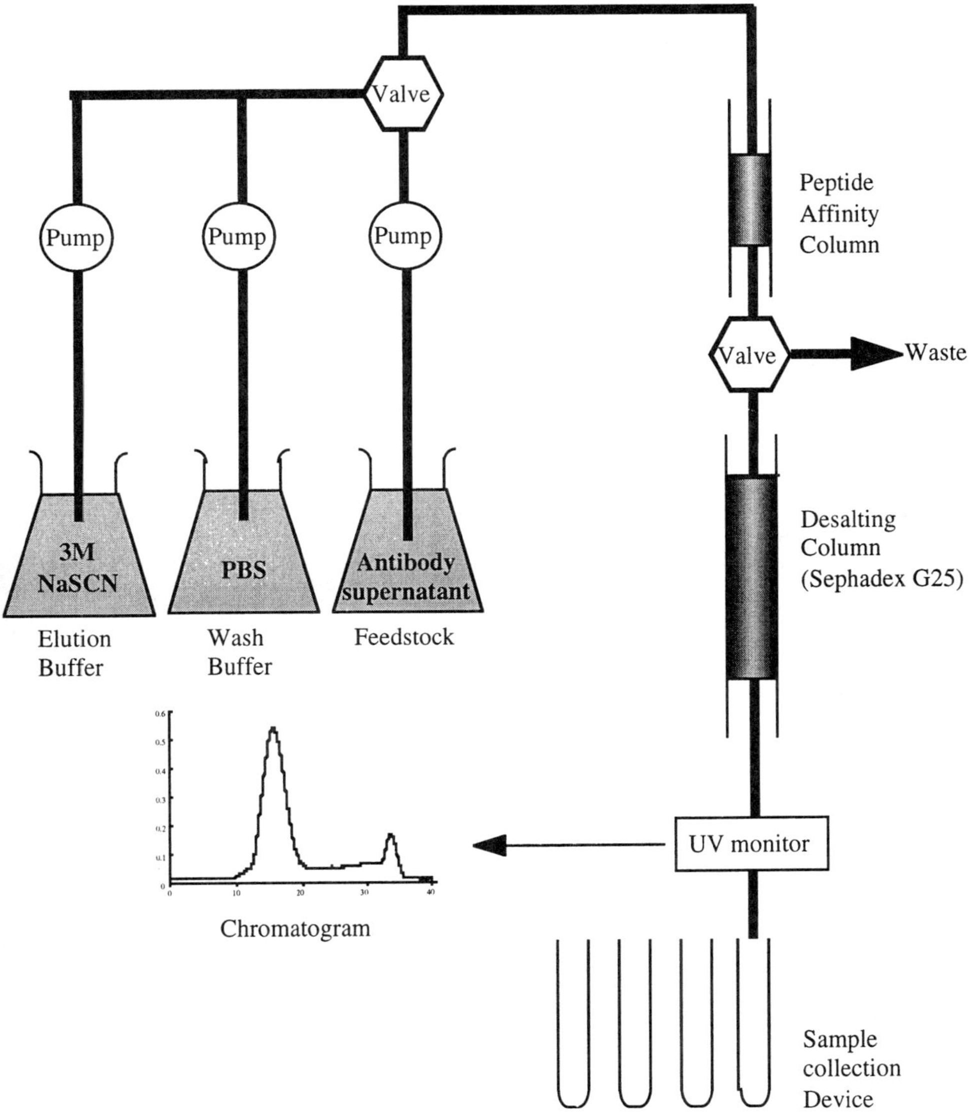

Fig. 3. Schematic representation of the apparatus and reagents needed for the purification of antibodies by peptide epitope affinity chromatography.

Finally, it may be owing to the very nature of the mucins that such a wealth of information has been provided by the techniques described. The VNTR provides a convenient short sequence on which to base peptide synthesis strategies. In addition, most murine antimucin antibodies analyzed to date have been shown to define short linear determinants. It is unlikely that all other proteins and antibodies will be so accommodating.

Table 1
Suitable Amino Acid Side Chain Protecting Groups for Solid Phase Peptide Synthesis on Pins

Side chain protecting group	Amino acid
t-Butyl ether	S, T, Y
t-Butyl ester	D, E
t-Butoxycarbonyl	K, H, W
2,2,5,7,8-Pentamethylchroman-6-sulfonyl	R
Trityl	C

2. Materials

2.1. Preparation of BSA-Peptide Conjugates

1. Conjugation buffer: sodium hydrogen carbonate buffer (0.1 *M*, pH 8.4).
2. BSA: crystalline, greater than 96% pure.
3. Glutaraldehyde: when used as a crosslinker must be freshly distilled or high commercial grade (Sigma, Poole, UK).
4. Dialysis buffer: sodium chloride 1% (w/v).

2.2. Solid-Phase Peptide Synthesis on Pins

All reagents used in solid-phase peptide synthesis should be of the highest available purity (analytical reagent grade or better) unless stated otherwise.

1. Mulitpin Peptide Synthesis Kit (Chiron Mimotopes, Clayton, Victoria, Australia).
2. Amino acids: All amino acids recommended for use with the Multipin Peptide Synthesis Kit have their α-amino group protected with the 9-fluorenylmethyloxycarbonyl (Fmoc) group. **Table 1** appropriate side chain protecting groups. Alternatively, protected amino acid esters may be used. These have the advantage of requiring no prior activation. However, they are prone to decomposition with prolonged storage and are best stored at –20°C.
3. Activators**:** The activation of protected amino acids with diisopropylcarbodiimide (DIC)/1-hydroxybenzotriazole (HOBt) is recommended, but other coupling reagents can be used.
4. N,N-Dimethylformamide (DMF): DMF used in peptide coupling procedures must be pure and free from amines. Several methods may be used to purify DMF (*see* **Note 16**).
5. Piperidine 20% v/v: used for Fmoc deprotection. Piperidine should be redistilled before use and made up to a 20% (v/v) solution in DMF.
6. Bromophenol blue: used as an indicator of coupling efficiency. Stock reagent is prepared by dissolving 33.5 mg of bromophenol blue in 5 mL of DMF. This should be diluted 1:200 for working concentration.
7. Acetylation mixture: DMF, acetic anhydride and triethylamine in a 50:5:1 (v/v/v) ratio.
8. Side chain deprotection mixture: trifluoroacetic acid, ethanedithiol, and anisol in a 38:1:1 (v/v/v) ratio.
9. Final wash solution**:** acetic acid 0.5% (v/v) in methanol/water (1:1, v/v).
10. Other reagents: methanol (MeOH), purified water.

2.3. ELISA Testing Procedure

1. Phosphate buffered saline (PBS), 0.01 *M*, pH 7.2 (1.34 g of $Na_2HPO_4{\cdot}2H_2O$, 0.39 g of $NaH_2PO_4{\cdot}2H_2O$, and 8.5 g of NaCl made up to 1 L with distilled water) is used as the buffer base for most of the following buffer reagents.

2. Blocking buffer: 2% (w/v) BSA, 0.1% (v/v) Tween-20, and 0.1% sodium azide in 0.01 *M* PBS.
3. Conjugate diluent: 1% (v/v) sheep serum, 0.1% (v/v) Tween-20, and 0.1% sodium caseinate (USB, Bioscience, Cambridge, UK) in 0.01 *M* PBS.
4. Citrate phosphate buffer: 17.8 g of $Na_2HPO_4{\cdot}2H_2O$ and 16.8 g of citric acid monohydrate made up to 1 L with distilled water, pH 4.0.
5. 2,2'-azino-bis[3-ethylbenz-thiazoline-6-sulfonic acid] (ABTS) substrate solution (Sigma): 0.5 mg/mL in citrate phosphate buffer with hydrogen peroxide (35% w/w) added to give a final concentration of 0.01% (w/v).
6. Disruption buffer: Sodium dihydrogen orthophosphate (0.1 *M*) pH 7.2, containing sodium dodecyl sulfate (SDS) (0.1% w/v). β-Mercaptoethanol (5 mL) is added immediately prior to use.

2.4. Purification of Antibodies Using Peptide Affinity Chromatography

1. Affinity support: CNBr-activated Sepharose 4B (Pharmacia, Uppsala, Sweden).
2. Equilibration buffer: 0.01 *M* PBS with azide (PBSA), pH 7.2 (1.34 g of $Na_2HPO_4{\cdot}2H_2O$, 0.39 g of $NaH_2PO_4{\cdot}2H_2O$, and 8.5 g of NaCl made up to 1 L with distilled water) with sodium azide 0.02% (w/v) added as a preservative.
3. Wash buffer: 0.5 *M* NaCl, pH 7.2, in distilled water.
4. Elution buffer: 3 *M* NaSCN, pH 7.2, in distilled water.
5. Desalting column: Sephadex G25 (Pharmacia).

3. Methods

3.1. Preparation of BSA-Peptide Conjugates (see Notes 1–3)

1. Dissolve BSA (10 mg) in 3 mL conjugation buffer in a clean glass vial.
2. Dissolve peptide (10 mg) in 1 mL conjugation buffer.
3. To the BSA solution, add 1 mL of peptide solution and 10 µL of glutaraldehyde. Then seal and agitate on a roller for 4 h at room temperature.
4. The conjugate is finally dialysed against sodium chloride (1%) for 48 h at 4°C.

3.2. Solid-Phase Peptide Synthesis on Pins (see Notes 4–11, and 17)

The Multipin Peptide Synthesis Kit (Chiron Mimotopes) contains derivatized gears, stems, 8 × 12 format pin holders, and reaction trays. In addition, it contains all the software needed for creating a synthesis schedule, running dispensing aids, and reading and plotting assay results.

3.2.1. Creating a Synthesis Schedule

The method described by Geysen et al. *(8)* for linear epitope scanning requires that many different peptides be synthesized simultaneously. To plan and execute a manual synthesis schedule for creating hundreds of peptides simultaneously is extremely time-consuming and fraught with the possibility of errors. Fortunately, computer software is available to generate a synthesis schedule based on any given protein sequence with any of the manipulations described (Chiron Mimotopes). These schedules calculate the weights and volumes of the various reagents required on each day of the synthesis and then instruct the operator where in the 96-well reaction tray each amino acid should go (*see* **Notes 5–8**).

3.2.2. Peptide Synthesis

The peptide synthesis procedure consists of cycles of N-terminal deprotection, washing and coupling until the desired peptides have been assembled, followed by side chain deprotection. The synthesis schedule provides details of the amounts of amino acids and activators required. It is advisable to weigh out all these reagents before beginning a synthesis since this is the most time-consuming step of the procedure. All steps are carried out at room temperature unless stated otherwise.

1. The appropriate number of gears required for that synthesis on that day should be removed from storage and assembled on to the block according to the synthesis schedule. It is important that only gears requiring deprotection in the next cycle of synthesis be added to the block.
2. Achieve deprotection of the amino terminus by immersing the pins in a bath containing 20% piperidine for 20 min. The piperidine solution should cover the gears. The pins are then washed as follows:
   a. DMF to cover gears for 2 min.
   b. MeOH (complete submersion) for 2 min.
   c. MeOH to cover gears for 2 min (three times).

   The pins are then allowed to air-dry in an acid-free fume hood for a minimum of 30 min.
3. Prepare HOBt and DIC solutions by dissolving in the appropriate amount of DMF (*see* **Note 13**). The addition of bromophenol blue to the HOBt to give a final concentration of 0.05 m*M* as an indicator of coupling efficiency is optional. The volume of HOBt solution specified on the synthesis schedule must be added to each amino acid to dissolve it fully before adding the specified amount of DIC.
4. Dispense amino acid solutions into a 96-well reaction tray according to the synthesis schedule. The recommended order of activating and dispensing amino acids is as follows: A D E F G I L M P S T V Y W Q N K C H R. Care should be taken to ensure that the amino acids are dispensed into the correct wells. Dispensing aids are now available that consist of a bank of LED lights set out in a microtiter plate format. Lights are lit beneath the reaction tray to indicate which wells should contain which amino aid. The dispensing aid is driven by the synthesis schedule software.
5. Place the block of Fmoc-deprotected pins into the reaction tray in the correct orientation. Place the tray into a polystyrene box to reduce evaporative losses and avoid contamination and leave to incubate for at least 4 h.
6. When the coupling reaction is complete, the blue colouration of bromophenol blue should have disappeared. The pins are then washed as follows:
   a. MeOH to half the pin height for 5 min.
   b. air-dry for 2 min.
   c. DMF to half the pin height for 5 min.

   The next cycle of peptide synthesis can begin immediately with Fmoc deprotection.
7. When the required peptides have been synthesized, deprotect and wash the free amino termini as described in **step 2**. The amino terminus may then be acetylated to remove the charge associated with a free amino terminus (if required) by incubating the pins in a reaction tray containing acetylation mixture at 150 µL/well for 90 min in an enclosed environment. Then wash the pins in MeOH for 15 min and then air-dry.
8. To achieve side chain deprotection, incubate the pins in a bath of side chain deprotection mixture for 2.5 h. Next, immerse in a final wash solution for 1 h, rinse twice in MeOH for 2 min each, and air-dry overnight. The pins are now ready for ELISA testing.

3.3. ELISA Testing Procedure (see Notes 7 and 12)

Antibody binding to peptides on pins is measured using an indirect ELISA procedure in which the solid phase on which the test antibody is captured is the peptide-coated gear, and the presence of the test antibody is reported using an enzyme-labeled secondary antibody. The enzyme catalyzes the reaction of ABTS substrate to its colored product, which can be measured using a spectrophotometer. The degree of color change is proportional to the amount of test antibody bound to the peptide on the gear.

Before antibody testing begins, the newly synthesized pins should be tested for nonspecific binding to the enzyme-labeled secondary antibody of choice. This is achieved by carrying out **steps 1** and **4–6**. Antibodies may now be tested as follows, with all incubations and washing steps performed at room temperature unless otherwise stated.

1. First precoat the pins in blocking buffer in order to minimize nonspecific binding to the gear. To achieve this, immerse the pins in a microtiter plate containing blocking buffer at 200 µL/well and incubate for 1 h with agitation.
2. Dilute the primary antibody to an appropriate concentration in blocking buffer and dispense into the wells of a microtiter plate at 200 µL/well. After removing from blocking buffer and flicking to remove excess buffer, incubate the pins in primary antibody at 4°C overnight.
3. Remove the pins from the microtiter plate and wash four times in a bath of PBS containing Tween-20 (0.1% v/v) for 10 min. Use fresh buffer for each wash.
4. Dilute an appropriate horseradish peroxidase-labeled secondary antibody conjugate (e.g., horseradish peroxidase-conjugated rabbit antimouse Ig is suitable for detecting murine primary antibodies) in conjugate diluent and dispense into the wells of a microtiter plate at 200 µL/well. Then incubate the washed pins in the secondary antibody solution for 1 h with agitation.
5. Wash the pins four times as in **step 3**. Prepare ABTS substrate solution immediately before use and dispense into the wells of a microtiter plate at 200 µL/well. Immerse the pins in the substrate solution in the correct orientation and allow to incubate for 45 min with agitation. The reaction can be stopped before the time is elapsed, if it appears that the reaction will give an optical density (OD) of 2 or greater, by removing the pins from the wells and then allowing the microtiter plate to shake for a further 15 min to allow full color dispersion. The OD of each well is determined spectrophotometrically at a wavelength of 405 nm.
6. Bound antibodies can be removed from the pins by sonication in disruption buffer at 60°C for 2 h, followed by repeated rinses in distilled water at 60°C and methanol (two times). The efficiency of the cleaning procedure should be tested by repeating **steps 4** and **5**. Absorbance levels above background indicate that antibody remains bound to the pins and further cleaning is required. Once the pins are clean, they should be sonicated for 30 min followed by rinsing as detailed just above (*see* **Note 17**).

3.4. Purification of Antibodies Using Peptide Affinity Chromatography (see Notes 12–15 and Fig. 3)

1. Prepare Sepharose-peptide affinity matrices and pack columns according to the manufacturer's instructions (*see* **Notes 12** and **13**).
2. Equilibrate the columns with 10 column volumes of PBSA at a flow rate of 1 mL/min.
3. Clarify hybridoma or bacterial culture supernatants by ultracentrifugation (40,000*g*, 1 h) and ultrafiltration (0.2 µm) and then store with 0.05% (w/v) sodium azide as a preservative.
4. Apply clarified supernatant to the column at a rate of 1 mL/min followed by washing with PBSA, to remove unbound material, until the trace from the ultraviolet monitor has returned to baseline.
5. **Optional:** Wash the column with 0.5 *M* NaCl (1 mL/min) to remove material that has bound to the column nonspecifically.

6. To achieve desorption of specifically bound, pure antibody, apply three column volumes of 3 *M* NaSCN to the column at a rate of 1 mL/min and finally return the column to PBSA (*see* **Notes 14** and **15**).
7. Antibody preparations desorbed using 3 *M* NaSCN must be desalted soon after elution from the affinity matrix. To achieve this, connect a gel filtration column containing a medium such as Sephadex G25 in series with the affinity column (**Fig. 3**).

4. Notes

1. Conjugation of peptides to carrier proteins can also be performed *in situ* in the well of a microtitier plate ***(11)***.
2. Synthetic peptide-carrier protein conjugates and synthetic branched-chain polypeptides provide highly characterized and reproducible sources of mucin-like antigenic material. However, bear in mind that these reagents are analogs of the natural antigen and do not possess the carbohydrate side chains that are a dominant characteristic of all mucins. Hence, results obtained in immunoassays, especially those involving the measurement of kinetic data, must be treated with caution.
3. The influence of carbohydrates on the recognition of peptide epitopes may be evaluated, at least to some extent, using synthetic glycopeptides rather than peptide alone. Glycopeptides can be produced by both chemical ***(12)*** and enzymatic ***(13,14)*** methods. These reagents have been of value in assessing the contribution of *O*-linked *N*-acetylgalactosamine (GalNAc) residues to mucin secondary structure and also in the studies to investigate the role of GalNAc residues in the binding of protein core antibodies ***(15)***. However, the glycosylation of mucin molecules is complex, and the production of higher-order synthetic mucin analogs with more than a single sugar at each glycosylation site is technically demanding.
4. The length of peptides synthesized seems to have no effect on the result obtained as demonstrated by two independent studies of anti-MUC1 protein core mAbs. One group used heptamers spanning the tandem repeat domain and overlapping each other by six amino acids ***(16)***, and the other used octamers overlapping each other by seven amino acids ***(17)***. Twelve antibodies were analyzed in total and the three that were analyzed in both studies gave identical minimum binding units.
5. In the omission analysis approach, a series of peptides are synthesized based on an epitope sequence. In each consecutive peptide, a single residue is omitted from the sequence. This allows the role of individual residues to be assessed. For example, an omission analysis series covering the immunodominant region of the MUC1 protein core may be synthesized. If an antibody were allowed to interact with this series of peptides, those sequences that produced a loss in binding compared with the parent sequence can be identified as part of a peptide in which an essential residue has been omitted. Antibody binding is maintained in those peptides in which the epitope is complete.
6. In the substitution analysis approach, each residue is replaced in turn with another amino acid. This amino acid is normally alanine, but for cases in which an alanine already exists at that substitution position, the residue can be replaced with glycine. The information provided when an antibody is allowed to react with this set of peptides is similar to that of omission analysis; however, in this case, the spatial arrangement of the respective epitope residues more closely resembles the native sequence.
7. RNET analysis offers the most critical and informative method for probing a peptide epitope in the MUC1 protein core. In this approach, a set of peptides with sequences based on a short minimum binding sequence or epitope are synthesized on the heads of

pins. Each residue within the epitope is systematically replaced by each of the other naturally occurring amino acids to provide a library of peptides with mutations at specific points. For example, for a tetrapeptide epitope consisting of the amino acids RPAP, the RNET would consist of four sets of 20 peptides with the general sequences: XPAP, RXAP, RPXP, and RPAX, where X is any naturally occurring amino acid.

8. Using RNET analysis, it is possible to identify residues that cannot be replaced by any other and thus are essential for antibody binding. Such residues almost certainly have a role in forming the correct three-dimensional structure of the epitope and may also provide the forces between epitope and paratope required to maintain binding. Conversely, it is possible to identify residues that are present in the epitope but can be replaced by other amino acids of similar size and charge. These can be assumed to play no direct role in antibody binding but may serve to restrict the conformation of the epitope so that the critical contact residues are presented in the optimal orientation for antibody binding.
9. RNET analysis has also provided information on the role of residues flanking the epitope that are not involved in antibody binding but are required for epitope presentation. Briggs et al. ***(18)*** studied the HMFG-2 antibody, which defines the minimum binding unit DTR, using a set of RNET peptides based on a PDTR tetrapeptide. These studies revealed that the antibody required the presence of the proline residue at position 1 for binding even though this residue was not found to be part of the minimum binding unit using Pepscan.
10. Care should be taken when interpreting the results of assays using dipeptides since ELISA signals from the best binding pairs may be low compared with those obtained for "complete" epitopes ***(19)*** In addition, the very nature of the library facilitates the binding of any antibody, including the second antibody conjugate used as a reporter molecule in the ELISA procedure. Careful control experiments must therefore be performed in order to allow differentiation of signals produced by binding of the test antibody and false-positive signals caused by the reporter antibody binding directly to the pins.
11. A dipeptide library may also be used to probe the specificity of antibodies against proteins whose primary structure is unknown as well as to identify peptide mimotopes for antibodies whose antigen is not protein in nature.
12. A column comprising of MUC1 immunodominant peptide conjugated to Sepharose 4B was able to adsorb in excess of 40 mg of IgG/mL of affinity matrix, resulting in a concentration factor of about 1000. Antibodies prepared in this way were found to be pure enough for administration to humans ***(9)***. A matrix such as this one has since been used for the purification of recombinant anti-MUC1 Fv fragments ***(20,21)*** as well as human anti-MUC1 antibodies from serum (Murray, A., unpublished findings).
13. A detailed study of the fine specificity of the reaction of an antibody with its antigen can lead to the design of peptide ligands with improved purification performance. For instance, it has been found that the C595 mAb has higher affinity for a synthetic MUC1 glycopeptide compared with the naked peptide. This information has been used to design a matrix consisting of a diglycosylated MUC1 glycopeptide linked to Sepharose, which demonstrates improved antibody yields in affinity chromatography experiments ***(12)***.
14. A variety of reagents can be used to achieve desorption of immunoreactive antibody. Consult standard chromatography text for further details.
15. NaSCN gradient elution provides a simple method for determining the relative affinities of a series of antibodies for a matrix or a single antibody for different matrices ***(12)***. The technique relies on the observation that a higher-affinity interaction would require a greater concentration of chaotrope to effect desorption of antibody. The use of an auto-

mated fast protein liquid chromatography system allows the production of accurate gradients that can be measured with a conductivity monitor. Manual gradient elution requires more effort and patience but can be achieved successfully with practice.

16. Distillation in vacuo results in high-purity DMF, which can be stored for up to 2 wk if kept under nitrogen in dark bottles at 4°C. Alternatively, stand over an activated molecular sieve (4 Å) for several days and then filter off the DMF. Use within 2 d.
17. A major advantage of the technique is that the peptides are covalently bound to the gears, so that harsh conditions such as sonication in SDS and mercaptoethanol can be used to remove bound antibodies after testing without affecting the peptide. Hence the peptide libraries can be reused to test other antibodies. In our experience, these immobilized peptide arrays have been used to test 50 or more antibodies with no loss in activity.

Reference

1. Allen, A. (1983) Mucus - a protective secretion of complexity.*TIBS*. **8,** 169–173.
2. Hayes, D. F., Zurawski, Jr., V. R. and Kufe, D. W. (1986) Comparison of circulating CA15-3 and carcinoembryonic antigen levels in patients with breast cancer. *J. Clin. Oncol.* **4,** 1542–1550.
3. Robertson, J. F. R., Pearson, D., Price, M. R., Selby, C., Blamey, R. W., and Howell, A. (1991) Objective measurement of therapeutic response in breast cancer using tumour markers. *Br. J. Cancer* **64,** 757–763.
4. Perkins, A. C., Symonds, I. M., Pimm, M. V., Price, M. R., Wastie, M. L., and Symonds, E. M. (1993) Immunoscintigraphy of ovarian carcinoma using a monoclonal antibody (In-111-NCRC-48) defining polymorphic epithelial mucin. *Nucl. Med. Commun.* **14**, 578-586.
5. Kunkler, R. B, Bishop, M. C., Green, D. J., Pimm, M. V., Price, M. R., and Frier, M. (1995) Targeting of bladder cancer with radiolabelled monoclonal antibody NCRC-48—a novel approach for intravesical therapy.*Br. J. Urol.* **75,** 9–17.
6. Biasson, L., Granowski, M., Carroll, M. J., Mather, S. J., Howell, R., Ellison, D., MacNeil, F. A., Wells, C. A., Carpenter, R. and Britton, K. E. (1998) 99mTc-labelled SM3 in the preoperative evaluation of axillary lymph nodes and primary breast cancer with change detection statistical processing as an aid to tumour detection. *Br. J. Cancer* **77,** 131–138.
7. Hudecz, F. and Price, M. R. (1992) Monoclonal antibody binding to peptide epitopes conjugated to synthetic branched chain polypeptide carriers. *J. Immuol. Meth.* **147,** 201–210.
8. Geysen, H. M., Rodda, S. J., Mason, T. J., Tirbbick, G., and Schoofs, P. J. (1987) Strategies for epitope analysis using peptide synthesis. *J. Immunol. Meth.* **102,** 259–274.
9. Price, M. R., Sekowski, M., and Tendler, S. J. B. (1991) Purification of anti-epithelial mucin monoclonal antibodies by epitope affinity chromatography. *J. Immunol. Methods* **139,** 83–90.
11. Price, M. R., Sekowski, M., Hooi, D. S. W., Durrant, L. G., Hudecz, F., and Tendler, S. J. B. (1993) Measurement of antibody binding to antigenic peptides conjugated in situ to albumin coated microtitre plates. *J. Immunol. Methods* **159,** 277–281.
12. Murray, A., Sekowski, M., Spencer, D. I. R., Denton, G. and Price, M. R. (1997) Purification of monoclonal antibodies by epitope and mimotope affinity chromatography. *J. Chromatog. A* **782**, 49–54.
13. Nishimori, I., Perini, F., Mountjoy, K. P., Sanderson, S. D., Johnson, N., Cerny, R. L., Gross, M. L., Fontenot, J. D., and Hollingsworth, M. A. (1994) *N*-Acetylgalactosamine glycosylation of MUC1 tandem repeat peptides by pancreatic tumour cell extracts. *Cancer Res.* **54**, 3738–3744.

14. Stadie, T. R. E., Chai, W., Lawson, A. M., Byfield, P. G. H., and Hanisch, F-G. (1995) Studies on the order and site specificity of GalNAc transfer to MUC1 tandem repeats by UDP-GalNAc:polypeptide *N*-acetylgalctosaminyltransferase from milk or mammary carcinoma cells. *Eur. J. Biochem.* **229**, 140–147.
15. Spencer, D. I. R., Price, M. R., Tendler, S. J. B., De Matteis, C. I., Stadie, T. and Hanisch, F.-G. (1996) Effect of glycosylation of a synthetic MUC1 mucin-core-related peptide on recognition by anti-mucin antibodies. *Cancer Lett.* **100**, 11–15.
16. Price, M. R., Hudecz, F., O'Sullivan, C., Baldwin, R. W., Edwards, P. M. and Tendler, S. J. B. (1990) Immunological and structural features of the protein core of human polymorphic epithelial mucin. *Mol. Immunol.* **27**, 795–802.
17. Burchell, J., Taylor-Papadimitriou, J., Boshell, M., Gendler, S., and Duhig, T. (1989) A short sequence, within the amino acid tandem repeat of a cancer associated mucin, contains immunodominant epitopes. *Int. J. Cancer* **44**, 691–696.
18. Briggs, S., Price, M. R., and Tendler, S. J. B. (1991) Immune recognition of linear epitopes in peptide fragments of epithelial mucins. *Immunology* **73**, 505–507.
19. Briggs, S., Price, M. R., and Tendler, S. J. B. (1993) Fine specificity of antibody recognition of carcinoma-associated epithelial mucins: antibody binding to synthetic peptide epitopes. *Eur. J. Cancer* **29A**, 230–237.
20. Davies, G. M., Bösze, S., Hudecz, F., Price, M. R., and Tendler, S. J. B. (1994) Characterisation of a recombinant Fv fragment of anti-MUC1 antibody HMFG1. *Cancer Lett.* **82**, 179–184.
21. Denton, G., Sekowski, M., Spencer, D. I. R., Hughes, O. D. M., Murray, A., Denley, H., Tendler, S. J. B., and Price, M. R. (1997) Production and characterisation of a recombinant anti-MUC1 scFv reactive with human carcinomas. *Br. J. Cancer* **76**, 614–621.
22. Petrakou, E., Graves, C. R. L., von Mensdorff-Pouilly, S., Robertson, J. F. R., and Price, M. R. (1997) Preliminary studies on the binding of human autoantibodies to the MUC1 antigen. *Int. J. Oncol.* **11**, 902.

13

Mucin Domains to Explore Disulfide-Dependent Dimer Formation

Sherilyn L. Bell and Janet F. Forstner

1. Introduction

The viscoelastic properties needed for the protective functions of secretory mucins are in part conditional on the capacity of mucin macromolecules to form linear polymers stabilized by disulfide bonds. The individual mucin monomers have a distinctive structure, consisting of a long central peptide region of tandem repeat sequences, flanked by cysteine-rich regions at each end, which are presumed to mediate polymerization. Secretory mucins contain approx 60–80% carbohydrate, with extensive *O*-glycosylation in the central tandem repeat regions, and *N*-linked oligosaccharides in the peripheral regions *(1)*.

The ability of mucin peptides to form large polymers, combined with their extensive posttranslational glycosylation and sulfation, results in complexes that reach molecular masses in excess of 10,000 kDa *(2)*. This leads to difficulties in resolving mucins by sodium dodecyl sulfate-polyacrylamide gel electrophoresis (SDS-PAGE) or agarose gel electrophoresis, because both unreduced and reduced mucin samples are capable of only limited movement. A related difficulty inherent in analyzing mucins lies in their strong negative charge owing to sialic acid and sulfate content *(3)*. Migration through polyacrylamide gels becomes more influenced by charge than by mass. The result is that interpretations of the size of mucin from electrophoretic mobility are not as straightforward as with other proteins.

Hypotheses concerning the regions of secretory mucins that could be involved in the initial dimer formation have centered on the terminal cysteine-rich, poorly glycosylated domains. Indirect evidence that these domains are involved has been shown by treating mucins with proteolytic enzymes and reducing agents that act on these regions, and noting a resultant decrease in the size of mucin and gel formation *(4–6)*. Intriguingly, more indirect evidence was found when database searches identified a functionally unrelated protein, von Willebrand factor (vWF) (which also forms S-S–dependent polymers), to exhibit a mucinlike pattern of cysteine residue align-

From: *Methods in Molecular Biology, Vol. 125: Glycoprotein Methods and Protocols: The Mucins*
Edited by: A. Corfield © Humana Press Inc., Totowa, NJ

ments in its N- and C-terminal domains *(7,8)*. The function of vWF to cause aggregation of platelets is also dependent on its ability to polymerize via disulfide bonds. More important for the present report, the DNA encoding vWF was expressed successfully in heterologous cells and shown to undergo S-S–dependent dimer and multimer formation ***(9,10)***. The minimum component necessary for initial dimerization is a region of approx 150 amino acids at the C-terminal end ***(11)***. A similar vWF "motif" has now been recognized near the C-terminus of several secretory mucins, including frog integumentary mucin FIM-B.1 *(7)*, bovine submaxillary mucin ***(12)***, porcine submaxillary mucin ***(13)***, human and rat intestinal mucin MUC2 ***(8,14)***, human tracheobronchial mucin MUC5AC *(15)*, human gall bladder mucin MUC5B ***(16,17)***, and human gastric mucin MUC6 *(18)*. Since these mucins are known to form oligomers in vivo, their shared C-terminal motif may also be involved in forming mucin dimers. Dimerization of mucin molecules represents the crucial first step in the transformation of individual mucin molecules into gel structures.

In this chapter, we describe a domain construct and expression approach used to examine directly whether the C-terminal domain of rat intestinal Muc2 is capable of dimerization through its cysteine residues. This method avoids, to a large extent, the numerous difficulties of dealing with full-length mucins, since the domain peptide is expressed as a relatively small, less glycosylated monomer or dimer. The principle is that DNA encoding the domain of interest is ligated to a known epitope sequence (for detection by a suitable antibody), and to a signal peptide sequence (to ensure secretion), by recombinant polymerase chain reaction (PCR) strategies. The resulting construct is ligated to an expression vector for transfection into heterologous cells. Once the expected peptide has been translated and processed, dimerization by disulfide bond formation is shown by comparing the sizes of immunoreactive, thiol-reduced and nonreduced products in cells and media by SDS-PAGE and Western blotting. The use of specific antibodies to various regions of the domain can provide assurance that the domain is expressed in an intact form or, alternatively, has been proteolytically processed during dimerization. After establishing the dimerization capability of the isolated C-terminal domain of rat Muc2, we describe methods for examining the role of glycosylation in dimerization by manipulating the system with inhibitors of glycosylation and/or deglycosylating enzymes.

2. Materials

2.1. Synthesis of Domain Constructs Using Recombinant PCR

1. PCR reagents: buffer, dNTPs, $MgCl_2$, *Taq* DNA polymerase obtained from Perkin Elmer (Foster City, CA).
2. Primer 1 (*see* **Fig. 1**). This is a sense primer containing an *Xba*I site to facilitate cloning, and also encoding part of the rat Muc2 signal peptide:
   a. 5'-CGTCTAGAATGGGGCTGCCACTAGCTCGCCTGGTGGCT-3'.
3. Primer 2 (*see* **Fig. 1**). The antisense primer containing signal peptide and "linker" sequence to be paired with primer 1:
   a. 5'-CACAGTTAGATTCCAGCCCTTGGCTAAGGCCAGGACTAGGCACACAG-3'.
4. Primer 3 (*see* **Fig. 1**). This primer specifies the "linker" sequence and primes the 5' end of the target domain of rat Muc2:
   a. 5'-GGCTTGGAATCTAACTGTGAAGTTGCTGC-3'.

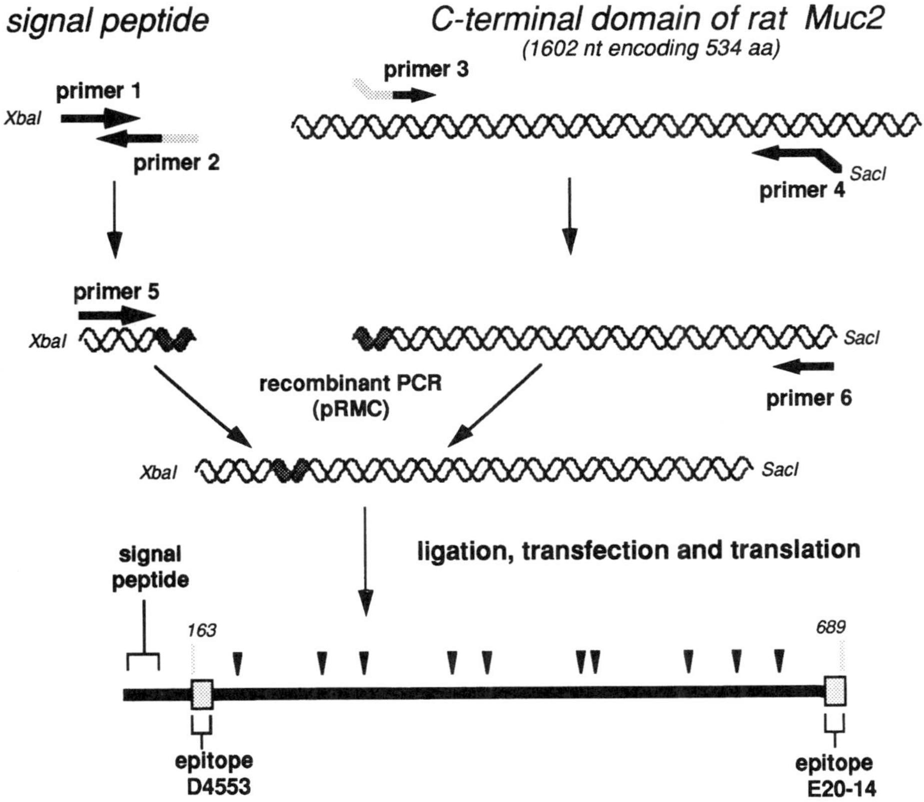

Fig. 1. Schematic showing the synthesis of construct pRMC and its expression in COS cells. The protocol is described in **Subheadings 3.1.–3.3.** Oligonucleotide primers are designed that facilitate the synthesis and joining of DNA sequence coding for the signal peptide and carboxyl-terminal 534 amino acids of rat Muc2 via recombinant PCR. The resulting construct is then subcloned into the expression vector pSVL for expression in COS cells.

5. Primer 4 (*see* **Fig. 1**). The antisense primer encoding the 3' end of rat Muc2 with an additional *Sac*I site sequence for cloning:
   a. 5'-CGAGCTCCTATCACTTCCTTCCTAGAAGCCG-3'.
6. Clone MLP-3500, which encodes the C-terminal 1121 amino acids of rat Muc2, was used as the DNA template *(8)*.
7. Outer primers 5:
   a. 5'-CGTCTAGAATGGGGCTGC-3'.
8. Antisense primer 6 (*see* **Fig. 1**):
   a. 5'-CGAGCTCCTATCACTTCC-3'.
9. Thermal cycler such as Perkin Elmer DNA Thermal Cycler 480.

2.2. Ligation of Construct to Transfection Vector

1. Invitrogen TA cloning kit (Invitrogen, Carlsbad, CA).
2. Transfection vector pSVL available from Pharmacia (Uppsala, Sweden).

3. Restriction enzymes *Sac*I, *Xba*I with supplied incubation buffer(s).
4. T4 DNA ligase and ligase buffer (Boehringer Mannheim, Mannheim, Germany).
5. Subcloning efficiency competent DH5α *Escherichia coli* in 50-µL aliquots.
6. Convection incubator maintained at 37°C.
7. DNA maxiprep columns (Qiagen, Chatsworth, CA).
8. Luria broth (LB) plates containing 50 µg/mL of ampicillin ***(19)***.
9. Agarose gels containing 0.5 µg/mL of ethidium bromide.
10. *Hin*dIII-digested DNA λ markers for size and quantity estimation.
11. 1X TAE: 0.04 *M* Tris base, 1 m*M* EDTA, 1.14 mL/L glacial acetic acid.

2.3. Transfection

1. COS-1 or COS-7 cell line obtained from American Type Culture Collection, (Rockville, MD).
2. Dulbecco's modified Eagle's medium (Gibco-BRL, Gaithersburg, MD) supplemented with 10% fetal bovine serum (FBS) (CanSera, Etobicoke, ON, Canada) and with 100 U/mL of penicillin and 100 µL of streptomycin (Gibco-BRL).
3. Hemocytometer.
4. Lipofectamine (Gibco-BRL).
5. Cell culture incubator to maintain an atmosphere of 37°C and 5% CO_2.
6. Transfection efficiency reporter such as pCMVßGAL or luciferase systems.

2.4. Harvesting of Transfected Cells

1. 2X Laemmli SDS sample buffer: 125 m*M* Tris-HCl, pH 6.8, 20% glycerol, 4% SDS, 0.005% bromophenol blue, plus or minus 1,4-Dithiothreitol (DTT) to give a final concentration of 10 m*M*.
2. Filter concentrators such as Centricon with a molecular weight cutoff of 30 kDa.
3. Phenylmethylsulfonylfluoride (PMSF).
4. Beckman J2-21 centrifuge, JA-20.1 rotor (Beckman Instruments, Palo Alto, CA).

2.5. SDS-PAGE and Western Blot Analysis

1. Tris-glycine polyacrylamide gels (precast) (Novex, San Diego, CA).
2. Gel electrophoresis apparatus (Novex Xcell II Mini Cell).
3. Gel running buffer: 25 m*M* Tris, 192 m*M* glycine, 0.1% SDS, pH 8.3.
4. Prestained protein standards.
5. Transfer buffer: 25 m*M* Tris, 192 m*M* glycine, 20% methanol, pH 8.3, store at 4°C.
6. Transfer apparatus such as Mini Trans-Blot (Bio-Rad, Hercules, CA).
7. Nitrocellulose membrane, cut to the size of the gel.
8. Blotting paper (0.33 mm).
9. Tris-buffered saline (TBS): 20 m*M* Tris base, 137 m*M* NaCl, final pH 7.6, and TBS with 0.1% Tween-20 added (TBST).
10. Blocking solution: 3% bovine serum albumin (BSA) in TBS.
11. Primary incubation solution: 1:1000 dilution (v/v) of rabbit polyclonal antibody raised against the deglycosylated C-terminal "link" glycopeptide ***(20)*** or against synthetic peptides D4553 corresponding to a 14 amino acid segment in the mucin domain, and E20-14, corresponding to the C-terminal 14 amino acids of the mucin domain ***(21)*** (*see* **Fig. 1**) in TBS and 0.1% BSA.
12. Secondary incubation solution: 1:10,000 dilution of goat antirabbit IgG alkaline phosphatase conjugate in TBS and 0.1% BSA.
13. Alkaline phosphatase detection system: equilibration solution (100 m*M* Tris-HCl, 100 m*M* NaCl, 50 m*M* $MgCl_2$, pH 9.5; store at 4°C), reaction solution (same as equilibration but

with the addition of 6.6 mg of 4-nitro blue tetrazolium chloride and 1.65 mg of 5-bromo-4-chloro-3-indolyl-phosphate), and stop solution (10 m*M* Tris-HCl, 1 m*M* EDTA, pH 8.0). Nitro blue tetrazolium (NBT) and 5-bromo-4-chloro-3-indolyl-phosphate (BCIP) are available from Boehringer Mannheim at 100 and 50 µg/µL concentrations, respectively.

2.6. The Role of Glycosylation

1. Tunicamycin (Sigma, St. Louis, MO) in DMEM, filter sterilized.
2. 20 m*M* benzyl 2-acetamido-2-deoxy-α-D-galactopyranoside (benzyl-α-GalNAc) (Sigma), filter sterilized.
3. Peptide-N^4-(acetyl-ß-glycosaminyl) asparagine amidase (*N*-glycosidase F, EC 3.5.1.52) (Boehringer Mannheim).
4. Nonidet P-40.
5. *N*-acetylneuraminidase from *Vibrio cholerae* (EC 3.2.1.18), (Boehringer Mannheim).
6. Lysis buffer: 50 m*M* Tris-HCl, pH 7.5, 150 m*M* NaCl, 1% Nonidet P-40, and 1X Complete™ protease inhibitor cocktail (Boehringer Mannheim).
7. Protein A-Sepharose (Boehringer Mannheim).
8. Immunoprecipitation buffer: 20 m*M* Tris-HCl, pH 8.0, containing 0.1 *M* NaCl, 1 m*M* Na_2EDTA, 0.1 m*M* PMSF, with and without 0.5% Nonidet P-40.
9. Neuraminidase incubation buffer: 40 m*M* Tris-HCl, 4 m*M* $CaCl_2$, pH 7.8.

3. Methods

3.1. Synthesis of Rat Muc2 Domain Construct (pRMC) Using Recombinant PCR

Figure 1 shows the general scheme of the synthesis of construct pRMC. Two DNA constructs are synthesized encoding the entire rat Muc2 signal peptide and the C-terminal 534 amino acids, respectively. Each construct is synthesized to contain an identical region (**Fig. 1**, shaded area) at which a single strand of one construct is able to complement the other during the annealing cycle of the second PCR step *(22)*, resulting in a recombinant product (pRMC) that can be cloned, using the incorporated restriction sites, into a suitable vector such as pSVL. The detailed procedure is as follows:

1. Synthesize DNA fragments to be joined via recombinant PCR. The 5' fragment will encode the 534 amino acids of the C-terminal end of rat Muc2 and a 3' *Sac*I restriction site for subcloning. This domain of Muc2 can be detected by the antibodies anti-d-link, anti-D4553, and anti-E20-14 *(8,20)*, which span the length of the pRMC peptide product (**Fig. 1**).
2. Make up 100 µL of PCR samples containing 10 µL of 10X polymerase buffer, 2.5 m*M* $MgCl_2$, 400 µ*M* dNTPs, 200 ng of primers 1 and 2 or 3 and 4, 1 ng of template DNA (with only primers 3 and 4), and 5 U of *Taq* polymerase. Perform a PCR program in a suitable thermal cycler at 94°C for 5 min, followed by 30 cycles at 94°C for 1 min, 60°C for 1 min, and 72°C for 1 min, ending with an extension period of 72°C for 7 min.
3. Couple the signal peptide and RMuc2 PCR products. Make up 100-µL reaction mixtures containing 10 µL of 10X polymerase buffer, 2.5 m*M* $MgCl_2$, 400 µ*M* dNTPs, and approx 50 ng of each PCR product as templates. Denature the DNA in a thermal cycler for 1 min at 94°C, followed by a 5-min incubation at 55°C to allow both templates to anneal. Add 5 U of *Taq* polymerase, and incubate the reaction at 72°C to allow extension of the linked templates. Add 2 µg each of primers 5 and 6, and continue the PCR with 30 cycles at 94°C for 30 s, 50°C for 1 min and 72°C for 2 min, ending with a 72°C extension for 7 min.

4. Check the PCR product for size and purity by electrophoresis on a 1% agarose gel. The construct may be verified by sequencing at this point or after insertion into a vector.

3.2. Ligation of Construct to the Transfection Vector

TA cloning may be necessary if direct ligation is unsuccessful. In cases where subcloning is impeded by incomplete restriction enzyme digestion, a specialized vector containing 3' deoxythymidine (T) overhangs can be utilized to ligate the PCR product generated by Taq polymerase (which leaves complementary deoxyadenosine or A, overhangs). The ligated TA overhangs can then be digested without difficulty. Alternatively, it can be assumed from the beginning that TA cloning will be used, which would result in a small change in primer design. Specifically, it would not be necessary to incorporate *Sac*I and *Xba*I sites into the beginning primers, because these sites already exist in the TA cloning vector.

1. Using fresh recombinant PCR product (less than 1 d old), perform TA cloning as outlined in the protocol of the supplier (Invitrogen). Pick isolated white colonies and check for the correct insert by restriction digest of DNA minipreps with *Sac*I and *Xba*I.
2. Add 5 µL of recombinant PCR product or positive TA clone DNA to 3 µL of H_2O, 1 µL of incubation buffer, and at least 1 U of *Sac*I to a microfuge tube. Incubate for 1 h in a 37° C water bath. Add at least 1 U of *Xba*I to the reaction and incubate for a further hour at 37°C. Inactivate the enzymes according to the manufacturer's instructions.
3. Purify the *Sac*I/*Xba*I restriction digest product. We used electroelution from the agarose fragment ***(19)*** in TAE buffer for 1 h followed by precipitation in 0.1 vol of 3 *M* sodium acetate and 2 vol of ethanol. Electrophorese on a 1% agarose gel and quantitate the sample. Use sufficient quantities of pSVL and insert DNA to result in an approx 1:3 molar ratio, respectively. Bring the ligation reaction to 8 µL with H_2O, and then add 1 µL of the 10X ligation buffer supplied with the enzyme and 1 µL (1 U) of T4 DNA ligase. Incubate the ligation reaction at 16°C for 16 h.
4. Thaw competent DH5α cells on ice, and to each 50-µL aliquot add 2 µL of the ligation reaction mixture (or H_2O for mock-transformation control), stir gently with a pipet tip, and incubate the cells on ice for 30 min.
5. Heat shock the reactions in a 37°C water bath for 20 s, and then incubate them on ice for 2 min.
6. Add 950 µL of LB to each reaction tube and incubate on a shaker at 37°C for 1 h.
7. Add 50 and 200 µL from each tube of transformed cells to warmed LB plus ampicillin plates. Distribute the cells over the plate evenly with a sterile glass spreader, allow plates to dry, and leave in a 37°C convection incubator for 16 hr.
8. After confirming that no growth has occurred on control plates of mock-transformed DH5α, pick at least five positive, isolated colonies. Check DNA minipreps of these positive colonies by restriction digestion with *Sac*I and *Xba*I. The positive clones may be sequenced at this point to confirm that the correct sequence is contained within the plasmid.
9. Prepare large-scale amounts of transfection-quality construct pRMC and control vector pSVL by resin purification such as with the Qiagen Maxiprep kit.

3.3. Transfection

The transfection system uses the expression vector pSVL, which is ideally suited for transfection into COS cells owing to the production of T-antigen, which increases pSVL expression ***(23)***.

1. Twenty-four hours before the transfection, seed COS cells into 3.5-cm tissue culture dishes at a density of 8×10^6 cells, as counted by a hemocytometer.
2. Pilot experiments are necessary to establish the ideal ratio of DNA to Lipofectamine. For this system, add 2 μg of pRMC or pSVL control to 100 μL of DMEM. Add 10 μL of Lipofectamine to 100 μL of DMEM in a separate tube.
3. Add the contents of the first tube containing DNA to the second tube containing Lipofectamine, and incubate the tubes at room temperature for 20 min. While the incubation is ongoing, remove the medium from all dishes of COS cells and wash with 1 mL of DMEM (containing no antibiotics or FBS). Drain the dishes thoroughly.
4. Add 800 μL of DMEM to the DNA-Lipofectamine mixture, pipet to mix, and add to the dish of COS cells. Leave all dishes in the tissue culture incubator at 37°C for 5 h.
5. Replace the mixture with 2 mL of DMEM plus 10% FBS (do not include antibiotics). Leave the dish in the tissue culture incubator for 43 h.
6. Remove serum-containing DMEM and wash COS cells thoroughly with DMEM without serum. Replace medium with 2 mL of DMEM without serum and incubate at 37°C for another 24 h. If it is critical to collect conditioned media from the entire incubation period, OptiMEM (Gibco-BRL) may be substituted for DMEM to avoid later problems associated with SDS-PAGE involving FBS (e.g., distortion of lower mobility proteins during electrophoresis and antibody crossreactivity).

3.4. Harvesting of Transfected Cells

1. After establishing the success of transfection with a reporter system, remove COS cells transfected with pRMC and pSVL control vector from the incubator, collect all medium, and wash the cells with 1 mL of phosphate-buffered saline (PBS).
2. Add 60 μL of 2X Laemmli sample buffer to each dish, and quickly scrape the lysates into microfuge tubes. The lysates may be stored at –70°C at this time.
3. Add PMSF to the conditioned medium to give a concentration of 0.1 m*M*, and concentrate using spin columns, such as Centricons, to attain a final volume of approx 100 μL. Medium may be stored at –70°C at this time.

3.5. SDS-PAGE and Western Analysis

The transfected COS cells are harvested for separation of lysate components on polyacrylamide gels. The percentage of polyacrylamide used will vary with the size of the expected product and the desired separation. Most experiments in the present study were performed using 8% gels.

1. Boil the cell lysates for 3 min to reduce viscosity, divide each sample equally between two microfuge tubes, and add DTT to one of each pair of samples to give a final concentration of 10 m*M*. Boil the samples for a further 2 min, briefly spin to collect droplets, and load each sample into a well of an 8% polyacrylamide gel.
2. Add 15 μL of 2X Laemmli sample buffer to 15-μL aliquots of conditioned media (two samples for each vector transfected) and reduce one of each pair with DTT (final concentration 10 m*M*). Boil, spin, and pipet samples into wells of an 8% polyacrylamide gel.
3. Add prestained molecular mass standards to one well and 2x Laemmli sample buffer to any remaining empty wells (*see* **Note 3**).
4. Electrophorese at 125 V for 100 min, or until the bromophenol blue dye reaches the bottom of the gel.
5. Transfer the separated proteins from the gel to a nitrocellulose membrane at a constant voltage of 100 V for 1 h using buffer conditions suitable for the transfer apparatus; the

Bio-Rad Mini Trans-blot requires buffer maintained at 4°C. The transfer setup is detailed in **ref. *19***.

6. Remove and submerge the membrane in blocking solution. Incubate on a rotator at room temperature for 1 h.
7. Wash the membrane five times in TBST at room temperature, 5 min per wash.
8. Incubate the membrane for 16 h at 4°C in primary incubation solution. We have used the anti-d-link, anti-D4553, and anti-E20-14 antibodies (*see* **Subheading 2.5., step 11**) in our experiments, with equivalent results. Wash the membrane as described in **step 5**.
9. Incubate the membrane for 1 h in secondary incubation solution. Wash the membrane as in **step 5**.
10. After a 3-min equilibration, develop the membrane in reaction solution containing NBT and BCIP until specific bands are visible. Keep all membranes protected from light during the color development.
11. Submerge the membrane in stop solution and dry on blotting paper. Compare control and experimental samples (**Fig. 2**) for the appearance of immunospecific bands that increase in mobility (i.e., decrease in size) upon treatment with DTT (*see* **Note 4**).

3.6. The Role of Glycosylation

3.6.1. Tunicamycin

1. Repeat **Subheadings 3.1.–3.5.** with the addition of 0.5 µg/mL of tunicamycin to the DMEM used in **Subheading 3.3. steps 5** and **6**. Harvest and analyze cells and conditioned media as outlined previously (see Note 5).

3.6.2. N-Glycosidase F

1. Repeat **Subheadings 3.1.–3.3.** inclusive.
2. Collect and concentrate the conditioned medium as described in **Subheading 3.4., step 3**. Add 1 U of *N*-glycosidase F (or H_2O as a control) to each 15-µL aliquot and incubate 16 h at 37°C in a water bath. Analyze by SDS-PAGE as described in **Subheading 3.5., steps 2–10**.
3. Wash the COS cells in PBS, scrape the cells into 500 µL of PBS and microfuge the suspension at 4°C to form a pellet. Resuspend the pellet in 50 µL of 1% SDS, and boil 1 min to denature.
4. Dilute the suspension with Nonidet P-40 to give final concentrations of SDS and Nonidet P-40 of 0.1 and 0.5 %, respectively. *N*-glycosidase F is deactivated by higher concentrations of SDS.
5. Add 1 U of *N*-glycosidase F to each cell lysate and medium sample (H_2O to controls) and incubate for 16 h at 37°C in a water bath.
6. Vacuum-dry the samples and reconstitute in 60 µL of 2X Laemmli sample buffer.
7. Analyze the conditioned media and cell lysates by SDS-PAGE and Western blotting (Fig. 3) as described in **Subheading 3.5.** (*see* **Note 6**).

3.6.3. Benzyl-α-GalNAc

1. Repeat **Subheadings 3.1.–3.5.** with the addition of benzyl-α-GalNAc at a final concentration of 2 m*M* to each test dish of COS cells. Harvest and analyze cell lysates and conditioned media as outlined previously (*see* **Note 7**).

3.6.4. N-Acetylneuraminidase

1. Repeat **Subheadings 3.1.–3.3.** inclusive.
2. Collect and concentrate the conditioned medium as described in **Subheading 3.4., step 3**. Add 10 mU of *N*-acetylneuraminidase from *V. cholerae* (or H_2O as a control) to each 50-µL

aliquot and incubate 16 h at 37°C in a water bath. Analyze by SDS-PAGE as described in **Subheading 3.5., steps 2–10**.
3. Harvest COS cells in 200 µL of lysis buffer and microfuge at 4°C to remove cell debris.
4. Incubate the remaining supernatant with 4 µL of antibody (anti-D4553 was used in this study) for 1.5 h at 4°C followed by a further 2.5-h incubation at the same temperature with the addition of 50 µL of Protein A-Sepharose. Isolate the immunoprecipitates by centrifugation at 4°C for 5 min followed by washing the resultant pellet three times with 500 µL of immunoprecipitation buffer (the third wash without Nonidet P-40).
5. Analyze the conditioned media and cell immunoprecipitates by SDS-PAGE and Western blotting (**Fig. 3**) as described in **Subheading 3.5.** (*see* **Note 8**).

4. Notes

1. The procedures outlined above for specific domain expression in heterologous cells demonstrate that the C-terminal region of rat Muc2 can form disulfide-dependent dimers, and that *N*-glycosylation plays a significant role in dimerization. A general schematic reflecting the results using these methods is shown in **Fig. 4**. The reader is referred to Perez-Vilar et al. *(25)* for an earlier application of domain expression to show dimerization of the C-terminal domain of porcine submaxillary mucin. The domain approach was necessitated by the virtual impossibility of studying detailed structural changes in molecules as large and as viscous as typical full-length secretory mucins. As has been true for the elucidation of vWF physiology and disease-associated mutants of vWF, the expression of individual mucin domains and their secretory fate holds promise of enlarging our understanding of structure-function relationships of mucins in vivo.
2. Since the mucin domain forms dimers, the present approach could be extended to include larger constructs and/or other domains, e.g., constructs encoding both C- and N-terminal domains, to test whether dimers can expand into larger S-S–linked oligomers or multimers. Various embellishments are also possible, including the addition to constructs of commercially available tag epitopes at selected areas of the domain and immunoprecipitation of translated products with specific antiepitope antibodies. Crosslinking agents could also be added to examine the possibility that nonmucin proteins bind to specific mucin domains. The kinetics of posttranslational modifications of the domain and its secretion could be studied by performing pulse-chase experiments using radioactive precursors added to transfected cells. The strategic use of inhibitors during cell incubations has the potential to reveal information about the pre- and post-Golgi movement of mucin domains along the secretory pathway. Finally, truncation or site-specific mutagenesis of the initial domain constructs could be introduced to explore the role of selected amino acids, such as cysteines required for dimerization, or *N*-glycosylation sites involved in mediating the correct folding for dimerization.
3. We have noted some discrepancies in mobility versus size correlations of different commercial batches and sources of molecular mass standards, particularly in the range above 150 kDa. Thus domain product sizes should be viewed as relative rather than absolute.
4. **Figure 2A,B** shows the results of this protocol for cell lysates and media. Nonspecific bands may be present and can be eliminated from consideration using the pSVL vector control lanes for comparison (**Fig. 2A,B**, lanes 1 and 3). The sizes of specific bands reported herein are given with reference to Bio-Rad prestained molecular mass standards. (Estimated dimer sizes are higher, by up to 30 kDa using Novex standards, whereas monomeric size approximations remain unchanged.) In cell lysates (**Fig. 2A**), an immunospecific band is seen at approx 150 kDa (lane 2), which is replaced with an 88-kDa band

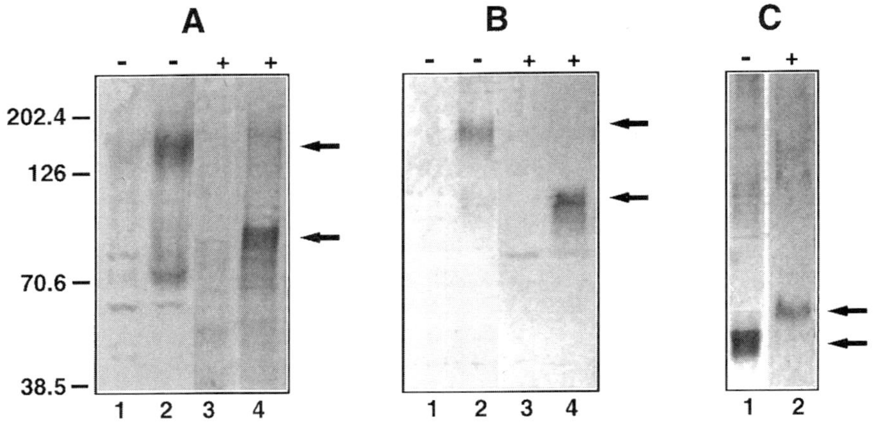

Fig. 2. SDS PAGE analysis of COS-1 cell translation products after transfection with pRMC, as outlined in **Subheading 3.5.** Western blots were performed using antibody to the deglycosylated "link" glycopeptide. "+" lanes refer to samples reduced with 10 m*M* DTT. **(A)** Thirty microliters of pRMC- (lanes 2 and 4) and pSVL control-transfected (lanes 1 and 3) cell lysates. Reduction produces a shift in mobility from a 150-kDa dimer to an 88-kDa monomer. **(B)** A similar shift, but of larger species, is seen using 30-μL aliquots of concentrated conditioned media from pRMC- (lanes 2 and 4) and pSVL-transfected (lanes 1 and 3) COS-1 cells. DTT reduction of samples (denoted +) causes a shift from 165 to 100 kDa. **(C)** Culture of COS-1 cells in 0.5 μL/mL of tunicamycin followed by SDS-PAGE analysis of the cell lysates reveals a highly mobile band, presumably tightly folded, at 45 kDa (lane 1), which yields a 62-kDa deglycosylated monomer on reduction. No dimer product is detected in lysates, and no monomer or dimer is detected in conditioned medium (not shown).

on thiol reduction (lane 4). These represent the mucin domain dimer and reduced monomer, respectively. In nonreduced samples there is also a band at 73 kDa. The identity of this species is not yet clear, but may represent misfolded or immature monomeric domain products. It converts to 88 kDa with reduction (**Fig. 2A**, lane 4). In the cell medium (**Fig. 2B**), lane 2 shows a secreted immunospecific product with a mobility equivalent to about 165 kDa (dimer) that disappears on reduction, leaving a 100-kDa species (lane 4) (monomer). Note that the apparent molecular masses of the secreted (medium) dimer and monomer are larger than the corresponding cell lysate species (i.e., 165 vs 150 for the dimer, 100 vs 88 for the reduced monomer). The explanation lies in a late glycosylation step, as noted in **Subheading 3.6.**

5. As seen in **Fig. 2C**, cells treated with tunicamycin exhibit a 45-kDa product (presumably a highly folded, nonglycosylated monomer) under nonreducing conditions (lane 1), and a 62-kDa band on reduction (lane 2). The calculated size of the expected translation product is 59 kDa (534 amino acids), which corresponds well with the observed 62-kDa band. No dimer form appears, indicating a prerequisite for *N*-glycosylation in domain dimerization. Because no bands were observed in the medium (not presented), it is clear that the nonglycosylated product cannot be secreted.
6. Results are shown in **Fig. 3A**, in which a mobility shift of the untreated control (lane 1) from 165 kDa to a band at 145 kDa (lane 2) is observed after treatment with *N*-glycosidase F.

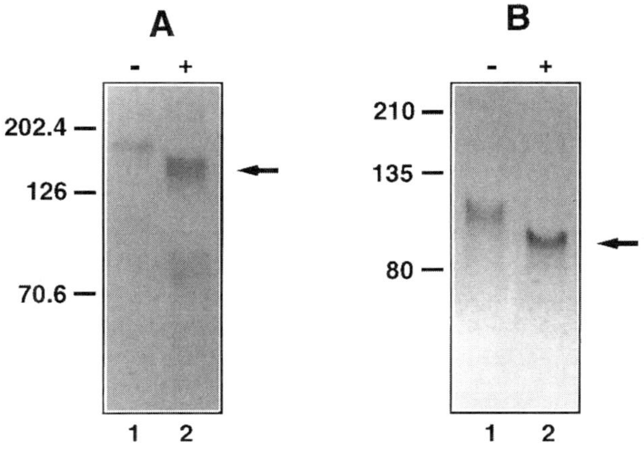

Fig. 3. COS-1 cells were transfected with pRMC and conditioned media incubated with (+) or without (–) 1 U/15 μL of *N*-glycosidase F **(A)** or 10 mU/50 μL of N-acetylneuraminidase **(B)**. In **(A)** the dimer shifted from 165 to 145 kDa. Gel **(B)** shows that the reduced monomer shifted from 100 to 88 kDa.

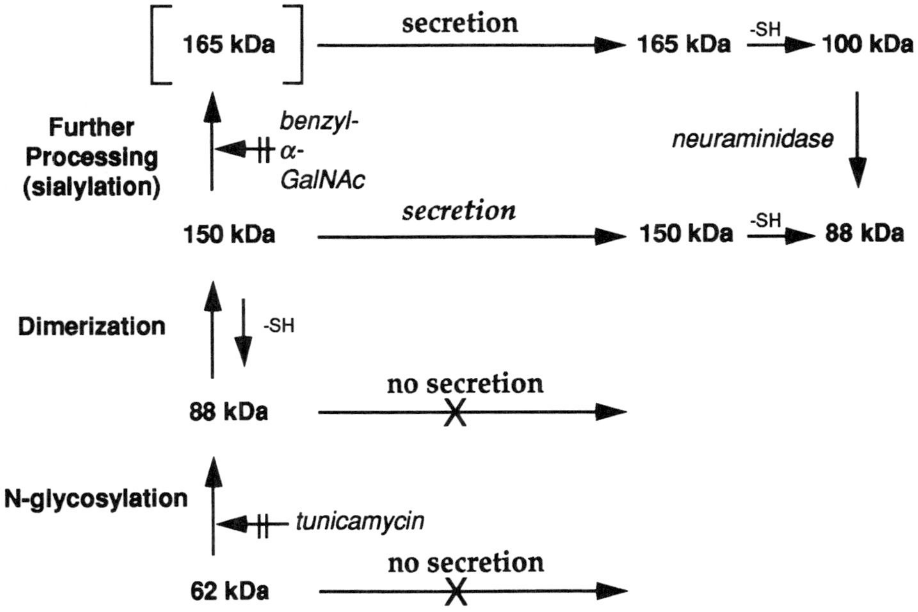

Fig. 4. Schematic showing the proposed interpretation of domain-processing events. A newly formed peptide of about 62 kDa undergoes *N*-glycosylation cotranslationally. The latter is inhibited by tunicamycin. N-glycosylation is mandatory for the 88-kDa monomer to dimerize via S-S bonds to form a 150-kDa dimer. Although the dimer is capable of being secreted, benzyl-α-GalNAc and neuraminidase experiments reveal that a further processing step occurs (sialylation) to form a slower-mobility dimer (165 kDa) that is secreted rapidly and thus detected only in the conditioned medium. The secreted dimer can be reduced to a 100-kDa monomer and desialylated to an 88-kDa species.

7. Benzyl-α-GalNAc has been reported to inhibit sialylation of glycoproteins *(24)* (*see* Chapter 22). Cell lysates gave products equivalent to those shown in **Fig. 2A**, lanes 2 and 4 (i.e., a dimer of 150 kDa, and a reduced monomer of 88 kDa). Media samples also showed dimers of 150 kDa and reduced monomers of 88 kDa (not presented). Thus, benzyl-α-GalNAc prevents the normal change (an apparent increase in molecular mass) of domain dimers when they are secreted.
8. Digestion of cell lysates and media with *N*-acetylneuraminidase did not change the mobility in cell lysates but decreased the medium-reduced monomer from 100 to 88 kDa (Fig. 3B, lane 2). Thus neuraminidase and benzyl-α-GalNAc both produce the same effect, namely a loss of sialic acid from secreted domain products. Collectively the deglycosylation experiments suggest that sialic acid is added to domain dimer oligosaccharides in the cells at a late stage of processing, immediately before secretion into the medium. The addition of sialic acid negative charge is responsible for the apparent increase in size of the media dimers relative to cell dimers (**Fig. 2A,B**).

Acknowledgments

Financial support for this study was obtained from the Canadian Cystic Fibrosis Foundation. Antibodies and advice were generously provided by members of the Forstner laboratory, Dr. Gongqiao Xu, and Dr. Ismat Khatri. DNA sequencing was performed by the Biotechnology Service Centre, The Hospital for Sick Children, Toronto, Canada.

References

1. Forstner, J. F. and Forstner, G. G. (1994) Gastrointestinal mucus, in *Physiology of the Gastrointestinal Tract* (Johnson, L. R., ed.), Raven Press, New York, pp. 1255–1283.
2. Allen, A. and Pearson, J. P. (1993) Mucus glycoproteins of the normal gastrointestinal tract. *Eur. J. Gastroenterol. Hepatol.* **5,** 193–199.
3. Tytgat, K. M. A. J., Swallow, D. M., Van Klinken, B. J.-W., Büller, H. A., Einerhand, A. W. C., and Dekker, J. (1995) Unpredictable behaviour of mucins in SDS/polyacrylamide-gel electrophoresis. *Biochem. J.* **310,** 1053,1054.
4. Pearson, J. P., Kaura, R., Taylor, W., and Allen, A. (1982) The composition and polymeric structure of mucus glycoprotein from human gallbladder bile. *Biochim. Biophys. Acta* **706,** 221–228.
5. Bell, A. E., Sellers, L. A., Allen, A., Cunliffe, W. J., Morris, E. R., and Ross-Murphy, S. B. (1985) Properties of gastric and duodenal mucus: effect of proteolysis, disulfide reduction, bile, acid, ethanol, and hypertonicity on mucus gel structure. *Gastroenterology* **88,** 269–280.
6. Sellers, L. A., Allen, A., Morris, E. R., and Ross-Murphy, S. B. (1987) Mechanical characterization and properties of gastrointestinal mucus gel. *Biorheology* **24,** 615–623.
7. Probst, J. C., Gertzen, E.-M., and Hoffmann, W. (1990) An integumentary mucin (FIM-B.1) from *Xenopus laevis* homologous with von Willebrand factor. *Biochemistry* **29,** 6240–6244.
8. Xu, G., Huan, L.-J., Khatri, I. A., Wang, D., Bennick, A., Fahim, R. E. F., Forstner, G. G., and Forstner, J. F. (1992) cDNA for the carboxyl-terminal region of a rat intestinal mucin-like peptide. *J. Biol. Chem.* **267,** 5401–5407.
9. Verweij, C. L., Hart, M., and Pannekoek, H. (1987) Expression of variant von Willebrand factor (vWF) cDNA in heterologous cells: requirement of the pro-polypeptide in vWF multimer formation. *EMBO J.* **6,** 2885–2890.

10. Voorberg, J., Fontijn, R., van Mourik, J. A., and Pannekoek, H. (1990) Domains involved in multimer assembly of von Willebrand factor (vWF): multimerization is independent of dimerization. *EMBO J.* **9,** 797–803.
11. Voorberg, J., Fontijn, R., Calafat, J., Janssen, H., Mourik, J. A. V., and Pannekoek, H. (1991) Assembly and routing of von Willebrand factor variants: the requirements for disulfide-linked dimerization reside within the carboxy-terminal 151 amino acids. *J. Cell Biol.* **113,** 195–205.
12. Bhargava, A. K., Woitach, J. T., Davidson, E. A., and Bhavanandan, V. P. (1990) Cloning and cDNA sequence of a bovine submaxillary gland mucin-like protein containing two distinct domains. *Proc. Natl. Acad. Sci. USA* **87,** 6798–6802.
13. Eckhardt, A. E., Timpte, C. S., Abernethy, J. L., Zhao, Y., and Hill, R. L. (1991) Porcine submaxillary mucin contains a cystine-rich, carboxyl-terminal domain in addition to a highly repetitive, glycosylated domain. *J. Biol. Chem.* **266,** 9678–9686.
14. Gum, J. R., Jr., Hicks, J. W., Toribara, N. W., Rothe, E.-M., Legace, R. E., and Kim, Y. S. (1992) The human MUC2 intestinal mucin has cysteine-rich subdomains located both upstream and downstream of its central repetitive region. *J. Biol. Chem.* **267,** 21,375–21,383.
15. Lesuffleur, T., Roche, F., Hill, A. S., Lacasa, M., Fox, M., Swallow, D. M., Zweibaum, A., and Real, F. X. (1995) Characterization of a mucin cDNA clone isolated from HT-29 mucus-secreting cells. *J. Biol. Chem.* **270,** 13,665–13,673.
16. Desseyn, J.-L., Aubert, J.-P., Seuningen, I. V., Porchet, N., and Laine, A. (1997) Genomic organization of the 3' region of the human mucin gene MUC5B. *J. Biol. Chem.* **272,** 16,873–16,883.
17. Keates, A. C., Nunes, D. P., Afdhal, N. H., Troxler, R. F., and Offner, G. D. (1997) Molecular cloning of a major human gall bladder mucin: complete C-terminal sequence and genomic organization of MUC5B. *Biochem. J.* **324,** 295–303.
18. Toribara, N. W., Ho, S. B., Gum, E., Gum, J. R., Jr., Lau, P., and Kim, Y. S. (1997) The carboxyl-terminal sequence of the human secretory mucin, MUC6. *J. Biol. Chem.* **272,** 16,398–16,403.
19. Sambrook, J., Fritsch, E. F., and Maniatis, T. (1989) *Molecular Cloning: A Laboratory Manual*, Cold Spring Harbor Laboratory, Cold Spring Harbor, New York.
20. Khatri, I., Forstner, G., and Forstner, J. (1993) Preparation of polyclonal antibodies to native and modified mucin antigens, in *Methods in Molecular Biology, Vol. 14: Glycoprotein Analysis in Biomedicine* (Hounsell, E. F., ed.), Humana Press, Totowa, NJ, pp. 225–235.
21. Xu, G., Forstner, G. G., and Forstner, J. F. (1996) Interaction of heparin with synthetic peptides corresponding to the C-terminal domain of intestinal mucins. *Glycoconj. J.* **13,** 81–90.
22. Higuchi, R. (1990) Recombinant PCR, in *PCR Protocols* (Innis, M., Gelfand, D. H., and Sninsky, J. J., eds.), Academic Press, Toronto, pp. 177–183.
23. Sprague, J., Condra, J. H., Arnheiter, H., and Lazzarini, R. A. (1983) Expression of a recombinant DNA gene coding for the vesicular stomatitis virus nucleocapsid protein. *J. Virol.* **45,** 773–781.
24. Delannoy, P., Kim, I., Emery, N., De Bolos, C., Verbert, A., Degand, P., and Huet, G. (1996) Benzyl-N-acetyl-α-D-galactosaminide inhibits the sialylation and the secretion of mucins by a mucin secreting HT-29 cell subpopulation. *Glycoconj. J.* **13,** 717–726.
25. Perez-Vilar, J., Eckhardt, A. E., and Hill, R. L. (1996) Porcine submaxillary mucin forms disulfide-bonded dimers between its carboxyl-terminal domains. *J. Biol. Chem.* **271,** 9845–9850.

VI

Carbohydrate Structural Analysis of Mucins

14

Monosaccharide Composition of Mucins

Jean-Claude Michalski and Calliope Capon

1. Introduction

Mucin oligosaccharides are constructed by monosaccharide addition to form common cores. This architecture limits the number of constituent monosaccharides. Monosaccharides commonly found in mucins may be divided into neutral (galactose [Gal]; fucose [Fuc]), hexosamines (*N*-acetylgalactosamine [GalNAc]; *N*-acetylglucosamine [GlcNAc]), and acidic compounds (sialic acids [NeuAc]). Additive heterogeneity comes from the possible substitution with aglycone residues such as sulfate, phosphate, or acetate groups. Prior to their analysis, monosaccharides must be released from the oligosaccharide chain by acidic hydrolysis. Monosaccharide composition can also be achieved on free oligosaccharide-alditols released from the native glycoprotein by reductive alkaline treatment (β-elimination). In this case, GalNAc is converted into *N*-acetylgalactosaminitol (GalNAc-ol). Different methods are available for the analysis of monosaccharides depending mainly on the amount of material available. Several techniques, such as gas-liquid chromatography (GLC) or high-performance liquid chromatography (HPLC), allow both quantitative and qualitative analysis of monosaccharide mixtures. Other chromatographic or electrophoretic procedures are described herein, but these only allow a rapid qualitative analysis of samples. Single separated monosaccharides may be further identified by physicochemical methods such as mass spectrometry (MS) or nuclear magnetic resonance.

1.1. Release and Identification of Sialic Acids

Sialic acids constitute a family of nine-carbon carboxylated sugars found in the external position on glycan chains. The diversity of sialic acids is generated by the presence of various substituents present on carbon 4, 5, 7, 8, and 9. The substituent on carbon 5 can be an amino, an acetamido, a glycolyl-amido, or a hydroxyl group and defines the four major types of sialic acids: neuraminic acid (NeuAc), *N*-acetylneuraminic acid (Neu5Ac), *N*-glycolylneuraminic acid (Neu5Gc), and keto-deoxynonulosonic acid (Kdn), respectively. Substituents of the hydroxyl groups present on

From: *Methods in Molecular Biology, Vol. 125: Glycoprotein Methods and Protocols: The Mucins*
Edited by: A. Corfield © Humana Press Inc., Totowa, NJ

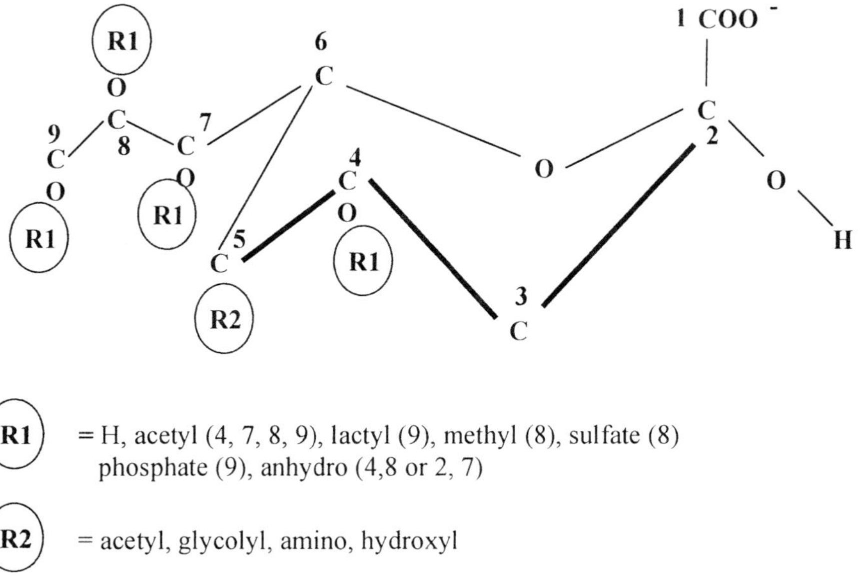

Fig. 1. The sialic acids. The nine-carbon backbone common to all known sialic acids may be substituted by R1 or R2 substituents, giving a family of more than 30 different compounds.

carbons 4, 7, 8, and 9 can be acetyl, lactyl, methyl, sulfate, or phosphate, anhydro forms can also occur (**Fig. 1**) *(1,2)*. Most of the substituents, largely *O*-acetyl groups are quite labile during acid or alkaline hydrolysis methods generally utilized for the release of monosaccharides. Consequently, the study of sialic acid must be generally considered independently of other monosaccharides. The study of sialic acid modifications has been attempted after release and purification by improving the methods to avoid any destruction, and is achieved either with low concentrated acid solutions *(3)* or with enzymatic hydrolysis.

Many techniques for detection and quantification of sialic acids have been described *(1)*. These techniques differ widely in the initial purification of sialic acids from other biological contaminants. One of the most widely used assays is the detection of free Neu5Ac and Neu5Gc acids by the thiobarbituric acid assay (TBA). Free sialic acids react with periodate under acidic conditions to produce β-formylpyruvic acid, which condenses with TBA to produce a purple chromogen (λ_{max} = 549 nm). The assay is sensitive to 1 nmol, but 2-deoxy-sugars interfere because they also condense with TBA to give a chromophore with a slightly lower λ_{max} (532 nm). Powell and Hart *(4)* have introduced an HPLC adaptation of the periodate–TBA assay sensitive to 2 pmol, and requiring no prior purification of released sialic acids. The characterization of released sialic acids can be achieved by chromatography: thin-layer chromatography, GLC *(3)*, or HPLC *(5–8)*. The last technique has higher sensitivity and resolving power. We have reported the HPLC separation of sialic acid quinoxalinones *(8)* that allows the detection of sialic acids at the femtomole level.

1.2. Analysis of Monosaccharides by GLC

GLC methods for identification of monosaccharides are powerful and extremely sensitive. Detection is usually by means of a flame ionization detector (FID), but sensitivity may be increased by coupling the gas chromatograph to an MS instrument (electron or chemical impact). Prior to analysis, monosaccharides must be released by hydrolysis of the oligosaccharides or the glycoproteins and converted to a volatile derivative ***(9,10)***.

1.3. Separation of Monosaccharides by HPLC

HPLC has been widely used because of the advantages of allowing rapid and direct quantification of underivatized or derivatized samples and the ability to characterize samples through coelution with samples of known structures or through retention time comparison. Separation methods are based on anion-exchange **(11)**, size exclusion ***(12)***, ion suppression ***(13)***, reversed-phase ***(14)***, and, most recently, high-performance anion-exchange chromatography (HPAEC).

HPAEC takes advantage of the weakly acidic nature of carbohydrates to give highly selective separations at high pH using strong anion-exchange pellicular resins ***(15)***. In HPAEC, strong alkaline solutions, usually NaOH, are used as eluent. Under these conditions, the hydroxyl groups of carbohydrates are converted to oxyanions with pK*a* values in the range of 12–14. The anomeric hydroxyl group of the reducing sugar is more acidic than the others but each of the hydroxyl groups is characterized by a different p*Ka* value ***(16)***; thus, the modification of some of the hydroxyl groups should greatly influence the elution positions (separation of anomeric and positional isomers) ***(17,18)***. Monosaccharides released from glycoproteins by the previously mentioned hydrolysis methods can be rapidly separated in less than 30 min. Because their molar responses are different, a calibration curve must be established for each monosaccharide. When coupled with pulsed amperometric detection (PAD), HPAEC allows direct quantification of underivatized monosaccharides or carbohydrates at low picomole levels (10–50 pmol) with minimal sample preparation and purification. PAD utilizes a repeating sequence of three potentials. The most important potential is E1, the potential at which the carbohydrate oxidation current is measured. Potential E2 is a more positive potential that oxidizes the gold electrode and completely removes the carbohydrate oxidation products. The third potential, E3, reduces the oxidized surface of the gold electrode in order to allow detection during the next cycle at E1. The three potentials are applied for fixed periods referred to as t_1, t_2, and t_3.

1.4. Electrophoretic Separation of Monosaccharides

Since the early 1990s, capillary electrophoresis has become a good alternative and rapid procedure for analytical separation of microquantities of carbohydrate compounds including monosaccharides ***(19)***. Separation of native monosaccharides is generally difficult owing to the lack of ionized groups and to their low extinction coefficients, which do not allow direct ultraviolet (UV) absorbance detection. Consequently, separation generally requires precolumn derivatization with reagents that contain a suitable chromophoric or fluorophoric group in order to facilitate separation and increase the sensitivity of detection. As described under HPLC, the most common tagging methods are based on the reductive amination procedure, wherein the reducing end of the sugar reacts with the primary amino group of the chromophore ***(20)***.

Different chromophores such as 2-aminopyridine *(20)*, 8-aminonaphthalene-1,3,6-trisulfonic acid (ANTS) **(21)**, ethyl-4-aminobenzoate, and 4-aminobenzonitrile *(22)*, have been used for electrophoretic separation of monosaccharides.

2. Materials

2.1. Release and Identification of Sialic Acids

1. 1000 mol wt cutoff dialysis tube (Bioblock Scientific, Illkirch, France).
2. Dowex AG 50 W × 8 (H^+) (Bio-Rad, Hercules, CA).
3. Dowex AG 3 × 4A ($HCOO^-$) (Bio-Rad).
4. Neuraminidase from *Vibrio cholerae* or *Clostridium perfringens* (Boehringer Mannheim, Indianapolis, IN).
5. HPLC equipment with fluorescent detector.
6. Lichrosorb RP 18 HPLC column (5-μm resin, 250 × 4.6 mm) equipped with an RT30-4 Lichrosorb RP18, 7-μm guard cartridge (Merck, Darmstadt, Germany).
7. Stock solution: 2.35 mL of phosphoric acid (85%), and 28.1 g of sodium perchlorate in 1 L of distilled water.
8. Working solution: water:methanol:2X buffer stock (2:3:5).
9. 1,2-Diamino-4,5-methylene dioxybenzene (DMB) (Merck).
10. C18 column (250 × 4.6 mm, particle size 5 μm) (Beckman, Fullerton, CA).
11. DMB–sialic acid HPLC solvents
    a. Solvent A: methanol:water (7:93 v/v).
    b. Solvent B: acetonitrile:methanol:water (11:7:82 v/v/v).

2.2. Analysis of Monosaccharides by GLC

1. Gas-liquid chromatograph fitted with an FID.
2. Magnesium turnings (Acros Organics, Geel, Belgium).
3. Sodium chloride and sulfuric acid (Sigma, St. Louis, MO).
4. Meso-inositol (Sigma).
5. Silver carbonate (Sigma).
6. Acetic anhydride (Sigma).
7. Heptane (Acros Organics, Sunnyvale, CA).
8. Bis-silyltrifluoroacetamide (BSTFA) (Pierce, Austin, TX).
9. Silicone OV 101 (BP1 phase, SGE).
10. Sodium borohydride (Merck).
11. Pyridine (Merck).
12. Dichloromethane (Merck).
13. Silicone BP 70 (SGE).
14. Helium gas (Air Liquide, Paris, France).

2.3. HPLC Separation Using Amino-Bonded Silica

1. HPLC apparatus equipped with a gradient system.
2. Refractive index detector.
3. Kromasil–NH_2 5 μm column (250 × 4.6 mm) (Alltech, Deerfield, IL).

2.4. Reversed-Phase HPLC of Pyridylamino-Monosaccharides

1. Analytical ODS (C18) 5-μm HPLC column (4.6 × 250 mm) (Zorbax, Interchim, Montluçon, France).
2. 0.25 *M* sodium citrate buffer, pH 4.0, containing 1% acetonitrile (Merck).

2.5. HPAEC-PAD

All eluents and chemical products must be of the highest purity available.

1. Gradient pump module (Dionex Bio-LC apparatus, Sunnyvale, CA).
2. A model PAD-2 detector equipped with a gold working electrode. The following pulse potentials and durations are used for detection: E1 = 0.05 V (t_1 = 360 ms); E2 = 0.70 V (t_2 = 120 ms); E3 = –0.50 V (t_3 = 300 ms) The response time is set to 3 s.
3. Eluent Degas module to sparge and pressurize the eluents with helium (Dionex).
4. Postcolumn with a DQP-1 single-piston pump (Dionex).
5. CarboPac PA-1 column (4 × 250 mm) (Dionex).
6. CarboPac PA-1 guard (4 × 50 mm) (Dionex).
7. CarboPac MA-1 column (4 × 250 mm) (Dionex).
8. CarboPac MA-1 guard (4 × 50 mm) (Dionex).
9. 18 *M* Ω deionized water (Milli-Q Plus System, Millipore, Bedford, MA).
10. NaOH 50% solution with less than 0.1% sodium carbonate (Baker, Deventer, The Netherlands).
11. Anhydrous sodium acetate (Merck).
12. Acetic acid (glacial, HPLC grade; Merck).
13. Eluents containing sodium acetate should be filtered through 0.45-µm nylon filters (Millipore) prior to use.
14. Solvents for separation of neutral monosaccharides, hexosamines, and uronic acids (*see* **Subheading 3.3.3.2.**).
    a. Eluent 1: Deionized water.
    b. Eluent 2: 25 m*M* NaOH and 0.25 m*M* sodium acetate.
    c. Eluent 3: 200 m*M* NaOH and 300 m*M* sodium acetate.
    d. Eluent 4: 125 m*M* NaOH and 10 m*M* sodium acetate.
15. Solvents for HPAEC separation of sialic acids (*see* **Subheading 3.3.3.3.**).
    a. Eluent 1: Deionized water.
    b. Eluent 2: 5 m*M* NaOAc.
    c. Eluent 3: 5 m*M* acetic acid (glacial, HPLC grade; Merck).
16. Solvents for separation of a mixture of unreduced and reduced monosaccharides (*see* **Subheading 3.3.3.4.**).
    a. Eluent 1: Deionized water.
    b. Eluent 2: 1.0 M NaOH.
17. Neu5Ac and Neu5Gc acid (Sigma).
18. A mixture of sialic acids released from bovine submaxillary gland mucin (BSM) (Sigma) (*see* **Subheding 3.1.1.**).

2.6. Electrophoretic Separation of Monosaccharides

1. Capillary zone electrophoresis apparatus fitted with a UV detector (Beckman).
2. Capillary tube (50 µm id × 65 cm) (Beckman). A part of the polyimine coating on the capillary tube is removed by burning at a distance of 15 cm from the cathode, to allow UV detection.
3. 2-Aminoacridone (AMAC) (Lambda Fluoreszentechnologie GmbH, Graz, Austria) made up to 0.1 *M* in acetic acid:dimethylsulfoxide (DMSO, Acros Organics, Sunnyvale, CA) (3:17 v/v). The solution is stored at –70°C.
4. 1 *M* sodium cyanoborohydride (Merck) in water. This solution is made fresh for each experiment.

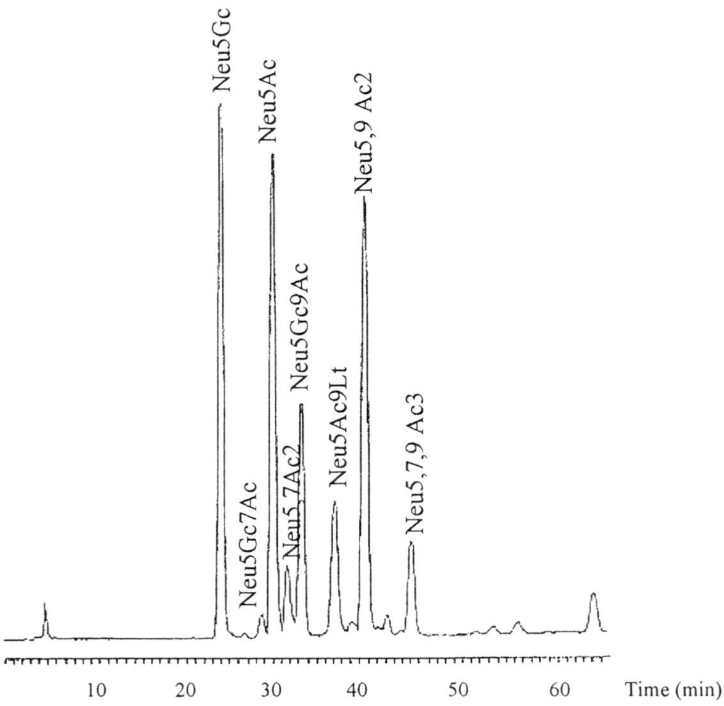

Fig. 2. HPLC separation of sialic acid quinoxalinones obtained after mild acid hydrolysis of BSM. Ac, acetyl; Lt, lactyl; Gc, glycolyl.

3. Methods

3.1. Release and Identification of Sialic Acids

Figure 2 illustrates the separation of the different sialic acid species obtained after hydrolysis of BSM. The different sialic acids may be characterized according to their specific retention times. Additionally, each sialic acid may be characterized by MS analysis *(8)*.

3.1.1. Chemical Hydrolysis of Sialic Acids

1. Suspend 1–10 mg of mucins in 5 mL of 2 *M* acetic acid in a Teflon-capped reaction tube.
2. Hydrolyze for 5 h at 80°C.
3. Dialyze the solution for 24 h against 20 vol of water (1000 mol wt cutoff tubing).
4. Lyophilize the diffusate. Direct analysis can be made at this stage.
5. Further purify sialic acids as follows:
   a. Redissolve the dialysate in 1 mL of water.
   b. Load the sample on a Dowex AG 50W × 8 (H^+) (Bio-Rad) column (10 mL).
   c. Wash the column with 100 mL of water.
   d. Lyophilize the effluent.
   e. Resuspend the lyophilysate in 1 mL of water.

f. Load the sample on a Dowex AG 3 × 4A ($HCOO^-$) (Bio-Rad) column (1 mL).
g. Wash the column successively with 7 mL of 10 m*M* , 7 mL of 1 *M* and 7 mL of 5 *M* formic acid.
h. Pool the fractions and lyophilize.

3.1.2. Enzymatic Release of Sialic Acid

1. Resuspend 1–10 mg of mucin in 2 mL of 100 m*M* HEPES-KOH, pH 7.0, 150 m*M* NaCl, 0.5 m*M* $MgCl_2$, and 0.1 m*M* $CaCl_2$.
2. Add 200 mU/mL of *V. cholerae* or 40 mU/mL *C. perfringens* enzyme.
3. Introduce the solution in a dialysis tube (1000 mol wt cutoff) and dialyze against 5 mL of the same solvent at 37°C overnight.
4. Collect the filtrate and purify the sialic acid as in **Subheading 3.1.1.**

3.1.3. TBA-HPLC Quantification of Sialic Acids

3.1.3.1. TBA Reaction

1. Sialic acids are released from mucins by mild acid hydrolysis as described in **Subheading 3.1.1.** The TBA assay is performed essentially according to Warren *(23)*.
2. Place 40 µL of free sialic acid solution (10–100 pmol/100 µL in water) in an Eppendorf tube.
3. Add 20 µL of sodium periodate (128 mg of sodium metaperiodate, 1.7 mL of phosphoric acid, and 1.3 mL of water).
4. After 20 min at room temperature, add slowly 0.1 mL of 10% sodium arsenite in 0.1 *N* H_2SO_4, and 0.5 *M* Na_2SO_4.
5. When the solution appears yellow-brown, gently vortex the tubes.
6. Add 0.6 mL of 0.6% TBA (0.6 g of TBA [Sigma] in 0.5 *M* Na_2SO_4 [Merck]).
7. After mixing, cap the tubes and heat at 100°C for 15 min.
8. Chill the tubes on ice and centrifuge before HPLC analysis.

3.1.3.2. HPLC Analysis (Fig. 3)

1. Equilibrate the column in the working solution.
2. Elute in the isocratic mode at a flow rate of 1 mL/min.
3. Run UV detection at 549 nm.
4. Quantify the sialic acid by integrating the surface of the sialic acid. Obtain the chromophore peak calibration curve with pure sialic acid solution (1–5 µg of sialic acid in 40 µL of water).
5. Wash the column extensively with 50% acetonitrile in water after use.

3.1.4. Characterization and Quantification of Sialic Acids by HPLC

3.1.4.1. Derivatization with DMB

1. Heat sialic acid samples released by mild hydrolysis in 7 m*M* DMB, 0.75 *M* β-mercaptoethanol, and 18 m*M* sodium hydrosulfite in 1.4 *M* acetic acid (100–200 µL) for 2.5 h in the dark.
2. Inject 10 µL of the reaction mixture on the C18 column.

3.1.4.2. Elution by HPLC

1. Equilibrate the column in 65% solvent A–35% solvent B.
2. Elute using a linear gradient from 65% A/35% B to 100% B over 60 min followed by isocratic elution by 100% B for 10 min at a flow rate of 1 mL/min.
3. Achieve on-line fluorescent detection at an emission wavelength of 448 nm and excitation wavelength of 373 nm with a response time of 0.5 s.

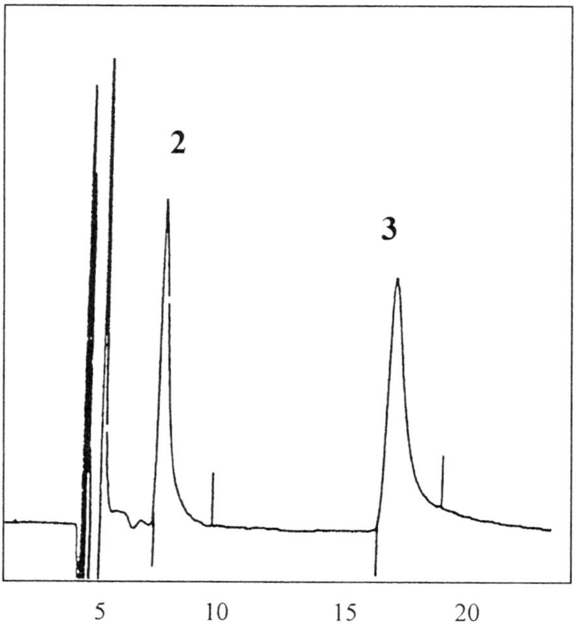

Fig. 3. HPLC analysis of TBA chromophores. 2, NeuAc chromophore; 3, deoxyhexose chromophore.

3.2. Analysis of Monosaccharides by GLC

Complete hydrolysis of oligosaccharide chains may be obtained using concentrated acid solutions.

3.2.1. Trifluoroacetic Hydrolysis

1. Dissolve the oligosaccharide-alditol sample or the native glycoprotein in 0.5 mL of a 4 *M* solution of trifluoroacetic acid (TFA) or a mixture of formic acid:water:TFA (3:2:1 v/v/v).
2. Heat at 100°C for 4 h in Teflon-capped tubes.
3. After hydrolysis, remove the acid by repeated evaporation under reduced pressure. Evaporation is completed by the addition of ethanol.

3.2.2. Formic Acid–Sulfuric Acid Hydrolysis

1. Dissolve oligosaccharide-alditols or native glycoproteins in 0.5 mL of 50% aqueous formic acid and hydrolyze for 5 h at 100°C in a Teflon-capped tube.
2. Repeat **step 1** using 0.25 *M* aqueous sulfuric acid for 18 h at 100°C.
3. Neutralize the hydrolysate with barium carbonate powder, filter, and concentrate to dryness (*see* **Note 1**).

3.2.3. Methanolysis

Methanolysis is a widely used method for hydrolysis of both oligosaccharides and native glycoproteins.

3.2.3.1. Preparation of Methanol–HCl Reagent

1. Obtain anhydrous methanol by refluxing with magnesium turnings (1 h) followed by distillation in a dry all-glass apparatus.
2. Generate gaseous HCl by slow addition (10–20 drops/min) of sulfuric acid to 250 g of solid NaCl.
3. Dry the hydrogen chloride gas through moisture traps containing concentrated sulfuric acid.
4. Then bubble hydrogen chloride gas through anhydrous methanol for 3 to 4 h.
5. Standardize the methanol–HCl reagent to 0.5 *M* by titration with NaOH and dilution with anhydrous methanol (*see* **Note 2**).

3.2.3.2. Methanolysis of Oligosaccharide or Mucus Glycoprotein (24)

1. Freeze-dry carefully in Teflon-capped tubes (complete dehydration of samples is the main condition of success) amounts of glycoproteins or oligosaccharides corresponding to 10 µg of total sugar to which 1 µg of mesoinositol is added as an internal standard.
2. Add 0.5 mL of methanol–HCl reagent.
3. Heat at 80°C for 24 h.

3.2.4. GLC Analysis of Monosaccharides as Their Methylglycoside (25) *Trimethylsilylated Derivatives*

GLC analysis allows the determination of the monosaccharide composition of glycans with amounts of total sugars not exceeding 1 µg (15–20 pmol of glycoproteins). It consists of the methanolysis of previously purified glycoprotein, followed by a re-N-acetylation and a trimethylsilylation leading to trimethylsilylation-derivatives.

3.2.4.1. Methanolysis and Derivatization

1. Mix amounts of purified glycoproteins corresponding to 0.5 µg of total sugars with 200 µL of 0.5 *M* methanol-HCl mixture for 24 h at 80°C.
2. After cooling the tube, neutralize the acidic solution by adding silver carbonate to give a pH of 6.0-7.0 as controlled with pH paper.
3. Re-*N*-acetylate by adding 10 µL of acetic anhydride and keep overnight at room temperature.
4. Centrifuge at 2000*g* for 5 min and collect the supernatant.
5. To eliminate fatty acid methyl esters, wash the methanolic phase two times with 200 µL of heptane (remove the upper phase).
6. Dry the methanolic lower phase under a stream of nitrogen.
7. Trimethylsilylate with 20 µL of BSTFA in the presence of 10 µL of pyridine for 1 h at room temperature.
8. Apply 1–5 µL of the solution of trimethylsilylated methylglycosides to GLC.

3.2.4.2. Gas Chromatography Conditions

A typical GLC chromatogram of TMS derivatives is given in **Fig. 4**. Neutral monosaccharides generally provide several peaks corresponding to pyrano, furano, α, and β forms.

1. Use a FID gas chromatograph and a glass solid injector (moving needle).
2. Use a capillary column (25 m × 0.33 mm) of silicone OV 101.
3. Use carrier gas helium at a pressure of 0.5 bar.
4. Program the oven temperature from 120 to 240°C at 2°C/min.
5. Use injector and detector temperatures of 240 and 250°C, respectively.

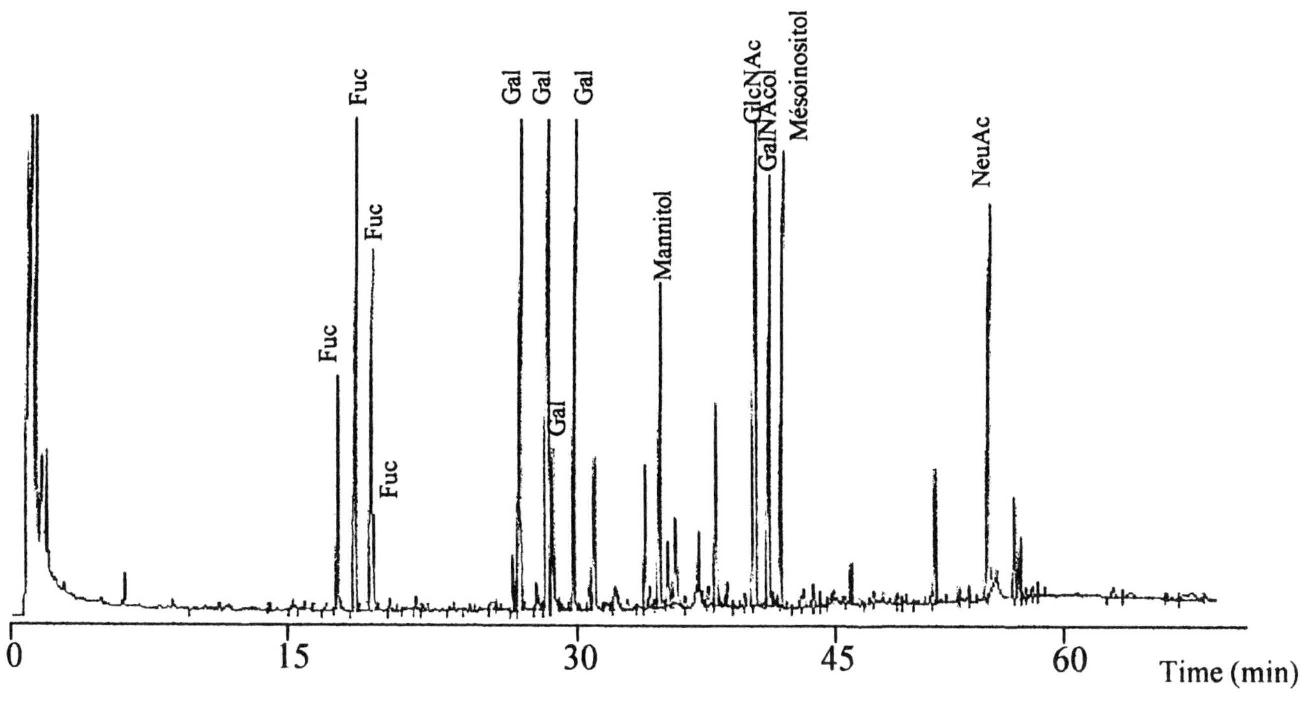

Fig. 4. GLC separation of trimethylsilylated deriviatives of monosaccharides released by methanolysis from β-eliminated intestinal mucin oligosaccharide-alditols.

3.2.5. GLC Analysis of Monosaccharides as Alditol Acetates (26) (Note 3)

3.2.5.1. Hydrolysis and Derivatization

1. Hydrolyze an amount of glycoprotein corresponding to 10 μg of total sugars with 100 μL of 4 *M* TFA in the presence of 2 μg of mesoinositol used as internal standard at 100°C for 4 h in glass tubes fitted with a Teflon screw cap (*see* **Note 4**).
2. After cooling, evaporate the solution and place the tube in a vacuum dessicator over P_2O_5.
3. Reduce the liberated monosaccharides for 1 h at room temperature with 100 μL of a solution of sodium borohydride (2 mg/mL of 0.05 *M* ammonia solution).
4. Destroy the excess sodium borohydride by adding of a 20% acetic acid solution until reaching pH 4.0. Eliminate borate by codistillation with methanol (three times) in a rotary evaporator.
5. Take up the residue in 0.5 mL of water and freeze-dry.
6. Peracetylate the reduced monosaccharides by adding of 50 μL of pyridine and 50 μL of acetic anhydride, and leave overnight at room temperature.
7. Remove the excess of reagent under a stream of nitrogen and take up the residue in 50 μL of dichloromethane containing 1% of acetic anhydride.

3.2.5.2. Gas Chromatography Conditions

A typical chromatogram is presented in **Fig. 5**.

1. Use a gas chromatograph equipped with FID detector.
2. Use a glass capillary column (12 m × 0.22 mm) of silicone BPX70.
3. Use: helium at a pressure of 0.6 bar as the carrier gas.
4. Program the oven temperature from 150 to 230°C and 3°C/min and then 230 to 250°C at 5°C/min.
5. Use injector and detector temperatures of 240 and 250°C, respectively.
6. Use an injection volume of 1 μL.

3.3. Separation of Monosaccharides by HPLC

3.3.1. Separation of Native Monosaccharides by HPLC Using Amino-Bonded Silica (Kromasil-NH_2)

A typical separation diagram is given in **Fig. 6**.

1. Inject 10 μL of a monosaccharide mixture (1 mg/mL [w/v]) on to a 5-μm Kromasil-NH2 column (250 × 4.6 mm).
2. Elute with acetonitrile:water (75:25 v/v) at a flow rate of 1 mL/min

3.3.2. Separation of Pyridylamino Derivatives of Monosaccharides by Reverse-Phase HPLC (27) (see Note 5)

Figure 7 presents a typical profile.

1. Hydrolyze oligosaccharides as previously described (*see* **Subheading 3.2.**).
2. Evaporate off the solvent and dissolve the residue in 100 μL of coupling reagent (prepared by dissolving 100 mg of 2-aminopyridine in 50 μL of acetic acid and 60 μL of methanol).
3. Heat the reaction mixture at 90°C for 30 min.
4. Evaporate the reaction mixture under nitrogen with the addition of toluene to remove excess reagent.
5. Dissolve the pyridylamino monosaccharides in water.

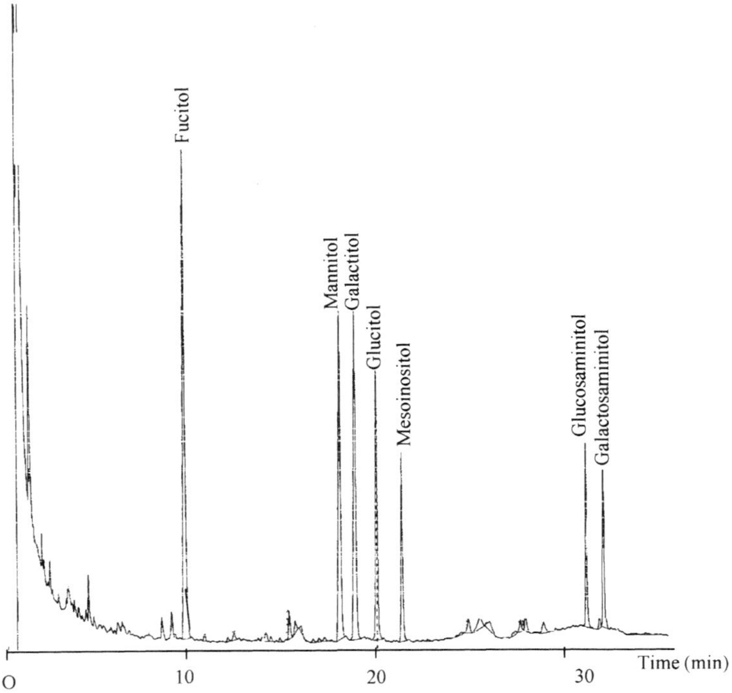

Fig. 5. Quantitative determination of monosaccharides by GLC in the itol-acetates form.

6. Apply samples to an analytical Ultrasphere ODS (C18) 5-µm HPLC column (4.6 × 250 mm) (Zorbax).
7. Elute with 0.25 *M* sodium citrate buffer, pH 4.0, containing 1% acetonitrile at a flow rate of 1.5 mL/min.
8. Detect by fluorescence with an excitation wavelength of 320 nm and an emission wavelength of 400 nm.

3.3.3. HPAEC Analysis of Monosaccharides (***Notes 6–11***)

3.3.3.1. HPAEC Separation of Neutral Monosaccharides and Hexosamines (*28*)

1. Filter 1 L of Milli-Q water and transfer to bottle 1 as eluent 1. Prepare 50 m*M* sodium hydroxide solution by suitable dilution of the 50% sodium hydroxide solution. Connect the sparge lines and degas for 30 min.
2. Disconnect the sparge lines and pressurize bottles 1 and 2.
3. Run 100% of each solvent through the lines to remove any bubbles.
4. Connect the Carbopac PA-1 column with the guard column.
5. Apply an isocratic run at a flow rate of 1.0 mL/min (*see* **Table 1**).
6. Start the system and wait until a stable backpressure is attained.
7. Turn on the PAD detector and wait for stabilization (15 min).
8. Check the baseline and collect chromatographic data by using an integrator.
9. After stabilization and the isocratic run, inject 10 µL of a mixture of standards containing 250 pmol of neutral monosaccharides and hexosamines. Determine the response factors using the areas.

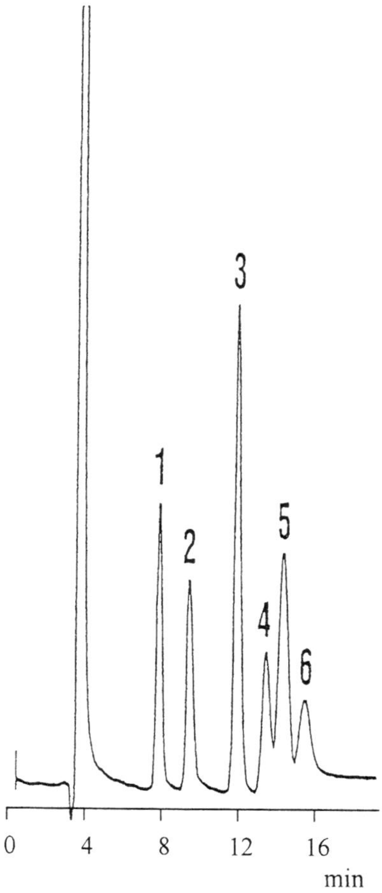

Fig. 6. Separation of native monosaccharides on an amino-bounded column. 1, Rhamnose; 2, xylose; 3, glucose; 4, mannose; 5, sucrose (internal standard); 6, galactose.

10. Check for good resolution between monosaccharides, their molecular responses, and the correct retention times (<30 min/run) before injecting the sample.
11. Inject the sample via a Rheodyne valve equipped with a 200- to 500-µL sample loop.
12. Quantify each monosaccharide in the sample by comparison with the injected area of standards.

3.3.3.2. HPAEC Separation of Neutral Monosaccharides, Hexosamines, and Uronic Acids *(28)*

1. Add a postcolumn reservoir and deliver NaOH (300 m*M*) via a mixing tee at a flow rate of 1 mL/min.
2. Follow the steps described in **Subheading 3.3.3.1.**
3. After each run, equilibrate with the starting eluent in order to yield highly reproducible retention times for the monosaccharides.

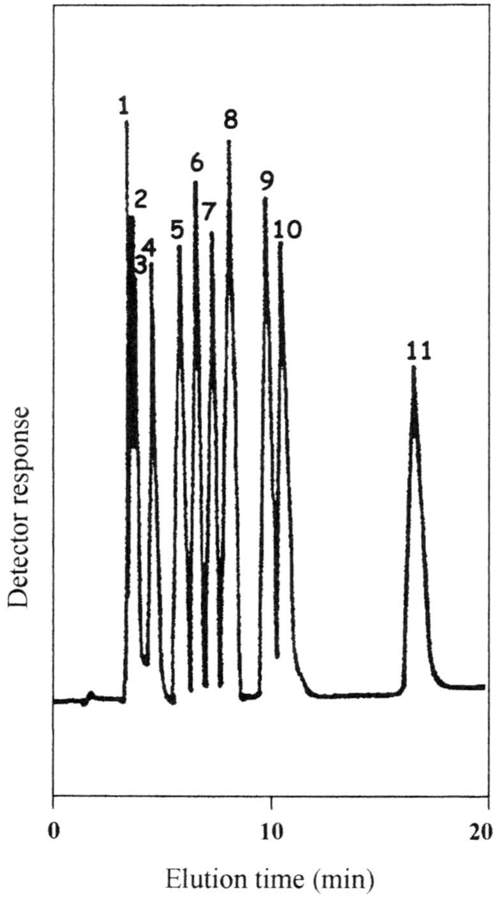

Fig. 7. HPLC separation of pyridylamino derivatives of monosaccharides by reversed-phase HPLC Ultrasphere ODS (4.6 × 250 mm). 1, PA-Gal; 2, PA-Glc; 3, PA-Man; 4, PA-Rib; 5, PA-Fuc; 6, PA-Rham; 7, PA-ManNAc; 8, PA-deoxy-Rib; 9, PA-GlcNAz; 10, PA-GalNAc; 11, PA-NeuAc.

4. Inject 10 µL of the mixture of standards containing 250 pmol each of the neutral monosaccharides, hexosamines, and the required uronic acids.
5. Apply the gradient in **Table 2** with eluents 1–4 at a flow rate of 1 mL/min.
6. Quantify the monosaccharides contained in the sample.

3.3.3.3. HPAEC Separation of Sialic Acids *(7)*

1. Connect a postcolumn solution of 300 m*M* sodium hydroxide at a flow rate of 1 mL/min
2. Follow **steps 1–8** as described in **Subheading 3.3.3.1.**
3. Inject 10 µL of a mixture containing 50 pmol of each sialic acid and check the resolution. The total running time is below 30 min.
4. Load the sample and start the run.
5. Apply a two-step run at a flow rate of 1.0 mL/min (*see* **Table 3**).

Table 1
Elution Gradient for the Separation of Neutral Monosaccharides and Hexosamines

Time (min)	Eluent 1 (%)	Eluent 2 (%)
0	94	6
30	94	6

Table 2
Elution Gradient for the Separation of Neutral Monosaccharides, Hexosamines, and Uronic Acids

Time (min)	Eluent 1 (%)	Eluent 2 (%)	Eluent 3 (%)	Eluent 4 (%)
0	89	11	0	0
20	89	11	0	0
50	39	11	50	0
60	39	11	50	0
63	0	0	0	100
70	0	0	0	100

Table 3
Elution Gradient for the Separation of Sialic Acid

Time (min)	Eluent 1 (%)	Eluent 2 (%)	Eluent 3 (%)
0	0	100	0
5	0	100	0
35	0	50	50
36	0	0	100
46	0	0	100

3.3.3.4. HPAEC Separation of a Mixture of Unreduced and Reduced Monosaccharides *(29)*

1. Follow the different steps described in **Subheading 3.3.3.1.** using a CarboPac MA-1 (4 × 250 mm) column with a CarboPac MA-1 (4 × 50 mm) guard column.
2. Inject 10 µL of a mixture containing 250 pmol of each monosaccharide used as a standard.
3. Apply a gradient at a flow rate of 0.6 mL/min (*see* **Table 4**).
4. If necessary use a second gradient to separate GalNAc-ol and GlcNAc-ol (*see* **Table 5**).
5. The following pulse and durations can be used: E_1 = 0.05 V (300 ms), E_2 = 0.60 V (120 ms), and E_3 = –0.80 V (60 ms).

3.4. Electrophoretic Separation of Monosaccharides

3.4.1. Capillary Zone Electrophoresis of Monosaccharides

Separation and characterization of monosaccharides by capillary zone electrophoresis as the borate complexes of *N*-2-pyridylglycamines (**Fig. 8**).

1. Release monosaccharides by acidic hydrolysis (*see* **Subheading 3.2.**).
2. Carry out precolumn derivatization with 2-aminopyridine by the reductive amination method (**Subheading 3.3.2.**). Freshly prepare the reagent solution by dissolving 10 mg of

Table 4
Elution Gradient for the Separation of Unreduced and Reduced Monosaccharides

Time (min)	Eluent 1 (%)	Eluent 2 (%)
0	60	40
2	60	40
22	45	55
25	45	55

Table 5
Elution Gradient for the Separation of GalNAc-ol and GlcNAc-ol

Time (min)	Eluent 1 (%)	Eluent 2 (%)
0	90	10
10	90	10
15	40	60
20	40	60

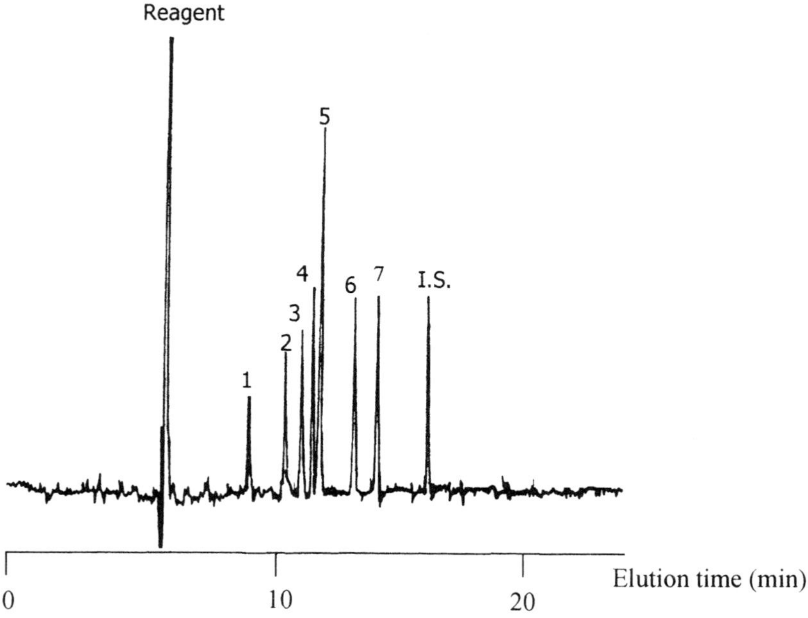

Fig. 8. Capillary zone electrophoresis separation of N-2-pyridylglycamines from different monosaccharides. 1, GalNAc; 2, Xyl; 3, GlcNAc; 4, Glc; 5, Ara; 6, Fuc; 7, Glc; I.S., internal standard cinnamic acid.

sodium cyanoborohydride in 1 mL of methanol containing 2-aminopyridine (10% w/v) and acetic acid (10% v/v).
3. Heat for 2 h at 50°C.
4. Injection 1 µL.
5. Run UV detection at 240 nm.
6. Use an applied potential of 15.0 kV.
7. Use a carrier buffer of 200 m*M* borate buffer, pH 10.5 (dissolve a pellet of potassium hydroxide in boric acid solution, to adjust the pH at the indicated value).

3.4.2. Analysis of Monosaccharides by Fluorophore-Assisted Carbohydrate Electrophoresis (FACE)

A new method of carbohydrate analysis has been introduced in which mono- or oligosaccharides having a reducing end group are labeled covalently with a fluorophore ANTS or AMAC *(30)* and are separated with high-resolution polyacrylamide gel electrophoresis. Monosaccharides released either enzymatically from oligosaccharides by sequential glycosidases digestion or by acidic hydrolysis can be analyzed in this way. Subpicomolar quantities of individual monosaccharides can be detected using an imaging system such as a digitizing camera based on charge-coupled devices or a Polaroid photographic camera. **Figure 9** shows a classic electrophoregram obtained for monosaccharides.

3.4.2.1. Polyacrylamide Gel Electrophoresis of AMAC-Labeled Monosaccharides *(31)*

1. Hydrolyze 50 µg of glycoprotein using 200 µL of 4 *M* TFA at 100°C for 4 h to release neutral monosaccharides and hexosamines (*see* **Note 12**).
2. Lyophilize the hydrolysate.
3. Add 5 µL of 0.1 M solution of AMAC in glacial acetic acid:DMSO (3:17 v/v) and 5 µL of a 1 *M* solution of $NaCNBH_3$ in water.
4. Incubate the mixture at 37°C for 15 h.
5. Lyophilize and dissolve in 6 *M* urea solution (20 pmol monosaccharides/µL) containing 0.05% (w/v) bromophenol blue. Store labeled monosaccharides at –70°C.
6. Electrophoresis buffer is 0.1 *M* Tris-borate buffer, pH 8.3. Electrophorese samples of 1 to 2 µL on 20% (w/v) polyacrylamide gels containing 0.67 % (w/v) *N-N'*-methylene-bisacrylamide. The final concentrations of *N,N,N',N'*-tetramethylenediamine and ammonium persulfate are 0.1% (v/v) and 0.1% (w/v), respectively. The gel dimensions are 140 mm id by 0.5 mm thick. Electrophorese samples are initially at 100 V for 30 min, then at 200 V for 30 min, and finally at 500 V for 90 min.
7. Use bromophenol blue as a marker, stop electrophoresis when the bromophenol blue reaches the anode.

3.4.2.2. Gel Imaging

The electrophoretic band patterns may be viewed in three ways: by visual inspection, by photography, and by electronic image using a suitable camera system.

1. Place the gels on a UV transilluminator with maximum emissions at wavelengths of either 254 or 365 nm.
2. Photograph the gels through a yellow filter.

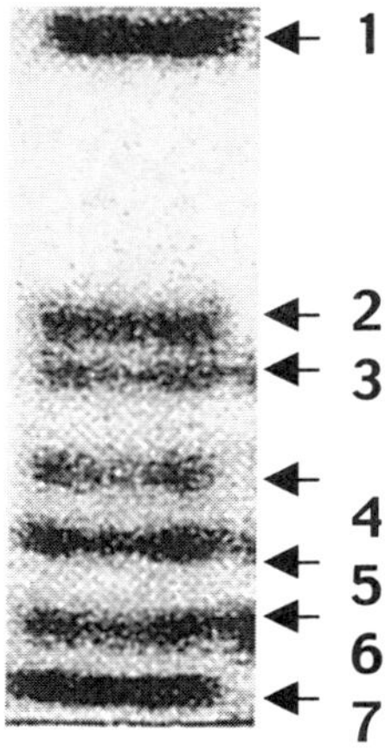

Fig. 9. FACE separation of AMAC-labeled monosaccharides.

4. Notes

1. Using both sulfuric acid and TFA for hydrolysis, sialic acids are destroyed and *N*-deacetylation of acetylhexosamines is observed.
2. The methanol-HCl reagent should be stored at –20°C in hermetically closed containers and its stability checked by titration at least monthly.
3. With the trimethylsilylated derivatives, neutral monosaccharides divide into at least four glycosides, resulting from anomeric and ring isomerization, each of which gives a peak on the chromatogram. Thus, in complex mixtures containing several monosaccharides, the multiplicity of peaks may lead to overlapping such that accurate determination cannot be achieved. To avoid this problem the alditol-acetates method can be utilized, in which each monosaccharide gives rise to a single peak.
4. Sialic acids are destroyed under the hydrolysis conditions used for alditol acetates.
5. Precolumn labeling of monosaccharides with chromophores or fluorophores allows their separation by reversed-phase HPLC and increases the detection level. Chromophores can be introduced either via the hydroxyl groups or by reaction of the aldehyde group with hydrazino or amino reagents. **Table 6** gives a list of different chromophores that may be used. The most commonly used method is based on reductive amination, wherein first the reducing end of a saccharide reacts with the primary amino group of the chromophore to form a Schiff base, which is subsequently reduced to a secondary amine ***(32)***. Using UV or fluorescent detection, quantities of compounds can be detected with picomole range.
6. The retention time of the monosaccharides is closely related to the p*Ka* of their hydroxyl groups.
7. Separation of hexoses is possible in HPAEC despite similarities in their overall molecular size and the number of hydroxyl groups. An isocratic elution with sodium hydroxide allows the separation of all known neutral monosaccharides and hexosamines. Monosaccharides are separated in the following order: Fuc, GalNAc, Gal, GlcNAc, Xyl, and Man, with detection in the picomole range ***(17)***. The elution order of Gal and GalNAc can be interchanged according to the eluent concentration.
8. Alditols devoid of the more acidic anomeric hydroxyl group are eluted much earlier than their reducing counterparts. The decrease in the elution volumes makes the separation of monosaccharide alditols much more difficult. Oligosaccharide-alditols released by

Table 6
Different Derivatization Reagents Used for HPLC Separation of Monosaccharides

Type of reagents	Detection	Sensitivity	HPLC	Reference
Benzoyl	UV 254 nm	1 ng	C18 (CH_3CN / H_2O)	***(33,34)***
p-Bromo-benzoyl;	UV 250 nm	1 ng	Silica; dichloromethane:ethyl acetate (25:4)	***(35)***
naphtoyl-benzoyl	UV260–254 nm	300 ng	Silica; ethyl acetate:hexane *(1:5)*	***(36)***
Dns-hydrazide	Fluorescence	2–5 pmol	C18	***(37)***
DABS hydrazide	Fluorescence	2–5 pmol	C18 (80% acetone:0.08 *M* CH_3COOH)	***(37,38)***
Fmoc hydrazide	Fluorescence	1 pmol	C18 (CH_3CN:AcCOOH)	***(39)***
Dansyl-hydrazide	425 nm	10 pmol	C18	***(40)***
1-phenyl-3-methyl-5-pyrazolone	UV 254 nm	1 pmol	C18 (CH_3CN 13%:phosphate buffer pH 7.0)	***(41,42)***
Pyridylamine	Fluorescence	1 pmol	C18 (0.25 *M* sodium citrate:1%, pH 4.0, CH_3CN)	***(27)***
Dansyl	Fluorescence	10 pmol	C18 (CH_3CN/H_2O)	***(43)***
DMB for sialic acids	Fluorescence	0.1 pmol	C18 (CH_3CN:MEOH:H2O) Per vol. 9 / 7 / 84	***(6)***
Anthranilic acid	Fluorescence	0.1 pmol	1-Butyl amino:phosphoric acid:tetrahydrofuran:acetonitrile:water	***(44)***
Phenyliso thiocyanate	UV	1 pmol	Ammonium acetate:acetonitrile:methanol:water	***(45)***
p-Aminobenzoïc ethyl ester	UV	1 pmol	Sodium acetate pH 4.5:acetonitrile:methanol	***(46)***

reductive elimination from mucus contain a terminal GalNAc-ol. This monosaccharide elutes earlier than the same monosaccharide with a reducing terminal group. After hydrolysis, monosaccharides and GalNAc-ol can be separated in a one-step elution using a CarboPac MA-1 column *(29)*.

9. Phosphorylated monosaccharides have been successfully chromatographed by HPAEC *(47)* as sulfated monosaccharides *(48)*.
10. Elution of acidic monosaccharides and sialic acids requires much stronger conditions with different gradients than neutral and basic monosaccharides: 0.1 *M* NaOH and 50 m*M* sodium acetate *(7)*.
11. Under alkaline elution conditions, the O-acetyl groups are removed rapidly, so the different sialic acids must be separated without introduction of base in the eluent. Because the detector works only at high pH, postcolumn addition of NaOH to the effluent before the entry into the PAD detector is required *(7)*.
12. In this case, sialic acids are destroyed and *N*-acetylhexosamines are deacetylated. Re-*N*-reacetylation may be achieved by dissolving the hydrolysate in 0.2 mL of saturated sodium bicarbonate and by adding 20 µL of acetic anhydride. For sialic acid release, hydrolysis is conducted with 0.1 *M* TFA at 80°C for 1 h.

References

1. Schauer, R. and Kamerling, J. P. (1997) Chemistry, Biochemistry and Biology of Sialic acids. *New Comprehens. Biochem.* **29b**, 243–402.
2. Varki, A. (1992) Diversity in the sialic acids. *Glycobiology* **2**, 25–40
3. Schauer R. (1987) Analysis of sialic acids. *Methods Enzymol.* **138**, 132–161.
4. Powell, L. D. and Hart, G. W. (1986) Quantification of picomole levels of N-acetyl- and N-glycolyl neuraminic acids by HPLC adaptation of the thiobarbituric acid assay. *Anal. Biochem.* **157**, 179–185.
5. Hara, S., Takemori ,Y., Yamaguchi, M., Nakamura, M., and Ohkura, Y. (1987) Fluorometric high performance liquid chromatography of N-acetyl- and N-glycolyl neuraminic acids and its application to their microdetermination in human and animal sera, glycoproteins and glycolipids. *Anal. Biochem.* **164,** 138–145.
6. Hara, S., Yamaguchi, M., Takemori, Y., Furihata, K., Ogura, H., and Nakamura, M. (1989) Determination of mono-O-acetylated N-acetylneuraminic acids in human and rat sera by fluorometric high performance liquid chromatography. *Anal. Biochem.* **179**, 162–166.
7. Manzi, A. E., Diaz, S., and Varki, A. (1990) High-performance liquid chromatography of sialic acids on a pellicular resin anion-exchange column with pulsed amperometric detection: a comparison with six other systems. *Anal Biochem* **188**, 20–32.
8. Klein, A., Diaz, S., Ferreira ,I., Lamblin, G., Roussel, P., and Manzi, A. E. (1997) New sialic acids from biological sources identified by a comprehensive and sensitive approach: liquid chromatography – electrospray ionization – mass spectrometry (LC – ESI – MS) of SIA quinoxalinones. *Glycobiology* **7**, 421–432.
9. Michalski, J. C. (1995) Analysis of monosaccharides by GLC and HPLC in *Methods on Glycoconjugates – A Laboratory Manual*, (Verbert, A., ed.), Harwood, pp. 31–40.
10. Chaplin, M. F. (1994) Monosaccharides in *Carbohydrate Analysis, A Practical Approach*, 2nd ed., (Chaplin, M. F. and Kennedy, J. F., eds.), IRL Press, pp. 1–42.
11. Mellis, S. J. and Baenziger, J. U. (1983) Size fractionation of anionic oligosaccharides and glycopeptides by high performance liquid chromatography. *Anal. Biochem.* **134,** 442–449.
12. Natowicz, M. and Baenziger, J. U. (1980) A rapid method for chromatographic analysis of oligosaccharides on Bio-Gel P4. *Anal. Biochem.* **105**, 159–164.

13. Green, E. D. and Baenziger, J. U. (1986) Separation of anionic oligosaccharides by high performance liquid chromatography. *Anal. Biochem.* **158**, 42–49.
14. Hase, S., Ikenaka, T., and Matsushima, Y. (1978) Structure analyses of oligosaccharides by tagging of the reducing end sugars with a fluorescent compound. *Biochem. Biophys. Res. Commu.*, **85**, 257–263.
15. Hardy, M. R., Townsend, R. R., and Lee, Y. C. (1988) Monosaccharide analysis of glycoconjugates by anion-exchange chromatography with pulsed amperometric detection. *Anal Biochem* **170**, 54–62.
16. Roberts, E. J., Wade, C. P., and Rowland, S. P. (1971) Effect of base concentration upon the reactivities of the hydroxyl groups in methyl D-glucopyranosides, *Carbohydr. Res.* **17**, 393–399.
17. Lee, Y. C. (1990) High-performance anion-exchange chromatography for carbohydrate analysis. *Anal. Biochem.* **189**, 151–162.
18. Lee, Y. C. (1996) Carbohydrate analyses with high-performance anion-exchange chromatography. *J. Chromatogr. A.* **720**, 137–149.
19. Oefner, P. J. and Chiesa, C. (1994) Capillary electrophoresis of carbohydrates. *Glycobiology* **4,** 397–412.
20. Honda, S., Iwase, S., Makino, A., and Fujiwara, S. (1989) Simultaneous determination of reducing monosaccharides by capillary zone electrophoresis as the borate complexes of N-2-pyridyl-glycamines. *Anal. Biochem.* **176,** 72–77.
21. Chiesa, C. and Horvath, C. (1993) Capillary zone electrophoresis of malto-oligosaccharides derivatized with 8-amino-naphtalene-1,3,6-trisulfonic acid. *J. Chromatogr.* **645**, 337–352.
22. Schwaiger, H., Oefner, P. J., Huber, C., Grill, E., and Bonn, G. K. (1994) Capillary zone electrophoresis and micellar electrokinetic chromatography of 4-aminobenzonitrile carbohydrate derivatives. *Electrophoresis* **15**, 941–952.
23. Warren, L. (1959) The thiobarbituric assay of sialic acids. *J. Biol. Chem.* **234**, 1971–1975.
24. Zanetta, J. P., Breckenridge, W. C., and Vincendon, G. (1972) Analysis of monosaccharides by gas-liquid chromatography of the O-methyl glycosides as trifluoroacetate derivatives. Application to glycoproteins and glycolipids. *J. Chromatogr.* **69**, 291–304.
25. Chaplin, M. F. (1982) A rapid and sensitive method for the analysis of carbohydrate components in glycoproteins using gas-liquid chromatography. *Anal. Biochem.* **123**, 336–341.
26. Henri, R. J., Blakeney, A. B., Harris, P. J., and Stone, B. A. (1983) Detection of neutral and amino sugars form glycoproteins and polysaccharides as their alditol-acetates. *J. Chromatogr.* **256**, 419–427.
27. Takemoto, H., Hase, S., and Ikenaka ,T. (1985) Microquantitative analysis of neutral and amino sugars as fluorescent pyridylamino-derivatives by high performance liquid chromatography. *Anal. Biochem.* **145**, 245–250.
28. Manzi, A. E. and Varki, A. (1994) Compositional analysis of glycoproteins in *Glycobiology, A Practical Approach,* (Fukuda M., ed.), IRL Press, pp. 27–77.
29. Karlsson, N. G. and Hansson, G. C. (1995) Analysis of monosaccharides composition by mucin oligosaccharide-alditols by high performance anion-exchange chromatography. *Anal. Biochem.* **224**, 538–541.
30. Jackson, P. (1994) The analysis of fluorophore labelled glycans by high-resolution polyacrylamide gel electrophoresis. *Anal. Biochem.* **216**, 243–252.
31. Jackson, P. (1991) Polyacrylamide gel electrophoresis of reducing saccharides labelled with the fluorophore 2-aminoacridone subpicomolar detection using an imaging system based on a cooled charge-coupled device. *Anal. Biochem.* **196**, 238–244.
32 . Borch, R. F., Bernstein, M. D., and Durst, H. D. (1971) The cyanohydridoborate anion as a selective reducing agent. *J. Am. Chem. Soc.* **93**, 2897–2904.

33. Jentoft, N. (1985) Analysis of sugars in glycoproteins by high-pressure liquid chromatography. *Anal. Biochem.* **148**, 424–433.
34. Gisch, D. J. and Pearson, J. D. (1988) Determination of monosaccharides in glycoproteins by reverse-phase high performance liquid chromatography on 2. 1 mm narrowbore columns. *J. Chromatogr.* **443**, 299–308.
35. Golik, J., Liu, M., Divoni, M., Furakawa, J., and Nakanishi, K. (1983) Characterization of methylglycosides at the pico to nano-gram level. *Carbohydr. Res.* **118**, 135–146.
36. White, C. A. and Kennedy, J. F. (1979) Analysis of derivatives of carbohydrates by high-pressure liquid chromatography. *Carbohydr. Res.* **76**, 1–10.
37. Anderson, J. M. (1986) Fluorescent hydrazides for high performance liquid chromatography, determination of biological carbonyls. *Anal. Biochem.* **152,** 146–153.
38. Muramoto, K., Goto, R., and Kamiya, H. (1987) Analysis of reducing sugars as their chromophoric hydrazones by high performance liquid chromatography. *Anal. Biochem.* **162,** 435–442.
39. Zhang, R. E., Cao, Y. L., and Hearn, M. W. (1991) Synthesis and application of fmoc-hydrazine for the quantitative determination of saccharides by reverse-phase high performance liquid chromatography in the low and subpicomole range. *Anal. Biochem.* **195,** 160–167.
40. Lin, J. K. and Wu, S. S. (1987) Synthesis of dansyl hydrazine and its use in the chromatographic determination of monosaccharides by thin-layer and high performance liquid chromatography. *Anal. Biochem.* **591,** 1320–1326.
41. Honda, S., Akao, E., Suzuki, S., Okada, M., Kakahi, K., and Mukamura ,J. (1989) High performance liquid chromatography of reducing carbohydrates as strongly ultraviolet absorbing and electrochemically sensitive 1-phenyl-3-methyl-5-pyrazolone derivatives. *Anal. Biochem.* **180,** 351–357.
42. Fu, D. and O'Neill, R. A (1995) Monosaccharide composition analysis of oligosaccharides and glycoproteins by high performance liquid chromatography. *Anal. Biochem.* **227,** 377–384.
43. Alpenfelsm W. F. (1981) A rapid method for the determination of monosaccharides as their dansyl hydrazones by high performance liquid chromatography. *Anal. Biochem.* **114,** 153–157.
44. Annumulam K. R. (1994) Quantitative determination of monosaccharides in glycoproteins by high performance liquid chromatography with highly sensitive fluorescence detection. *Anal. Biochem.* **220,** 275–283.
45. Spiro, M. J. and Spiro, R. G. (1992) Monosaccharide determination of glycoconjugates by reverse-phase by high performance liquid chromatography of their phenylthiocarbamyl derivatives. *Anal. Biochem.* **204,** 152–157.
46. Kwon, H. and Kim, J. (1993) Determination of monosaccharides in glycoproteins by reverse-phase by high performance liquid chromatography. *Anal. Biochem.* **215,** 243–252.
47. Hernandez, L. M., Ballon, L., Alvarado, E., Gillea-Castro, B. L., Burlingame, A. L., and Ballon, C. E. (1989) A new *Saccharomyces cerevisiea* mnn Mutant N-linked oligosaccharide structure. *J. Biol. Chem.* **264**, 11,849–11,856.
48. Hemmrich, S., Bertozzi, C. R., Leffler, H., and Rosen, S. D. (1994) Identification of the sulfated monosaccharides of GlyCAM-1 an endothelial derived ligand for L-selectin. *Biochemistry* **33**, 4820–4829.

15

O-Linked Oligosaccharide Chain Release and Fractionation

Elizabeth F. Hounsell

1. Introduction

O-linked chains of glycoproteins have classically been released by alkaline borohydride degradation, in which mild alkali (0.05 *M* OH^-) is used to cause β-elimination from the β carbon of serine (R-H) or threonine ($R\text{-}CH_3$) in the protein backbone.

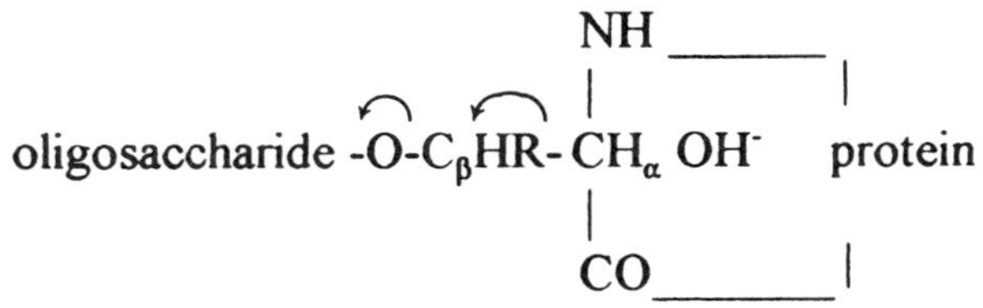

To inhibit subsequent elimination around the glycosidic ring of the linkage monosaccharide, and on backward down the oligosaccharide if linked at C-3, the reaction is carried out in the presence of 1 *M* sodium borohydride to give simultaneous reduction of the reducing sugar formed after elimination. An advantage of this technique is that there is now a large database of nuclear magnetic resonance (NMR) chemical shifts for mucin-type oligosaccharide alditols *(1,2)* that can be searched in a computer-assisted way *(3)* for structural identification. The disadvantage of this technique is that the resulting alditol is not capable of undergoing a reductive amination procedure for coupling to a sensitive flourescent label or for polyvalent coupling to lipid or protein for subsequent immunoassay. However, it is possible to reoxidize selectively the alditol using periodate, which results in a new aldehyde being formed for reductive amination, and, indeed, this is the basis for a method for structural analysis by thin-layer chromatography-mass spectometry (MS) *(4)*. Note, however, that for subsequent biological or immunological assay, any branching at GalNAc-Ser/Thr, often found in mucins, is destroyed by this procedure.

From: *Methods in Molecular Biology, Vol. 125: Glycoprotein Methods and Protocols: The Mucins*
Edited by: A. Corfield © Humana Press Inc., Totowa, NJ

If branching is not present, a useful cleavage can be obtained by the enzyme *O*-glycanase (α-*N*-acetylgalactosaminidase). This enzyme has the disadvantage that only relatively simple oligosaccharides are released with any reproducibility *(5,6)*; the disaccharide Galβ1-3GalNAcα1-Ser/Thr is the best substrate. Hydrazinolysis is probably the preferred technique for release of all oligosaccharides in both the presence and the absence of sulfate and/or sialic acids, if the conditions are optimized. It may be a good idea to carry out this reaction in triplicate with the alkaline borohydride degradation and *O*-glycanase in order to ensure that the complete picture of a mucin at a structural level is obtained. Note that if the protein is required for analysis, rather than the oligosaccharide, mild alkali in the absence of borohydride *(7)* or hydrolysis with trifluoromethanesulfonic acid (TFMSA) can be used to keep the protein intact.

After the release of oligosaccharide alditols, these can be fractionated by normal- or reversed-phase high-performance liquid chromatography (HPLC) *(8)* or anion-exchange chromatography, e.g., high-performance anion-exchange chromatography (HPAEC) on two CarboPac PA-100 columns (Dionex Camberley, Surrey, UK) in series *(9)*. Sulfated and sialylated oligosaccharide alditols can also be separated by these techniques; however, for better results the reversed-phase HPLC should be by porous graphitized carbon (PGC) in 0.1% trifluoroacetic acid (TFA) ***(10,11)***, and the normal-phase HPLC should be in the presence of buffers *(9)*. Sulfated oligosaccharide alditols are retained well on the HPLC columns *(12)*. For reducing oligosaccharides, PGC is again a good alternative for both neutral and anionic oligosaccharides, which have similar retention times to each other and to their alditols. On HPAEC *(13)*, the reducing oligosaccharides will be retained significantly longer than the alditols, to give improved chromatography. Sulfated oligosaccharides may now be retained too long on a CarboPac PA-100 column. Although HPAEC with pulsed amperometric detection (PAD) is a sensitive technique, additional sensitivity can be obtained by flourescent labeling. The usual labels, 2-aminopyridine *(14,15)* and 2-aminobenzamide *(15,16)* cause sulfated oligosaccharides to be retained too long on reversed phase or PGC column. For neutral or sialylated oligosaccharides, separation is primarily achieved by reversed-phase and normal-phase chromatography, often as a two-dimentional map ***(14,15)***, or on weak ion exchange, e.g., GlycoSep C™ (Oxford GlycoSciences, Abingdon, Oxford, UK) *(17)*. Neutral 2-aminobenzamide (2-AB) oligosaccharides (naturally occuring or after desialylation) can be analyzed by gel filtration in water as eluent (usually Bio-Gel P4 chromatography) to give molecular size estimation. This is particularly useful to follow exoglycosidase digestions to obtain additional sequence information. More information about these techniques can be obtained from **refs.** ***18–20***. The use of electrospray ionization MS coupled with collision-induced dissociation MS is another possibility for analysis at the glycopeptide level *(21)*. The internet provides a source for predicting potential *O*-glycosylation sites on proteins at http://www.cbs.dtu.dk/netOglyc/cbsnetOglyc.html.

2. Materials

2.1. Alkaline Borohydride Degradation (β-Elimination) and Chromatography

1. 1 *M* $NaBH_4$ in 0.05 *M* NaOH made up fresh.
2. Glacial acetic acid.

3. Methanol.
4. Cation-exchange column Dowex 50W X8 H^+ form.
5. Phenyl boronic acid (PBA) Bond Elut columns (Jones Chromatography, Hengoed, UK) activated with MeOH *(22)*.
6. 0.2 *M* NH_4OH.
7. 0.01, 0.1, and *0.5 M* HCl.
8. HPLC apparatus fitted with an ultraviolet detector (approx 1 nmol of mono- and oligosaccharides containing *N*-acetyl groups can be detected at 195–210 nm) and pulsed electrochemical detector (oligo- and monosaccharides ionized at high pH can be detected at picomole level).
9. Columns: reversed-phase C_{18}, amino bonded silica, PGC (Hypercarb 7µ, Hypersil Ltd., Runcorn Cheshire, UK), CarboPac PA100 and CarboPac PA1 (Dionex Camberley).
10. Eluents for HPLC: HPLC grade water, acetonitrile, 0.1% aqueous TFA; acetonitrile containing 0.1% TFA, ammonium formate.
11. Eluents for HPAEC: 12.5 *M* NaOH (BDH, Poole, Dorset, UK) diluted fresh each day to 200, 100, 80, and 1.5 m*M*. After chromatography and detection, salt needs to be removed by a Dionex micromembrane suppressor or by cation-exchange chromatography before further analysis, e.g., by methylation.
12. Eluent A: 0.08 *M* NaOH.
13. Eluent B: 0.5 *M* sodium acetate (Aldrich, Poole, Dorset, UK) in 0.08 *M* NaOH.

2.2. Reoxidation

1. Sodium periodate, analytical reagent grade.
2. Imidazole (Sigma, Poole, Dorset, UK) 40 m*M* adjusted to pH6.5 with HCl
3. Butan-2,3-diol.

2.3. Release of the Core 1 Disaccharide (Galβ1-3GalNAcα–) from Mucins

1. Bio-Spin® chromatography columns (Bio-Rad, Hercules CA), or home made columns suitable for 1.5-mL microcentrifuge tubes packed with 0.8 mL of Bio-Gel P30 acrylamide gel matrix . Store at 4°C in 0.15 *M* sodium chloride-17.5 m*M* sodium citrate, pH 7.0, containing 0.2% w/v sodium azide as preservative.
2. *O*-Glycanase®; *Streptococcus (Diplococcus) pneumoniae* Endo-α-N-acetylgalactosaminidase (EC 3.2.1.907) (Oxford GlycoSciences). Store at –20°C for up to 6 mo, but avoid repeated freeze-thawing.
3. 0.1 *M* sodium citrate-phosphate, pH 6.0; make up with HPLC-grade water.
4. Dowex 50X-12 H^+-form resin.
5. Sephadex GM25.

2.4. Hydrazinolysis

1. 0.5-mL screw-capped V-bottomed Reacti-vials™ (Pierce and Warriner, Chester, UK).
2. Anhydrous hydrazine (Pierce and Warriner).
3. Whatman CF-11 cellulose chromatography medium.
4. Reagent A: butanol:ethanol:acetic acid, 4:1:0.5 (v/v/v).
5. Reagent B: butanol:ethanol:water, 4:1:1 (v/v/v).
6. Reagent C: acetic anhydride:methanol, 2:5 (v/v).
7. Reagent D: 0.2 *M* sodium acetate.
8. Sep-Pak C_{18} cartridge (Waters, Watford, UK).
9. Alternatively, the Glycorelease *N*- and *O*-Glycan recovery kit (Oxford GlycoSciences) can be used.

2.5. Isolation of the Protein by TFMSA Destruction of Oligosaccharides

1. Anhydrous TFMSA *(23)*.
2. Reagents as specified in Glyco Free™ Deglycosylation kit (K500, Oxford, GlycoSciences).

2.6. 2-AB Fluorescence Labeling and Size Exclusion Chromatography (SEC)

1. Signal labeling kit (K404,Oxford GlycoSciences).
2. Dowex AG50 X 12 (H^+ form).
3. Dowex AG1 X 8 (acetate form).
4. RAAM 2000 Glycosequencer (Oxford GlycoSciences) or Biogel P4 column (100 × 2 cm) in a water jacket at 55°C and HPLC pump with refractive index and fluorescence detectors.
5. GlycoSepH™ and GlycoSepC™ HPLC column (Oxford GlycoSciences).

3. Methods

3.1. Alkaline Borohydride Degradation (β-Elimination)

1. Release *O*-linked chains by treatment with 0.05 *M* sodium hydroxide in the presence of 1 *M* $NaBH_4$ or $NaB[^3H]_4$ for 16 h at 50°C.
2. Degrade excess $NaBH_4$ or $NaB[^3H]_4$ by the careful addition with the sample on ice of glacial acetic acid (to pH 7.0) or acetone (1 mL/100 mg of $NaBH_4$) followed by repeated evaporation with methanol.
3. Desalt on a cation-exchange column and analyze by reversed-phase HPLC (**Subheading 3.6.2.**) or HPAEC (**Subheading 3.7.1.**).
4. Or, for microscale identification of the presence of alditols, dissolve the sample in 200 µL 0.2 *M* NH_4OH and add to the top of a PBA minicolumn prewashed with MeOH, water, and 0.2 *M* NH_4OH.
5. Wash the PBA column with 2 × 100 µL 0.2 *M* of NH_4OH and 2 × 100 µL of water.
6. Specifically elute the alditols in 1 *M* acetic acid.
7. Evaporate the sample and reevaporate with 2 × 100 µL of water.

3.2. Oxidation of Oligosaccharide Alditols

1. To the dry alditols, add twice the molar ratio of sodium periodate in imidazole buffer (*see* **Note 1**).
2. Oxidize in the dark at 0°C for 5 min.
3. Destroy excess oxidant with butan-2,3-diol (two times the molar excess over periodate) for 40 min at 0°C in the dark.

3.3. Release of Core 1 Disaccharide (Galβ1-3GalNAcα–) from Mucin Using Endo-α-N-Acetylgalactosaminidase (O-Glycanase)

3.3.1. The Removal of Glycerol from O-Glycanase (see Note 2)

1. Invert a Bio-Spin P30 polyacrylamide BioGel column (0.8 mL column volume) several times and allow the buffer to drain by gravity.
2. Wash the column with 300 µL of 0.01 *M* citrate-phosphate, pH 6.0, place in a collection tube, and centrifuge for 2 min at 1100*g*. Repeat four times.
3. Make up 6 mU *O*-Glycanase to 100 µL with 0.01 *M* citrate-phosphate, pH 6.0, and load the sample carefully and directly to the center of the column, drop wise (*see* **Note 3**).
4. Centrifuge for 4 min at 1100*g* and collect the excluded *O*-Glycanase.

5. Pass the excluded *O*-Glycanase through a second Bio-Spin column to maximize glycerol removal.

3.3.2. Hydrolysis of Galβ1-3GalNAcα- from mucin using O-Glycanase

1. Reconstitute mucin (100 μg) with 90 μL of 0.1 *M* citrate-phosphate, pH 6.0, containing 100 μg/mL of bovine serum albumin and 0.02% (w/v) sodium azide. Mix well.
2. Add 0.6 mU of deglycerolyated *O*-Glycanase (10 μL).
3. Incubate for 18 h at 37°C (*see* **Note 4**).
4. Load reactions on Dowex 50X-12H^+ form resin (3 × 0.5 cm column) and elute with three column vol of HPLC-grade water.
5. Collect the effluent and eluent and then pool (2.5-mL volume).
6. Load onto a PD10-Sephadex GM25 column and elute with HPLC-grade water to isolate liberated disaccharide from intact mucin.

3.4. O-Linked Oligosaccharide Release by Hydrazinolysis

1. Dry salt-free glycoprotein into a V-bottomed Reacti-vial and remove from the lyophilizer immediately before the reaction is due to commence.
2. Using a clean, dry, acid-washed glass pipet, transfer 100 μL of anhydrous hydrazine to the vial and cap immediately. Incubate at 60°C for 5 h (*see* **Note 5**).
3. Allow the Reacti-vial to cool to room temperature, and transfer the reaction mixture to a 1-mL cellulose (Whatman CF-11) microcolumn washed with reagent A.
4. Wash the column with 3 × 1mL reagent B.
5. Re-*N*-acetylate the glycans on the column by adding 1.4 mL of reagent C for 30 min at room temperature.
6. Wash the column with 4 × 1 mL of reagent B followed by 1 mL of methanol (*see* **Note 6**).
7. Elute the oligosaccharides with 2 × 1 mL of reagent D.
8. Complete the re-*N*-acetylation with 0.1 mL of acetic anhydride for 30 min at room temperature.
9. Wash a Sep-Pak C_{18} cartridge with 2 mL of methanol and 2 mL of H_2O.
10. Transfer the sample containing *O*-glycans to the cartridge and collect the eluate. Elute the remaining glycans with 0.5 mL of H_2O.

3.5. TFMSA Treatment

1 Make up a 3 g/mL solution of TFMSA in anisole and cool in dry ice/ethanol
2. Add 10 times the weight of TFMSA in anisole to the lyophilized material in a teflon screw-capped vial standing on a bed of ice.
3. Incubate at 0°C for 6–16 h with occasional vigorous shaking.
4. With the vial on ice, add 1 vol of cold anhydrous diethyl ether and then add this mixture to 1 vol of a frozen slush of aqueous pyridine.
5. Warm the solution to room temperature and extract with ether.
6. Collect the aqueous phase containing the peptide with partial glycosylation depending on the reaction time.
7. Alternatively, follow the instructions in the GlycoFree kit.

3.6. HPLC of 2-AB-Labeled Oligosaccharides

3.6.1. Preparative HPLC on a GlycoSep C HPLC Column

1. Wash the column with water for 30 min at a flow rate of 0.4 mL/min.
2. Wash the column with acetonitrile for 30 min at a flow rate of 0.4 mL/min.

3. Wash the column with 0.5 *M* ammonium acetate, pH 4.5, for 30 min at a flow rate of 0.4 mL/min (*see* **Note 7**).
4. Rewash the column with water for 30 min at a flow rate of 0.4 mL/min.
5. Equilibrate the column in 75% aqueous acetonitrile at a flow rate of 0.4 mL/min.
6. Inject the 2-AB-labeled sample in 75% aqueous acetonitrile with fluorescence detection using an excitation λ = 330 nm and an emission λ = 420 nm.
7. Elute the sample with the following gradient with fraction collection at a flow rate of 0.4 mL/min (*see* **Note 8**):
   a. 75% acetonitrile for 5 min
   b. 62.5% acetonitrile over the next 25 min
   c. 60% over the next 30 min
   d. Back to 75% at 60 min for the total run

3.6.2. Porous Graphitized Carbon Chromatography

1. Wash the column with water for 30 min at a flow rate of 0.75 mL/min (*see* **Note 9**).
2. Wash the column with acetonitrile for 30 min at a flow rate of 0.75 mL/min.
3. Equilibrate the column in 0.2% TFA in 20% aqueous acetonitrile at a flow rate of 0.75 mL/min (*see* **Note 10**).
4. Inject the 2-AB-labeled sample in water, with fluorescence detection using an excitation λ = 330 nm and an emission of λ = 420 nm (*see* **Note 11**).
5. Elute the sample with the following gradient at a flow rate of 0.75 mL/min with water/acetonitrile containing 0.2% TFA:
   a. 20% acetonitrile for 2 min
   b. 40% acetonitrile over 33 min
   c. 30% acetonitrile over another 35 min
   d. Back to 20% acetonitrile over the next 4 min
6. Store the column in 75:25 (v/v) acetonitrile:water.

3.7. HPAEC of Released Oligosaccharides

3.7.1. HPAEC Analysis of Alkaline-Borohydride–Released Mucin Oligosaccharide Alditols

1. Inject an aliquot (50 and 100 μL) of released mucin oligosaccharide alditols, containing 320 ng of D-melibiose as internal standard, onto the two CarboPac PA-100 columns (connected in series) equilibrated in eluent A.
2. Elute with an increasing gradient of eluent B at a flow rate of 0.7 mL/min.
3. Profile the chromatogram obtained using a range of well-characterized human oligosaccharide alditol standards.
4. Reequilibrate the columns before each subsequent sample application.

3.7.2. HPAEC Analysis of O-Glycanase-Released Galβ1-3GalNAc

1. Inject a 1-μg aliquot of the *O*-Glycanase-released product, containing 320 ng of D-melibiose as internal standard, onto the two CarboPac PA100 columns (connected in series) equilibrated in eluent A.
2. Elute with an increasing gradient of eluent B at a flow rate of 0.7 mL/min.
3. Quantify the Galβ1-3GalNAc released using a calibration curve for standard amounts of Galβ1-3GalNAc (*see* **Note 12**).
4. Reequilibrate the column before each subsequent sample application (*see* **Note 13**).

3.8. Fluorescence Labeling with 2-AB

1. Dry salt free glycans into a 0.5-mL Eppendorf tube.
2. Prepare labelling reagent of 70% DMSO 30% glacial acetic acid containing 0.25 *M* of 2-AB and 0.1 *M* of $NaCNBH_3$ (*see* **Note 14**).
3. Add 5 µL of labeling reagent and incubate the sample at 65°C for 2 h.
4. Centrifuge sample briefly.
5. Transfer to a hydrophillic separation disk (supplied with labeling kit) washed with 1 mL of water, lml 30% acetic acid, and 1 mL of acetonitrile.
6. Load the sample onto disk and leave for 15 min (*see* **Note 15**).
7. Wash the tube with 100 µL acetonitrile and add to disk.
8. Wash disk with 1 mL of acetonitrile followed by 5 × 1 mL 4% water in acetonitrile.
9. Elute the sample with 3 × 0.5mL of water (*see* **Note 15**).
10. Dry the sample to about 100 µL.
11. Prepare 150 µL of AG50-X12 resin in a microcolumn and wash with 5 mL 1.5% triethylamine in water followed by 3 × 1 mL water (see Note 16).
12. Add 150 µL of AG1 X8 (acetate form) to the microcolumn taking care not to disturb the AG50 resin.
13. Wash with 0.5 mL of water.
14. Load the sample in 100 µL water and elute with 4 × 0.4 mL of water.
15. Filter the sample through a 0.45-µm filter and dry for further analysis.

4. Notes

1. Standard *O*-linked chains, i.e., those having the linkage

   \
   6
   GalNAcαl-Ser/Thr
   3
   /

   are cleaved by periodate at the C-4—C-5 bond, thus giving a characteristic product for chains linked at the C-3 and/or C-6.
2. The efficient removal of glycerol (added as an enzyme stabilizer) from commercial *O*-Glycanase is necessary for the subsequent HPAEC analysis of hydrolyzed Galβ1-3GalNAcα-, because glycerol was found to be highly PAD active *(20)*. Deglycerolated *O*-Glycanase, however, does not store well, so it is best to remove the glycerol on the day of use.
3. The optimum volume of sample is between 50 and 100 µL. Applying more or less results in poor recovery of desalted sample. Likewise, sample directed down the side of the gel also results in poor recovery.
4. 1 mU of *O*-glycanase releases approx 500 ng of Gaβ1-3GalNAc from 1 µg of antifreeze glycopeptides and 50 ng from 1 µg of asialofetuin *(20)*.
5. The reaction should be incubated in either an oven or a heating block but not in a water bath. Because of the highly toxic and flammable nature of hydrazine, all manipulations that involve its use should be carried out in a fume cupboard with additional skin and eye protection. The evaporator used should also be vented into a fume cupboard; if this includes a pump, the pump should be left on overnight. For the release of *N*-linked glycans incubate at 95°C for 5 h.

6. If the column is not washed with methanol, the final eluent will consist of an immiscible butanol/water mixture, which will prove difficult to dry.
7. To ensure that high-purity eluents are always obtained, it is recommended that ammonium acetate be obtained by titrating the relevant acid (e.g., 0.5 *M* HPLC-grade acetic acid) to the relevant pH with HPLC-grade ammonium hydroxide. Ammonium formate may also be used as an eluent, at similar pH values.
8. The precise elution gradient can be varied to suit the diversity of oligosaccharides being studied. The pH of the ammonium acetate will greatly affect the resolution and retention and can be tailored to suit the analytes being investigated.
9. This flow rate is for a 100 × 4.6 mm Hypercarb S column. The smaller Glycosep H column should not be run at flow rates in excess of 0.5 mL/min.
10. Great care should be taken to ensure thorough equilibration of the column prior to injecting samples because these columns are quite sensitive to changes in organic-phase composition.
11. The use of 0.1% TFA will cause quenching of fluorescence detection.
12. To optimize detector sensitivity and avoid baseline drift at low NaOH concentrations, the addition of 0.3 *M* NaOH postcolumn (i.e., before it enters the detector) is required to increase pH to ≥12. Use of a thinner electrode gasket (0.005 in.) gives the best signal-to-noise ratio. Flow rate at which the postcolumn reagent is added should be reproducible between runs and must be the gradient pump flow; a pressure of 112–114 psi (eluent vessel must be able to withstand higher pressure) should achieve this flow rate with no significant increase in gradient pump pressure. Beware of air bubbles in the beaded mixing coil, which prevent efficient pH increase. **Note:** Always turn off the gradient pump before turning off the postcolumn eluent stream ***(20)***.
13. Retention times will steadily decrease unless the column is regularly regenerated with strong alkali. Regenerate the CarboPac PA-100 anion-exchange column for 5 min with 20% 0.5 *M* sodium acetate – 15 μ*M* NaOH and then re-equilibriate for 30 min with 15 μ*M* NaOH before each subsequent sample application.
14. In addition to introducing 2-AB groups, reductive amination can also be used to introduce other fluorescent labels, such as 2-aminopyridine ***(24)*** or 8-amino naphtalene-1,3,6-trisulfonic acid ***(25)***, and also to couple to protein or lipid (*see* **Subheading 1.**). Here the commercial labeling reagent is sufficient to label up tp 50 nmol of oligosaccharide. If poor sloubility of the reductant ($NaCNBH_3$) is observed, this can be improved by the addition of 10 μL of water to the labeling mixture prior to adding it to the samples.
15. Care should be taken so that the flow rate through the disk is approx 1 drop/s and that air bubbles do not form below the disk. Air bubbles can be removed by gentle pressure on the disk; however, it is difficult to remove them completely.
16. Desalting may also be carried out by Bio-Gel P2 chromatography eluted in water.

Acknowledgment

The author wishes to thank Gail Evans for preparation of the manuscript and the UK Medical Research Council for funding.

References

1. Hounsell, E. F. (1995) ^{1}H-NMR in the structural and conformational analysis of oligosaccharides and glycoconjugates, in *Progress in NMR Spectroscopy*, vol. 27 (Emsley, J. W., Feeney, J., and Sutcliffe, L. H., eds.), Elsevier, Kidlington, Oxford, UK, pp. 445–474.

2. Hounsell, E. F., Davies, M. J., and Renouf, D. V. (1996) O-linked protein glycosylation structure and function. *Glycoconj. J.* **13,** 19–26.
3. Hounsell, E. F. and Wright, D. J. (1990) Computer assisted interpretation of ^{1}H-NMR spectra in the structural analysis of oligosaccharides. *Carbohydr. Res.* **205,** 19–29.
4. Stoll, M. S., Hounsell, E. F., Lawson, A. M., Chai, W., and Feizi, T. (1990) Microscale sequencing of O-linked oligosaccharides using periodate oxidation of alditols, coupling to phosphatidylethanolamine dipalmitoate and TLC-MS analysis of the resulting neoglycolipids. *Eur. J. Biochem.* **189,** 499–507.
5. Iwase, H., Ishii, I., Isihara, X., Omura, S., and Hotta, K. (1988) Release of oligosaccharides possessing reducing end N-acetylgalactosamiine from mucus glycoprotein in *Streptomyces* sp. OH-11242 culture medium through action of endo-type glycosidase. *Biochem. Biophys. Res. Commun.* **151,** 4222–4228.
6. Fan, J.-Q., Yamamoto, K., Matsumoto, Y., Hirabayashi, Y., Kumagai, H., and Tochikura, T. (1990) Action of endo-α-N-acetylgalactosaminidase from *Alcalgenes* sp on amino acid-O-glycan: comparison with the enzyme from *Diplococcus pneumoniae. Biochem. Biophys. Res. Commun.* **169,** 751–757.
7. Hounsell, E. F., Jones, N. J., and Stoll, M. S. (1985) The application of high performance liquid chromatography to the purification of oligosaccharides containing neutral and acetamido sugars. *Biochem. Soc. Trans.* **13,** 1061–1064.
8. Hounsell, E. F. (1986) HPLC of carbohydrates, in *HPLC of Small Molecules.* (Lim, C. K., ed.), IRL Press, pp. 49–68.
9. Campbell, B. J., Davies, M. J., Rhodes, J. M., and Hounsell, E. F. (1993) Separation of neutral oligosaccharide alditols from human meconium using high pH anion exchange chromatography. *J. Chromatogr.* **622,** 137–146.
10. Davies, M. J., Smith, K. D., Carruthers, R. A., Chai, W., Lawson, A. M., and Hounsell, E. F. (1993) The use of a porous graphitised carbon (PGC) column for the HPLC of oligosaccharides, alditols and glycopeptides with subsequent mass spectrometry analysis. *J. Chromatogr.* **646,** 317–326.
11. Davies, M. J. and Hounsell, E. F. (1996) Comparison of separation modes for high-performance liquid chromatography of glycoprotein- and proteoglycan-derived oligosaccharides. *J. Chromatogr.* **720,** 227–234.
12. Lo-Guidice, J-M., Herz, H., Lamblin, G., Plancke, Y., Roussel, P., and Lhermitte, M. (1997) Structures of sulfated oligosaccharides isolated from the respiratory mucins of a non-secretor (O, Le(a+b–)) patient suffering from chronic bronchitis. *Glyco. J.* **14,** 113–125.
13. Davies, M. J. and Hounsell, E. F. (1998) HPLC and HPAEC of oligosaccharides and glycopeptides, in *Glycoscience Protocols* (Hounsell, E. F., ed.), Humana, Totowa, NJ, pp. 79–100.
14. Kakehi, K., and Honda, S. (1993) Analysis of carbohydrates in glycoproteins by high-performance liquid chromatography and high-performance capillary electrophoresis, in *Glycoprotein Analysis in Biomedicine* (Hounsell, E. F., ed.), Humana, Totowa, NJ, pp. 69–80.
15. Routier, F. H., Hounsell, E. F., Rudd, P. M., Takahashi, N., Bond, A., Axford, J. S., and Jefferis, R. Quantitation of human IgG glycoforms isolated from rheumatoid sera: A critical evaluation of chromatographic methods. *J. Immunol. Methods* **213,** 113–130.
16. Guile, G. R., Rudd, P. M., Wing, D. R., Prime, S. B., and Dwek, R. A. (1996) A rapid high resolution HPLC method for separating glycan mixtures and analysing oligosaccharide profiles. *Anal. Biochem.* **240,** 210–226.
17. Guile, G. R., Wong, S. Y. C., and Dwiek, R. A. (1994) Analytical and preparative separation of anionic oligosaccharides by weak anion-exchange high-performance liquid chromatography on an inert polymer column. *Anal. Biochem.* **222,** 231–235.

18. Hounsell, E. F. (ed.) (1998) *Glycoanalysis Protocols*, 2nd ed., Humana Press, Totowa, NJ.
19. Paramonov, N. and Hounsell, E. F. (2000) Application of HPLC to analysis of oligosaccharides of glycoproteins, in *HPLC of Small Molecules*, 2nd ed. (Perret, D., Flanagan, R., and Hounsell, E. F., eds.), Oxford University Press, Oxford, UK.
20. Campbell, B. J. and Rhodes, J. M. (1998) Purification of gastrointestinal mucins and analysis of their O-linked oligosaccharides, in *Glycoanalysis Protocols*, vol. 76 (Hounsell, E. F., ed.), Humana Press, Totowa, NJ.
21. Greis, K. and Hart G. (1998) Analytical methods for the study of O-GlcNAc glycoproteins and glycopeptides, in *Glycoanalysis Protocols*, vol. 76 (Hounsell, E. F., ed.), Humana Press, Totowa, NJ.
22. Stoll, M. S. and Hounsell, E. F. (1988) Selective purification of reduced oligosaccharides using a phenylboronic acid bond elute column: potential application in HPLC, mass spectrometry, reductive amination procedures and antigenic/serum analysis. *Biomed. Chromatogr.* **2,** 249–253.
23. Gerken, T. A., Owens, C. L. and Pasumarthy, M. (1996) Determination of the site-specific O-glycosylation pattern of the porcine submaxillary mucin tandem repeat glycopeptide. *J. Biol. Chem.* **272,** 9709–9719.
24. Natsuka, S. and Hase S. (1998) Analysis of N– and O-Glycans by Pyridylamination, in *Glycoanalysis Protocols*, vol. 76 (Hounsell, E. F., ed.), Humana Press, Totowa, NJ.
25. Klock, J. C. and Starr, C. M. (1998) Polyacrylamide gel electrophoresis of fluorophore-labelled carbohydrates from glycoproteins, in *Glycoanalysis Protocols*, vol. 76 (Hounsell, E. F., ed.), Humana Press, Totowa, NJ.

16

Structural Analysis of Mucin-Type *O*-Linked Oligosaccharides

André Klein, Gérard Strecker, Geneviève Lamblin, and Philippe Roussel

1. Introduction

The carbohydrate moiety of mucin is characterized by the presence of oligosaccharides linked to the peptide backbone by an *O*-glycosidic linkage between an *N*-acetylgalactosamine residue and a hydroxylated amino acid (serine or threonine). These linkages are alkali labile and the carbohydrate chains can be released as oligosaccharide-alditols by a β-elimination, with NaOH in the presence of $NaBH_4$. The structures of carbohydrate chains found in mucins can be as simple as the disaccharide NeuAc α2→6GalNAc in ovine submaxillary mucin and as complex as the ones found in human respiratory or salivary mucins, in which several hundred different carbohydrate chains exist *(1,2)*. This diversity is generated (1) by the different monosaccharides constituting the glycans, generally fucose, galactose, *N*-acetylgalactosamine, *N*-acetylglucosamine, and *N*-acetyl neuraminic acid, but also other monosaccharides such as ketodesoxynonulosonic acid or *N*-glycolylneuraminic acid, and, finally the occurance of sulfation of galactose and *N*-acetylglucosamine *(3,4)*; and (2) by the difference in length, in branching, and by the occurrence of all the different possible linkages between the constituting monosaccharides. The diversity of mucin-type oligosaccharides can be extreme. For example, 88 oligosaccharides have been isolated from the respiratory mucins of a single individual *(5–10)* and more than 150 have been isolated from the jelly coat from eggs of different species of amphibians *(11–15)*.

This chapter gives an overall idea of the strategy of elucidation of primary structure of glycans, the most current nuclear magnetic resonance (NMR) techniques used in structure determination, and some of the mass spectrometry (MS) techniques available for the glycobiologist.

1.1. Strategy

The amount of pure oligosaccharide (or of a mixture of two compounds, or three at the most) and the facilities that are available in the laboratory environment will define

From: *Methods in Molecular Biology, Vol. 125: Glycoprotein Methods and Protocols: The Mucins*
Edited by: A. Corfield © Humana Press Inc., Totowa, NJ

the strategy. Two types of techniques are used: a destructive technique, MS and a nondestructive technique, NMR. **Figure 1** summarizes the sequence of techniques. When the amount of oligosaccharide permits the use of NMR first, especially if two-dimentional (2D) NMR can be performed, MS is only used to confirm the structure if it is a novel one; in most cases, it is not needed. MS is quite useful when there is not enough material for NMR or when the mixture studied is too complex.

1.2. Nuclear Magnetic Resonance

NMR spectroscopy constitutes the most suitable method for the structure determination of carbohydrate chains. This method was introduced in the 1970s and rapidly received a large application for analyzing the sequence of *N*-acetyllactosamine- and oligomannosidic-type glycans.

Originally, and before the development of 2D NMR spectroscopy, the method was limited to one-dimentional ^{1}H-NMR spectroscopy, which has led to the concept of "structural-reporter groups." In these conditions, depending on the field of spectrometer (600–300 MHz), 20–100 nmol will constitute a sufficient amount of material to apply a "finger-print" method. Nevertheless, the method is restricted to compounds which are members of a series of closely related sequences, as it is happily the case for most of *O*-glycans.

When the material is available in the range of 0.1–5 µmol, the *de novo* structural elucidation of the sequences can be easily deduced from the compilation of data furnished by various homo and heteronuclear 2D NMR methods.

1.3. Mass Spectrometry

MS has become an indispensable tool for the determination of carbohydrate structures. The information provided by this methodology ranges from the accurate molecular weight determination to the complete primary structure with a sensitivity such that only picomoles of oligosaccharides are necessary. These remarkable advances have been made possible with the appearence of novel methods of ionization such as fast atom bombardment ionization (FAB), electrospray ionization (ESI), and matrix-assisted laser desorption ionization (MALDI).

2. Methods

2.1. Nuclear Magetic Resonance

2.1.1. Proton-NMR as a Fingerprinting Method

The proton-NMR method was developed by Vliegenthart and colleagues during the 1970s and essentially applied to the structure determination of *N*-glycans of the *N*-acetyllactosamine and oligomannoside type ***(16)***. More recently, a similar procedure was summarized for the primary structural analysis of oligosaccharide-alditol released from mucin-type *O*-glycosylproteins ***(17)***. This method is based on the recognition of some atom resonances that constitute probes for representative structural elements. These structural-reporter groups resonate outside the bulk constituted by the nonanomeric protons. ^{1}H-NMR structural-reporter group signals correspond to the following atom resonances: anomeric protons; GalNAc-ol H-2, H-4, H-5 and H-6'

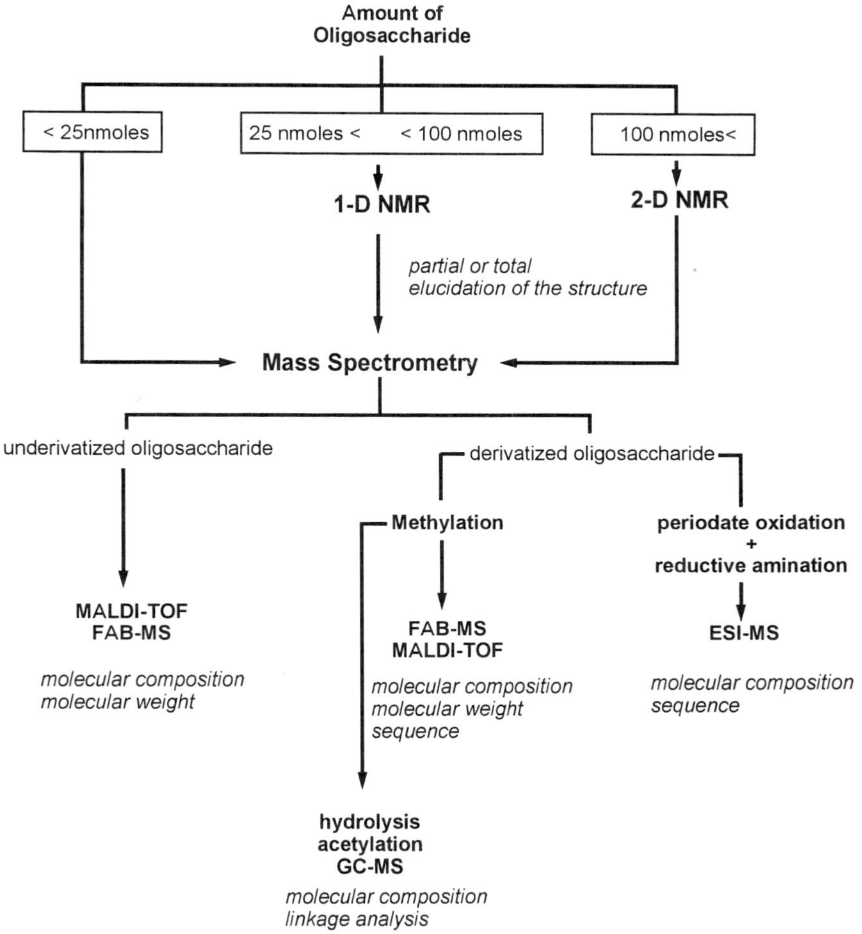

Fig. 1. Strategy of elucidation of oligosaccharide primary structure.

atoms; Gal H-3 and H-4 atoms; Fuc H-5 and H-6 atoms; NeuAc H-3ax and H-3eq atoms; and CH_3 of the acetamido groups.

The first step of spectrum analysis consists of the identification of the core region (**Table 1**), based on the characteristic chemical shifts of the H-2 and H-5 atom resonances of the GalNAc-ol unit. Moreover, the quadruplet of the H-6' signal of GalNAc-ol is upfield shifted out of the bulk at $\delta \sim 3.50$ ppm in the case of an O-6 substitution with sialic acid. The presence of α-2,3- or α-2,6-linked sialic acid is clearly shown by the respective chemical shift of the H-3ax and H-3eq signals of the monosaccharides. The H-3ax and H-3eq resonances of the α-2,3-linked NeuAc are systematically downfield shifted, compared to the corresponding signals of α-2,6-linked NeuAc.

The attachment of NeuAc at O-3 of a Gal unit causes downfield shifts of the Gal structural-reporter groups, as clearly indicated in **Fig. 2**, in which the NMR spectra of asialo and sialo glycans are compared (compounds N-1 and A-1).

Table 1
Chemical Shifts of GalNAc-ol Residues Characteristic of the Nature of Oligosaccharide-Alditol Cores

	Gal(β1–3)**GalNAc-ol**	GlcNAc(β1–3)**GalNAc-ol**	Gal(β1–3)[GlcNAc(β1–6)]**GalNAc-ol**	NeuAc(α2–6)**GalNAc-ol**
H-2	4.393	4.286	4.391	4.246
H-3	4.063	3.995	4.069	3.846
H-4	3.506	3.546	3.468	3.411
H-5	4.193	4.141	4.277	4.020
H-6'	3.628	ND	ND	3.532

	Gal(β1–3)[NeuAc(α2–6)]**GalNAc-ol**	GlcNAc(β1–3)[NeuAc(α2–6)]**GalNAc-ol**	GlcNAc(β1–6)**GalNAc-ol**
H-2	4.378	4.260	4.242
H-3	4.055	3.984	3.841
H-4	3.534	ND	3.379
H-5	4.244	4.185	4.021
H-6'	3.486	3.490	3.933

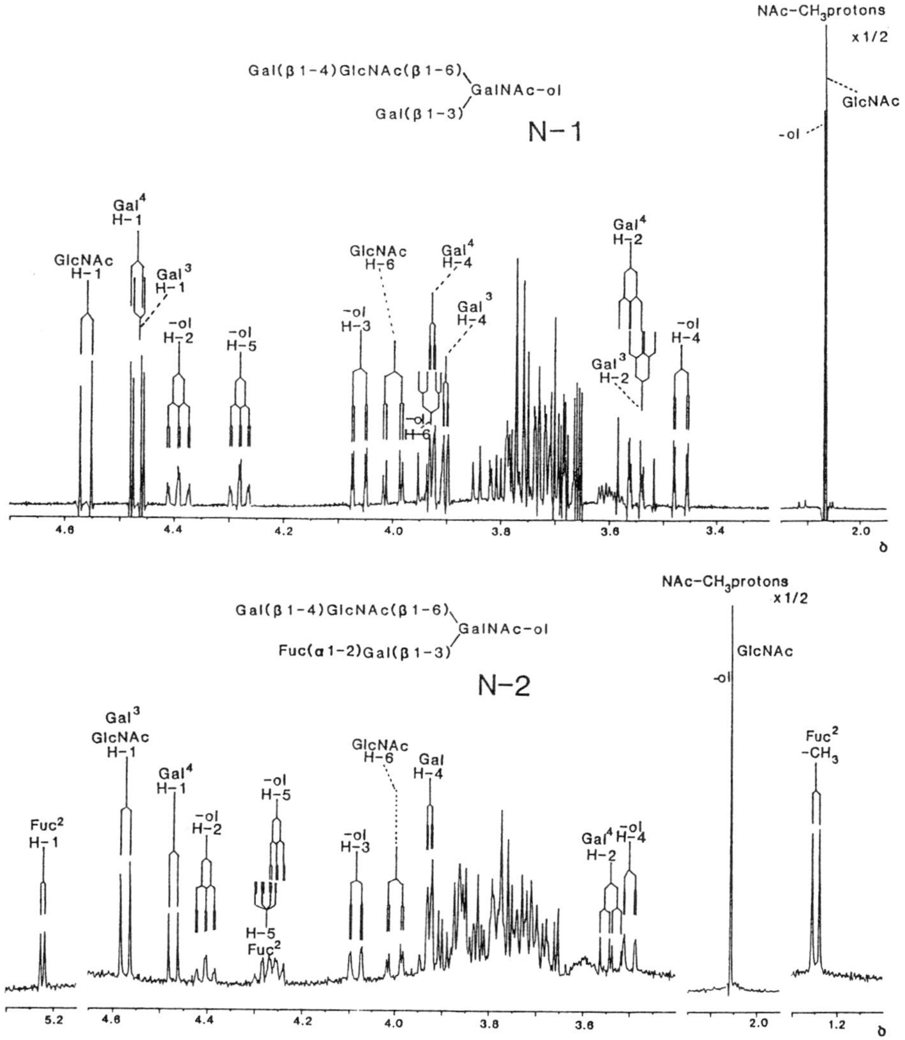

Fig. 2. ^{1}H-NMR spectra of three oligosaccharide-alditols. N-1, basic structure devoid of fucose and sialic acid residue; N-2, downfield shift of Gal[3] H-1 owing to the α-1,2 fucose; A-1, downfield shift of Gal[3] H-1 and H-3 owing to the α-2,3 sialylation. The first superscript after the abbreviated name of a monosaccharide residue indicates to which position of the adjacent monosaccharide it is glycosidically linked (e.g., Gal[4] in the case of Galβ1→4 GlcNAcβ1→).

Fucose units can be easily identified according to the presence of methyl resonances at ~ 1.2 ppm. The α-1,2 (H), α-1,3 (Le^x), and α-1,4 (Le^a) linkages are deduced from the position of H-1, H-5, and H-6 resonances. For instance, the Le^x and Le^a epitopes can be characterized on the basis of their H-1 (Le^x: δ ~ 5.13–5.14; Le^a : δ ~ 5.02–5.05), and H5 (Le^x: δ ~ 4.80–4.85; Le^a: δ ~ 4.86–4.88) resonances, whereas the

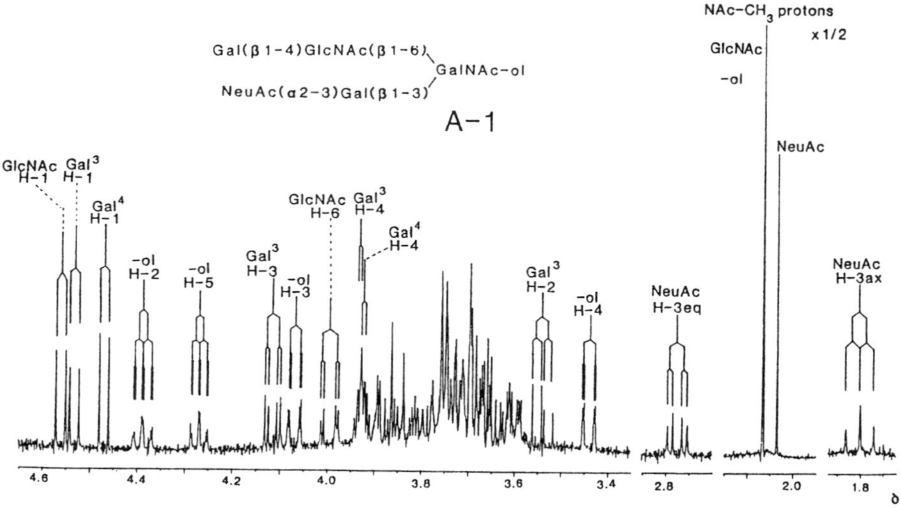

Fig. 2. Continued.

H-5 signal of α-1,2-linked Fuc is observed at δ ~ 4.25–4.34 ppm. These NMR data imply, respectively, the presence of type 2 (Galβ1-4GlcNAc) and type 1 (Galβ1-3GlcNAc) backbone structures.

The analysis of compounds with a higher complexity in backbone sequence cannot be developed in some pages, and such analyses often call for additional chemical (methylation analysis) or physical (MS, nuclear Overhauser effect [NOE] measurements) operations.

Figures 3 and **4** give examples of complex backbones. Compounds were analyzed by methylation analysis, which precises the location of the hydroxyl groups implicated in the glycosidic linkages. The comparison of the spectra N-4B, N-4A, N-6B, and N-8 clearly indicates the presence of the same sequence Gal(β1-4)GlcNAc(β1-6) GalNAc-ol, as shown by the NMR parameters of Gal^4 and $GlcNAc^6$. This comparison between N-1, A-1, and A-3 (**Fig. 2**) also gives the chemical shift of the sialylated and terminal Gal unit O-3 linked to GalNAc-ol. Consequently, the upper branch of compound A-3 is composed of two Gal and two GlcNAc units. Other models (not shown here) also possess the sequences Gal(β1-3)GlcNAc(β1-3)Gal(β1-4)GlcNAc or Gal(β1-4)GlcNAc(β1-3)Gal(β1-4)GlcNAc, with additional Fuc or NeuAc attached to the terminal galactose units. Since the presence of these peripheral monosaccharides affects essentially the chemical shifts of the terminal galactose, the anomeric protons can be assigned step-by-step.

The compilation of 168 NMR spectra which are included in the review of Kamerling et al. *(17)* clearly shows that a spectrum is unique and can be used as an "identity card." This review presents the carbohydrate chains in a logical order and gives a classification according to the nature of the core, the nature of the backbone, and the peripherical monosaccharides. Actually, the nature of the backbone can be often di-

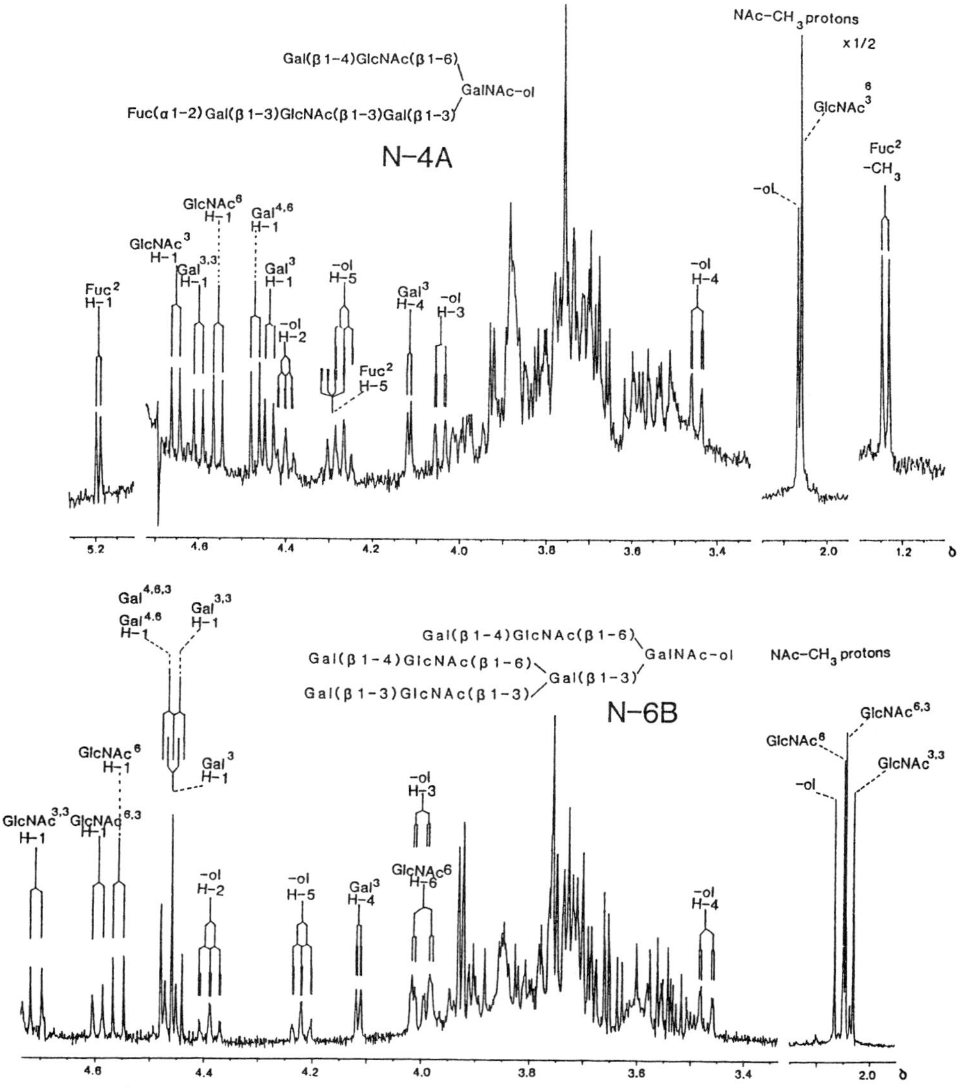

Fig. 3. ^{1}H-NMR spectra of oligosaccharide-alditols with the common tetrasaccharide N-1 (*see* **Fig. 2**) variously substituted.

rectly deduced from the linkage of the fucose units, as discussed previously. When the presence of these structural elements has been established, the number of possibilities decreases deeply, and a survey of the corresponding class of oligosaccharide-alditols furnishes rapidly the structure of the compound. Most of the *O*-glycans that constitute the carbohydrate moiety of mucins isolated from human tissues have now been described, but completely new structures can be isolated from other biological sources. If a sufficient amount of material is available, *de novo* structural elucidation of glycan sequences should be performed by 2D NMR spectroscopy.

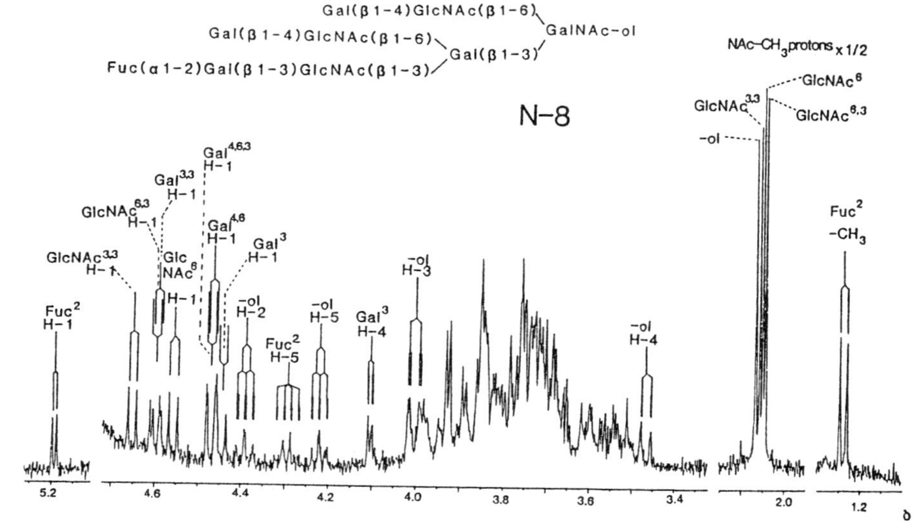

Fig. 3. Continued.

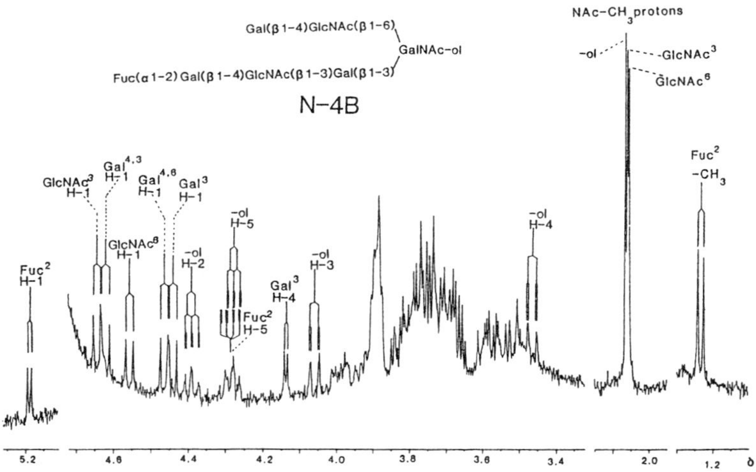

Fig. 4. ^{1}H-NMR spectrum of oligosaccharide-alditols with the common tetrasaccharide N-1 (*see* **Fig. 2**) variously substituted.

2.1.2. *De novo* Structural Elucidation of Glycan Sequences by NMR Spectroscopy

Homonuclear correlated spectroscopy (COSY) provides information on directly coupled protons with regard to coupling constants. Starting from the anomeric proton, the H-2, H-3, H-4, and so on, atom resonances, which were masked in the bulk, can be assigned. Nevertheless, such an assignment is generally difficult when the cross peaks

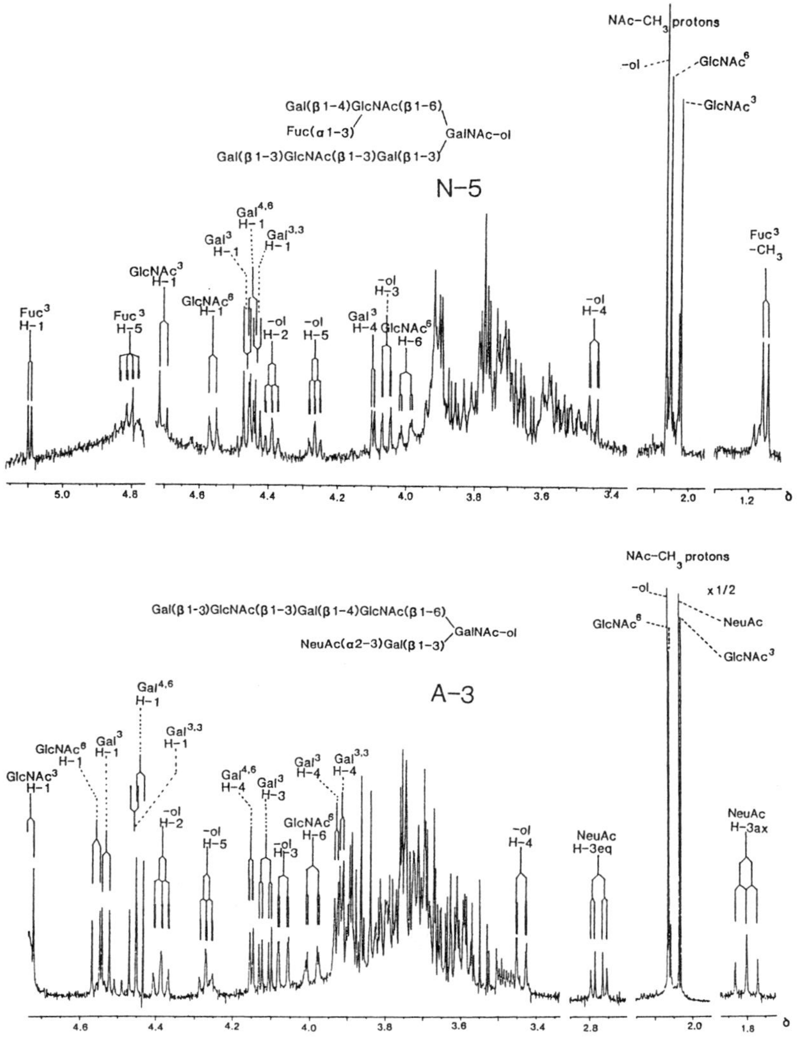

Fig. 4. Continued.

occur around the diagonal. Two-dimensional total correlation spectroscopy (2D ^{1}H-TOCSY) can be used to characterize all the proton resonances. A transfer of magnetization from H-1 to H-6 is observed in the case of the α– and β-*gluco* configuration ($J_{1,2}$–$J_{4,5}$ ~ 8 Hz), wheareas the small $J_{4,5}$ that characterizes the *galacto* configuration interrupt the assignment at the H-4 resonance. Relayed COSY spectra have the advantage of successively assigning the H-2 (COSY), H-3 (one-step-relayed COSY), and H-4 (two-steps-relayed COSY) atom resonance of the carbohydrate units.

The example given in **Fig. 5** clearly shows the presence of two β-Gal, one α-Gal and one GlcNAc units. For *N*-acetylglucosamine, H-2, H-3, and H-4 signals are trip-

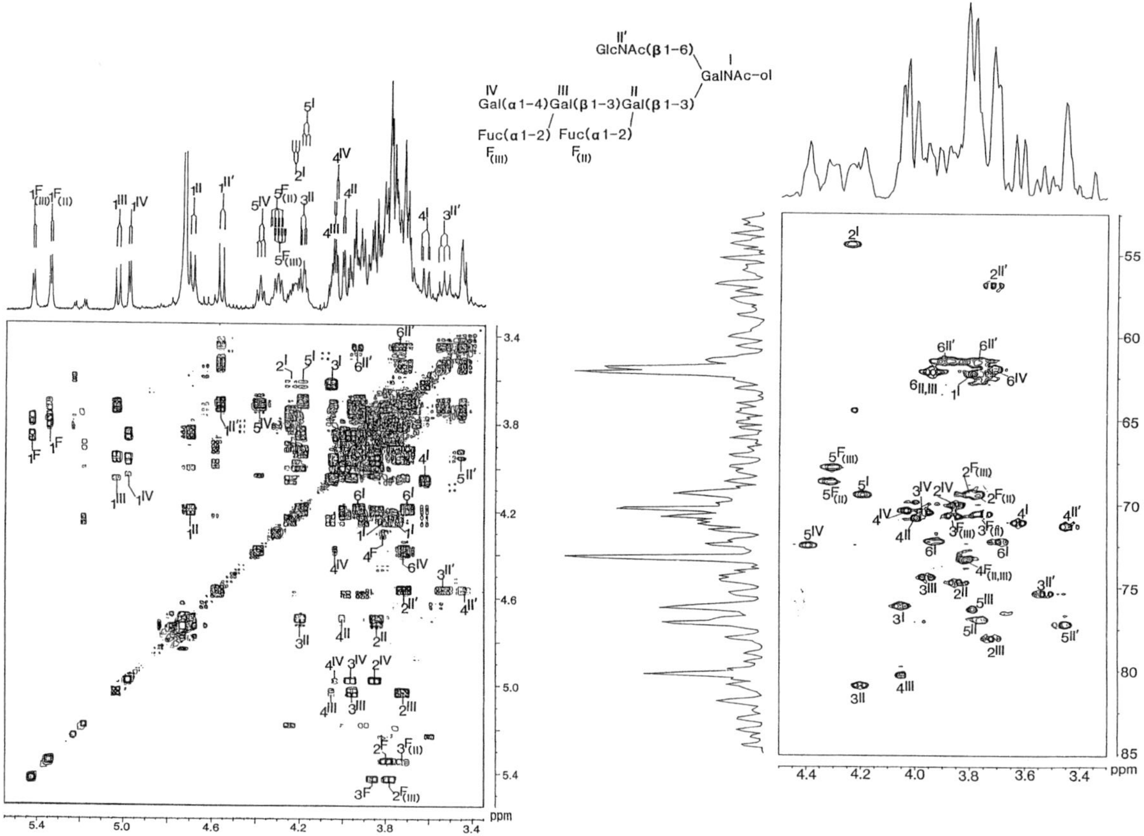

Fig. 5. COSY **(left)** and HMQC **(right)** NMR spectra of an heptasaccharide-alditol isolated from the oviducal mucin of *Xenopus laevis*. 2^{II}, proton H-2 of monosaccharide unit II. For the COSY spectrum, starting from the anomeric proton (i.e., 1^{III}) the protons 2^{III}, 3^{III}, and 4^{III} are

lets (transaxial hydrogens), whereas the same resonances of β-galactose are triplet, pseudo-doublet, and pseudo-singlet, respectively. Fucose, which possesses an α-*galacto* configuration, is easily distinguished from α-galactose owing to the overlap of the correlations H-6→H-5→H-4 and H-1→H-2→H-3→H-4.

The heteronuclear multiple quantum-coherence spectroscopy (HMQC) relies on the ^{1}H and ^{13}C, which are directly attached. The position of the glycosidic linkage is clearly observed owing to a strong downfield shift (4–10 ppm) that affect the substituted carbon. In the example depicted in **Fig. 5**, the crosspeak 4^{II} observed at ~ 70 ppm corresponds to an unsubstituted carbon, in opposite to the signal 4^{III} at 80 ppm.

The most elegant method for establishing the exact sequence of the oligosaccharide is indisputably the heteronuclear multiple-bond correlation spectroscopy (HMBC) which relies the ^{1}H and ^{13}C via their $^{3}J_{H,C}$ coupling. Unfortunately, the method is too insensitive for being used systematically. In the example given in **Fig. 5** these expected connectivities should be observed: $1^{IV}{\rightarrow}3^{IV}$, 5^{IV}, 4^{III}; $1^{III}{\rightarrow}3^{III}$, 5^{III}, 3^{III}; $1^{II}{\rightarrow}3^{II}$, 5^{II}, 3^{I}, and so on.

2D Nuclear Overhauser effect spectroscopy (NOESY) or rotating-frame NOE spectroscopy (ROESY) is generally used for establishing the sequence of carbohydrate chains. Since the strongest NOE is not always between the protons connected to the linkage, the method may fail to establish the position of the glycosidic substitution. **Figure 6** describes such an assignment. A preliminary methylation analysis has shown the presence of one terminal Gal, two *O*-3-substituted GalNAc, two O-4-substituted Gal, and one *O*-3-substituted GalNAc-ol. The COSY experiment indicates that Gal and GalNAc have β and α configuration, respectively. The exact assignment of the five anomeric protons resulted from the connectivities $1^{IV}/3^{V}$, $1^{V}/4^{IV}$, $1^{IV}/3^{III}$, $1^{III}/4^{II}$, and $1^{II}/3^{I}$.

Fig. 5. *(continued)* successively assigned. The shape of the correlation peaks (triplet for H-2, pseudo-doublet for H-3, pseudo-singlet for H-4) allows the estimatation of the coupling constant: L (large) for $J > 6$ Hz; S (small) for $J < 4$ Hz. For the monosaccharide unit III, we observe $J_{1,2} > 6$ Hz; $J_{2,3} > 6$ Hz; $J_{3,4} < 4$ Hz; $J_{4,5} < 1$ Hz. These values demonstrate the following configuration of the ring protons: H-1, axial; H-2, axial; H-3, axial; H-4, equatorial, which are characteristic of a galactose residue in a β configuration. With the same demonstration, units II, II, and IV possess, respectively the β-Gal, β-Glc, and α-Gal configuration. Hexose and *N*-acetylhexosamine are discriminated according to their H-2 resonances, strongly deshielded in the case of *N*-acetylhexosamine. For the HMQC spectrum, a comparison with the corresponding COSY spectrum allows the assignment of the ^{13}C resonances (i.e. 4^{III} represents the correlation peak between Gal III ^{1}H-4/^{13}C-4). For the β-Gal units, the ^{13}C-atom resonances 2^{II}, 3^{II}, 2^{III}, and 4^{III} are deshielded from 3 to 10 ppm, as compared to the standard values observed for β-methylgalactoside. On the contrary, the signal 3^{III} and 4^{II} possess normal values (nonsubstituted hydroxyl group at position 3 and 4, respectively).

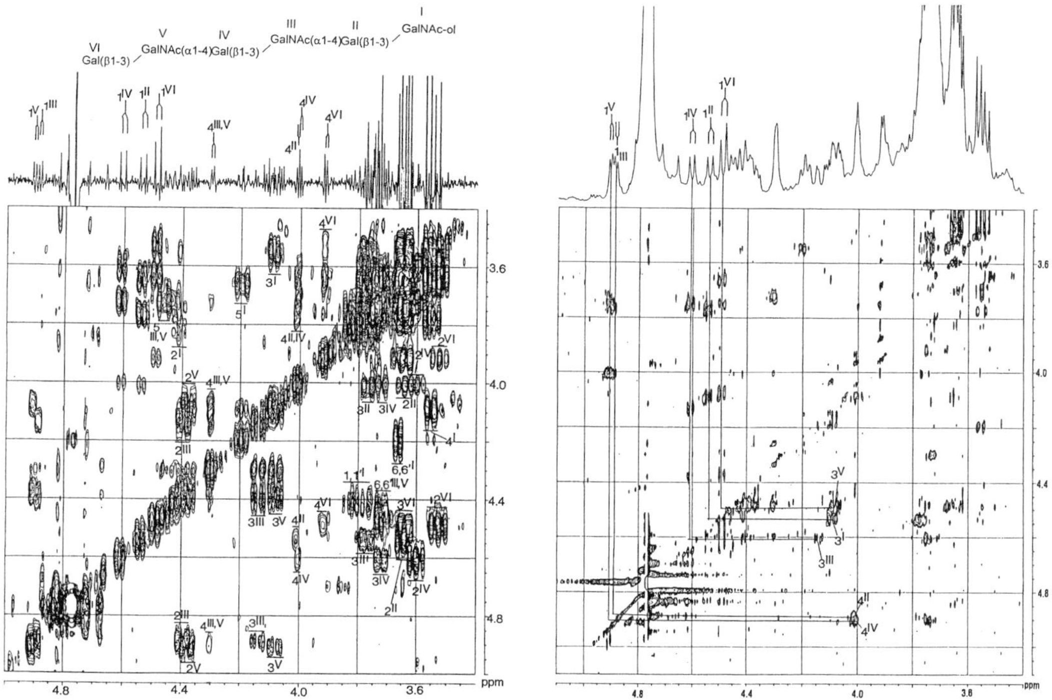

Fig. 6. COSY **(left)** and ROESY **(right)** NMR spectra of a hexasaccharide-alditol isolated from the oviducal mucin of *Rana palustris.* 2^{II}, proton H-2 of monosaccharide unit II. For the COSY spectrum, the carbohydrate units were identified on the basis of the set of the vicinal coupling constants as described in **Fig. 5**. For the ROESY spectrum, the correlations $1^{VI} \rightarrow 3^{V}$, $1^{V} \rightarrow 4^{IV}$, $1^{IV} \rightarrow 3^{III}$, $1^{III} \rightarrow 4^{II}$, and $1^{II} \rightarrow 3^{I}$ clearly indicate the sequence of the oligosaccharide and confirm the assignment of each anomeric proton.

2.2. Mass Spectometry

2.2.1. FAB-Mass Spectometry

2.2.1.1. Principle

The sample containing 20–100 pmol of analyte is mixed with the matrix on the target. Bombardment of this sample/matrix mixture in the ionization source with an energetic beam of atoms such as argon or xenon forms ions that are extracted from the surface of the target, accelerated, and then mass analyzed in the spectrometer. The FAB ionization method can be applied to analysis of native or derivatized oligosaccharides.

2.2.1.2. Methods

Oligosaccharides must be prepared in a desalted form to remove all sodium or other metal from the sample in order to enhance the production of the $[M+H]^+$ or $[M-H]^-$ ion. For experiments in which the M+Na ion is desired, the appropriate salt can be added to the sample. The matrix used is either glycerol for native oligosaccharides or thioglycerol for permethylated samples *(18)*. An alkaline matrix can be used such as glycerol:water:triethanolamine (3:2:1, v/v/v) and will amplify the signal for sulfated or sialylated compounds.

Native oligosaccharides exhibit poorer response than their derivatized forms. Carbohydrate permethylation and peracetylation are the most commonly used procedures. The derivatization of the oligosaccharide presents many advantages : it increases the sensitivity, the desalting step is rendered easier, and the fragmentation patterns of the derivatives are well defined *(18)*.

NaOH permethylation according to Ciucanu and Kerek *(19)* is a simple and efficient procedure. Dried oligosaccharides are first dissolved in dimethyl sulfoxide. Then finely powdered NaOH is added with methyliodine, the reaction mixture is placed in an ultrasonic bath for 1 to 2 h at room temperature, and the reaction is stopped by careful addition of water. The permethylated oligosaccharide is extracted by chloroform or by solid-phase extraction procedure over a Sep-Pak® cartridge (Waters, Milford, MA). For neutral oligosaccharide-alditols, this is the method of choice. The methylated oligosaccharide-alditols are dissolved in methanol containing sodium acetate (0.1%) and loaded on the metal target with thioglycerol as a matrix *(2)*. If sufficient material is available, the methylated sample can be further analyzed after methanolysis and acetylation to identify the methylated derivatives *(20)*.

The NaOH methylation procedure is too strong for sulfated oligosaccharide. These samples require a modified Hakomori procedure *(21)*, which preserves the sulfate residue. The oligosaccharides are dissolved in methyl sulfoxide to which is sequentially added a solution of methylsulfinylmethanide (sodium hydride with methyl sulfoxide) and methyliodine at room temperature. The reaction is stopped by water, and the methylated oligosaccharides are purified over a C18, Sep-Pak cartridge *(22)*.

2.2.1.3. Results

Most informative signal are given by the pseudomolecular ions corresponding to $(M+H)^+$, $(M+Na)^+$ when recorded in the positive mode and $(M-H)^-$ in the negative

Table 2
Calculation of Molecular Weight of Oligosaccharides[a]

Compounds	Mass incrementation in native oligosaccharides	Mass incrementation in permethylated oligosaccharides
Hexose	180	250
Desoxyhexose	164	220
N-acetylhexosamine	221	291
N-acetylhexosaminitol	223	307
N-acetylneuraminic acid	309	407
Glycosidic linkage	–18	–46

[a]The nominal value of *m/z* is given excluding the fractional mass increment of each atom; e.g., H is counted as 1 instead of 1.008. The molecular weight is obtained by the sum of the respective molecular weight of each constituting monosaccharide minus the weight corresponding to the number of glycosidic linkages.

mode. Molecular ions give information on the chemical composition of the oligosaccharides if they are composed of the generally constituting monosaccharides (hexose, *N*-acetylhexosamine, deoxyhexose, *N*-acetylneuraminic acid). The increments of masses per residue are given in atomic mass units in **Table 2**, to this sum the mass carresponding to the number of linkage has to be substracted.

When sufficient material is available, fragmentation of the oligosaccharide in the source may occur. With underivatized oligosaccharides, the fragmentation patterns are ambiguous, rendering the determination of the sequence uncertain. Permethylated oligosaccharides have a pattern of breakdown that results from the cleavage of the glycosidic bonds. Different pathways of fragmentation have been described, in which the charge is retained either on the reducing or on the non-reducing end of the oligosaccharide ***(18,23)***. When ring cleavages are observed, they are more difficult to assign but give off the linkage position. Finally, secondary ions, resulting from the preferential elimination of substituents on a certain position (i.e., 3-position of an *N*-acetylhexosamine) complete the structural identification of the oligosaccharide.

2.2.2. MALDI-Mass Spectometry

2.2.2.1. Principle

The sample containing 1–100 pmol of analyte is mixed with the matrix and loaded on a sample slide. The function of the matrix is to isolate the sample molecules from each other and to absorb photons from an ultraviolet laser, usually a 337-nm N_2 laser. Absorption of the laser energy by the matrix leads to a fast ejection of molecular material. Ionization takes place by proton transfer from matrix ions to neutral analyte molecules. This mild ionization occurs in a gas phase resulting from the energy of the laser at the surface of the slide. MALDI is usually combined with a time-of-flight (TOF) mass spectrometer.

2.2.2.2. Methods

Oligosaccharide samples (10–50 pmol dissolved in 1 mL of water) and the matrix solution (1 mL of a solution of 2, 5-dihydroxybenzoic acid [12 mg/mL] in meth-

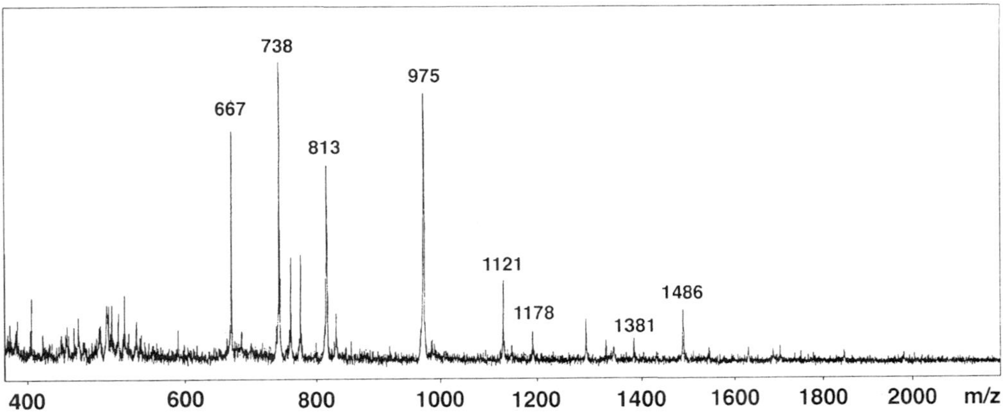

Fig. 7. MALDI-TOF of a total fraction of acidic oligosaccharide-alditols isolated from a human colon carcinoma cell line mucin produced in nude mice as xenograph *(24)*.

anol:water, 80:20), are placed and dried on the mass spectrometer target. NaCl can be added to enrich the Na^+ concentration and produces M+Na ions. For di- or triacidic fractions, 3-aminoquinolein (10 mg/mL in acetonitrile:water, 80:20) is used, and the ions are analyzed in the negative mode *(24)*.

2.2.2.3. Results

Pseudomolecular ions corresponding either to $(M+H)^+$, $(M+Na)^+$ when recorded in the positive mode or $(M-H)^-$ in the negative mode are observed. As in FAB-MS, molecular ions give information on the chemical composition of the oligosaccharide if they are composed of the generally constituting monosaccharides (hexose, *N*-acetylhexosamine, deoxyhexose, *N*-acetylneuraminic acid). **Figure 7** gives a MALDI-TOF mass spectrum of a total fraction of acidic oligosaccharide-alditols isolated from a human colon carcinoma cell line mucin produced in nude mice as xenograph. Twenty-two mg of mucin were used for total structural determination, of which 50 pmol of saccharides were used to obtain the spectrum. The chemical composition of nine oligosaccharide could be deduced from this analysis. **Table 3** gives the interpretation of the spectrum.

2.2.3. ESI-Mass Spectometry

2.2.3.1. Principle

In ESI-MS, analytes are ionized at atmospheric pressure directly from a flowing liquid stream. The sample solution enters the ES chamber at a flow rate between 1 and 50 µL/min through a capillary that is maintained at a high potential, and the solution emerges from the capillary as an aerosol of charged droplets. Solvent evaporation through a nitrogen flux results in the formation of charged sample molecules, and the

Table 3
MALDI-TOF MS of Monosulfated Oligosaccharides[a]

m/z	HSO_3	Deoxyhexose	Hexose	*N*-acetylhexosamine	*N*-acetylhexosaminitol
667	1	–	1	1	1
738	1	1	1	1	–[b]
813	1	1	1	1	1
829	1	–	2	1	1
975	1	1	2	1	1
1121	1	2	2	1	1
1178	1	1	2	2	1
1381	1	1	2	3	1
1486	1	2	3	2	1

[a]All *m/z* values correspond to pseudomolecular ions owing to loss of one H. The nominal value of *m/z* is given excluding the fractional mass increment of each atom; e.g., H is counted as 1 instead of 1.008. The error in mass assignment was about ±1/1000.

[b]Instead of an *N*-acetylhexosaminitol, the *m/z* value indicated the presence of a group of mass 131 that corresponded to a by-product of β-elimination in which the *N*-acetylgalactosaminitol has been degraded in a hexene-tetrol *(24)*.

ions produced are then directed into the high-vacuum region of the mass spectrometer. The compatibility with low flow rate allows coupling with high-performance liquid chromatography (HPLC) or capillary electrophoresis.

2.2.3.2. Methods

A recent technique developed by Morelle et al. *(25)* allows microsequencing of oligosaccharide-alditols. It consists of three steps: mild periodate oxidation of the oligosaccharide-alditols, coupling of the oxidized fragments to 2-aminopyridine, and separation and characterization of the derivatives by ESI-MS/MS.

First, the mild periodate oxidation (10 nmol of oligosaccharide-alditols disssolved in an imidazole buffer, and reacted with 50 μmol of IO_4^- at 0°C for 30 min) cleaves the *N*-acetylgalactosaminitol between the carbons 4 and 5 and also the exocyclic chain of sialic acid when it is not 8 or 9 substituted. The oligosaccharide-alditol is cleaved in two fragments, one corresponding to the 3-branch and the other to the 6-branch. Second, the resulting aldehydes are coupled by reductive amination to 2-aminopyridine: in a first step, the two oxidized fragments react with 2-aminopyridine to form a Schiff base that is reduced by borane dimethylamine complex. The excess reagents are removed by evaporation and extraction. Third, the pyridylamino derivatives are separed by HPLC with an octadecylsilane column and collected. The samples are dissolved in methanol:water:acetic acid at a concentration of 50 pmol/μL and injected in the electrospray source.

3.2.3.3. Results

After periodate oxidation, the cleavage of the *N*-acetylgalactosaminitol results in two fragments: one corresponds to the 3-branch and the other to the 6-branch of the

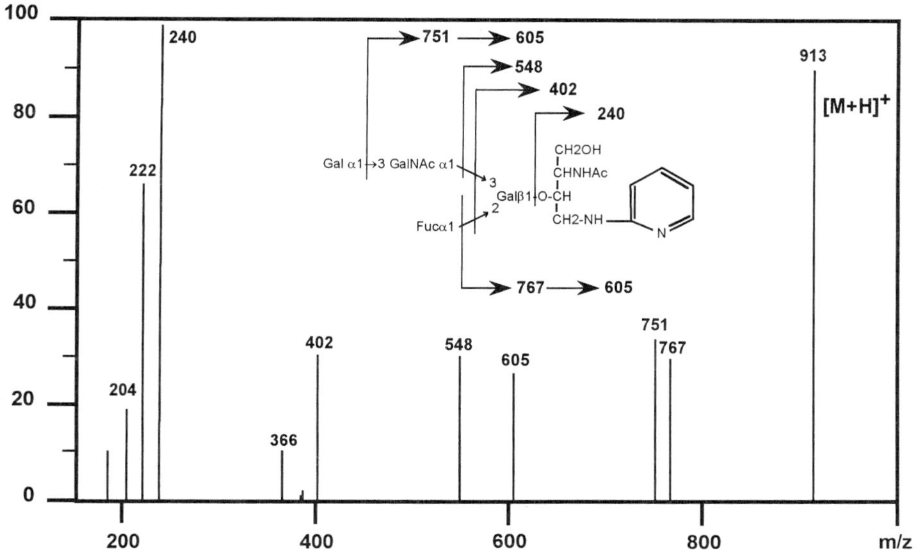

Fig. 8. ESI-MS of the 3-branch, an oligosaccharide-alditol, after mild periodate oxidation and derivatization with 2-aminopyridine.

oligosaccharide. These two fragments differ by the number of carbons present on the aglycon: four carbons on the 3-branch, the two other carbons of the hexosaminitol on the 6-branch. The separation of each branch is achieved on a C18 column and each branch is studied by ESI-MS.

The positive-ion ES of the pyridylamino-oligosaccharide gives a spectrum with $[M+H]^+$ and $[M+Na]^+$, with $[M+H]^+$ predominating. Molecular ions formed by ES ($[M+H]^+$ and $[M+Na]^+$) are submitted to collision-induced dissociation in the second quadrupole region of the mass spectrometer. In this type of fragmentation, the glycosidic cleavages are predominent. In **Fig. 8**, the cleavage of the glycosidic linkages of the 3-branch of an oligosaccharide alditol gives the ions at m/z 751, 548, 402, and 240, the position of the fucose residue on the first galactose residue is indicated by the ions at *m/z* 767, 605, and 548. The ion at *m/z* 366 corresponds to a preferential cleavage at the internal *N*-acetylhexosamine residue with the charge retained on the nonreducing end; further fragmentation of this ion gives the fragment at *m/z* 204 that correponds to an oxonium ion of an *N*-acetylgalactosamine residue.

References

1. Lamblin, G., Lhermitte, M., Klein, A., Houdret, N., Scharfman, A., Ramphal, R., and Roussel, P. (1991) The carbohydrate diversity of human respiratory mucins: a protection of the underlying mucosa? *Am. Rev. Respir.Dis.* **144,** S19–S24.
2. Klein, A., Carnoy, C., Wieruszeski, J. M., Strecker, G., Strang, A. M., van Halbeek, H., Roussel, P., and Lamblin, G. (1992)The broad diversity of neutral and sialylated oligosaccharides derived from human salivary mucins *Biochemistry* **31,** 6152–6165.

3. Lo-Guidice, J. M., Wieruszeski, J. M., Lemoine, J., Verbert, A., Roussel, P., and Lamblin, G.(1994) Sialylation and sulfation of the carbohydrate chains in respiratory mucins from a patient with cystic fibrosis. *J. Biol. Chem.* **269,** 18,794–18,813.
4. Lo-Guidice, J. M., Herz, H., Lamblin, G., Plancke, Y., Roussel, P., and Lhermitte, M. (1997) Structures of sulfated oligosaccharides isolated from the respiratory mucins of a non-secretor (O, Le^{a+b-}) patient suffering from chronic bronchitis. *Glycoconjugate J.* **14,** 113–125.
5. Klein, A., Lamblin, G., Lhermitte, M., Roussel, P., Breg, J., van Halbeek, H., and Vliegenthart, J. F. G. (1988) Primary structure of neutral oligosaccharides derived from respiratory-mucus glycoproteins of a patient suffering from bronchiectasis, determined by combination of 500-MHz ^{1}H-NMR spectroscopy and quantitative sugar analysis.1-Structure of 16 oligosaccharides having the Galβ(1→3)GalNAc-ol core (type 1) or the Galβ1→3 [GlcNAcβ1→6]GalNAc-ol core (type 2).*Eur. J. Biochem.* **171,** 631–642.
6. Klein, A., Carnoy, C., Lamblin, G., Roussel, P., Van Kuik, J. A., De Waard, P., and Vliegenthart, J. F. G. (1991) Isolation and structural characterization of novel neutral oligosaccharide-alditols from respiratory-mucus glycoproteins of a patient suffering from bronchiectasis: 1. Structure of 11 oligosaccharides having the GlcNAcβ(1→3) Galβ(1→4)GlcNAcβ(1→6)GalNAc-ol structural element in common. *Eur. J. Biochem.* **198,** 151–168.
7. Klein, A., Carnoy, C., Lamblin, G., Roussel, P, Van Kuik, J. A., De Waard, P., and Vliegenthart, J. F. G. (1993) Isolation and structural characterization of novel sialylated oligosaccharide-alditols from respiratory-mucus glycoproteins of a patient suffering from bronchiectasis. *Eur. J. Biochem.* **211,** 491–500.
8. Breg, J., van Halbeek, H., Vliegenthart, J. F. G., Klein, A., Lamblin, G., and Roussel, P. (1988) Primary structure of neutral oligosaccharides derived from respiratory-mucus glycoproteins of a patient suffering from bronchiectasis, determined by combination of 500-MHz ^{1}H-NMR spectroscopy and quantitative sugar analysis. 2-Structure of 19 oligosaccharides having the GlcNAcβ(1→3)GalNAc-ol core (type 3) or the GlcNAcβ1→3 [GlcNAcβ1→6]GalNAc-ol core (type 4). *Eur. J. Biochem.* **171,** 643–654.
9. Van Halbeek, H., Breg, J., Vliegenthart, J. F. G., Klein, A., Lamblin, G., and Roussel, P. (1988) Isolation and structural characterization of low-molecular-mass monosialyl oligosaccharides derived from respiratory-mucus glycoproteins of a patient suffering from bronchiectasis. *Eur. J. Biochem.***177,** 443-460.
10. van Kuik, J. A., de Waard, P., Vliegenthart, J. F. G., Klein, A., Carnoy, C., Lamblin, G., and Roussel, P. (1991) Isolation and structural characterization of novel neutral oligosaccharide-alditols from respiratory-mucus glycoproteins of a patient suffering from bronchiectasis: 2. Structure of 12 hepta to nonasaccharides, six of which possess the GlcNAcβ(1→3)[Galβ(1→4) GlcNAcβ (1→6)]Galβ(1→3) GalNAc-ol common structural element. *Eur. J. Biochem.* **198,** 169–183.
11. Strecker, G., Wieruszeski, J.-M., Michalski, J. C., Alonso, C., Leroy, Y., Boily, B., and Montreuil, J. (1992) Primary structure of neutral and acidic oligosaccharide-alditols derived from the jelly coat of the *Mexican axolotl.* Occurence of oligosaccharide with fucosyl α(1-3) fucosyl (α1-4)-3deoxy-D-glycero-D-galacto-nonulosonic acid and galactosyl (α1-4) [fucosyl(α1-2)]galactosyl(β1-4)N-acetylglucosamine sequence. *Eur. J. Biochem.* **207,** 995–1002.
12. Strecker, G., Wieruszeski, J.-M., Fontaine, M. D., and Plancke, Y. (1994) Strucutre of the major neutral oligosaccharide-alditols released from the egg jelly coat of *Axolodt maculatum.* Characterization of the carbohydrate sequence GalNAc(β1-4) [Fuca(1-3)]GlcNAcβ(1-3/6).*Glycobiology* **4,** 605–609.

13. Strecker, G., Wieruszeski, J.-M., Plancke, Y., and Boilly, B. (1995) Primary structure of 12 neutral oligosaccharide alditols released from the jelly coats of the anouran *Xenopus laevis* by reductive β-elimination. *Glycobiology* **5,** 137–146.
14. Morelle, W. and Strecker, G. (1997) Structural analysis of the oligosaccharide-alditols released by reductive β-elimination from the jelly coat of *Rana utricularia* eggs. *Biochem. J.* **321,** 879–887.
15. Morelle, W. and Strecker, G. (1997) Structural analysis of oligosaccharide-alditols released by reductive b-elimination from oviductal mucins of *Bufo bufo*: characterization of the carbohydrate sequence Galα(1-3)GalNAcα(1-3)[Fucα(1-2)]Gal. *Glycobiology* **7,** 777–790.
16. Vliegenthart, J. F. G., Dorland, L., and van Halbeek, H. (1983) High resolution 1H nuclear magnetic resonance spectroscopy as a tool in the structural analysis of carbohydrates related to glycoproteins. *Adv. Carbohydr. Biochem.* **41,** 209–374.
17. Kamerling, J P. and Vliegenthart, J. F. G. (1992) High-resolution ^{1}H-nuclear magnetic resonance spectroscopy of oligosaccharide-alditols released from mucin-type *O*-glycoproteins, in *Biological Magnetic Resonance* (Erliner, L. J. and Reben, J., eds.), vol 10, pp. 1–287.
18. Dell, A. (1987) FAB mass spectrometry of carbohydrates *Adv. Carbohydr. Chem. Biochem.,* **45,** 20-71.
19. Ciucanu, I. and Kerek,F. (1984) A simple and rapid method for the permethylation of carbohydrates *Carbohydr. Res.* **131,** 209–217.
20. Fournet, B., Strecker, G., Leroy, Y., and Montreuil, J. (1981) Gas-liquid chromatography and mass spectrometry of methylated and acetylated methyl glycosides. Application to the structural analysis of glycoprotein glycans. *Anal. Biochem.* **116,** 489–502.
21. Hakomori, S. I. (1964) A rapid permethylation of glycolipid, and polysaccharide catalysed by methylsulfinyl carbanion in dimethyl sulfoxide. *J. Biochem.* **55,** 205–208.
22. Dell, A., Morris, H. R., Greer, F., Redfern, J. M., Rogers, M. E., Weisshaar, G., Hiyama, J., and Renwick, A. G. C.(1991) Fast-atom-bombardment mass spectrometry of sulfated oligosaccharides from ovine lutropin *Carbohydr. Res.* **209,** 33–50.
23. Domon, B. and Costello, C. E. (1988) A systematic nomenclature for carbohydrate fragmentations in FAB-MS/MS spectra of glycoconjugates *Glycoconjugate J.* **5,** 397–409.
24. Capon, C., Wieruseski, J. M., Lemoine, J., Byrd, J. C., Leffler, H., and Kim Y. S. (1997) Sulfated Lewis X determinants as a major structural motif in glycans from LS174T-HM7 human colon carcinoma mucin. *J. Biol. Chem.* **272,** 31,957–31,968.
25. Morelle, W., Lemoine, J., and Strecker, G. (1998) Structural analysis of *O*-linked oligosaccharide- alditols by electrospray tandem mass spectrometry after mild periodate oxidation and derivatization with 2-aminopyridine. *Anal. Biochem.* **259,** 16–27.

17

Measurement of Sulfate in Mucins

Mathew J. Harrison and Nicolle H. Packer

1. Introduction

The sulfation of the terminal sugar residues of mucins is a common and extensive posttranslational event that greatly influences the ultimate viscoelastic properties of mucin. Highly sulfated and/or sialylated mucins comprise a considerable proportion of the mucous layers of the gastrointestinal, respiratory, and reproductive tracts, and have been demonstrated to be associated with some pathological conditions and pathogenesis *(1–4)*. The precise biological roles for glycan sulfation are largely unknown; however, several groups have demonstrated discrete biological roles for specific instances of glycan sulfation of non-mucin glycoproteins. These roles include the control of the circulatory half-life of human luteinizing hormone, symbiotic interactions of leguminous plants and nitrogen-fixing bacteria, and the targeting of lymphocytes to lymph nodes *(5)*.

Historically, efforts to assess qualitatively and quantitatively the extent of glycoprotein sulfation have been based on the metabolic incorporation of ^{35}S. However, this method is expensive and painstaking to perform. Sulfate and phosphate are currently indistinguishable by normal modes of mass spectrometric analysis (79.96 and 79.97 Dalton, respectively), although collision-induced fragmentation electrospray mass spectrometry (MS) has been used to distinguish between phosphate and sulfate on a fungal glycoprotein-cellobiohydrolase I *(6)*. Others have also used tandem MS to assign sugar-sulfate linkages on rat MUC2 *(7)*.

Here we present a chromatographic method for the identification and quantitation of sulfation on mucin. This method is comparatively cheap and fast to perform, and can be automated to provide fast-screen or high-throughput analysis. It is particularly easy to implement for groups who are currently performing mucin sugar analysis, since much of the necessary chromatographic equipment and the hydrolysis method are similar for both methods. Although this method is not particularly sensitive by today's analytical standards (>1 nmol), it provides fast and reproducible quantitation of sulfate for highly sulfated biomolecules such as mucins.

From: *Methods in Molecular Biology, Vol. 125: Glycoprotein Methods and Protocols: The Mucins*
Edited by: A. Corfield © Humana Press Inc., Totowa, NJ

2. Materials

2.1. Solvents and Reagents

1. Water, carbonate-free (*see* **Note 1**), and sparged with a constant flow of argon.
2. 0.1 *M* NaOH, made up with decarbonated water, and similarly sparged with argon.
3. At least 2 L of 50 m*M* sulfuric acid (*see* **Note 2**).
4. HCl, concentrated (Aristar, BDH Laboratory Supplies, Poole, Dorset, UK).
5. 1 m*M* 2-sulfo-*N*-acetylglucosamine, 3-sulfo-*N*-acetylglucosamine, and 6-sulfo-*N*-acetylglucosamine, in water (*see* **Note 3**).

2.2. Equipment

1. High-performance liquid chromatography (gradient controller, binary pump) (*see* **Note 4**).
2. AS11 (4 × 250 mm) ion-exchange column (Dionex) (*see* **Note 5**).
3. AMMS II postcolumn ion suppressor (Dionex, Corp., Sunnyvale, CA).
4. Conductivity detector (Dionex).
5. Heating block, set to 100°C, or boiling water bath.
6. Vacuum centrifuge (Savant Speedvac™, Savant Instruments, Farmingdale, NY).

3. Methods

3.1. Sample Hydrolysis

1. Transfer an appropriate volume (*see* **Note 6**) of each desalted sample (*see* **Note 7**) to a separate screw-top (or otherwise tightly sealable) Eppendorf tube, and dry completely in a Speedvac.
2. Resuspend each sample in 50 µL of 4 *M* HCl and mix thoroughly by vortexing.
3. Transfer to a 100°C heating block and hydrolyze for precisely 4 h.
4. After briefly centrifuging to spin condensate to the bottom of the tube, transfer the hydrolyzed samples to a vacuum centrifuge and dry completely. When dry, add 50 µL of water to each sample and dry again, in order to reduce the levels of chloride (expunged as gaseous HCl).

3.2. Analysis of Sulfate by Ion Chromatography

3.2.1. Chromatography

Figure 1 presents typical elution positions of ionic species using this method.

1. Attach the AS11 ion column to the high-performance liquid chromatograph. Attach the ion suppressor downstream of the column, and the conductivity cell downstream of the ion suppressor, according to the manufacturer's recommendations. Attach the neutralizing counterflow (50 m*M* sulfuric acid) reservoir to the ion suppressor.
2. The initial chromatographic conditions are 5 m*M* NaOH (95% water, 5% 0.1 *M* NaOH) at a flow rate of 1 mL/min. Sulfuric acid is used as the neutralizing countercurrent in the ion suppressor at a flow rate of about 3 mL/min (*see* **Note 8**).
3. Resuspend dry, hydrolyzed samples in an appropriate (25–100 µL) volume of water. A typical injection volume is 25 µL.
4. Perform chromatography over the following curved gradient:
   a. 0 to 10 min: injection at t = 0, curved gradient (curve = 8) from 5 m*M* NaOH (95% water, 5% 0.1 *M* NaOH) to 30 m*M* NaOH (70% water, 30% 0.1 *M* NaOH).
   b. 10 to 11 min: isocratic gradient (curve = 5) at 30 m*M* NaOH.
   c. 11 to 15 min: curved gradient (curve = 1) from 30 m*M* NaOH to 5 m*M* NaOH.
   d. 16 to 25 min: re-equilibration in 5 m*M* NaOH.

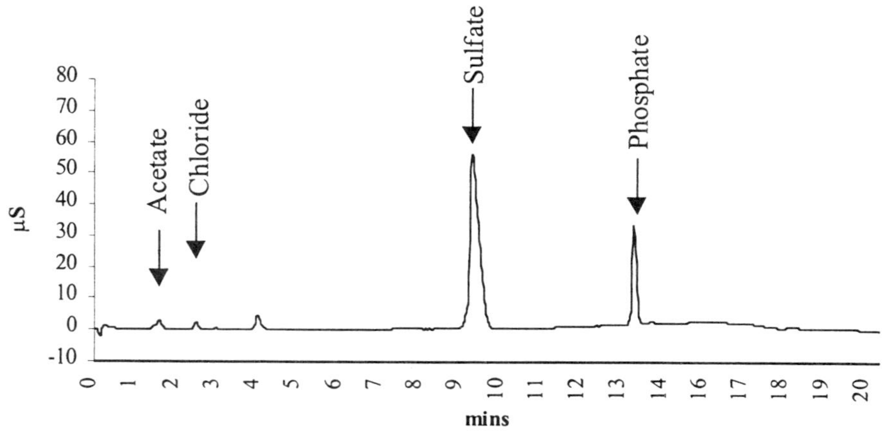

Fig. 1. Elution positions of acetate (CH_3COO^-), chloride (Cl^-), sulfate (HSO_4^-), and phosphate ($H_2PO_3^-$) ions (*see* also **Note 11**). Chromatogram displays 10 nmol each of phosphate and sulfate, and trace amounts of chloride and acetate. Note the difference in detector response between equivalent amounts of sulfate and phosphate.

3.2.2. Quantitation of Sulfate

Released inorganic sulfate can be quantitated by comparison to an external control, consisting of monosaccharide sulfates that are hydrolyzed at the same time (*see* **Note 9**).

1. Take 3 and 15 (total) nmol of a mixture of monosaccharide sulfates and hydrolyze as in **Subheading 3.1.**
2. Calculate the amount of sulfate from the area of the sulfate peak as follows:

$$\text{No. nmol sulfate} = (\text{Area}_{\text{mucin sulfate peak}} - \text{Area}_{\text{-ve control sulfate peak}})/\text{Area}_{\text{1 nmol sulfate}}$$

3. **Figure 2** shows a sample analysis of human MUC2 glycopeptides A and B (kindly supplied by I. Carlstedt, University of Lund, Sweden). An average value of 79 nmoles of sulfate from 50 mg of mucin glycopeptides was established (*see* **Note 10**).

4. Notes

1. Carbonate ions (CO_3^{2-}) elute at a retention time close to that of sulfate. It is necessary to decarbonate water by boiling in a 2-L conical flask covered with an upturned beaker. After 5–10 min of rapid boiling, transfer the hot flask to an ice bath or similar, and sparge the decarbonated water with a constant stream of argon until the water is approximately at room temperature. Make up the hydroxide eluent, taking care to minimize exposure to air.
2. Large volumes of 50 m*M* sulfuric acid are consumed in the neutralization of the postcolumn flow by the ion suppressor before analytes can be detected by the conductivity cell. Allow approx 250 mL/h of chromatography.
3. Note that carbohydrates are hygroscopic and must be thoroughly dried before weighing and making up as standards. Standards of 1 m*M* each of 2-, 3-, and 6-sulfo-N-acetylglucosamine in water may be aliquoted and stored at –20°C, and used as required.

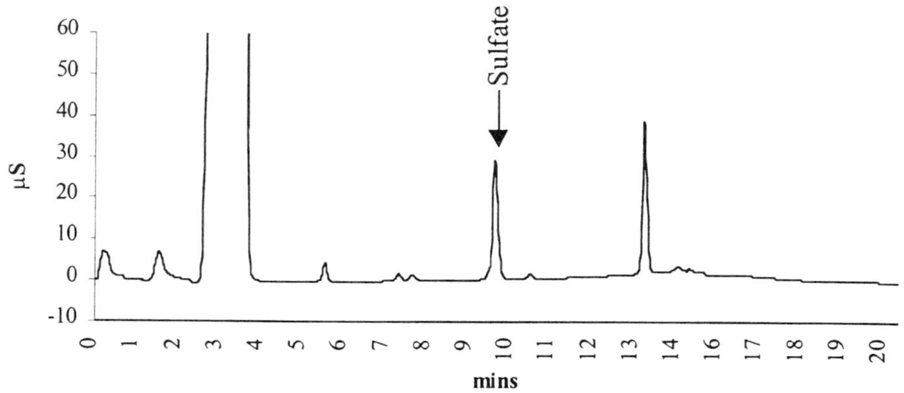

Fig. 2. Quantitation of sulfation of glycopeptides A and B from human MUC2. The chromatogram shown corresponds to half of an original hydrolysis amount of 50 μg.

4. Our system for sulfate analysis uses the same chromatographic system as that for sugar analysis, in which the ion chromatography column and high-performance anion-exchange chromatography (HPAEC) column are connected to a conductivity cell and pulsed amperometric detector, respectively. Since both modes of chromatography use the same solvents, either ion chromatography or monosaccharide-HPAEC is achieved by simply switching the column inlet to the desired column and selecting the appropriate detector.
5. The Dionex AS11 ion-exchange column was originally marketed as a strong anion-exchange (Omni-Pac™ PAX-100) column and was used extensively for the analysis of inorganic ions.
6. The limit of detection of sulfate by ion chromatography as described here is about 1 nmol. Since mucins are often highly sulfated (i.e., the number of moles of sulfate >> the number of moles of protein), the amount of protein required for analysis will often be quite small. For example, 10 pmol of a highly sulfated mucin (>100 mol sulfate/mol of protein) would be adequate for a confident analysis, whereas the sulfation of a protein that contains a unique sulfation site (1 mol sulfate/mol of protein) would not be detectable with <1 nmol of protein. Because the molecular weight of mucins is difficult to determine, the molar amount of the mucin is usually unknown. However, if we generalize and estimate an average molecular weight of 1 MDalton, then 10–100 μg of mucin should be hydrolyzed in the first instance.
7. Sample preparation and handling prior to hydrolysis are important factors affecting the accuracy and precision of sulfate analysis. Naturally, samples must be free from buffers that contain phosphate and/or sulfate (i.e., no phosphate-buffered saline!), as well as sulfate-phosphate-containing detergents (e.g., sodium dodecyl sulfate), chaotropes (e.g., thiourea), and reducing agents (e.g., mercaptoethanol). Ideally, desalting into water should be performed as a matter of course prior to sulfate analysis. For proteins that have poor solubility in water, consider desalting into partially organic solvents such as 50% acetonitrile or 50% isopropanol, or volatile buffers such as ammonium acetate or ammonium bicarbonate. If desalting is not possible or cannot reliably be performed, sulfate may be determined subtractively by direct injection of a known volume of unhydrolyzed sample in order to determine the amount of sulfate present as free (unbound) sulfate. This value is then subtracted from the value obtained after hydrolysis of an equivalent volume.

8. A flow rate of approx 3 mL/min of 50 m*M* sulfuric acid is sufficient to produce a stable baseline over the course of the separation method. Note that the flow rate of sulfuric acid to the ion suppressor does not need to be precise; in our laboratory the flow of sulfuric acid to the suppressor is regulated by positive pressure from an air cylinder. Upward baseline drift is diagnostic of an insufficient flow rate of sulfuric acid to the ion suppressor. However, an excessive flow rate of sulfuric acid to the suppressor is not responsible for downward baseline drift.
9. To quantitate sulfate, it is necessary to calibrate the analysis to an external control that is hydrolyzed under the same conditions. Adding an inorganic ion as an internal standard risks inaccuracies from ionic contaminants. Since there are differences in sulfate yield from sugar sulfates of different linkage, even after 4 h of hydrolysis (linkage lability: 6- > 2- > 3-), we recommend including two external controls that contain an equimolar mixture of 2-, 3- and 6-sulfo-N-acetylglucosamine in water: a 3 nmol control (1 nmol of each sugar sulfate), and a 15 nmol control (5 nmol of each sugar sulfate). Alternatively, an adequate quantitation can be obtained by using an external control of hydrolyzed inorganic Na_2SO_4. The use of only inorganic sulfate as a calibrant will give a slightly lower sulfate quantitation than the use of monosaccharide sulfates. Note that it is also necessary to include a negative control, in which a volume of water is hydrolyzed under the same conditions, and any reagent sulfate contamination can be subtracted.
10. Since the molecular weight of mucins is difficult to determine, the number of moles of sulfate may be expressed either as per nanomole of total amino acids, as determined by amino acid analysis (*see* Chapter 10), or as per nanomoles of total monosaccharides, as determined by HPAEC-pulsed amperometric detection.
11. Note that in this chromatographic method, monosaccharide-6-phosphates elute with similar retention times to those of inorganic sulfate, although under the hydrolysis conditions described, monosaccharide phosphates quantitatively release to phosphate. Note also that inorganic phosphate is clearly resolved by this chromatography and can also be quantitated. The hydrolysis of MUC2 glycopeptides (**Fig. 2**) showed the presence of inorganic phosphate, which was not present in the unhydrolyzed sample. Monosaccharide-sulfates have not been observed to elute from the column under the method described here.

Acknowledgments

This work was supported by an Australian Proteome Analysis Facility Industry Research Development scholarship, Macquarie University. The authors acknowledges the support of the National Health and Medical Research Council, Australia.

References

1. Corfield, A. P., Myerscough, N., Bradfield, N., Corfield, C. D. A., Gough, M., Clamp, J. R., Durdey, P., Warren, B. F., Bartolo, D. C. C., King, K. R., and Williams, J. M. (1996) Colonic mucins in ulcerative colitis: Evidence for loss of sulfation. *Glycoconj. J.* **13,** 809–822.
2. Ishikawa, N., Shi, B. B., Khan, A. I., and Nawa, Y. (1995) Reserpine-induced sulphomucin production by goblet cells in the jejunum of rats and its significance in the establishment of intestinal helminths. *Parasite Immunol.* **17,** 581–586.
3. Lo-Guidice, J. M., Herz, H., Lamblin, G., Plancke, Y., Roussel, P., and Lhermitte, M. (1997) Structures of sulfated oligosaccharides isolated from the respiratory mucins of a non-secretor (O, Le-a+b-) patient suffering from chronic bronchitis. *Glycoconj. J.* **14,** 113–125.
4. Veerman, E. C. I., Bank, C. M. C., Namavar, F., Appelmelk, B. J., Bolscher, J. G. M., and Nieuw Amerongen, A. V. (1997) Sulfated glycans on oral mucin as receptors for *Helicobacter pylori. Glycobiology* **7,** 737–743.

5. Hooper, L. V., Manzella, S. M., and Baenziger, J. U. (1996) From legumes to leukocytes: biological roles for sulfated carbohydrates. *FASEB J.* **10,** 1137–1146.
6. Harrison, M. J., Nouwens, A. S., Jardine, D., Zachara, N. E., Gooley, A. A., Nevalainen, H., and Packer, N. H. *Eur. J. Biochem.* **256,** 119–127.
7. Karlsson, N. G., Johansson, M. E., Asker, N., Karlsson, H., Gendler, S. J., Carlstedt, I., and Hansson, G. C. (1996) Molecular characterization of the large heavily glycosylated domain glycopeptide from the rat small intestinal MUC2 mucin. *Glycoconj. J.* **13,** 823–831.

VII

Mucin Biosynthesis

18

Biosynthesis of Mucin Cell and Organ Culture Methods for Biosynthetic Study

Anthony P. Corfield, Neil Myerscough, Alexandra W. C. Einerhand, B. Jan-Willem Van Klinken, Jan Dekker, and Christos Paraskeva

1. Introduction

The study of the biosynthesis has been greatly assisted by the use of cultured cells and tissue explants in short-term culture. Cells are available from a wide range of tissue sources, and this chapter focuses on the use of intestinal cells and tissue. Human colonic cell lines have been widely used in biosynthetic studies and the relationship of some lines to stages in the adenoma-carcinoma sequence is of particular interest, allowing study of the expression of mucin during the development and progression of disease ***(1–3)***. Recently the importance of proliferation, differentiation, and apoptosis has attracted attention to the use of culture systems for the study of cell behavior in normal and disease processes ***(4,5)***. In the same way, tissue obtained from patients at surgery or as biopsies can be placed in short-term primary or organ culture to study similar changes in disease ***(6,7)***.

Improvements in the study of glycoproteins, especially mucins, have been achieved through the use of defined human mucosal cells that can be grown in long-term culture ***(1,2)***. Radioactive tracer methods allow relatively small numbers of cells and tissue fragments (biopsies) to be analyzed, and cell culture also gives access to larger amounts of the mucins produced by the individual cell lines ***(8,9)***.

Each of the model systems described for the study of the synthesis and secretion of mucin has distinct advantages and drawbacks. Ideally we would wish for a clonal cell line, which has all the typical characteristics of the mucin-producing cells in vivo. Because no such cell line exists for any of the mucin producing cells, we must settle for either tissue explants, in which the mucin-producing cells have retained their natural tissue context, or isolated clonal cells from carcinomas. The first system is naturally short-lived, which affects reproducibility, whereas the latter system is far more reproducible; however, these cells will have irreversible genetic changes that distin-

From: *Methods in Molecular Biology, Vol. 125: Glycoprotein Methods and Protocols: The Mucins*
Edited by: A. Corfield © Humana Press Inc., Totowa, NJ

guish them from the cells in vivo. On the other hand, the cell lines may consist of only one cell type, which can be an advantage.

Study of the colon has the advantage that the sequence of changes during carcinogenesis in the colonic epithelial cells has been documented particularly well. This implies that cells can be isolated from each stage of malignant growth: starting from healthy tissue, from which normal epithelial cells can be brought into primary culture and display only minimal growth in vitro, via the adenoma stage that maintains intermediary characteristics to the full blown carcinoma cells, which usually grow quite well ***(10)***. This sequence of changes in the epithelium allows precise studies of how the changes in cells influence the mucins that are produced. Several of the end products of malignant transformation of colonic epithelium have yielded valuable cell lines, such as the goblet cell-like LS174T and the enterocyte-like Caco-2 cell lines, which are widely used to study the synthesis and regulation of mucin in detail ***(11,12)***. The PC/AA cell lines, originally derived from a single, large, colonic tubular adenoma, have been used at early premalignant, intermediate premalignant, adenocarcinoma, and mucinous carcinoma stages ***(1,2,8,9,13)***. Furthermore, the HRA-19 colorectal cell line ***(3)*** is valuable because it can be cloned to give all three differentiation pathways to columnar, goblet, and endocrine cells.

The methods described here cover the use of cellular systems for the study of the biosynthesis of mucin which includes cultured cells (primary cultures and cell lines) and organ culture. The data refer to human colorectal cells and tissue, but similar systems and principles apply to a wide range of other tissues where mucins are produced.

2. Materials

1. Human gastrointestinal (GI) biopsies or surgical tissue from stomach (corpus and antrum), duodenum, jejunum, ileum, colon (ascending, transverse, descending, sigmoid, and rectum), or gallbladder. Tissue obtained at surgery is dissected to isolate the mucosal layer and cut into small sections of approx 2–4 mm^3. Endoscopically obtained biopsies are used without further manipulations.
2. GI tissue explants of rat or mouse from stomach (corpus and antrum), duodenum, jejunum, ileum, colon (proximal, mid, and distal), or gallbladder (only in mouse). Full thickness explants, including the muscle layers are used cut to give segments of 2–4 mm^2 of mucosal surface.
3. Cell lines:
   a. Swiss 3T3 cells obtained from the American Tissue Type Culture Collection (ATCC, Manassas, VA, cat. no. CCL92).
   b. LS174T, colonic adenocarcinoma cell line (the parental line obtained from ATCC, cat. no. CL 188).
   c. Caco-2, colonic adenocarcinoma cell line (from Prof. G. J. Strous, Laboratory for Cell Biology, Utrecht, The Netherlands).
   d. A431, epidermoid carcinoma cell line (ATCC, cat. no. CRL 1555).
4. Fetal bovine serum (FBS). Batch testing is essential.
5. Cell culture media
   a. Standard growth medium: Dulbecco's modified Eagles medium (DMEM) containing 2 m*M* glutamine, 1 µg/mL of hydrocortisone sodium succinate, 0.2 U/mL of insulin, 100 IU/mL of penicillin, 100 mg/mL of streptomycin, and 20% of FBS.

   b. Washing medium: The same as the standard growth medium, but with 5% FBS, double the concentration of penicillin and streptomycin, and 50 μg/mL of gentamicin.
   c. Digesting solution: The same as the washing medium but containing 5% FBS together with 1.5 mg/mL of collagenase (Worthington type 4, Lorm Diagnostics, Reading, UK) and 0.25 mg/mL of hyaluronidase (type 1; Sigma, Poole, UK).
   d. 3T3 Conditioned medium: DMEM is supplemented with 10% FBS, 2 m*M* glutamine, 100 IU/mL of penicillin, 100 μg/mL of streptomycin, and put onto 24-h postconfluent 3T3 cell layers for 24 h. After conditioning, the medium is filtered through a 0.2-mm filter (Nalgene, Milton Keynes, UK) and further supplemented to give 20% FBS, 1 μg/mL of hydrocortisone sodium succinate, and 0.2 U/mL of insulin.
   e. Cell growth medium: Dulbecco's modified Eagles medium (DMEM, Gibco/BRL, Parsley, Scotland) containing 100 IU/mL of penicillin, 100 μg/mL of streptomycin, 3.7 g/L of $NaHCO_3$, and nonessential amino acids (sterile 100X stock solution, Gibco/BRL), supplemented with 20% of FBS (LS174T cells) or 10% of FBS for Caco-2 and A431 cells.
   f. Mouse thymocyte culture medium: RPMI medium (Gibco/BRL) supplemented with 10% FBS, 2 m*M* of glutamine, 100 IU/mL of penicillin, 100 μg/mL of streptomycin and 2 m*M* of glutamine.
6. Collagen (human placental type 4; Sigma) at 1 mg/mL is prepared in 1 part glacial acetic acid to 1000 parts sterile tissue culture grade distilled water and, and stored at 4°C.
7. Dispase solution: Dispase (grade 1; Boehringer, Lewes, UK) is prepared in DMEM containing 10% FBS, sterile filtered, and stored at –20°C.
8. MF-1 mice (Olac, Bicester, UK).
9. Trypsin 0.1% by weight in 0.1% EDTA in PBS.
10. Acridine orange (Sigma).
11. Organ culture medium:
    a. Grid cultures: DMEM, containing 10% FBS, 10 m*M* of sodium bicarbonate, 2 m*M* of glutamine, 100 μg/mL of streptomycin, 50 IU/mL of penicillin, 50 μg/mL of gentamicin, and 20 m*M* of HEPES, pH7.2
    b. Submerged cultures: Eagle's minimal Essential medium (EMEM, Gibco/BRL), supplemented with: non-essential amino acids (sterile 100X stock solution, Gibco/BRL), 100 IU/mL of penicillin, 100 μg/mL of streptomycin, 2 m*M* of L-glutamine. Incubate under 95% O_2/5% CO_2.
12. Carbogen gas: 5% O_2/5% CO_2.
13. Culture dishes (Costar, Cambridge MA).
14. Airtight capped, transparent tubes (3–5 mL) with round bottom (Sarstedt, Nümbrecht, Germany).
15. Water bath at 37°C.

3. Methods

3.1. Preparation of Collagen Coated Culture Flasks and Swiss 3T3 Feeder Cells (see Notes 1–3)

1. To prepare collagen-coated flasks, coat tissue culture flasks (T2525 cm^2) with a thin layer of collagen solution (**Subheading 2., item 6**; 0.2 mg/flask), and allow to dry at room temperature in a laminar flow hood for 2–4 h (*see* **Note 4**).
2. Grow Swiss 3T3 cells (**Subheading 2., item 1**) on collagen on plastic tissue culture flasks in DMEM containing 10% fetal calf serum until they are 24 h postconfluent.

3. Lethally irradiate the cells with 60 Kgray (6 mrad) of radiation, or treat with 10 µg/mL of mitomycin C (Sigma) for 2 h.
4. Wash the cells and produce a single-cell suspension using a Pasteur pipet (wide aperture to avoid shearing of cells). The cells can either be used immediately as feeders or be stored at 4°C as a single-cell suspension for up to 4 d (*see* **Note 5**).

3.2. Primary Culture-Enzyme Digestion (see Notes 3, 6, and 7)

1. Wash the tumor specimens (adenoma and carcinoma) four times in washing medium (**Subheading 2., item 5b**) and cut with surgical blades to approx 1 mm^3 in a small volume of the same medium.
2. Wash the tissue four times in washing medium, and place in digestion solution (**Subheding 2., item 5c**). Pellet cells by centrifugation at 300*g* for 5 min. Roughly put 1 cm^3 in 20–40 mL of solution.
3. Rotate at 37°C overnight (12–16 h).
4. Mix the suspension by using a Pasteur pipet to improve the separation of the epithelial elements from the stroma resulting from enzymatic digestion.
5. Filter the suspension through 50-mm mesh nylon gauze, or repeatedly allow to settle out under gravity and collect the pellets. The large clumps of cells and epithelial tubules (organoids that contain the majority of the epithelial cells) are separated from the single cells (mostly from the blood and stroma) and cell debris.
6. Wash the cell pellets three times, and place in culture on collagen-coated T25 flasks in the presence of Swiss mouse 3T3 feeders (approx 1×10^4 cells/cm^2) at 37°C in a 5% CO_2 in air incubator ***(13)***. In some situations, 3T3-conditioned medium (**Subheading 2., item 5d**) can be used instead of adding mouse 3T3 cells directly to cultures (*see* **Note 5**).

3.3. Long-Term Culture of Adenoma Cell Lines (see Note 8)

1. Prepare culture conditions for adenoma cell lines as previously described for primary cultures.
2. Carry out passage of adenoma cells as clumps of cells using sufficient dispase just to cover the cells, and incubate for approx 30 min at 37°C. Remove the cells as a sheet, and pipet with a Pasteur pipet to remove them from the flask and to break up the sheets into smaller clumps of cells (13).
3. Wash the clumps of cells and replate under standard culture conditions. Reattachment of cells may take several days, and during medium changing, any floating clumps of cells must be centrifuged and replated with fresh medium.

3.4. Long-Term Culture of Clonal Carcinoma Cell Lines, Including LS174T, Caco-2, and A431 Cell Lines (see Notes 8–10)

1. Grow the carcinoma cell lines in tissue culture plastics without collagen coating and 3T3 feeders in DMEM supplemented with 10% FBS and 1 m*M* of glutamine (**Subheading 2., item 5e**).
2. Carry out passage as single cells using 0.1% trypsin in 0.1% EDTA (*see* **Notes 9** and **10**).

3.5. Apoptotic and Differentiating Cells

3.5.1. Isolation and Identification of Apoptotic and Differentiating Cells (see Note 11)

1. During routine culture of cell lines, remove floating cells with the medium and pellet by centrifugation. Most cell lines give rise to floating cells, the majority of which undergo apoptosis.

2. Stain the cells with acridine orange at 5 µg/mL in PBS, which stains the DNA and allows visualization of the condensed chromatin of apoptotic cells. Stain cells for 10 min and then observe with a fluorescent microscope using narrow-band FITC excitation (excitation wavelength, 450–490 nm; and barrier filter, 520–560 nm). Count at least 300 cells.
3. Extract DNA from 10^6 cells and electrophorese on 2% (w/v) agarose gel containing 0.1 mg/mL of ethidium bromide at 40 V until the dye front has migrated 3 to 4 cm. Run the DNA from an equivalent number of attached cells as a control, and use dexamethasone treated mouse thymocytes as a positive control for DNA laddering (*see* **Subheading 3.5.2.** and **Note 11**).

*3.5.2. Preparation of Thymocyte Cell Cultures (see **Note 11**)*

1. Sacrifice MF-1 mice (**Subheading 2., item 8**) at 2–3 mo, dissect out the thymus and release thymocytes by pressing through sterile gauze.
2. Suspend the thymocytes in thymocyte culture medium (**Subheading 2., item 5f**) at a density of 10^6/mL and treate with 10^{-7} *M* of dexamethasone.
3. Collect samples after 18 h incubation with dexamethasone.

3.6. Organ Culture of GI Tissue

3.6.1. Grid Cultures **(6)** *(see **Note 2**)*

1. Place the biopsies or explants, singly or up to six per dish, on lens tissue placed over a stainless steel grid in culture dishes with a central well containing 2 mL of medium (**Subheading 2., item 11a**). The orientation of the tissue is with the luminal surface uppermost.
2. Place dishes in an incubation oven connected with a continuous supply of carbogen gas (**Subheading 2., item 12**) at 37°C.

3.6.2. Submerged Cultures **(14,15)** *(see **Notes 2, 12–15**)*

1. Place the biopsy or explants, up to a maximum of six, in a small, airtight capped tube in 100 µL of EMEM (**Subheading 2., item 11b**) (*see* **Notes 16** and **17**).
2. Blow carbogen gas into the tube and seal the carbogen gas atmosphere by replacing the cap.
3. Place the tube in a 37°C water bath.

4. Notes

1. The production and secretion of mucins by cultured colonic cells should be examined using cells at different stages of confluency, because the stages of growth may alter the differentiation properties of the cells and therefore the amount and type of mucin produced.
2. When using organ and primary cultures, it is important to consider the heterogenous nature of the cell types in the tissue (i.e., stromal elements and lymphoid cells in addition to the colonic epithelial cells). It may not be clear which cell type is producing the glycoproteins. Identification of specific cell-located glycoprotein expression can be further examined using histological methods with chemical, lectin, or antibody stains or *in situ* hybridization to identify the cellular origin of the mucin of interest (Chapters 3 and 27).
3. The primary culture techniques described can be used for normal adult colon. However, these are not as reproducible as those with the adenomas and carcinomas, and there are more problems from contaminating stromal elements. There are at present no normal adult colonic epithelial cell lines, only adenoma ***(1)*** and carcinoma cell lines such as PC/JW/F1 ***(13)***, HT29 ***(10)***, LS174T (goblet cell like), and Caco-2 (enterocyte like) ***(11,12)***.
4. Collagen-coated flasks are necessary to obtain efficient attachment of primary cultures and some adenomas and carcinomas to the flasks, and to retain the optimum differenti-

ated characteristics of the cells *(13)*. "Tissue culture-treated" culture plates (Costar) (**Subheading 2., item 13**) can be used successfully for established carcinoma cell lines without collagen coating.

5. The use of 3T3 cell feeders requires controls to determine which cell type is producing the glycoproteins of interest. This can be achieved using 3T3-conditioned medium in which the production of mucin is being assessed, or removing the 3T3 feeders from the flask once the epithelium has grown. In addition, the 3T3 cells should be examined for the production of mucin under the culture conditions used.
6. Colorectal adenomas invariably need digestion with enzymes because of their organization into well-differentiated glandular structures. With carcinomas, it is possible to adopt a nonenzymatic approach with surgical blades to release small clumps of tumour cells that can be cultured *(16)*.
7. When using all colonic cell lines, it is important to check the true colonic epithelial nature using a battery of markers, including antikeratin antibodies, ultrastructural analysis showing the presence of desmosomes, and other colonic differentiation markers *(10)*. This verification is especially important with primary cultures and newly derived lines, but may also be important when culture conditions induce changes in cell behavior.
8. Although many tumor cell lines, especially colon carcinomas, can be grown in simple media without 3T3 feeders and collagen coating, the colonic cells retain better differentiated phenotypes when using the more complex culture conditions described for primary cultures.
9. LS147T, Caco-2, and A431 cells are cultured in DMEM as described under **Subheading 2., item 5e** at 37°C and 5% CO_2. Caco-2 and A431 cells are passaged (trypsinised) and split 1:4 and the medium is changed every 2 to 3 d. LS174T cells are split 1:2 and the medium changed daily.
10. The three cell lines LS174T, Caco-2 and A431, together produce all gastrointestinal mucins known at present: LS174T produces MUC1, -2, -5AC, -5B, and -6 *(12)*; Caco-2 produces MUC1 and -3 *(12)* and A431 produces MUC4 (Van Klinken, Einerhand, and Dekker, unpublished results). They serve as excellent cell lines for the isolation of the corresponding mucin precursors as detailed in chapters 20 and 21. In addition, the PC/AA cell lines produce only MUC 1 and MUC2 at early passage, but at later passage and in later premalignant and malignant stages, they show *de novo* expression of MUC5AC, MUC5B, and MUC6 (Corfield, Myerscough, and Paraskeva, unpublished results).
11. Proliferating cells in the bottom of the colonic crypts migrate to the upper half of the crypts, where they differentiate. These differentiated cells migrate to the top of the crypt, and there is evidence that they die by apoptosis and that apoptosis may be the terminal stage of differentiation. The relationship among proliferation, differentiation and apoptosis may be studied using the culture system described here. We have shown that cultured colonic normal adenoma and carcinoma cells spontaneously die by apoptosis in vitro and that the levels of apoptosis can be modulated by dietary short-chain fatty acids (butyrate, acetate, and propionate) and bile acids *(5)*. During routine culture of colonic epithelial cells, some cells detach from the flask and float in the medium *(4)*. These cells contain condensed chromatin, which can be detected with acridine orange staining. In addition, characteristic DNA laddering resulting from internucleosomal fragmentation can be seen after analysis of total cellular DNA. Dexamethasone treated mouse thymocytes are a convenient source of cells to use as a positive control for DNA laddering.
12. A mucosal biopsy is taken by endoscopy from a human individual with an otherwise intact organ. A tissue explant is obtained from surgically resected human or animal tissue. For human tissue, the mucosa can be relatively easily dissected from the muscle

layer due to its size. In animal tissues, smaller sections of the entire wall of the organ are used, because these tissues are more fragile and dissection is difficult. The presence of the muscle layer does not appear to affect the synthesis of epithelial mucins.

13. Submerging the tissue segments in the culture medium is particularly advantageous when the secretion of mucin is studied. The grid technique results in the accumulation of a mucous gel layer on the top of the tissue during prolonged incubations and requires careful collection as a separate fraction. Upon submersion, the mucins are secreted into the medium, although some of the mucin may remain adherent to the epithelial tissue. The submerged culture system, under basic nonstimulated conditions, leads to the recovery of 20% of the total MUC2 in the medium during a 4-h culture of human colonic biopsies. This percentage is very reproducible when compared for a large number of patients (Van Klinken, Einerhand, and Dekker, unpublished results).
14. The submerged tissue culture technique has the advantage that the mucins are more efficiently labeled during metabolic labeling experiments compared to the grid culture system. In other words a higher incorporation of radioactive precursor is found in the mucins prepared using the submerged technique if the same concentration of precursor is used in both culture systems. Moreover, the labeling is more economical because similar labeling levels can be achieved with smaller (5%) amounts of precursor owing to the smaller volume (0.1 vs 2 mL). Because prolonged incubations will certainly deplete the medium of nutrients the submerged technique in small volumes (0.1 mL/explant) is only applied to the metabolic pulse labeling of tissue. The grid culture system is thought to mimic the in vivo state more closely, with retention of the secreted mucus at the tissue surface. Although lower levels of incorporation into mucins are found, these are quite adequate for comparative studies in disease ***(6,7)***.
15. Standard media usually contain sodium bicarbonate as primary buffer (2.2 g/L), and are used only with a 5% CO_2 atmosphere. Alternatively, the bicarbonate buffer can be replaced with the nontoxic buffer HEPES, which enables incubation under an air atmosphere and does not require CO_2. GI tissue is extremely sensitive to hypoxic conditions; therefore, the use of an oxygen atmosphere (carbogen gas: 95% O_2/5% CO_2) is preferred to the 95% air atmosphere.
16. Temperature control by a water bath is preferred to an incubator because the heat exchange is much quicker and the temperature is prone to only very slight fluctuations. This is particularly relevant for pulse/chase experiments because these involve relatively short incubation periods and require rapid temperature equilibration (*see* Chapters 20 and 21).
17. If the incubation is part of a pulse/chase experiment, the EMEM should be depleted of the compound used as a precursor to label the mucin (*see* Chapter 19, **Subheading 3.1.**, and Chapters 20 and 21).

References

1. Williams, A. C., Browne, S. J., Manning, A. M., Hague, A., van der Stappen, J. W. J., and Paraskeva, C. (1993) Biological consequences of the genetic changes which occur during human colorectal carcinogenesis. *Sem. Cancer Biol.* **4,** 153–159.
2. Williams, A. C., Harper, S. A. and Paraskeva, C. (1990) Neoplastic transformation of a human colonic epithelial cell line: Experimental evidence for the adenoma to carcinoma sequence. *Cancer Res.* **50,** 4724–4730.
3. Kirkland, S. (1988) Clonal origin of columnar, mucous and endocrine lineages in human colorectal epithelium. *Cancer* **61,** 1359–1363.
4. Hague, A., Manning, A. M., Hanlon, K., Huschtscha, L. I., Hart, D. and Paraskeva, C. (1993) Sodium butyrate induces apoptosis in human colonic tumour cell lines in a p53

independent pathway: Implications for the possible role of dietary fibre in the prevention of large bowel cancer. *Int. J. Cancer* **55,** 498–505.

5. Hague, A., Elder, D. J. E., Hicks, D. J. and Paraskeva, C. (1995) Apoptosis in colorectal tumour cells: Induction by the short chain fatty acids butyrate, propionate and acetate and the bile salt deoxycholate. *Int. J. Cancer* **60,** 400–406.
6. Corfield, A. P., Warren, B. F., Bartolo, D. C. C., Wagner, S. A., and Clamp, J. R. (1992) Mucin changes in ileoanal pouches monitored by metabolic labelling and histochemistry. *Brit. J. Surgery* **79,** 1209–1212.
7. Probert, C. S. J., Warren, B. F., Perry, T., Mackay, E. H., Mayberry, J. F., and Corfield, A. P. (1995) South Asian and European colitics show characteristic differences in colonic mucus glycoprotein type and turnover. Potential identification of a lower risk group for severe disease and cancer. *Gut* **36,** 696–702.
8. Vavasseur, F., Dole, K., Yang, J., Matta, K. L., Corfield, A. P., Myerscough, N., Paraskeva, C., and Brockhausen, I. (1994) O-glycan biosynthesis in human colonic cells during progression to cancer. *Eur. J. Biochem.* **222,** 415–424.
9. Corfield, A. P., Clamp, J. R., Casey, A. D., and Paraskeva, C. (1990) Characterization of a sialic acid-rich mucus glycoprotein secreted by a premalignant human colorectal adenoma cell line. *Int. J. Cancer* **46,** 1059–1065.
10. Laboisse, C. L. (1989) Differentiation of colon cells in culture, in *The Cell and Molecular Biology of Colon Cancer* (Augenlicht, L. H., eds.), CRC Press, Boca Raton, FL, pp. 27–43.
11. Van Beers, E. H., Al, R. H., Rings, E. H. H. M., Einerhand, A. W. C., Dekker, J., and Büller, H. A. (1995) Lactase and sucrase-isomaltase gene expression during Caco-2 cell differentiation. *Biochem. J.* **308,** 769–775.
12. Van Klinken, J.-W., Oussoren, E., Weenink, J.-J., Strous, G. J., Büller, H. A., Dekker, J. and Einerhand, A. W. C. (1996) The human intestinal cell lines Caco-2 and LS174T as models to study cell-type specific mucin expression. *Glycoconjugate J.* **13,** 757–768.
13. Paraskeva, C., Buckle, B. G., Sheer, D. and Wigley, C. B. (1984) The isolation and characteristics of colorectal epithelial cell lines at different stages in malignant transformation from familial polyposis coli patients. *Int. J. Cancer* **34,** 49–56.
14. Dekker, J., Van Beurden-Lammers, W. M. O., and Strous, G. J. (1989) Biosynthesis of gastric mucus glycoprotein of the rat. *J. Biol. Chem.* **264,** 10,431–10,437.
15. Van Klinken, B.J.W., De Bolos, C., Büller, H.A., Dekker, J. and Einerhand, A.W.C. (1997) Biosynthesis of mucins (MUC2-6) along the longitudinal axis of the gastrointestinal tract. Am. J. Physiol. 273, G296-G302
16. Leibovitz, A., Stinson, J. C., McComb, W. B., McCoy, C. E., Mazur, K. C., and Mabry, N. D. (1976) Classification of human colorectal adenocarcinoma cell lines. *Cancer Res.* **36,** 3562–3569.

19

Metabolic Labeling Methods for the Preparation and Biosynthetic Study of Mucin

Anthony P. Corfield, Neil Myerscough, B. Jan-Willem Van Klinken, Alexandra W. C. Einerhand, and Jan Dekker

1. Introduction

Radiolabeling methods have been introduced into the study of the biology of mucin for several reasons *(1–3)*. In many instances, the biochemical analysis of mucins in any form may be limited owing to the small amounts of mucosal tissue available, of cells from culture systems, and the difficulty in obtaining normal material for comparison *(3–6)*. The use of radiolabeling in direct assessment of the biochemistry of the metabolism of mucins, in particular their biosynthesis, is well suited to these techniques in the same way they have been adopted for other proteins and glycoconjugates. It is currently of special interest in evaluating the different stages in the maturation, aggregation, and secretion of mucin.

Separation methods for mucins have relied on the properties of these molecules, typically their buoyant density on density gradients, high molecular weight on gel filtration, their charge on ion-exchange chromatography and combination of molecular size and charge on agarose gel electrophoresis *(7–10)*. These methods have been applied to microscale radiolabelled mucins *(2,3)*, and to larger amounts of mucins from cell culture or resected tissue (*3,11*, *see* Chapters 1, 2, and 7). Although these separations yield pure fractions of mucin in most cases, contamination with proteoglycans and nucleic acids needs to be controlled. Identification and elimination of these contaminants may be necessary depending on the data on the mucin required.

This chapter describes the methods to radioactively tag mucins by metabolic labelling. Since these techniques require living, mucins producing cells, they will closely follow the protocols for optimal cell and tissue culture as described in Chapter 18. Continuous culture (4–96 h) of cell lines or tissue in the presence of radiolabeled monosaccharides, amino acids, or sulfate will lead to the accumulation of radioactively labeled mature mucins in the cells and the culture medium. This chapter further concentrates on the subsequent isolation and detailed analysis of these mature mucins.

From: *Methods in Molecular Biology, Vol. 125: Glycoprotein Methods and Protocols: The Mucins*
Edited by: A. Corfield © Humana Press Inc., Totowa, NJ

However, the possibilities of metabolic labeling to study minute amounts of mucin can also be applied to the mucin precursors, i.e., to the biosynthesis of the mucin polypeptide and other early steps in mucin biosynthesis such as *N*-glycosylation and early *O*-glycosylation. In addition, the dynamics of each step during synthesis and secretion can be studied. Experiments to identify mucin precursors and the dynamics of processing steps require short labeling periods (15–60 min), referred to as pulse labeling, and ensuing incubations to follow the processing of the labeled molecules with time (4–6 h), referred to as chase incubations. Since these mucin precursors are present in only very small amounts relative to the mature mucin, the isolation of these biosynthetic intermediates requires immunoprecipitation, which is described in Chapters 20 and 21.

2. Materials

1. A source of mucin-producing cells. These can be biopsies, explants or cell lines as described in Chapter 18.
2. Radioactively labelled precursors:
   a. L-(^{35}S)methionine/L-(^{35}S)cysteine, (Promix™, Amersham, Amersham, UK), a mixture containing 65% (^{35}S)methionine and 25% (^{35}S)cysteine, specific activity 1000 Ci/mmol (37,000 GBq/mmol), concentration 10 mCi/mL (370 MBq/mL).
   b. L-(^{3}H)threonine (Amersham): specific activity 5–20 Ci/mmol (185–740 GBq/mmol), concentration is 1 mCi/mL (37 MBq/mL).
   c. D-(6-^{3}H)galactose (Amersham): specific activity 20–40 Ci/mmol (740–480 GBq/mmol), concentration is 1 mCi/mL (37 MBq/mL).
   d. D-(^{3}H)glucosamine (Amersham): specific activity 20–40 Ci/mmol (740 GBq/mmol), concentration is 1 mCi/mL (37 MBq/mL).
   e. Sodium (^{35}S)sulfate (in aqueous solution, code SJS-1; Amersham) specific activity 1050 Ci/mmol (38,850 GBq/mmol), concentration 2 mCi/mL (74 MBq/mL).
3. Media for metabolic pulse-labeling (15–60 min):
   a. Eagle's minimal essential medium (EMEM) (Gibco-BRL, Paisley, Scotland) without L-methionine and L-cysteine for labeling with Promix (*see* **item 2a**), supplemented with nonessential amino acids, 100 IU/mL of penicillin, 100 µg/mL of streptomycin, and 2 m*M* of L-glutamine.
   b. EMEM without L-threonine (Gibco/BRL), supplemented with nonessential amino acids, 100 IU/mL of penicillin, 100 µg/mL of streptomycin, and 2 m*M* of L-glutamine.
   c. EMEM with low glucose (Gibco/BRL), (50 mg/mL instead of 1000 mg/mL) for labeling with D-(6-^{3}H)galactose, supplemented with nonessential amino acids, 100 IU/mL of penicillin, 100 µg/mL of streptomycin, and 2 m*M* of L-glutamine.
   d. EMEM without sulfate for labeling with (^{35}S)sulfate supplemented with nonessential amino acids, 100 IU/mL of penicillin, 100 µg/mL of streptomycin, and 2 m*M* of L-glutamine. This medium is not available commercially, and all compounds are reconstituted from concentrated stock solutions (Gibco/BRL) except that $MgSO_4$ is replaced with an equimolar solution of $MgCl_2$. Streptomycin should also be avoided because this is supplied as the sulfate form.
4. Medium for chase incubations: standard EMEM (Gibco/BRL) supplemented with nonessential amino acids, 100 IU/mL of penicillin, 100 µg/mL of streptomycin, and 2 m*M* of L-glutamine.
5. Guanidine hydrochloride, approx 7 *M* stock solution, prepared as follows:

   a. Dissolve guanidine hydrochloride (grade 1; Sigma, Poole, UK) in high purity (e.g., Milli Q) water or phosphate-buffered saline (PBS) to give a concentration of approx 7 or 8 *M* and stir for 24 h at room temperature.
   b. Add 10 g/L of activated charcoal and stir at 4°C for 24 h.
   c. Filter through two layers of Whatman No. 1 filter paper.
   d. Add a further 10 g/L of activated charcoal, stir overnight at room temperature.
   e. Filter through two layers of Whatman No. 1 filter paper and then pass through 0.2-μm Millipore filter (three times).
   f. Monitor the concentration by refractive index.
6. Dithiothreitol (DTT) (Sigma).
7. Sodium iodoacetamide (Sigma).
8. PBS/inhibitor cocktail in PBS and 6 *M* guanidine hydrochloride. 1 m*M* of phenylmethylsulfonylfluoride (PMSF), 5 m*M* of EDTA, 0.1 mg/mL of soybean trypsin inhibitor (STI), 5 m*M* of *N*-ethylmaleimide, 10 m*M* of benzamidine, and 0.02% of sodium azide. Prepare inhibitor cocktail fresh as required. This solution is made up from stocks of concentrated guanidine hydrochloride as in **item 5**.
9. Cesium chloride (CsCl) (Sigma).
10. Sepharose CL-2B (Pharmacia, Milton Keynes, UK). Use 1 × 30 cm or 2.5 × 80 cm in all-glass columns equilibrated in 4 *M* guanidine hydrochloride/PBS or 10 m*M* Tris-HCl, pH 8.0.
11. Agarose gels for electrophoresis, 0.8 or 1% (w/v). Agarose (SeaKem LE agarose, Flowgen, Sittingbourne, UK) is made up at 0.8 or 1% in the running buffer. Gels 15 × 15 cm are run in a standard submarine horizontal electrophoresis apparatus.
12. Buffers for electrophoresis and vacuum blotting.
   a. Running buffer for agarose gel electrophoresis: 40 m*M* Tris-acetate 1 m*M* EDTA, pH 8.0, containing 0.1% of sodium dodecyl sulfate (SDS) (add from 10% SDS stock).
   b. Sample buffer for agarose gel electrophoresis: 40 m*M* Tris -acetate 1 m*M* EDTA, pH 8.0, containing 0.1% of SDS with 10% of glycerol and 1% of bromophenol blue.
   c. Vacuum-blotting buffer: 3.3 *M* sodium citrate, pH 7.0, containing 3 *M* NaCl.
13. Markers for electrophoresis; Rainbow markers, high molecular weight range (Amersham), maximum 200 kDa (myosin) and IgM, 990 kDa (Sigma).
14. Immobilon P (polyvinyldene difluoride, PVDF) membrane (Millipore, Watford, UK).
15. High molecular weight (>20 kDa) polyethylene glycol, PEG 20,000 (Sigma)
16. Periodic acid-Schiff reagent (PAS) commercial solution (Sigma).
17. Precipitation buffer: 95% ethanol/1% sodium acetate cooled to –70°C.
18. Proteoglycan degrading enzymes and incubation buffers. Protease inhibitors may be included in the incubation buffers (*see* **Notes 1** and **2**).
   a. Chondroitinase ABC from Proteus vulgaris (Boehringer Mannheim, Lewes, UK). Incubation buffer: 250 mM Tris-HCl, 176 mM sodium acetate, 250 mM sodium chloride, pH 8.0.
   b. Hyaluronidase from bovine testis (Sigma). Incubation buffer: PBS.
   c. Heparinase types II and III from Flavobacterium heparinum, (Sigma), Incubation buffer: 5 mM sodium phosphate, 200 mM NaCl, pH 7.0.
19. Sephadex G100 (Pharmacia, Milton Keynes, UK). Use 30 × 1 cm all-glass columns equilibrated and run in 10 m*M* of Tris-HCl, pH 8.0.
20. Ultraturrax homogenizer (Jahnke and Kunkel, Stauffen, Germany).
21. Ultracentrifuge routinely capable of 100,000*g* for up to 72 h.

3. Methods

3.1. Metabolic Labeling

3.1.1. Continuous Labeling of Mature Mucins (see **Notes 1** and **2**)

3.1.1.1. Continuous Labeling Using Gastrointestinal (GI) Cells

1. Cells are cultured in standard growth medium as described in Chapter 18.
2. Radioactive precursors (**Subheading 2., item 2**) are added: D-(^{3}H)glucosamine, 370–1850 kBq, L-(^{3}H)threonine, 370–1850 kBq, (^{35}S)sulphate 185–1850 kBq (*see* **Notes 3–5**).
3. Cells are incubated under standard conditions for 4–96 h.

3.1.1.2. Continuous Labeling Using GI Tissue

1. Culture tissue samples on grids with standard growth medium under conditions described in Chapter 18.
2. Add radioactive precursors: D-(^{3}H)glucosamine, 370–740 kBq, L-(^{3}H)threonine, 370–740 kBq or (^{35}S)sulphate 925 kBq. In dual labelling experiments the ratio (^{3}H):(^{35}S) 370 kBq:925 kBq is used with colonic tissue (*see* **Notes 3–5**).
3. Incubate biopsies and explants for up to 24 h under standard conditions described in Chapter 18.

3.1.2. Pulse/Chase Labeling of Mucin Precursors and Mature Mucins

3.1.2.1. Pulse/Chase Labeling Using Cell Lines, Particularly LS174T, Caco-2, and A431

1. Culture cells as described in Chapter 18.
2. Remove the medium and wash the cells with sterile PBS at 37°C and apply the appropriate medium (EMEM) lacking the compound to be used as precursor (**Subheading 2., item 3**). Incubate for 45 min.
3. Add the radioactively labeled compound for 30–60 min (**Note 6**).
4. Remove the medium containing the label, wash in sterile PBS at 37°C and add standard EMEM (**Subheading 2., item 4**).
5. Incubate for 1–20 h.

3.1.2.2. Pulse/Chase Labeling Using GI Tissue

1. Use the submerged technique for tissue culture (Chapter 18). Place the tissue segments (one per tube) immediately after excision in the appropriate EMEM to deplete the compound to be used as label. Incubate for 30 min in 100 µL per tube to deplete this compound.
2. Add 100 µCi (3700 kBq) of the radioactive compound and incubate for 15–60 min (*see* **Note 9**).
3. Remove the medium containing the label. Wash the tissue once in 500 µL of EMEM at 37°C (**Subheading 2., item 4**).
4. Add 500 µL of EMEM at 37°C per test tube, and chase incubate for maximally 6 h.
5. After pulse labeling or chase incubation, homogenize the tissue for immunoprecipitation to allow immunoisolation of mucins, as described in Chapter 20.

3.2. Collection of Secreted and Cellular Material

3.2.1. Collection of Radioactive Fractions After Metabolic Labeling in Cell Culture (see **Note 1**)

1. Collect the medium and wash the cells with a further 5 mL of fresh nonradioactive medium. Medium and washings are combined and mixed 1:1 (v/v) with 6 *M* guanidine/PBS inhibitor cocktail (**Subheading 2., items 5** and **8**).

2. Irrigate the flasks with 5 mL of 6 *M* guanidine hydrochloride in PBS/inhibitor cocktail (*see* **Note 1**) containing 10 m*M* DTT at room temperature, and scrape the cells off with a cell scraper.
3. Wash the cells twice with 6 *M* guanidine hydrochloride PBS/inhibitor cocktail containing 10 m*M* DTTl. Collect the cells by low-speed centrifugation and pool the total washings.
4. Adjust the DTT washings to a 2.5X molar excess with sodium iodoacetamide and incubate for 15 h at room temperature in the dark (*see* **Note 10**).
5. Dialyze the medium and DTT wash material extensively at room temperature against three changes of 6 *M* guanidine hydrochloride in PBS.
6. Homogenize the washed cell pellet in 1 mL of 6 *M* guanidine hydrochloride PBS/inhibitor cocktail with an Ultraturrax for 10 s at maximum setting, on ice.
7. Centrifuge the homogenate at 100,000*g* for 60 min, and decant the supernatant. Resuspend the membrane fraction in 1 mL of 6 *M* guanidine hydrochloride PBS/inhibitor cocktail.

*3.2.2. Collection of Radioactive Fractions After Metabolic Labeling in Organ Culture. (*see ***Note 3)***

1. After incubation, remove the medium from the central well and wash the tissue and dish with 1 mL of PBS. Pool the medium and washings, and dialyze against three changes of 1 L of 6 *M* guanidine hydrochloride PBS over 48 h at room temperature (*see* **Note 2**).
2. Homogenize the tissue on ice in 1 mL of PBS/inhibitor cocktail or 6 *M* guanidine hydrochloride PBS/inhibitor cocktail in an all-glass Potter homogenizer, ensuring complete disruption of the mucosal cells (about 20 strokes). Remove connective tissue, if present, which resists disruption and sediments as large fragments.
3. Centrifuge the homogenate at 12,000*g* for 5 min at 4°C, and separate the supernatant soluble fraction from the membrane pellet.
4. Resuspend the membrane pellet in 1 mL of 6 *M* guanidine hydrochloride PBS/inhibitor cocktail.

3.3. Separation of Mucins from the Fractions Obtained After Culture (see Note 11)

3.3.1. Density Gradient Centrifugation

1. Make up the samples from the fractions prepared according to **Subheadings 3.2.1.–3.2.2.** in 4 *M* guanidine hydrochloride/PBS to a concentration of approx 1–5 mg/mL related to protein concentration, or containing a suitable amount of radioactivity (e.g., >10,000 cpm) for subsequent analytical techniques. Add solid CsCl to give a density of about 1.4 g/mL, and stir at room temperature for 15 h.
2. Load the samples into centrifuge tubes and centrifuge at 100,000 g for at least 48 h at 10°C to obtain a CsCl density gradient.
3. Aspirate 0.5-mL samples from the top of the tube by pipet, or drain from the bottom after piercing the tubes. Weigh the samples to obtain the density of each fraction.
4. Slot-blot aliquots (5–50 mL, after dilution if necessary) of each fraction onto Immobilon P membrane and visualize with a carbohydrate stain (*see* **Note 12**) *(6)*. Quantify the results using a densitometer *(7)*. For the radioactive samples cut out each slot-blot and place in scintillation cocktail for quantitation (*see* **Note 13**). Analysis by sodium dodecyl sulfate-polyacrylamide gel electrophoresis (SDS-PAGE) can be used (*see* **Note 14**).
5. Pool the mucin-containing fractions located at densities between 1.30 and 1.55 g/mL.

3.3.2. Gel Filtration (see **Notes 11** and **15**)

1. Make up samples in 10 m*M* Tris-HCl, pH 8.0, or 4 *M* guanidine hydrochloride/PBS buffer to give concentrations of 1–5 mg/mL of protein or by radioactivity (e.g., >5000 cpm), and load onto columns of Sepharose CL-2B. Elute the column with the same buffer and collect fractions (1–5 mL).
2. Identify mucin-containing fractions as described in **Subheading 3.3.1., step 4** (*see* **Notes 12–14**).
3. Pool the mucin-containing fractions identified in excluded or included volumes.

3.3.3. Agarose Gel Electrophoresis and Vacuum Blotting (see **Notes 16** and **17**)

1. Mix samples containing 10–500 mg mucin or >10,000cpm with 50 mL sample buffer and load onto horizontal 0.8–1% agarose gels. Run Rainbow markers and IgM as migration markers (*see* **Note 17**).
2. Run the gels at 20 V for 18 h, or 20 mA for 14 h at room temperature.
3. Remove the gels after electrophoresis and stain directly using Coomassie blue stain (*see* **Note 18**) or submit to vacuum blotting. Alternatively, dry the gels and use immediately for autoradiography or fluorography to visualize the radioactive mucins (*see* **Note 19** and Chapter 20).
4. Blot gels onto Immobilon P membranes using a standard apparatus in vacuum-blotting buffer for 2 h at 40 mbar. The success of the transfer can be visualized immediately if Rainbow markers are included on the gels.
5. Probe the blotted membranes with chemical stains (e.g., PAS), lectin conjugates or antibodies using standard techniques for visualization (*see* **Note 13**). Radioactive components can also be detected on the same blots (*see* **Note 19**).

3.3.4. Concentration of Mucin Samples After Purification (see **Note 20**)

1. Dialyze the samples against 4–5 changes of 5 L of distilled water or against two changes of 1 L of 4 *M* guanidine hydrochloride. Salt free samples can be freeze dried to concentrate (*see* **Note 20**).
2. Dialyze the samples against high molecular weight polyethylene glycol 20,000 (30%) in distilled water or 4 *M* guanidine hydrochloride. Samples for ion-exchange can then be dialyzed against suitable urea containing buffers with 0.05% CHAPS.
3. Mix a solution containing mucin (1–5 mg or >10,000cpm) with 4 vols of precipitation buffer cooled to –70°C and leave for 45 min at –70°C. Return to room temperature for centrifugation at 12,000*g* for 20 min and collect the pellet.

3.3.5. Identification and Removal of Proteoglycans (see **Notes 21** and **22**)

1. Mix samples containing 0.5–2.0 mg of mucin or >10,000 cpm with the followng:
   a. Chondroitinase ABC (5 U/mL) and incubate for 16 h at 37°C in incubation buffer (**Subheading 2., item 18a**).
   b. Hyaluronidase (10 mg/mL) and incubate for 16 h at 37°C in incubation buffer (**Subheading 2., item 18b**).
   c. Heparinase II and III (100 mU of each enzyme) and incubate for 16 h at 37°C in incubation buffer (**Subheading 2., item 18c**).

   Carry out control incubations under the same conditions without enzyme.
2. Stop the incubations by addition of 1-mL of 10 ml Tris-HCl, pH 8.0, and load the products onto columns of Sephadex G100, eluting with the same buffer and collecting 1-mL fractions.

3. Test the individual fractions for radioactivity or carbohydrate by slot-blotting or by colorimetric analysis (*see* **Note 22**). Analysis by SDS-PAGE can also be used (*see* **Note 14**).

4. Notes

1. A protease inhibitor cocktail is needed to avoid the degradation of mucins by bacterial enzymes and in cell homogenates. STI and PMSF are the most important for colonic tissue.
2. We have evidence that mucins will be degraded during dialysis, and other prolonged procedures under circumstances of insufficient protease inhibition *(19)*. During dialysis always be sure to use adequate precautions against degradation. It is convenient to dialyze against 4–6 *M* guanidine hydrochloride at 4°C, which will inhibit nearly all enzymatic activities. Dialysis against specific protease inhibitors such as PMSF and STI is also very expensive. Note that the presence of guanidine hydrochloride will not irreversibly eliminate the degrading enzymes by itself. On removal of the guanidine hydrochloride, e.g., by dialysis or gel filtration, the enzymes may be reactivated.
3. In radiolabeling experiments, the total amount of mucin is often small, and significant losses owing to nonspecific adsorption onto plastic and glass vessels, silicone rubber, and dialysis tubing may occur. Treatment of all vessels and tubing with 1% Triton in PBS before use improves yields.
4. The continuous labeling techniques do not use an incubation period in which the radioactive label is depleted. The commonly used label glucosamine is not a normal component in standard media and can be regarded as a supplement. The pulse/chase protocol always starts with a 30- to 45-min period during which the compound chosen as precursor is first (partially) depleted from the medium. After the depletion period, the label is added to the same medium and incubation proccedes with no medium change.
5. The choice of radioactive precursor is important. Radioactive glucosamine is most commonly used because it is incorporated into *N*-acetyl-D-glucosamine, *N*-acetyl-D-galactosamine, and the sialic acids, major monosaccharide components of mucins. In tissues other than liver, labels such as mannose and fucose may be randomized to other monosaccharides before they are transferred to glycoproteins.
6. The metabolic rate of the cells should be considered for optimum labeling with the radioactive precursor during continuous labeling cultures. The type of isotope will govern the amount to be added, typically (^{14}C)- and (^{35}S)- precursors in the range 185–1850 kBq/experiment and (^{3}H) precursors in the range 370–1850 kBq/experiment. Short term labeling of 2-4h may not result in labelling of secreted material and could require higher doses of radioactive precursor (1.85–3.7 MBq/experiment). Longer labeling periods may reflect synthesis, catabolism and recycling of glycoproteins. Dual labeling experiments with, e.g., (^{35}S) and (^{3}H) need to be planned so that the relative incorporation of each isotope is readily detectable in the isolated product; thus, consideration of **Notes 5** and **7** is necessary. Organ culture experiments should be controlled by histochemical criteria to ensure the integrity of the tissue during the incubation period. Diseased tissue may show signs of degradation during acceptable culture times for normal samples.
7. In continuous-labeling cultures, increased incorporation of radioactive precursors may be achieved by reduction of the concentration of the same nonlabeled compound in the medium (monosaccharides or amino-acids) for the labeling period as described for the pulse/chase cultures (*see* **Notes 8** and **9**). However, this should be balanced against any changes in the growth of the cells or tissue under these "depleted" conditions and for the total culture period.

8. For the labeling of each cell type use 5 μCi (185 kBq) (^{35}S) amino acids or (^{35}S) sulfate/cm^2 cell culture. All cell cultures can be labelled in an air incubator with 5% CO_2 or in an airtight container flushed with 5% CO_2 /95% O_2 and placed in a 37°C incubator.
   a. Optimal labeling conditions for LS147T cells (4 cm^2 wells are best): Use only preconfluent or freshly confluent cultures because cells will detach progressively from the plastic after they reach confluence, and labeling efficiency deteriorates rapidly after confluency *(12)*. In addition cells with a low passage number must be used owing to deterioration of the labelling efficiency at later passages. An aliquot of 20 μCi (740 kBq) is added to each tissue culture well in 250 μL of depletion medium. Pulse labeling is performed for either 30 min ([^{35}S] amino acids) or 60 min ([^{35}S] sulfate).
   b. Optimal labeling conditions for Caco-2 (F25 flasks, 25 cm^2): Use cells that are at least 5–7 d confluent, which will give the most efficient labeling *(12)*. An aliquot of 125 μCi (4625 kBq) is added to 1 mL of depletion medium in each 25 cm^2 culture flask. Pulse labeling times are identical to LS174T cells.
   c. Optimal labeling conditions for A431 are identical to those for Caco-2, but labeling efficiency is highest in preconfluent cultures (Van Klinken, Einerhand, and Dekker, unpublished results)
9. For each of the following labels used in pulse/chase labeling experiments: [^{35}S]methionine/[^{35}S]cysteine (i.e., Promix™), [^{3}H]threonine, [^{35}S]sulfate, and [^{3}H]galactose—100 μCi (3700 kBq) of label is added per tube containing one tissue segment. Metabolic labeling has been performed for each of the following GI tissues of human, rat and mouse *(13–17)*: stomach, gallbladder (not in rat), duodenum, jejunum, ileum, caecum, ascending colon, transverse colon, descending colon, and sigmoid colon. For Promix and [^{3}H]threonine, pulse times are as follows: 30 min except for stomach (15 min) and for ileum (60 min). Stomach tissue is very efficiently labelled with labelled amino acids, whereas labeling of the ileum appears to be very inefficient. The pulse time for (^{35}S)sulfate and (^{3}H)galactose are: 60 min except for stomach and sigmoid colon (both 30 min). Again ileum is very inefficiently labeled with the latter compounds. The origin(s) of the differences in labeling efficiency among these organs is not known. It is essential that incubation commences within 10 min of removal from the body.
10. The intestinal mucins are present as adherent gels, which are not always soluble in concentrated guanidine hydrochloride alone *(18)*. To achieve complete solution, reduction, and alkylation are necessary. This leads to the formation of mucin subunits which can be identified on agarose gel electrophoresis *(8)*.
11. The sequence of purification steps in mucin purification is significant. If density gradient centrifugation is followed by gel filtration, any lower molecular weight subunits or degradation products may be identified.
12. The use of a general carbohydrate stain is useful to detect mucins on slot blots. The periodic acid Schiff stain (PAS) can be used and sensitivity is improved on the membranes as salt is eliminated (*see* **Note 13**). Lectins can also be used as general probes. Wheat germ agglutinin–horse radish peroxidase conjugate has been found to be a satisfactory and sensitive probe for mucins on slot blots *(20)*.
13. Owing to high salt concentrations in density gradient centrifugation experiments, many colorimetric assays and some radioactive scintillation cocktails are inefficient. Extensive dialysis of each fraction may allow colorimetric assays to be performed, but with small amounts of metabolically labeled material this results in significant losses (*see* **Note 3**). The slot blotting technique (**Note 12**) is more reliable and sensitive for colorimetric and radioactive detection.

14. After dialysis of the guanidine hydrochloride containing fraction of the CsCl density gradient against water, the fractions can also be analysed by SDS-PAGE ***(13–17,19)***. Typically, 4% SDS-PAGE gels are used. These gels can be PAS stained, or Western blotted with specific antibodies to detect the mucins (*see* also Chapter 20). If higher polyacrylamide concentrations are used (7.5% or 3–10% gradient gels) double staining of the gels with PAS and Coomassie Blue will clearly show the contaminating nonmucin proteins within each fraction. Alternatively, radioactively labeled mucins can be visualized directly by similar analysis on SDS-PAGE followed by fluorography, as described in Chapter 20.
15. Gel filtration is the most rapid and convenient method to obtain a mucin-enriched fraction from crude culture samples for comparative studies ***(1,2)***. It is also a starting point for preparation of native mucins for further purification and analysis (*see* **Note 11**). Automated fast-protein liquid chromatography (FPLC) systems can also be used to analyze samples in the same way, but are not as flexible in preparation of larger fractions of mucin.
16. The separation of mucins on agarose gel electrophoresis allows the largest mucin molecules to be analyzed, whereas SDS-PAGE systems may show incomplete migration of the sample into the gels ***(8)***.
17. Markers for agarose gel electrophoresis reflect the separation of molecules on the basis of their size and charge. This is in contrast to the normal conditions used for proteins and some glycoproteins on SDS-PAGE. Accordingly, the use of markers on agarose gel electrophoresis can give only an estimate of relative migration and not of molecular size.
18. Protein stains for mucins are usually very poor. Coomassie blue and silver stains frequently give negative results. Chemical stains for carbohydrate, lectin conjugates, or specific antibodies are the most useful.
19. If the mucins analysed by blotting procedures are radioactive, it is possible, after probing the blot with antibodies or lectins, to subject the membrane to autoradiography to colocalize the radioactive components. In this way a dual identification of alleged mucin bands can be made.
20. When possible, samples of mucin should be kept in solution, preferably in 4 *M* guanidine hydrochloride. Desalting and/or freeze-drying may result in the production of a residue that cannot be resolubilized. Concentration of mucin samples may be difficult and where samples are treated by such methods assessment of mucin loss is advisable. Mild precipitation methods at reduced temperature or with specific antibodies for smaller samples are best but require suitable controls for efficiency.
21. The proportion of proteoglycan in colonic cell and tissue samples is normally low but in cases of cell transformation or selection of subclones ***(21)*** care must be taken to control high molecular weight material with a buoyant density in the range of 1.35–1.60 g/mL for the presence of proteoglycans. In metabolic labeling experiments the presence of cells having a high turnover rate for proteoglycans, e.g., fibroblast hyaluronan synthesis, may significantly add to the proportion of labeled material isolated in "mucin" fractions. Both of these situations require the analysis of suspected mucin products using enzymatic degradation.
22. The assay of carbohydrate breakdown products of proteoglycan degradation must be carried out using liquid assay systems because these products are of low molecular weight and will not be detectable using membrane-blotting techniques. Assay for total carbohydrate or hexosamine is appropriate.

References

1. Corfield, A. P., Warren, B. F., Bartolo, D. C. C., Wagner, S. A., and Clamp, J. R. (1992) Mucin changes in ileoanal pouches monitored by metabolic labelling and histochemistry. *Brit. J. Surgery* **79,** 1209–1212.

2. Probert, C. S. J., Warren, B. F., Perry, T., Mackay, E. H., Mayberry, J. F., and Corfield, A. P. (1995) South Asian and European colitics show characteristic differences in colonic mucus glycoprotein type and turnover. Potential identification of a lower risk group for severe disease and cancer. ***Gut* 36,** 696–702.
4. Forstner, J. F. and Forstner, G. G. (1994) Gastrointestinal mucus, in *Physiology of the Gastrointestinal Tract.* (Johnson, L. R., eds.), Raven Press, New York, pp. 1245–1283.
5. Jass, J. R. and Roberton, A. M. (1994) Colorectal mucin histochemistry in health and disease: A critical review. *Pathology International.* **44,** 487–504.
3. Vavasseur, F., Dole, K., Yang, J., Matta, K. L., Corfield, A. P., Myerscough, N., Paraskeva, C., and Brockhausen, I. (1994) O-glycan biosynthesis in human colonic cells during progression to cancer. *Eur. J. Biochem.* **222,** 415–424.
7. Thornton, D. J., Carlstedt, I., and Sheehan, J. K. (1994) Identification of glycoproteins on nitrocellulose membranes and gels, in *Basic Protein and Peptide Protocols,* vol. 32 (Walker, J. M., eds.), Humana Press, Totowa, pp. 119–128.
8. Thornton, D. J., Holmes, D. F., Sheehan, J. K., and Carlstedt, I. (1989) Quantitation of mucus glycoproteins blotted onto nitrocellulose membranes. *Anal. Biochem.* **182,** 160–164.
9. Thornton, D. J., Howard, M., Devine, P. L., and Sheehan, J. K. (1995) Methods for separation and deglycosylation of mucin subunits. *Anal. Biochem.* **227,** 162–167.
6. Corfield, A. P., Clamp, J. R., Casey, A. D., and Paraskeva, C. (1990) Characterization of a sialic acid-rich mucus glycoprotein secreted by a premalignant human colorectal adenoma cell line. *Int. J. Cancer* **46,** 1059–1065.
10. Carlstedt, I. and Sheehan, J. K. (1984) Macromolecular properties and polymeric structure of mucus glycoproteins, in *Mucus and mucosa,* vol. 109, Pitman, London, pp. 157–172.
11. Thornton, D. J., Devine, P. L., Hanski, C., Howard, M., and Sheehan, J. K. (1994) Identification of two major populations of mucins in respiratory secretions. *Am. J. Respir. Crit. Care Med.* **150,** 823–832.
12. Van Klinken, J.-W., Oussoren, E., Weenink, J.-J., Strous, G. J., Büller, H. A., Dekker, J., and Einerhand, A. W. C. (1996) The human intestinal cell lines Caco-2 and LS174T as models to study cell-type specific mucin expression. *Glycoconjugate J.* **13,** 757–768.
13. Tytgat, K. M. A. J., Büller, H. A., Opdam, F. J. M., Kim, Y. S., Einerhand, A. W. C., and Dekker, J. (1994) Biosynthesis of human colonic mucin: Muc2 is the predominant secretory mucin. *Gastroenterology* **107,** 1352–1363.
14. Dekker, J., Van Beurden-Lamers, W. M. O., and Strous, G. J. (1989) Biosynthesis of gastric mucus glycoprotein of the rat. *J. Biol. Chem.* **264,** 10,431–10,437.
15. Dekker, J. and Strous, G. J. (1990) Covalent oligomerization of rat gastric mucin occurs in the rough endoplasmic reticulum, is N-glycosylation dependent, and precedes initial O-glycosylation. *J. Biol. Chem.* **265,** 18,116–18,122
16. Van Klinken, B. J. W., De Bolos, C., Büller, H. A., Dekker, J., and Einerhand, A. W. C. (1997) Biosynthesis of mucins (MUC2-6) along the longitudinal axis of the gastrointestinal tract. *Am. J. Physiol.* **273,** G296–G302.
17. Van Klinken, B. J. W., Dekker, J., Van Gool, S. A., Van Marle, J., Büller, H. A., and Einerhand, A. W. C. (1998) MUC5B is the prominent mucin in human gallbladder and is also expressed in a subset of colonic goblet cells. *Am. J. Physiol.* **274,** G871–G878.
18. Asker, N., Baekstrom, D., Axelsson, M. A. B., Carlstedt, I., and Hansson, G. C. (1995) The human MUC2 mucin apoprotein appears to dimerize before O-glycosylation and shares epitopes with the 'insoluble' mucin of rat small intestine. *Biochem. J.* **308,** 873–880.
19. Dekker, J., Van Beurden-Lammers, W.M.O., Oprins, A., and Strous, G.J. (1989) Isolation and structural analysis of rat gastric mucus glycoprotein suggests a homogenous protein backbone. *Biochem. J.* **260,** 717-723.

20. Ayre, D., Hutton, D. A., and Pearson, J. P. (1994) The use of wheat germ agglutinin to improve binding of heterogeneous mucin species to nitrocellulose membranes. *Anal. Biochem.* **219,** 373–375.
21. Ogata, S., Chen, A., and Itzkowitz, S. H. (1994) Use of model cell lines to study biosynthesis and biological role of cancer-associated sialosyl-Tn antigen. *Cancer Res.* **54,** 4036–4044.

20

Identification of Mucins Using Metabolic Labeling, Immunoprecipitation, and Gel Electrophoresis

B. Jan-Willem Van Klinken, Hans A. Büller, Alexandra W. C. Einerhand, and Jan Dekker

1. Introduction

Metabolic labeling of mucins is a powerful method for two reasons: (1) it lowers the detection limits of the mucins and their precursors considerably, and (2) it provides data on the actual synthesis of mucins in living cells. The produced radioactive mucins can be isolated and studied using biochemical methods, as described in Chapter 19, but these techniques apply basically to the study of mature mucins. In this chapter, we outline the methods for the immunoprecipitation of mucins, i.e., the immunoisolation of the mature mucins as well as their corresponding precursors. By applying metabolic labeling using amino acids and immunoprecipitation with the proper antibodies against the mucin polypeptide, it becomes possible to detect the earliest mucin precursor in the rough endoplasmic reticulum, to follow its subsequent complex conversion into a mature mucin, and to observe its storage and eventual secretion ***(1–3)***. Moreover, this antibody-based technique has the required specificity to discriminate the primary translation-product of each mucin gene. How mucin precursors can be distinguished is described in detail for each of the MUC-type mucins in Chapter 21.

The type of metabolic labeling used is known as pulse/chase labeling: the radioactive label is administered for a short period of time, followed by removal of the label and an extended incubation in absence of the radioactive label. By homogenization of the cells or tissue at various time points and subsequent immunoprecipitation by polypeptide-specific antibodies, we are able to follow the whereabouts of the mucins during this time course. Also, this protocol enables us to interfere with various steps of the cellular processing, giving us a unique angle at the diverse steps in the mucin biosynthesis (*see* Chapter 21).

From: *Methods in Molecular Biology, Vol. 125: Glycoprotein Methods and Protocols: The Mucins*
Edited by: A. Corfield © Humana Press Inc., Totowa, NJ

The mucin molecules can be caught in various stages of their synthesis. We use three different labels for pulse-labeling of mucins: (1) essential amino acids, which are incorporated into the polypeptide in the RER, (2) galactose, which is incorporated early in *O*-linked glycosylation in the medial and *trans*-Golgi apparatus (namely in core-type 1, 2, or 6 *O*-glycosylation), but galactose is also incorporated during chain elongation in backbone 1, 2, and 3 structures, and in chain termination in the form of αGal (*see* Chapters 14–17), and (3) sulfate, which is incorporated in the *trans*-Golgi stack and *trans*-Golgi network, as *O*-glycosylation elongation-terminator (*see* Chapters 14 and 17). Following the movements of the pulse-labeled mucins through the cellular compartments of the mucin-producing cells during the chase-incubations gives essential information about the dynamics of each step of the complex processes that eventually leads to secretion of a fully mature and functional mucin molecule ***(1–3)***.

2. Materials

1. Source of mucin-producing cells: These can be biopsies, tissue explants, or cell lines, which are cultured as described in Chapter 18.
2. Radioactively labeled glycoprotein precursors (Amersham, Little Chalfont, Buckinghamshire, UK), which are described in detail in Chapter 19:
   a. L-[^{35}S]methionine/[^{35}S]cysteine (Pro-Mix™).
   b. L-[^{3}H]threonine.
   c. D-[1-^{3}H]galactose.
   d. [^{35}S]sulfate.
3. Media (Gibco/BRL, Gaitersburg MD, USA) for metabolic pulse-labeling (15–60 min), as described in detail in Chapter 19:
   a. Eagle's minimal essential medium (EMEM) without L-methionine and L-cysteine.
   b. EMEM without L-threonine.
   c. EMEM with low D-glucose (50 μg/mL instead of 1000 μg/mL).
   d. EMEM without sulfate.
4. Medium for chase incubations: EMEM (Gibco/BRL), supplemented with nonessential amino acids, 100 IU/mL penicillin, 100 μg/mL streptomycin, and 2 m*M* L-glutamine.
5. Sodium dodecyl sulfate-polyacrylamide gel electrophoresis (SDS-PAGE) gels, 4% polyacrylamide running gels with 3% polyacrylamide stacking gel, according to the Laemmli system: prepared from stock solution with 30% (w/v) acrylamide and 0.8% (w/v) bisacrylamide, and SDS-PAGE apparatus (mini Protean II, Bio-Rad, Richmond CA).
6. SDS-PAGE sample buffer: for 5X concentrated buffer, 10% SDS (w/v), 5% (v/v) 2-mercaptoethanol, 50% glycerol (v/v), 625 m*M* Tris-HCl, pH 6.8, bromophenol blue to desired color.
7. Agarose electrophoresis gels and apparatus for analysis of mucins (*see* Chapter 19).
8. Amplify™ (Amersham).
9. X-ray film (Biomax-MR, Kodak, Rochester, NY).
10. Homogenization buffer: 50 m*M* Tris-HCl, pH 7.5, 5 m*M* EDTA, 1% (w/v) SDS, 1% (v/v) Triton X-100, 1% (w/v) bovine serum albumin (BSA), 10 m*M* iodacetamide, 100 μg/mL soybean trypsin inhibitor, 10 μg/mL pepstatin A, aprotinin 1 % (v/v) from commercial stock solution, 1 m*M* PMSF, 10 μg/mL leupeptin. (All reagents are from Sigma, St Louis, MO.)
11. Glass/Teflon tissue homogenizer, 5-mL model (Potter/Elvehjem homogenizer).

12. A Protein A-containing carrier to precipitate immunocomplexes. There are two alternatives:
    a. *Staphylococcus aureus* bacteria, formaldehyde-fixed (commercial preparation, consisting of a 10% (w/v) suspension in sterile PBS: IgGSorb, New England Enzyme Center, Boston MA).
    b. Protein A-Sepharose CL-4B: commercial suspension, consisting of a 50% (v/v) suspension of Sepharose beads in sterile solution (Pharmacia, Upsala, Sweden).
13. ImmunoMix (wash buffer for immunoprecipitations): 1% (w/v) Triton X-100, 1% (w/v) SDS, 0.5% (w/v) sodium deoxycholate, 1% (w/v) BSA (Boehringer, Mannheim, Germany), 1 m*M* PMSF in PBS.
14. PBS: 10-fold diluted.
15. 10% (v/v) acetic acid/10% (v/v) methanol in water.
16. Schiff's reagent for periodic acid-Schiff (PAS) staining (Sigma).

3. Methods (Note 1)

3.1. Immunoprecipitation of Mucins and Mucin Precursors (Note 2)

1. Label the cells or tissue of interest according to the pulse/chase protocol described in Chapter 19 (*see* **Note 3**). Use L-[^{35}S]methionine/ [^{35}S]cysteine or L-[^{3}H]threonine to label the polypeptide of the mucins, and use D-[1-^{3}H]galactose or [^{35}S]sulfate to label the mature mucin (*see* **Notes 4–6**).
2. After incubation, the tissue or cell culture is placed on ice to immediately stop the metabolic incorporation of the radiolabel.
3. The medium of chase-incubations is collected, and centrifuged at 12,000*g* for 5 min. The pellet is discarded. To the supernatant of chase-incubated tissue segments, add homogenization buffer up to an end volume of 1000 µL. For supernatants of chase incubated cell lines, add an equal volume of homogenization buffer. After thoroughly mixing, the sample is kept on ice until immunoprecipitation.
4. For cell cultures, the cell monolayer is washed once with ice-cold PBS, then 1 mL of homogenization buffer is added to the tissue culture flask or well, and the cells are collected using a cell scraper. The scraped cells are transferred to a glass/Teflon homogenizer. Tissue segments are washed once with ice-cold PBS, transferred to a glass/Teflon homogenizer using tweezers, and immediately 1 ml homogenization buffer is added. The cells or tissue are homogenized with 20 stokes of the homogenizer (*see* **Note 7**).
5. The homogenates are centrifuged three times at 12,000*g* for 5 min. After each centrifugation the clear supernatant is collected, and the pellets are discarded (**see Note 8**).
6. Take small aliquots (50–100 µL) of each homogenate and medium and add one-fourth volume of five-times concentrated Laemmli sample buffer. Heat in boiling water immediately for 5 min, and stored at –20°C until analysis (**Subheading 3.2.**).
7. Take aliquots of 100–1000 µL of the homogenate or the medium samples, and adjust to 1000 µL with homogenization buffer. Prepare vials containing the appropriate anti-mucin antibodies (*see* **Notes 9** and **10**). Centrifuge the homogenates at 12,000*g* for 5 min, and add the clear supernatant to the vials containing the antibodies.
8. Incubate 16 h at 4°C, under gentle agitation (head-over-head rotation).
9. Prepared new vials, containing sufficient protein A-containing carrier to precipitate all the IgG-containing immunocomplexes, either IgGSorb or protein A Sepharose (*see* **Note 11**). Wash these preparations once with 1 mL of ImmunoMix to clear any soluble protein A (*see* **Note 12**). Centrifuge the samples and add the clear supernatant to vials containing the washed IgGSorb or protein A Sepharose.
10. Incubate for 1 h, at 4°C under gentle agitation (head-over-head rotation).

11. Wash the immunocomplexes, which have now been bound to the protein A-containing carrier (*see* **Note 12**). Wash at room temperature three times with ImmunoMix, and then twice with, 10-fold diluted PBS. After the last wash, drain as much buffer from the pellets as possible. When protein A-Sepharose beads are used, the buffer can be removed most efficiently by suction through a syringe with a very fine hypodermic needle.
12. Add Laemmli sample buffer containing 5% 2-mercaptoethanol to the pellets: 20 µL 1x sample buffer to *S. aureus* pellets, and 15 µL 3X sample buffer to protein A-sepharose pellets. Mix thoroughly and incubate in boiling water for 5 min. Analyze directly or store at –20°C until analysis (**Subheading 3.2.**).

3.2. Analysis of Immunoprecipitated Mucins on Gel Electrophoresis

3.2.1. SDS-PAGE (see **Note 13**)

1. Prepare SDS-PAGE gels, according to standard procedures, with 3% acrylamide stacking gels and 4% polyacrylamide running gels.
2. Analyze the homogenates and the immunoprecipitated mucins on the SDS-PAGE gels (*see* **Note 14**). Run the appropriate very high molecular mass markers on the same gel (*see* **Note 15**).
3. Fix the gel in 10% acetic acid /10% methanol for at least 15 min, and stain the gel with periodic acid/schiff's reagent (PAS), to reveal the presence of mature mucins (*see* **Note 16**).
4. Incubate for exactly 10 min with Amplify, and dry the gel immediately on a gel dryer (*see* **Note 17**).
5. Expose the dried gel to X-ray film or to a PhosphorImager plate (*see* **Note 18**).

3.2.2. Agarose Electrophoresis (see **Note 19**)

1. Prepare 0.8% agarose gels, according to standard procedures (*see* Chapter 19).
2. Analyze the homogenates and the immunoprecipitated mucins on the agarose gels.
3. Place the agarose gel on a pre-wetted piece of 3MM paper, and dry the gel immediately on a gel dryer.
4. Expose the dried gel to X-ray film or to a PhosphorImager plate (*see* **Note 18**).

4. Notes

1. The methods for metabolic labeling and immunoprecipitation have been optimized for the use on gastrointestinal cell lines or tissue samples, particularly for each of the following gastrointestinal tissues of human, rat and mouse: stomach, gallbladder (not in rat), duodenum, jejunum, ileum, cecum, ascending colon, transverse colon, descending colon, and sigmoid ***(5–8,11,14,18,19)***, as well as for the following cell lines: LS174T, Caco-2, and A431 ***(9)***. As the protocol works for quite a number of tissues and cell lines, we feel confident that it will probably work for most, if not all, mucin-producing tissues and cell lines.
2. All procedures regarding homogenization and immunoprecipitation take place on ice, using ice-cold buffers and ice-cooled apparatus. The washing in ImmunoMix and tenfold diluted PBS is performed at room temperature. It proves essential to **never** freeze the samples prior to immunoprecipitation, as this will often result in degradation of the mucin-precursor.
3. The details regarding the use of the four radiolabels and the corresponding media to label each of the tissues and cell lines, mentioned in **Note 1**, are specifically described in Chapter 19. Each experiment comprises of one pulse-labeling and one or more closely timed chase incubations in the absence of radiolabel. After chase incubations the medium as well as the tissue are collected to study the presence of mucins.

4. The commercial Pro-Mix preparation, consists of a ^{35}S-labeled protein lysate of *E. coli*, which were grown in the presence of [^{35}S]sulfate as sulfur source in their medium. Of all ^{35}S-labeled compounds in Pro-Mix, 65% is L-[^{35}S]methionine and 25% is L-[^{35}S]cysteine, whereas 10% of the ^{35}S-containing compounds in the mixture are not specified (Amersham, Pro-Mix™ data sheet). However, if there is any free [^{35}S]sulfate, or metabolizable [^{35}S]sulfate-containing compounds, in Pro-Mix, this will not be incorporated as [^{35}S]sulfate into glycoproteins, as the incorporation of radiolabeled sulfate is very efficiently inhibited by the presence of a large excess of free nonlabeled sulfate in the medium. Commercially available, highly purified [^{35}S]methionine or [^{35}S]cysteine will work equally well as Pro-Mix. However, these reagents are far more expensive (about 10-fold), while in our experience they give very similar labeling efficiencies.
5. Application of [^{35}S]amino acids or [^{3}H]threonine will both yield radioactively labeled mucin precursors, labeled in their polypeptide chains. Most mucins are particularly rich in threonine (up to 35% of the amino acid composition), and therefore the essential amino acid threonine may seem a good candidate for polypeptide labeling. However, it appears that the ^{3}H-label, which emits a far weaker ß-radiation that ^{35}S, necessitates very long exposure times in autoradio- or fluorography (*see* also **Notes 6**, **17**, and **18**). It is our very consistent finding that, although less abundant in the amino acid composition of mucins, labeling with [^{35}S]methionine and/or [^{35}S]cysteine will yield mucin precursor bands that are far more easily detected than ^{3}H-labeled precursors. Thus, for the application in immunoprecipitation and analysis on electrophoresis ^{35}S-labeled amino acids are a far better alternative, allowing far shorter exposure times. The only notably exception is MUC1, which contains no methionine or cysteine in its extracellular, repeat-containing domain, and therefore can only be labeled with [^{3}H]threonine *(4)*.
6. The use of [^{3}H]galactose or [^{35}S]sulfate to label mature mucins gives practically indistinguishable results (e.g., **refs.** *2,3*). The incorporation of galactose in *O*-linked glycans starts earlier (medial to *trans*-Golgi) than the incorporation of sulfate (*trans*-Golgi and *trans*-Golgi network). Thus, the processing of the mucins in the Golgi apparatus is very fast and efficient *(3)*, as is commonly observed for other glycoproteins in cell biological studies. However, for very similar reasons as outlined above for the application of differently labeled amino acids, the ^{35}S-labeled sulfate will yield a far more intense signal in autoradio- or fluorography, simply due to its more intense ß-emission. Therefore, [^{35}S]sulfate is our usual choice to metabolically label mature mucins, as it allows relatively short exposure times (see also **Notes 17** and **18**).
7. Normally, SDS is included in the homogenization buffer to reduce nonspecific binding of proteins to the immunocomplexes that will form after the addition of antibodies in the ensuing steps of the protocol. However, it is known that some antibodies will not recognize their epitopes in the presence of SDS. For the use of polyclonal antisera, the inclusion of SDS in the homogenization buffer may result in a slightly lower yield of immunoprecipitated mucin, but the immunoisolated mucins will be considerably more pure than in the absence of SDS. Therefore, the use of SDS for polyclonal antisera is absolutely recommended. Monoclonal antibodies exist of only one type of immunoglobulin, and if this particular monoclonal antibody is unable to recognize its epitope in the presence of SDS, then SDS must be omitted from the homogenization buffer.
8. Upon homogenization tissue segments often give a quite considerable pellet, which mainly consists of muscle and connective tissue. It is however, absolutely essential that the supernatant, which is collected, is clear: immunoprecipitation is a precipitating technique, so anything that precipitates spontaneously during centrifugation (in later steps of the procedure) will inevitable contaminate the mucin preparation.

9. In this procedure, the antibodies present in one sample (particularly the IgG-fraction) will end up in one lane of the gels, which are used to analyze the samples. As a result, the maximal amount of antibody that can be added to one homogenate or medium sample is determined by the amount of antibody that will overload the lane of the gel, which will be used for analysis. In practice, when using 0.75- to 1.5-mm thick slab gels, the maximal amount of antibody is about 25 μL serum, or an equivalent amount of IgG, e.g., in the form of a monoclonal antibody or protein A-isolated IgG-fractions.
10. There are quite a number of anti-mucin antibodies available, which are specific for the polypeptide of each respective MUC-type mucin. It is very important to realize that only these anti-peptide antibodies will (1) give the immunoprecipitation its MUC-type mucin specificity and (2) enables us to recognize the precursor, which is not yet *O*-glycosylated. Further, there is an important distinction between antirepeat antibodies, which will recognize only the repeated amino acid sequences, which become masked upon *O*-glycosylation of the mature mucins. These type of antibodies will most likely only recognize the precursor, but not the cognate mature mucin. Antibodies directed against the unique, non-*O*-glycosylated regions of the polypeptide will be able to recognize both precursor and mature mucin, and all the intermediate forms if these may appear. The latter type of antibody is of course the antibody of choice to perform pulse/chase analysis, as only these antibodies are able to recognize all the subsequent forms of the mucin molecules that may appear. The antibodies which have proven specificities against peptide epitopes of specific MUC-type mucins are listed in **Table 1**.
11. Normally, if either 25 μL of antiserum or an equivalent amount of IgG is used (as indicated in **Note 8**), then the following amounts of protein A carriers are sufficient to precipitate all IgG-containing immunocomplexes: 50 μL of the (10%, w/v) IgGSorb suspension, and 25 μL of the (50%, v/v) protein A-Sepharose suspension.
12. The washing of IgGSorb is as follows: Centrifuge 30 s at 12,000*g* and remove the supernatant, but leave approx 50 μL of buffer above the pellet. This pellet is difficult to resuspend. Therefore, first resuspend the pellet in this small amount of remaining buffer by vigorous agitation (Vortex), before the addition of the next volume of ImmunoMix. The protein A-Sepharose beads, which are much larger than the *S. aureus* bacteria, can be resuspended in 5 s by Vortex, and then collected by 5 s centrifugation at 12,000*g*. The resulting pellet is very easily resuspended in buffer.
13. SDS-PAGE is the method of choice to identify and quantify mucin precursors (see also Chapters 6 and 21). Mucin precursors are relatively "normal" glycoproteins, with only a relatively small amount of *N*-glycosylation and no *O*-glycosylation, which will be separated by SDS-PAGE following the normal rules that govern mobility on these gels. The only disadvantage is the extremely large sizes of these mucin precursors (*see* Chapter 21), making it difficult to accurately assess their molecular masses. Mature mucins behave rather unpredictable on SDS-PAGE, as was discussed at length elsewhere ***(22,23)***. The mobility of the mature mucins is governed by their intrinsic negative charge rather than by their actual molecular mass. Moreover, on reducing SDS-PAGE most mature mucins migrate only a very small distance into the running gel, making distinction between the various mature mucin species rather difficult. Nevertheless, it is very important to note that the mobility of any mature mucin from a defined source is always highly reproducible. Therefore, the mobility of a particular mature mucin on SDS-PAGE can be used to establish its identity, but not as a means to assess its actual molecular mass ***(22,23)***.
14. Mucins and their precursors will only migrate small distances into the 4% running gel. The migrating distance may be improved by extending the running time, for instance to

Table 1
Antibodies Directed Against Mucin Polypeptides of MUC1–MUC6 for the Use in Immunoprecipitation of Mucins

Antibody[a]	Specificity[b]	Clonality	Antirepeat/ antiunique	Recognition[c]	Refs.
139H2	Human MUC1	Monoclonal	Antirepeat	p, m	*4*
Anti-HCM, anti-HCCM	Human MUC2 (r,m)	Polyclonal	Antiunique	p, m	*5,6*
Anti-RCM	rat MUC2 (h,m)	Polyclonal	antiunique	p, m	*7,8*
MRP	human MUC2 (r)	Polyclonal	Antirepeat	p	*5,7,9–12*
Anti-SI mucin	Human MUC2	Polyclonal	Antiunique	p, m	*12*
Anti-MUC2TR	Human MUC2 (r)	Polyclonal	Antirepeat	p	*13*
Anti-MCM	Mouse MUC2 (r,h)	Polyclonal	Antiunique	p, m	*8*
WE9	Human MUC2 (r,m)	Monoclonal	Antiunique	p, m	*5,7,8*
M3P	human MUC3	Polyclonal	Antirepeat	p	*9–11*
Anti-MUC4	Human MUC4	Polyclonal	Antirepeat	p	*11*
Anti-HGM	Human MUC5AC (r)	Polyclonal	Antiunique	p, m	*14–16*
Anti-RGM	rat MUC5AC (h)	Polyclonal	Antiunique	p, m	*2,3,11,16*
LUM5-1	Human MUC5AC	Polyclonal	Antiunique	p, m	*11,17*
Anti-HGBM	Human MUC5B	Polyclonal	Antiunique	p, m	*18,19*
Anti-MUC5B	Human MUC5B	Monoclonal	Antirepeat	p	*19,20*
Anti-MUC6.1	Human MUC6	Polyclonal	Antirepeat	p	*9,11,21*

[a]All antibodies listed are directed against specific mucin polypeptides. Moreover, each of the antibodies has proven its usefulness in immunoprecipitation of mucins in metabolic labeling experiments. The name of the antibody in this column corresponds to the name given in the first publication in which it was described. Specific immunoprecipitations using antipeptide antibodies to other alleged MUC-type mucins (e.g., MUC7 and MUC8) have not been described.

[b]The specificity is indicated against the primary antigen. The cross-reactivity against homologous mucins in other species is indicated in parenthesis: h, human; r, rat; m, mouse.

[c]The recognition of the mucin precursor (p) and of the mature mucin (m) is indicated.

1.5–2 times the time required for the dye-front to reach the end of the gel. Usually this can be done without significant loss of band tightness.

15. Mucin precursors usually have very high molecular masses in the range of 300–900 kDa (*see* Chapter 21). There is a very limited choice of markers with sizes in this molecular mass range. Options are several unreduced protein oligomers: thyroglobulin (660 kDa), ferritin (440 kDa), IgM (990 kDa), and mouse laminin (approx 900 kDa). We often use unreduced rat gastric mucin precursor, that gives bands of 300 kDa for the monomeric precursor, and 600 kDa for the dimeric precursor *(3)*. Most of these markers are thus not ideal, as they are all used in unreduced form, which may not be totally comparable with the more fully denatured precursor proteins that will result from reduction in the presence of SDS. In general, it remains difficult to establish the actual molecular mass of the mucin precursors. Nevertheless, the mobilities of the individual mucin precursors relative to these markers is highly reproducible, implying that the mucin precursors of each of the known MUC-type mucin genes can be identified by their relative mobility on SDS-PAGE (*see* Chapter 21).

16. The staining of mature mucins by PAS will help to identify the position of radiolabeled mature mucins on the gels by carefully overlaying the PAS stained gel by the corresponding X-ray film.
17. Amplify is a commercial water-based solution that acts primarily as a scintillation fluid, that will amplify the ß-emissions of the radiolabeled molecules. The result is a fluorograph rather than an autoradiograph. This form of fluorography will shorten the exposure times to X-ray film to about one-tenth relative to autoradiography.
18. Standard X-ray film (e.g., Fuji-RX) is not particularly sensitive. A more sensitive detection of ^{35}S, ^{14}C, and ^{33}P can be achieved by the use of Kodak Biomax-MR. However, the detection of ^{3}H is not improved by this more sensitive type of film. In contrast, the PhosphorImager only detects radioactivity directly (i.e., it works like autoradiography), and therefore the application of Amplify is of no consequence to the intensity of the signal when using this apparatus.
19. Agarose electrophoresis is particularly well suited to separate mature mucins, as this method allows far better separation of the mucins compared to SDS-PAGE *(19,23)*. However, like for SDS-PAGE, the mobility of mature mucins is both dependent on their molecular mass and on their intrinsic negative charge. Thus, agarose electrophoresis is well suited to identify particular species of mature mucins, but accurate estimates of molecular masses are not possible.

References

1. Strous, G. J. and Dekker, J. (1992) Mucin-type glycoproteins. *Crit. Rev. Biochem. Mol. Biol.* **27,** 57–92.
2. Dekker, J., Van Beurden-Lamers, W. M. O., and Strous, G. J. (1989) Biosynthesis of gastric mucus glycoprotein of the rat. *J. Biol. Chem.* **264,** 10,431–10,437.
3. Dekker, J. and Strous, G. J. (1990) Covalent oligomerization of rat gastric mucin occurs in the rough endoplasmic reticulum, is N-glycosylation dependent and precedes O-glycosylation. *J. Biol. Chem.* **265,** 18,116–18,122.
4. Hilkens, J. and Buijs, F. (1988) Biosynthesis of MAM-6, an epithelial sialomucin. *J. Biol. Chem.* **263,** 4215–4222.
5. Tytgat, K. M. A. J., Büller, H. A., Opdam, F. J. M., Kim, Y. S., Einerhand, A. W. C., and Dekker, J. (1994) Biosynthesis of human colonic mucin: Muc2 is the most prominent secretory mucin. *Gastroenterology* **107,** 1352–1363.
6. Tytgat, K. M. A. J., Opdam, F. J. M., Einerhand, A. W. C., Büller, H. A., and Dekker, J. (1996) MUC2 is the prominent colonic mucin expressed in ulcerative colitis. *Gut* **38,** 554–563.
7. Tytgat, K. M. A. J., Bovelander, F. J., Opdam, F. J. M., Einerhand, A. W. C., Büller, H. A., and Dekker, J. (1995) Biosynthesis of rat MUC2 in colon and its analogy with human MUC2. *Biochem. J.* **309,** 221–229.
8. Van Klinken, B. J. W., Duits, L. A., Verburg, M., Tytgat, K. M. A. J., Renes, I. B., Büller, H. A., Einerhand, A. W. C., and Dekker, J. (1997) Mouse colonic mucin as a model for human colonic mucin. *Eur. J. Gastroenterol. Hepatol.* **9,** A66 (abstract).
9. Van Klinken, B. J. W., Oussoren, E., Weenink, J. J., Strous, G. J., Büller, H. A., Dekker, J., and Einerhand, A. W. C. (1996) The human intestinal cell lines Caco-2 and LS174T as models to study cell-type specific mucin expression. *Glycoconjugate J.* **13,** 757–768.
10. Ho, S. B., Niehans, G. A., Lyftogt, C., Yan, P. S., Cherwitz, D. L., Gum, E. T., Dahiya, R., and Kim, Y. S. (1993) Heterogeneity of mucin gene expression in normal and neoplastic tissues. *Cancer Res.* **53,** 641–651.

11. Van Klinken, B. J. W., De Bolos, C., Büller, H. A., Dekker, J., and Einerhand, A. W. C. (1997) Biosynthesis of mucins (MUC2-6) along the longitudinal axis of the human gastrointestinal tract. *Am. J. Physiol.* **273,** G296–G302.
12. McCool, D. J., Forstner, J. F., and Forstner, G. G. (1994) Synthesis and secretion of mucin by the human colonic tumour cell line LS180. *Biochem. J.* **302,** 111–118.
13. Asker, N., Baeckström, D., Axelsson, M. A. B., Carlstedt, I., and Hansson G. C. (1995) The human MUC2 mucin apoprotein appears to dimerize before O-glycosylation and shares epitopes with the "insoluble" mucin of rat small intestine. *Biochem. J.* **308,** 873–880.
14. Tytgat, K. M. A. J., Klomp, L. W. J., Bovelander, F. J., Opdam, F. J. M., Van der Wurff, A., Einerhand, A. W. C., Büller, H. A., Strous, G. J., and Dekker, J. (1995) Preparation of anti-mucin polypeptide antisera to study mucin biosynthesis. *Anal. Biochem.* **226,** 331–341.
15. Klomp, L. W. J., Van Rens, L., and Strous G. J. (1995) Cloning and analysis of human gastric mucin cDNA reveals two types of conserved cysteine-rich domains. *Biochem. J.* **308,** 831–838.
16. Klomp, L. W. J., Van Rens, L., and Strous G. J. (1994) Identification of a human gastric mucin precursor N-linked glycosylation and oligomerization. *Biochem. J.* **304,** 693–698.
17. Sheehan, J. K., Thornton, D. J., Howard, M., Carlstedt, I., Corfield, A. P., and Paraskeva, C. (1996) Biosynthesis of the MUC2 mucin: Evidence for a slow assembly of fully glycosylated units. *Biochem. J.* **315,** 1055–1060.
18. Klomp, L. W. J., Delely, A. J., and Strous, G. J. (1994) Biosynthesis of human gallbladder mucin. *Biochem. J.* **304,** 737–744.
19. Van Klinken, B. J. W., Dekker, J., Van Gool, S. A., Van Marle, J., Büller, H. A., and Einerhand, A. W. C. (1998) MUC5B is the prominent mucin biosynthesized in the human gallbladder and is also expressed in a subset of colonic goblet cells. *Am. J. Physiol.* **274,** G871–G878.
20. Perini, M., Vandamme-Cubadda, N., Aubert J. P., Porchet, N., Mazzuca, M., Lamblin, G., Hersovics, A., and Roussel, P. (1991) Multiple apomucin translation products from human respiratory mucosa. *Eur. J. Biochem.* **196,** 321–328.
21. De Bolos, C., Garrido, M., and Real, F. X. (1996) Muc6 apomucin shows a distinct normal tissue distribution which correlates with lewis antigen expression in the stomach. *Gastroenterology* **109,** 723–734.
22. Tytgat, K. M. A. J., Swallow, D. M., Van Klinken, B. J. W., Büller, H. A., Einerhand, A. W. C., and Dekker, J. (1995) Unpredictable behavior of mucins in SDS/polyacrylamide-gel electrophoresis. *Biochem. J.* **310,** 1053,1054.
23. Van Klinken, B. J. W., Einerhand, A. W. C., Büller, H. A., and Dekker J. (1998) Strategic biochemical analysis of mucins. *Anal. Biochem.* **265,** 103–116.

21

Mucin Precursors

Identification and Analysis of Their Intracellular Processing

Alexandra W. C. Einerhand, B. Jan-Willem Van Klinken, Hans A. Büller, and Jan Dekker

1. Introduction

MUC-type mucins are generally very large glycoproteins. They are encoded by very large mRNAs, and possess polypeptides between 200 and more than 900 kDa (1). The only notable exception is MUC7, which is considerably smaller, i.e. the polypeptide is only 39 kDa ***(1)***. Without exception however, mucins are very heavily *O*-glycosylated: Up to 50-80% of their molecular mass is due to *O*-glycosylation ***(1,2)***. Moreover, potential *N*-glycosylation sites are found in virtually all mucin sequences, and in several MUCs *N*-glycosylation is actually demonstrated ***(1,2)***. Human MUC2 for instance contains 30 potential *N*-glycosylation sites, and if these are all used, the *N*-glycans together would constitute a molecular mass of about 60 kDa. It is only the very large size of the mature mucins, that makes the amount of *N*-glycosylation seem insignificant ***(3)***. Generally, the sizes of the mature mucins are difficult to estimate; The approximations run from 1 to 20 MDa for single mucin molecules, which hampers many forms of biochemical analysis ***(3)***. Also, the extensive glycosylation of mucins results in an intrinsically very heterogeneous population of mature mucins.

The detection of mucin precursors forms an attractive alternative to assess the expression of specific mucins and to quantify mucin synthesis. Each precursor of the MUC-type mucins can be identified by immunoprecipitation using specific anti-mucin polypeptide antibodies (*see* Chapter 20). Very importantly, each of these precursors can be identified on reducing SDS-PAGE by its distinct molecular mass ***(3–5)***. Thus, immunoprecipitation in combination with sodium dodecyl sulfate-polyacrylamide gel electrophoresis (SDS-PAGE) can be used to detect expression of individual MUC-type mucins with high specificity in homogenates of tissue or cell lines. The mucin precursor bands, recognizable on SDS-PAGE, can be quantified as sensitive measures of mucin biosynthesis (*see* Chapter 6).

From: *Methods in Molecular Biology, Vol. 125: Glycoprotein Methods and Protocols: The Mucins*
Edited by: A. Corfield © Humana Press Inc., Totowa, NJ

Biochemically and cell biologically, MUC-type mucin precursors can be recognized by a number of characteristics, which will help in their identification *(2,3)*. Like any glycoprotein, the MUC polypeptide is synthesized at the rough endoplasmic reticulum (RER) and cotranslationally *N*-glycosylated. The product of this initial stage of biosynthesis will be referred to as the mucin precursor. Then, the precursors will oligomerize through formation of disulfide bonds, and be transported to the Golgi apparatus, where they will be fully *O*-glycosylated and sulfated, as many of the *O*-glycans of mucins contain terminal sulfate (*see* Chapter 17). Mucins that have completed synthesis are referred to as mature mucins.

In this Chapter, we focus on the identification of each of the known MUC-type mucin precursors by immunoprecipitation using antipeptide antibodies. Moreover, a number of biochemical and cell biological assays will be described which establish the presence in the RER of each alleged MUC-type mucin precursor. These assays are based on the following characteristics of the mucin precursors *(1–3)*: (1) The precursors contain only high mannose *N*-glycans, (2) Most precursors form, over time, disulfide-linked dimers within the RER, (3) *O*-glycosylation of the precursors, and conversion of the *N*-linked glycans to complex *N*-glycans, occurs only after their transport to the Golgi apparatus, and (4) A clear precursor/product relationship exists, as a result of the conversion over time of the precursors into their cognate mature mucins. The described methods will help researchers in the field to recognize and quantify the precursors of the known MUC-type mucins, and we will provide appropriate control experiments to verify the specificity of each of these procedures. Moreover, these methods will help to allocate previously unidentified mucin precursors.

2. Materials

1. Source of mucin-producing cells, such as biopsies, tissue explants, or cell lines, which are cultured as described in Chapter 18.
2. Radioactively labeled essential amino acids (Amersham, Little Chalfont, Buckinghamshire, UK), described in detail in Chapter 19:
   a. L-(^{35}S)methionine/(^{35}S)cysteine (Pro-Mix™).
   b. L-(^{3}H)threonine.
3. Media (Gibco/BRL, Gaitersburg MD) for metabolic pulse-labeling and chase incubations, as described in detail in Chapter 19.
4. Homogenization buffer for immunoprecipitation, as described in Chapter 20.
5. Glass/Teflon tissue homogenizer, 5 mL model (Potter/Elvehjem homogenizer).
6. Anti-mucin antisera directed against the mucin-polypeptide of interest (see Chapter 20, **Table 1**).
7. Protein A-containing carrier to precipitate immunocomplexes, as described in Chapter 20.
8. ImmunoMix, as described in Chapter 20.
9. PBS: 10-fold diluted.
10. SDS-PAGE gels: 4% polyacrylamide running gels with 3% polyacrylamide stacking gel, as described in Chapter 20.
11. SDS-PAGE sample buffer containing 1% SDS and 5% (v/v) 2-mercaptoethanol.
12. SDS-PAGE sample buffer containing 1% SDS, without reducing agent.
13. 10% (v/v) acetic acid/10% (v/v) methanol in water.
14. Schiff's reagent for PAS staining (Sigma, St. Louis MO).

15. Amplify™ (Amersham).
16. X-ray film (Biomax-MR, Kodak, Rochester, NY).
17. Brefeldin A (BFA), stock solution, 1 mg/mL in water.
18. Tunicamycin (Calbiochem, La Jolla CA), stock solution, 1 mg/mL in 10 m*M* NaOH in water.
19. Carbonyl cyanide *M*-chlorophenylhydrazone (CCCP, Sigma), stock solution, 1 m*M* in ethanol.
20. Endoglycosidase H (Endo H, New England Biolabs, Beverly MA), 500,000 U/mL.
21. 10-times concentrated Endo H-buffer (New England Biolabs), containing 0.5 *M* sodium citrate (pH 5.5).
22. Peptide:*N*-glycosidase F (PNGase F, New England Biolabs), 1,000,000 U/mL.
23. 10-times concentrated PNGase F-buffer (New England Biolabs), containing 0.5 *M* sodium phosphate (pH 7.5).
24. Nonidet-40 (New England Biolabs), 10% in water.
25. 10-times concentrated denaturing buffer (New England Biolabs), containing 5% SDS and 10% 2-mercaptoethanol.
26. *Dolichos biflorus*-agglutinin (DBA) Sepharose CL-4B beads (Sigma).
27. DBA column buffer: PBS (pH 7.2), supplemented with 1% (v/v) Triton X-100, 1 m*M* phenylmethylsulfonyl fluoride, 50 μg/mL pepstatin A, 25 μg/mL leupeptin, 1% (w/v) BSA, 10 m*M* iodoacetamide, and 0.1% NaN_3.
28. *N*-acetyl-Galactosamine (GalNAc), 100 m*M* solution in the above mentioned DBA column buffer.
29. Freunds complete adjuvant (Difco, Detroit MI,).

3. Methods (Note 1)

3.1. Identification of the Precursors of MUC-Type Mucins by Their Distinct Molecular Masses Through Metabolic Labeling and Immunoprecipitation (Note 1)

1. Metabolically pulse-label the mucin-producing tissue or cells with radiolabeled essential amino acids (*see* Chapter 19).
2. Homogenize the samples and isolate the radiolabeled mucin precursor of interest by immunoprecipitation using specific antipolypeptide antibodies (*see* Chapter 20).
3. Analyze the immunoprecipitated mucin precursors on 4% SDS-PAGE using reducing sample buffer.
4. Identify the mucin precursor according to its apparent molecular mass, using the appropriate molecular mass markers and/or control samples (*see* **Notes 2–6**).

3.2. Relation of the Mucin Precursor to its Mature Form Revealed by Pulse/Chase Experiments (Notes 1, 7, and 8)

1. Metabolically pulse-label seven samples of mucin-producing tissue or cells using radiolabeled essential amino acids, as described in Chapter 19. Immediately homogenize one sample after pulse-labeling. The pulse-medium is discarded.
2. Chase-incubate the remaining six tissue or cell samples, homogenize one sample after 1, 2, 3, 4, 5, and 6 h, respectively, of chase incubation, and isolate the media of each respective chase sample.
3. Isolate the radiolabeled mucin of interest from the seven homogenates and the six media, respectively, by immunoprecipitation using antipolypeptide antibodies (*see* **Note 8**).
4. Analyze the immunoprecipitated mucin precursors on 4% SDS-PAGE using reducing sample buffer and the appropriate molecular mass markers (*see* **Notes 2–5**).

5. PAS-stain the gels to reveal the position of the mature mucins. Prepare fluorographs of the gels using Amplify and X-ray film.
6. Analyze the kinetics of disappearance of the precursor and the appearance of the mature mucin, and the appearance of the mature mucin in the medium (*see* **Note 9**).

3.3. Identification of the Mucin Precursors as RER-Localized Proteins (see Note 1)

3.3.1. Inhibition of Vesicular RER-to-Golgi Transport (see Note 10)

3.3.1.1. Inhibition of Vesicular RER-to-Golgi Transport by Brefeldin A (BFA) (see Note 11)

1. Treat seven samples of mucin-producing tissue or cells with BFA for 30 min under normal culturing conditions; 10 μg/mL for tissue, 0.1–2 μg/mL for cell lines (*see* **Note 12**).
2. Metabolically pulse-label the tissue or cells by radiolabeled essential amino acids, as described in Chapter 19 (*see* **Note 12**). Homogenize one sample immediately after the pulse-labeling.
3. Chase-incubate the six remaining samples of the tissue or cells in continued presence of BFA (identical concentrations as above), chase the samples for 1, 2, 3, 4, 5, and 6 h, respectively. Homogenize each sample immediately after its respective chase incubation. Also isolate and homogenize the media of the chase incubations.
4. Isolate the radiolabeled mucin precursor of interest from the homogenates and media by immunoprecipitation using anti-polypeptide antibodies (*see* Chapter 20).
5. Analyze the immunoprecipitated mucin precursors on 4% SDS-PAGE using reducing sample buffer. Compare the mobility of the mucin precursor bands in the BFA-treated samples to the precursor bands in a pulse/chase experiment under normal conditions, described in **Subheading 3.2.** (*see* **Note 13**). Perform DBA affinity chromatography to study initial *O*-glycosylation (*see* **Subheading 3.3.1.2.**).

3.3.1.2. DBA Affinity Chromatography to Detect Initial O-Glycosylation (see Note 14)

1. Perform this entire procedure at 4°C. Prepare a DBA-Sepharose column, and wash extensively with DBA column buffer.
2. Prepare a homogenate of [^{35}S]amino acids-labeled tissue or cells in DBA column buffer (Avoid the use of Tris). Apply this homogenate to the column, and elute with DBA column buffer. Collect the flow-through and store on ice.
3. Elute the terminal GalNAc-containing proteins from the column by 100 m*M* GalNAc in DBA column buffer. Collect the eluate and keep on ice.
4. Immunoprecipitate the mucin precursor from the flow-through (containing the nonbound proteins), and from the eluate (the GalNAc-containing proteins), as described in Chapter 20.
5. Analyze the presence of mucin precursor in both column fractions by reducing SDS-PAGE (*see* **Note 14**).

3.3.1.3. Inhibition of Vesicular RER-to-Golgi Transport by CCCP (see Note 15)

1. Metabolically pulse-label seven samples of mucin-producing tissue or cells by radiolabeled essential amino acids, as described in Chapter 19. Homogenize one sample immediately after the pulse-labeling. Discard the pulse-medium.
2. Chase-incubate the six remaining samples of the tissue or cells in the presence of CCCP (tissue; 10 μg/mL, cells; 0.1–1 μ*M*), and chase the samples for 1, 2, 3, 4, 5, and 6 h, respectively. Homogenize each sample immediately after its respective chase incubation. Also isolate and homogenize the media of the chase incubations.

3. Isolate the radiolabeled mucin precursor of interest from the homogenates and media by immunoprecipitation using anti-polypeptide antibodies (see Chapter 20).
4. Analyze the immunoprecipitated mucin precursors on 4% SDS-PAGE using reducing sample buffer. Compare the presence of the mucin precursor band in the homogenates to the pulse/chase experiment under normal conditions, described in **Subheading 3.2.** (*see* **Note 15**).

*3.3.2. Analysis of Disulfide Bond Formation of Mucin Precursors (*see ***Notes 1** *and* **16***)*

1. Perform a pulse/chase experiment on mucin-producing tissue or cells, using [^{35}S]amino acids, as described in **Subheading 3.2.**
2. Immunoprecipitate the mucins, as described in Chapter 20, until the second of the two wash steps in 10-fold diluted PBS.
3. Add the second aliquot (i.e. the last wash step) of 1 mL of 10-fold diluted PBS. Divide the resuspended pellet into two equal aliquots of 500 µL in separate vials. Centrifuge these two suspensions, and remove the buffer thoroughly.
4. Boil one pellet in sample buffer containing 5% 2-mercaptoethanol, and the duplicate pellet in sample buffer without reducing agent, and analyze these samples on SDS-PAGE (*see* **Notes 16–18**).

*3.3.3. Identification of Mucin Precursors as High Mannose N-Glycan Containing Glycoproteins (*see ***Note 1***)*

3.3.3.1. Characterization of *N*-Glycans by Endo H and PNGase Digestion (*see* **Note 19**)

1. Metabolically pulse-label a sample of mucin-producing tissue or cells using [^{35}S]amino acids, as described in Chapter 19. Immediately homogenize the sample after pulse-labeling.
2. Isolate the radiolabeled mucin precursor of interest from the homogenate by immunoprecipitation using antipolypeptide antibodies (*see* **Note 8**).
3. Endo H digestion: Add 10 µL denaturing buffer to the *S. aureus* or protein A Sepharose pellet, denature the sample for 5 min at 100°C. Cool to room temperature, add 1.2 µL Endo H-buffer and 500 U Endo H to the sample, and incubate 1 h at 37°C.
4. PNGase F digestion: Add 10 µL denaturing buffer to the *S. aureus* or protein A Sepharose pellet, denature the sample for 5 min at 100°C. Cool to room temperature, add 1.2 µL PNGase F-buffer and 1000 U PNGase F to the sample, and incubate 1 h at 37°C.
5. Add reducing Lemmli sample buffer to the digestion mixtures, and analyze the mucin precursors on 4% SDS-PAGE, using the appropriate molecular mass markers (*see* **Notes 2–5**, and **19**).

3.3.3.2. Inhibition of *N*-Glycosylation by Tunicamycin (*see* **Notes 20** and **21**)

1. Incubate one sample of mucin-producing tissue (50 µg/mL) or cells (5–20 µg/mL) for 3 h with tunicamycin. Perform a control incubation under identical conditions.
2. Metabolically pulse-label both samples of mucin-producing tissue or cells using [^{35}S]amino acids, as described in Chapter 19. Immediately homogenize the samples after pulse-labeling.
3. Isolate the radiolabeled mucin precursor of interest from the homogenate by immunoprecipitation using antipolypeptide antibodies (*see* **Note 8**).
4. Analyze the mucin precursors on 4% SDS-PAGE using reducing sample buffer, using the appropriate molecular mass markers (*see* **Notes 2–5**, and **20**).

3.4. Identification of Previously Unidentified Mucins Through Detection of Their Precursors (see Notes 1, 22, and 23)

1. Isolate mucins using density centrifugation on CsCl/guanidinium·HCl gradients (*see* Chapter 1). Thoroughly dialyze the isolated mucins against water.
2. Prepare a polyclonal antiserum in rabbits against the isolated mucins, using Freunds complete adjuvant *(8)*.
3. Metabolically pulse-label a sample of the mucin-producing tissue or cells from which the mucin was isolated using [^{35}S]amino acids, as described in Chapter 19. Immediately homogenize the sample after pulse-labeling.
4. Isolate the radiolabeled mucin precursors from the homogenate by immunoprecipitation using the polyclonal antiserum raised against the isolated mucins from this particular source.
5. Analyze the mucin precursors on 4% SDS-PAGE using reducing sample buffer, using the appropriate molecular mass markers (*see* **Notes 2–5**, **22**, and **23**).

4. Notes

1. Mucin precursors, because of their low abundance, can only be detected through metabolic labeling. All methods described in this chapter are based on the methods to culture tissue and cell lines (*see* Chapter 18), methods for metabolic labeling of the mucin precursors (*see* Chapter 19), and methods to specifically immunoprecipitate the mucin precursors (*see* Chapter 20).
2. Each precursor of the known human, rat or mouse MUC-type mucins can be distinguished by its unique apparent molecular mass by SDS-PAGE. These data are summarized in **Table 1**, which serves as a reference table to identify each known mucin precursor by SDS-PAGE (*see* also Chapter 20 for listed molecular mass markers).
3. The distribution of MUC2-MUC6 based on detection by immunoprecipitation of their respective precursors in gastrointestinal tissue and in cell lines are summarized in **Table 2**, which serves as reference table for mucin precursor synthesis in these organs and cells. MUC1 is not included, as it is expressed in virtually all epithelia at low levels, i.e., its expression is not tissue specific. Thus far, no data are available for other MUC-type mucins, like MUC7 and MUC8.
4. The information on the molecular masses of the mucin precursors of the rat and mouse is incomplete. However, the analogy to their human counterparts suggests that also in these species a clear distinction can be made between the various mucin precursors based on their molecular masses (**Table 1**).
5. Three cell lines are included for reference, which collectively produce the precursors of MUC1 through MUC6 (**Table 2**). These cell lines are available at low costs through the American Type Culture Collection (ATCC), and can be cultured as described in Chapter 19. The mucin precursors immunoprecipitated from these cell lines serve as excellent markers to detect these respective mucin precursors in other human mucin-producing sources. Moreover, immunoprecipitation of a particular mucin precursor from one of these cell lines can provide the proper positive control for the immunoprecipitation procedure of this particular mucin precursor from other sources.
6. MUCs often display genetic polymorphisms, which affect the number of tandemly repeated amino acid sequences *(1,2)*. Therefore, different individuals or cell lines may biosynthesize precursors of a particular MUC gene of slightly variable lengths. When immunoprecipitating precursors of a particular MUC, we sometimes observe distinct interindividual differences in the molecular masses of these MUC precursors (**Table 1**).

Table 1
Apparent Molecular Masses of MUC-Type Mucin Precursors as Determined by Immunoprecipitation and Reducing SDS-PAGE

Mucin	Species	Molecular mass[a]	References
MUC1	Human	160-400[b]	***6***
MUC2	Human	600[b]	***4,5,7–9***
Muc2	Mouse	600	***10***
rMuc2	Rat	600[b]	***11***
MUC3	Human	550[b]	***4,5***
MUC4	Human	>900	***4,5***
MUC5AC	Human	500	***4,5,12***
rMuc5AC	Rat	300[b]	***13–15***
MUC5B	Human	470	***16,17***
MUC6	Human	400	***4,5***

a The apparent molecular masses were estimated (expressed as kDa) after immunoprecipitation by reducing SDS-PAGE.

*b*These mucin precursors were shown to display interindividual heterogeneity, leading to small variations in the apparent molecular masses on reducing SDS-PAGE (*see* also **Note 6**).

Table 2
Distribution of Mucin Precursors in Human Gastrointestinal Tissues and in Cell Lines as Determined by Metabolic Labeling and Immunoprecipitation

Tissue	MUC2	MUC3	MUC4	MUC5AC	MUC5B	MUC6	Refs.
Stomach	–[a]	–	–	+++	–	+	***5,12***
Duodenum	++	++	–	–	–	–	***5***
Jejunum	++	++	–	–	–	–	***5***
Ileum	+	++	–	–	–	–	Unpub.[b]
Proximal colon	+++	–	+	–	+	–	***5***
Distal colon	+++	–	+	–	+	–	***5,7***
Gallbladder	–	+	–	–	+++	–	***16***
LS174T	+++	–	–	+	+	++	***4***
Caco–2	+	++	–	–	–	–	***4***
A431	–	ND[c]	++	ND	ND	ND	Unpub.[b]

*a*Per organ or cell line we have indicated, in a semi-quantitative manner, the relative amounts of mucin precursors: –, no expression ; +, detectable; ++, moderate expression; +++, strong expression.

*b*Data on human ileum and A431 cells; Van Klinken, B. J. W., Büller, H. A., Dekker, J., and Einerhand, A. W. C., unpublished.

*c*ND, not determined.

This phenomenon is best documented for MUC1 in which the variation in molecular mass of the precursors, produced from these different alleles, can be quite high: approx 160–310 kDa ***(6,18)***. However, for the other mucins the interindividual variations in the molecular masses of the mucin precursors are quite small. That is, there is variation in the exact position of the precursor band on reducing SDS-PAGE, and sometimes double bands can be observed in particular individuals. However, it is very important to note that these variations in apparent molecular mass are relatively small, and that they will not lead to any confusion regarding the identity of the immunoprecipitated mucin precursor.

7. For gastrointestinal tissues, over a period of up to 6 h, at 37°C under normal culture conditions, all precursor will be processed to mature mucin. For cell lines, like LS174T, this conversion may take longer (up to 24 h). In these experiments, the mature mucin can be recognized on SDS-PAGE by its molecular weight, by PAS-staining, and often by its heterogeneous appearance (smear). Also the position of the mature mucin on SDS-PAGE can be revealed by metabolic labeling of duplicate tissue or cell samples with [^{3}H]galactose or [^{35}S]sulfate (*see* Chapters 19 and 20).
8. Pulse/chase experiments will only reveal the precursor/product relationship of the mucin precursor and its cognate mature mucin if antibodies are used, which are able to recognize both the precursor as well as the mature mucin. Therefore, the antibodies used in these experiments must be able to recognize the mucin polypeptide in a manner independent of *O*-glycosylation (extensively described in Chapter 20).
9. Precursors are never present in the medium. If however, a known precursor is found in the medium, this can be taken as evidence of cell lysis during the experiment.
10. Inhibition of vesicular transport from the RER to the Golgi complex will lead to the accumulation of mucin precursors in the RER. This accumulation is generally accepted as evidence of RER localization *(2)*.
11. BFA is a fungal metabolite, which inhibits the anterograde vesicular transport from the RER to the Golgi complex, but not the retrograde transport of vesicles from the Golgi complex to the RER. This results in accumulation of RER-localized protein in the RER, but also in an enrichment within the RER with enzymes (like glycosyltransferases), which are normally present in the *cis*-Golgi cisternae *(2,22)*.
12. BFA is added to the medium during the 30 min period, which is used to deplete the compound to be used as label. During the metabolic pulse-labeling the medium is not changed, i.e., BFA remains present in the medium.
13. BFA will retain the mucin precursors in the RER. However, some enzymes involved in initial *O*-glycosylation are redistributed to the RER in the presence of BFA, resulting in initial *O*-glycosylation of these precursors. As a result, the precursor band will gradually transform over time into a smear, slightly above the normal precursor position on reducing SDS-PAGE (14,20). As BFA is a potent inhibitor of secretion, none of these partly *O*-glycosylated precursors will appear in the medium as secreted product *(14,20,22)*.
14. DBA has a high affinity for terminal GalNac residues. Therefore, the binding of mucin precursors to this lectin is taken as evidence that initial *O*-glycosidic α(1–0) GalNac addition to serine and threonine residues has occurred *(14)*. This initial *O*-glycosylation will occur in the presence of BFA, but not in the presence of CCCP *(14,20)*.
15. CCCP inhibits the oxidative phosphorylation in the mitochondria, resulting in a sharp drop in ATP levels in the cells. As the RER-to-Golgi transport is highly energy dependent, the addition of CCCP will almost instantaneously inhibit this transport. The presence of CCCP will lead to accumulation of all mucin precursors, formed in the pulse-labeling, in the RER *(14,20)*. Never, add CCCP prior to or during the pulse-labeling, as this will inhibit nearly all protein synthesis *(20)*.
16. Most mucin precursors form disulfide-bound dimers in the RER *(14,20)*. When we perform a pulse/chase experiment on tissue or cells with radiolabeled amino acids, and analyze the immunoprecipitated mucin precursors on nonreducing SDS-PAGE, we are able

to demonstrate, next to the monomeric precursor band, a band with a much higher apparent molecular mass than the monomeric mucin precursor. Reduction of parallel samples will show that radioactivity in this high molecular weight band can be retrieved as the monomeric mucin precursor on reducing SDS-PAGE, thus proving the dimerization of the mucin precursor. The pulse sample usually only contains only monomeric precursors, when analyzed on nonreducing SDS-PAGE. The precursor dimer appears during the chase-period (typically within 30–60 min), and shows clear precursor/product relationship with the monomeric precursor ***(14,20)***. It is advisable, to perform electrophoresis for extended time to ensure that all putative dimers enter the running gel ***(20)***.

17. The application of BFA or CCCP in pulse/chase experiments, as described in **Subheadings 3.3.1.1.** and **3.3.1.3.**, has no effect on the kinetics of oligomerization of the mucin precursors ***(14,20)***.
18. Care should be taken not to run samples with reducing and nonreducing sample buffer alongside on the same gel. The reduction of disulfide bonds is a fast process and the reducing agents (typically 2-mercaptoethanol) are highly diffusible compounds. Therefore, the risk exists that 2-mercaptoethanol will diffuse through the gel and reduce the disulfide bonds in nonreduced samples. If these samples are run on the same gel, at least one lane should be left unused in between.
19. *N*-linked glycans are added to RER-localized proteins in a conformation known as "high mannose" *N*-glycans. Upon transport through the Golgi apparatus these *N*-glycans are modified to "complex" *N*-glycans. The high mannose *N*-glycans can be split from the polypeptide by the action of Endo H. This enzyme is however not capable to release the complex form of these glycans. PNGase F releases all *N*-glycans, irrespective of their conformation. Thus, if a mucin precursor is demonstrated to contain only high mannose *N*-glycans this is taken as good evidence that this molecule is present within the RER ***(2–4,7,11,13,14)***. The sensitivity of the mucin precursors towards these enzymes is demonstrated on SDS-PAGE by an increase in mobility.
20. Tunicamycin inhibits the *N*-glycosylation completely, resulting in RER-localized polypeptides without any glycosylation. When mucin precursors are immunoprecipitated from tunicamycin-treated tissue or cells, this will yield the "naked" mucin polypeptide. Upon reducing SDS-PAGE this will give the most accurate indication of the molecular mass of the mucin polypeptide. Moreover, the position of this "naked" mucin polypeptide on reducing SDS-PAGE is identical to the position of Endo H- or PNGase F-digested mucin precursors, which can serve as appropriate evidence that the Endo H and/or PNGase F digestions have removed all *N*-glycans from mucin precursors (e.g., **ref. *13***).
21. The inhibition of *N*-glycosylation by tunicamycin slows down the process of oligomerization of the mucin precursors considerably ***(14,17,20)***. Since both *N*-glycosylation and oligomerization take place in the RER, this lends additional experimental evidence to the notion that the mucin precursors are actually present in the RER. To observe this inhibitory effect on oligomerization, pulse/chase experiments must be performed in the continuous presence of tunicamycin.
22. The procedures to isolate mucins from any given source and to prepare polyclonal antibodies against these intact mucins are described previously ***(8)***. Polyclonal antisera raised following this protocol are always specific for the unique, non-*O*-glycosylated polypeptide regions of the mucins, which are expressed in this particular mucin source. It has been demonstrated for many different tissues, that these antisera will be able to recognize the mucin precursors in the respective tissue or cells in metabolic labeling experiments ***(7–17)***. Thus, immunoprecipitation using these antisera on pulse-labeled tissue or cells

will reveal which mucins are expressed in this particular mucin source. As each mucin precursor can be identified by its unique mobility on reducing SDS-PAGE (**Table 1**), the identity of the immunoprecipitated mucin precursors can be established (*see* also Chapter 20).

23. An excellent example of the successful application of this method is the study of human gallbladder mucin. Human gallbladder mucin was isolated using CsCl/guanidinium.HCl density gradients, a polyclonal antiserum was raised, and the expression of mucin precursors was studied by metabolic labeling experiments ***(17)***. It appeared that the antiserum recognized only one mucin precursor with an apparent molecular mass of 470 kDa. By comparative immunoprecipitation analysis it appeared that this mucin precursor was not identical to the precursor of MUC1, 2, 3, 4, 5AC, 6, or 7, leading us to conclude that gallbladder mucin was either a novel mucin or MUC5B ***(4,21)***. Finally, using specific monoclonal antibodies to immunoprecipitate MUC5B precursor, we were able to show that the major human gallbladder mucin was identical to MUC5B ***(16)***.

References

1. Van Klinken, B. J. W., Dekker, J., Büller, H. A., and Einerhand, A. W. C. (1995) Mucin gene structure and - expression updated: protection versus adhesion. *Am. J. Physiol.* **269,** G613–G627.
2. Strous, G. J., and Dekker, J. (1992) Mucin-type glycoproteins. *Crit. Rev. Biochem. Mol. Biol.* **27,** 57–92.
3. Van Klinken, B. J. W., Einerhand, A. W. C., Büller, H. A., and Dekker, J. (1998) Strategic biochemical analysis of mucins. *Anal. Biochem.***265,** 103–116.
4. Van Klinken, B. J. W., Oussoren, E., Weenink, J. J., Strous, G. J., Büller, H. A., Dekker, J., and Einerhand, A. W. C. (1996) The human intestinal cell lines Caco-2 and LS174T as models to study cell-type specific mucin expression. *Glycoconjugate J.* **13,** 757–768.
5. Van Klinken, B. J. W., De Bolos, C., Büller, H. A., Dekker, J., and Einerhand, A. W. C. (1997) Biosynthesis of mucins (MUC2-6) along the longitudinal axis of the gastro-intestinal tract. *Am. J. Physiol.* **273,** G296–302.
6. Hilkens, J. and Buijs, F. (1988) Biosynthesis of MAM-6, an epithelial siaolomucin. *J. Biol. Chem.* **263,** 4215–4222.
7. Tytgat, K. M. A. J., Büller, H. A., Opdam, F. J. M., Kim, Y. S., Einerhand, A. W. C., and Dekker, J. (1994) Biosynthesis of human colonic mucin: Muc2 is the most prominent secretory mucin. *Gastroenterology* **107,** 1352–1363.
8. Tytgat, K. M. A. J., Klomp, L. W. J., Bovelander, F. J., Opdam, F. J. M., Van der Wurff, A., Einerhand, A. W. C., Büller, H. A., Strous, G. J., and Dekker, J. (1995) Preparation of anti-mucin polypeptide antisera to study mucin biosynthesis. *Anal. Biochem.* **226,** 331–341.
9. Tytgat, K. M. A. J., Opdam, F. J. M., Einerhand, A. W. C., Büller, H. A., and Dekker, J. (1996) MUC2 is the prominent colonic mucin expressed in ulcerative colitis. *Gut* **38,** 554–563.
10. Van Klinken, B. J. W., Duits, L. A., Verburg, M., Tytgat, K. M. A. J., Renes, I. B., Büller, H. A., Einerhand, A. W. C., and Dekker, J. (1997) Mouse colonic mucin as a model for human colonic mucin. *Eur. J. Gastroenterol. Hepatol.* **9,** A66 (abstract).
11. Tytgat, K. M. A. J., Bovelander, F. J., Opdam, F. J. M., Einerhand, A. W. C., Büller, H. A., and Dekker, J. (1995) Biosynthesis of rat MUC2 in colon and its analogy with human MUC2. *Biochem. J.* **309,** 221–229.
12. Klomp, L. W. J., Van Rens L., and Strous, G. J. (1994) Identification of a human gastric mucin precursor N-linked glycosylation and oligomerization. *Biochem. J.* **304,** 693–698.
13. Dekker, J., Van Beurden-Lamers, W. M. O., and Strous, G. J. (1989) Biosynthesis of gastric mucus glycoprotein of the rat. *J. Biol. Chem.* **264,** 10,431–10,437.

14. Dekker, J. and Strous, G. J. (1990) Covalent oligomerization of rat gastric mucin occurs in the rough endoplasmic reticulum, is N-glycosylation dependent and precedes O-glycosylation. *J. Biol. Chem.* **265,** 18,116–18,122.
15. Dekker, J., Van der Ende, A., Aelmans, P. H., and Strous, G. J. (1991) Rat gastric mucin is synthesized and secreted exclusively as filamentous oligomers. *Biochem. J.* **279,** 251–256.
16. Van Klinken, B. J. W., Dekker, J., Van Gool, S. A., Van Marle, J., Büller, H. A., and Einerhand, A. W. C. (1998) MUC5B is the prominent mucin biosynthesuzed in the human gallbladder mucin and is also expressed in a subset of colonic goblet cells. *Am. J. Physiol.* **274,** G871–G878.
17. Klomp, L. W. J., Delely, A. J., and Strous, G. J. (1994) Biosynthesis of human gallbladder mucin. *Biochem. J.* **304,** 737–744.
18. Swallow, D. M., Gendler, S., Griffiths, B., Corney G., Taylor-Papadimitriou J., and Bramwell, M. E. (1987) The human tumor-associated epithelial mucins are coded by an expressed hyperviable gene locus PUM. *Nature* **328,** 82–84.
19. Tytgat, K. M. A. J., Van der Wal, J. W. G., Büller, H. A., Einerhand, A. W. C., and Dekker, J. (1996) Quantitative analysis of MUC2 synthesis in ulcerative colitis. B*iochem. Biophys. Res. Comm.* **224,** 397–405.
20. Van Klinken, B. J. W., Einerhand, A. W. C., Büller, H. A., and Dekker J. (1998) The oligomerization of a family of four, genetically clustered, human gastrointestinal mucins. *Glycobiology* **8,** 67–75.
21. Van Klinken, B. J. W., Tytgat, K. M. A. J., Büller, H. A., Einerhand, A. W. C., and Dekker, J. (1995) Biosynthesis of intestinal mucins: MUC1, MUC2, MUC3, and more. *Biochem. Soc. Trans.* **23,** 814–818.
22. Klausner, R. D., Donaldson, J. G., and Lippincott-Schwartz (1992) Brefeldin A: Insights into the control of membrane traffic and organelle structure. *J. Cell Biol.* **116,** 1071–1080.

22

Inhibition of Mucin Glycosylation

Guillemette Huet, Philippe Delannoy, Thécla Lesuffleur, Sylviane Hennebicq, and Pierre Degand

1. Introduction

Mucins are secreted or membrane-bound large glycoproteins produced by epithelial cells of normal and malignant tissues. The secreted mucins are the major components of the mucous gel overlaying respiratory, gastrointestinal, or genital epithelia. Mucins constitute a family of extensively *O*-glycosylated glycoproteins (40–80% by weight) *(1,2)* encoded by a family of different MUC genes *(3)*. The oligosaccharide side chains substitute threonine or serine residues of tandemly repeated sequences in the core of the molecule.

The biochemical properties and functions of mucins are greatly dependent on their *O*-glycosylation state. In particular, mucins can display a role in cellular protection or cellular adhesion *(4)*. On the one hand, filamentous mucins highly substituted by negatively charged *O*-glycans act as a protective barrier for epithelial cells. On the other hand, the terminal oligosaccharides of mucins can interact with cellular or bacterial receptors and promote adhesion on the epithelial cells. Mucins display tissue-specific patterns of *O*-glycosylation. Alterations in the glycosylation of mucins commonly occur in many mucosal diseases. In particular, glycan epitopes of mucins are important markers in cancer *(5,6)*.

Studies on the regulation of the biosynthesis and secretion of mucins have been improved by the use of human mucosal cells, which can be grown in long-term culture *(7)*. To address the function of carbohydrates, inhibitors of *O*-glycosylation have been used in in vivo experiments on cultured cells *(8)*. For that purpose, aryl-*N*-acetyl-α-galactosaminides (GalNAcα-*O*-aryls) have been initially used as potential competitors of the glycosylation of *N*-acetylgalactosamine (GalNAc) residues linked to the core protein, since these sugar analogs were suitable substrates

From: *Methods in Molecular Biology, Vol. 125: Glycoprotein Methods and Protocols: The Mucins*
Edited by: A. Corfield © Humana Press Inc., Totowa, NJ

for UDP-GlcNAc:GalNAc-Rβ1, *3-N*-acetylglucosaminyltransferase ***(9)***. Actually, in in vitro experiments, these compounds inhibit the UDP-Gal:GalNAc-Rβ1,3-galactosyltransferase ***(8)***. GalNAcα-*O*-aryls (benzyl-, phenyl-, and *p*-nitrophenyl derivatives of *N*-acetylgalactosamine), when added in the medium of cultured cells, are metabolized within the cells and give rise to different internal derivatives. In vivo the resulting effects of GalNAcα-*O*-aryls on the *O*-glycosylation of mucins are different from the effects obtained in vitro ***(8,10–12)***. In this way, in mucin-secreting colon cancer cells such as the HM7 variant of LSI74T cells ***(8,10)*** and the HT-29 MTX subpopulation ***(11,12)***, GalNAcα-*O*-aryls are highly converted into the disaccharide Galβ1-3GalNAcα-*O*-aryls, but this conversion does not impair some significant β1,3-galactosylation of *O*-linked GalNAc to the core protein as well. The disaccharide-formed Galβ1-3GalNAcα-*O*–aryls have proved, on the contrary, to behave as a strong competitive inhibitor of the elongation of the mucin Galβ1-3GalNAcα sequences by *N*-acetylglucosa-minyltransferases, sialyltransferases, and fucosyltransferases ***(8,10–12)***. Hence, in in vivo experiments, GalNAcα-*O*-aryls mainly act as inhibitors of the elongation of the Galβ1-3GalNAcα sequence (T-antigen) of mucins.

The carbohydrate changes induced by GalNAcα-*O*-aryl treatments can be evaluated by different methods:

1. Mucins can be directly analyzed in the cell culture media or in the cell lysates by Western blotting using lectins and/or antibodies directed against carbohydrate epitopes.
2. Mucins can be isolated from cell lysates or culture media by the conventional procedures using ultracentrifugation on a cesium bromide gradient ***(13,14)***, and analyzed by carbohydrate composition or by (enzyme-linked immunosorbent assay (ELISA) with lectins and/or glycan epitope-specific antibodies.

The effects on mucin secretion can be estimated using metabolic labeling with [^{3}H]threonine or by histochemical staining. And, the intracellular metabolization of GalNAcα-*O*-aryls can be studied using metabolic labeling with [^{3}H]galactose and reversed-phase, high-performance liquid chromatography (HPLC).

2. Materials

1. Alcian blue (AB), pH 2.5: 0.1% (w/v) AB in 3% (v/v) acetic acid, pH 2.5.
2. Amplify (Amersham, Buchler Gmbh, Braunschweig, Germany).
3. Anti-digoxygenin (DIG) Fab fragment conjugated with alkaline phosphatase (Boehringer Mannheim, Germany).
4. Aryl-*N*-acetyl-α=D=galactosaminides (Sigma, St. Louis, MD).
5. Blocking solutions:
   a. Blocking solution 1: Tris-buffered saline (TBS), pH 7.5, containing 2% polyvinyl pyrrolidone K-30 (Aldrich).
   b. Blocking solution 2: TBS, pH 7.5, containing 0.5% blocking reagent (Boehringer Mannheim). Heat at 60°C for 1 h.
   c. Blocking solution 3: TBS, pH 7.5, containing 6% bovine serum albumin (BSA).
   d. Blocking solution 4: 0.01 *M* phosphate-buffered saline (PBS), pH 6.8, containing 1% (w/v) BSA.

6. Blotting buffers:
   a. Anode buffer: 0.3 *M* Tris HCl, pH 10.4, containing 20% methanol.
   b. Cathode buffer: 0.04 *M* 6-amino-*N*-hexanoic acid containing 20% methanol.
7. Cell disrupter Sonifer B-30 adapted with an exponential probe (Branson Sonic Power).
8. Continous flow radiochromatography detector Flo-One/Beta.
9. ECL detection kit (Amersham).
10. Eukitt (Kindler GMBN, Freeburg, Germany).
11. Hibar prepacked column RT 250-4, Lichrosorb RP-18, 5 µm (Merck, Darmstadt, Germany).
12. Incubation buffer:
    a. Incubation buffer 1: TBS, pH 7.5 containing 1% (w/v) BSA, 1% sodium dodecyl sulfate (SDS), 1% Triton X-100, 10% fetal calf serum (FCS).
    b. Incubation buffer 2: TBS, pH 7.5 containing 10% FCS.
13. Labeling :
    a. [^{3}H]-L-threonine (ICN, Costa Mesa, CA) 46 Ci/mmol, 1.7 GBq/mmol (50 mCi/mL).
    b. D-[6–^{3}H]galactose (Amersham): 20–40 Ci/mmol, 0.7–1.5 TBq/mmol (1 mCi/mL).
14. Lectin buffer: TBS, pH 7.5, containing 1 m*M* $MgCl_2$, 1 m*M* $MnCl_2$, and 1 m*M* $CaCl_2$.
15. Cell culture media:
    a. Standard growth medium: Dulbecco's modified Eagle's medium containing 2 m*M* glutamine, 100 UmL of penicillin, 100 mg µL of streptomycin, and 10% fetal bovine serum.
    b. Growth medium with GalNAcα-*O*-aryl: The medium **in ituma** is added with aryl-*N*-acetyl-α-D-galactosaminide.
    c. Threonine-free medium (Life Technologies).
    d. Low-glucose Dulbecco's minimal essential/H-16 medium (Gibco).
16. Molecular weight-markers are the high molecular range weight of Rainbow colored markers (Amersham), maximun 200 kDa (myosin).
17. Monoclonal anti-T (Thomsen-Friedenrich) antibody available from commercial suppliers.
18. Nuclear Red: 0.1% (w/v) nuclear red in 5% aluminum sulfate solution. Heat to dissolve and filter.
19. Paraformaldehyde: 4% (w/v) in 0.1*M* phosphate buffer, pH 7.4.
20. PBS:
    a. 0.01 *M* PBS, pH 6.8.
    b. 0.01 *M* PBS, pH 6.8, containing 0.1% (v/v) Tween-20.
21. Peroxidase-labeled antimouse IgG antibody (Jackson Immunoresearch Laboratories, West Grove, PA).
22. Peroxidase-labeled streptavidin.
23. RIPA buffer: 1 m*M* Tris-HCl, pH 8.0, 0.01 *M* NaCl, 0.1% SDS, 1% Triton X-100, 0.5% sodium deoxycholate, 1% phenylmethylsulfonylfluoride, 1 m*M* sodium ethylene diaminetetraacetate.
24. Stop solution: 1 *N* HCl.
25. Substrates:
    a. Substrate solution 1: 50 µL of 4-nitroblue tetrazolium chloride (77 mg mL in 70% dimethylformamide) and 37.5 µL of 5-bromo-4-chloro-3-indolyl-phosphate (50 mg mL in dimethylformamide) in 10 mL of 100 m*M* Tris-HCl, pH 9.5, 50 m*M* $MgCl_2$, and 100 m*M* NaCl.

b. Substrate solution 2: 1 × 10 mg tablet of *O*-phenylenediamine dihydrochloride (Sigma) in 10 mL of 0.01 *M* PBS, pH 5.5. Add 15 µL of 30% hydrogen peroxide immediately before use.

26. TBS: 50 m*M* Tris-HCl, 150 m*M* sodium chloride, pH 7.5.
27. X-O Kodak films (Amersham).
28. Whatman 3MM paper.
29. SDS-polyacrylamide gels for electrophoresis: stacking gels of 2% and running gels of 2–10% polyacrylamide gradient gel (acrylamide/bisacrylamide ratio of 37.5) are prepared at 2-mm thickness.

3. Methods

3.1. Cell Culture in the Presence of GalNAcα-O-benzyl

3.1.1. Conditions of Use of GalNAcα-O-benzyl

GalNAcα-*O*-benzyl is directly dissolved in the culture medium for 2 h at room temperature with continuous stirring. Then, the medium is sterilized by filtration. GalNAcα-*O*-benzyl may be used at various concentrations up to 10 m*M*. A range of different concentrations should be tested for the viability of the cells and the obtained effects. Indeed, the response is expected to be different according to the cell type, and in particular, its glycosyltransferase pattern.

GalNAcα-*O*-benzyl can be used in the following ways:

1. In a short treatment over a 24-h period. The cells are cultured in the standard medium up to their differentiation state into mucin-secreting phenotype, and then for 24 h in the medium enriched in GalNAcα-*O*-benzyl.
2. In a long time period treatment. The cells are cultured in the standard medium up to d 2 after seeding, and then in the medium enriched in GalNAcα-*O*-benzyl. The medium is changed daily with new medium containing GalNAcα-*O*-benzyl.

3.1.2. Metabolic Labeling

To study the effects of GalNAcα-*O*-benzyl on the secretion of mucins, cells are seeded on 6-well plates. For metabolic labeling with [^{3}H]Threonine, the current medium is substituted by threonine-free medium. [^{3}H]Threonine (50 µCi mL) is added in the threonine-free medium of control cells and in the threonine-free medium containing GalNAcα-*O*-benzyl of treated cells. Then, the culture media are collected and centrifuged and the cells are lysed in 1 mL of RIPA buffer and centrifuged. The supernatants are analyzed on 2–10% gels and autoradiography (*see* **Subheading 3.2.4.**).

To study the intracellular metabolism of GalNAcα-*O*-benzyl, cells are seeded into 6-well culture plates. For metabolic labeling with [6–^{3}H]galactose, the current medium is substituted by low glucose medium to facilitate the incorporation of the precursor. [6–^{3}H]Galactose (1 mCi mL) is added in the medium simultaneously with GalNAcα-*O*-benzyl for up to 72 h. After removing the culture media, cells are lysed by sonication in 1 mL of distilled water and centrifuged at 13,000*g*. The supernatants are collected, heat-denaturated, and filtered through 0.22 µm ultrafiltration units.

GalNAcα-*O*-benzyl derivatives are analyzed by reversed-phase HPLC using a Lichrosorb RP-18 column.

3.2. Visualization of Mucins after Electrophoresis

Samples of culture media or cell lysates are subjected to electrophoresis on 2–10% polyacrylamide gels in the presence of SDS.

3.2.1. Transfer to Nitrocellulose

Proteins are transferred from the gel to nitrocellulose for 1 h using a semi-dry electroblot apparatus. Six sheets of Whatman 3MM paper are immersed in the anode buffer and covered with the membrane and then with the gel, both previously rinsed in the anode buffer. The gel is then covered with six more sheets of Whatman 3MM paper that have been immersed in the cathode buffer. The transfer is carried out at 0.8 mA/cm^2.

3.2.2. Lectin Staining (see Notes 1–4)

All steps should be performed at room temperature with gentle shaking except the color development.

1. Wash the membrane three times in TBS for 5 min.
2. Incubate in blocking solution 1 for 2 h (*see* **Note 1**).
3. Wash the membrane three times in TBS for 5 min.
4. Incubate with DIG-labeled lectin (*see* **Note 2**) in lectin buffer for 1 h (*see* **Note 3**).
5. Wash the membrane in TBS (three times for 5 min).
6. Incubate in blocking solution 2 for 1 h.
7. Wash the membrane in TBS (three times for 5 min).
8. Incubate with anti-DIG Fab fragment conjugated with alkaline-phosphatase diluted 1000-fold in TBS (1 µg mL).
9. Wash the membrane in TBS (three times for 5 min).
10. Incubate the membrane in substrate solution 1 until color development (*see* **Note 4**).
11. Stop the reaction by immersion and gentle shaking in distilled water.
12. Dry the nitrocellulose membrane at room temperature.

3.2.3. Antibody Staining

All steps should be performed with gentle shaking.

1. Wash the membrane in TBS for 15 min.
2. Incubate the membrane in blocking solution 3 for 1 h at 37°C.
3. Wash the membrane twice in TBS containing 0.1% Tween-20 for 15 min.
4. Incubate the membrane at room temperature with the first antibody diluted 1000-fold in the incubation buffer 1 for 2 h.
5. Wash the membrane in TBS containing 0.1% Tween-20 (two times for 15 min).
6. Incubate the membrane at room temperature with the peroxidase-conjugated second antibody diluted 4000-fold in incubation buffer 2 for 2 h.

7. Wash the membrane in TBS containing 0.1% Tween-20 (two times for 15 min).
8. Incubate with the ECL solution for 1 min.
9. Expose to hyperfilm.

3.2.4. Autoradiography

1. Fix the gel overnight in 40% ethanol, 10% glycerol, 10% acetic acid (v/v/v).
2. Soak in Amplify for 20 min.
3. Dry on Whatman 3MM paper.
4. Expose to X-O Kodak film (Amersham).

3.3. ELISA of Purified Mucins

Figure 1 gives an example of ELISAs of purified mucins.

3.3.1. Coating

1. Solubilize the mucins in PBS buffer overnight at 4°C.
2. For coating, incubate different amounts of mucins (from 10 to 1000 ng) in wells of a 96-well plate overnight at 4°C, and then empty the plate.
3. Incubate the plate with blocking solution 4 at room temperature for 2 h.
4. Wash the plate in PBS (three time for 5 min.).

3.3.2. Lectin Staining

1. Incubate for 1 h with biotinylated lectin diluted in blocking solution 4 (from 1 to 10 µg mL).
2. Wash in PBS containing 0.1% (v/v) Tween-20.
3. Wash three times in PBS.
4. Incubate for 90 min with peroxidase-labeled streptavidin diluted in blocking solution 4 (2 µg mL).
5. Repeat **steps 2** and **3**.
6. Develop color with substrate solution 2.
7. Stop the reaction by adding 50 µL of 1 *N* HCl.
8. Read the optical density at 492 nm.

3.3.3. Antibody Staining

1. Incubate for 1 h with first antibody diluted in blocking solution 4.
2. Wash in PBS containing 0.1% (v/v) Tween-20.
3. Wash three times in PBS.
4. Incubate for 90 min with peroxidase-labeled second antibody at 0.2 µg mL in blocking solution 4.
5. Repeat **steps 2** and **3**.
6. Develop color with substrate solution 2.
7. Stop the reaction by adding 50 µL of 1 *N* HCl.
8. Read the optical density at 492 nm.

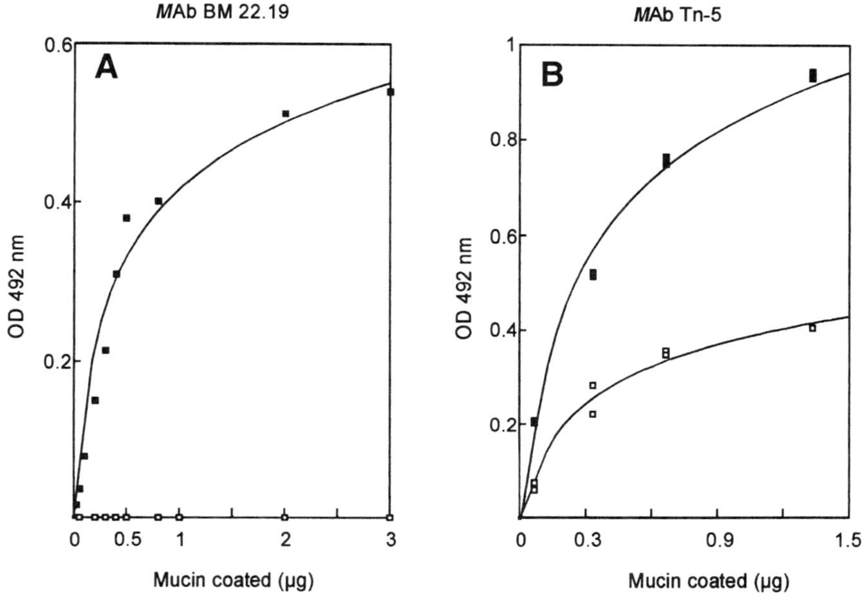

Fig. 1. Example of ELISAs of purified mucins from control (□), and from GalNAcα-*O*-benzyl–treated cells (■) with monoclonal antibody (MAb) BM 22.19 detecting T-antigen (**A**), and with MAb Tn-5 detecting Tn-antigen (**B**) ***(12)***.

3.4. AB Staining on Cryostat Sections of Cell Layer Rolls

3.4.1. Cryostat Sections of Cell Layer Rolls

Cells grown in 25-cm^2 flasks are rinsed twice in PBS, and dry scraped with a rubber policeman, and the cell pellet is frozen in liquid nitrogen. Cryostat sections (6 µm) of the cell pellet are performed and fixed using 4% paraformaldehyde (15 min), followed by washing in PBS (two times for 15 min). After drying, slides can be stored at –20°C ***(15)***.

3.4.2. AB Staining of Cryostat Sections

1. Wash slides in distilled water for 5 min.
2. Incubate in AB for 15–30 min.
3. Wash in distilled water (three times for 5 min).
4. Counterstain using nuclear red for 5 min.
5. Wash in distilled water (three times for 5 min).
6. Dehydrate in alcohol (70°, 95°, 100°) and toluol before mounting in Eukitt.

3.5. Analysis of GalNAcα-O-aryl Derivatives

3.5.1. Preparation of Samples

After continuous labeling with [6–^{3}H]Galactose, the culture media are collected from the 6-well culture plates and cells are washed three times with sterile PBS. Cells are then harvested and lysed in 1 mL of distilled water by sonication for 2 min using a cell disrupter adapted with an exponential probe. Cell lysates are centrifuged for 15 min at 13,000*g*. The supernatants are heat denatured (5 min at 100°C) and filtered through onto 0.22 µm ultrafiltration units prior the injection.

For the analysis of the GalNAcα-*O*-benzyl derivatives secreted in the culture media, medium samples are heat-denatured (5 min at 100°C), centrifuged for 15 min at 13,000*g*, and filtered through onto 0.22-µm ultrafiltration units prior to injection on an HPLC column.

3.5.2. Reversed-Phase HPLC Fractionation of GalNAcα-O-benzyl Derivatives

One hundred microliters of sample are injected at a flow rate of 1 mL min and eluted isocratically with distilled water for 10 min. An acetonitrile gradient is then applied by increasing the percentage of acetonitrile from 0 to 50% in 20 min. The percentage is still maintained at 50% for 15 min and the column is reequilibrated in water for 20 min prior to the next injection. Detection of the radioactive compounds is performed using a continuous flow radiochromatography detector. **Figure 2** gives an example of a separation profile.

3.5.3. Identification of GalNAcα-O-benzyl Derivatives

Identification is performed by the coinjection of [^{14}C]labeled radioactive standards (i.e., [^{14}C]Gal, [^{14}C]Galβ1-3GalNAcα-*O*-benzyl, [^{14}C]NeuAcα2-3Galβ1-3GalNAcα-*O*-benzyl) and dual-label scintillation counting.

4. Notes

1. The lectin staining protocol is an evolution of the Glycan Detection protocol from Boehringer Mannheim, provider of the DIG-labeled lectins. Several modifications have been made to optimize the initial procedure. The main change concerns the substitution of the first blocking solution (**step 1**) by polyvinyl pyrrolidone K-30 at 2% in TBS (blocking solution 1) prior to the incubation with DIG-labeled lectins. This change was made to decrease the background owing to the nonspecific binding of lectins to the blocking reagent, initially used for the saturation of the membrane after Western blotting. The blocking reagent (commercially available from Boehringer Mannheim, cat. no. 1 096 176) is used in a second step of blocking (blocking solution 2, **step 6**) prior to incubation with the nitrocellulose membrane with the anti-DIG Fab fragments.
2. The DIG-labeled lectins are used at the following concentrations: Amaranthin from *Amaranthus caudatus* (ACA-dig) 2.5 µg mL; *Maackia amurensis* agglutinin (MAA-dig) 5 µg mL; peanut (*Arachis hypogaea*) agglutinin (PNA-dig) 2 µg mL; *Sambucus nigra* agglutinin (SNA-dig) 2µg mL.
3. PNA recognizes T-antigen (Galβ1-3GalNAc-R) only when the disaccharide is unsubstituted by sialic acid, either linked on 3 position to Gal, or on 6 position to GalNAc. It is possible to visualize and estimate the total amount of T-antigen (sialylated or not) by total desialylation of mucins after transfer onto the nitrocellulose membrane. Desialylation of the blot is performed as follows: After **step 2,** incubate the membrane in a plastic bag

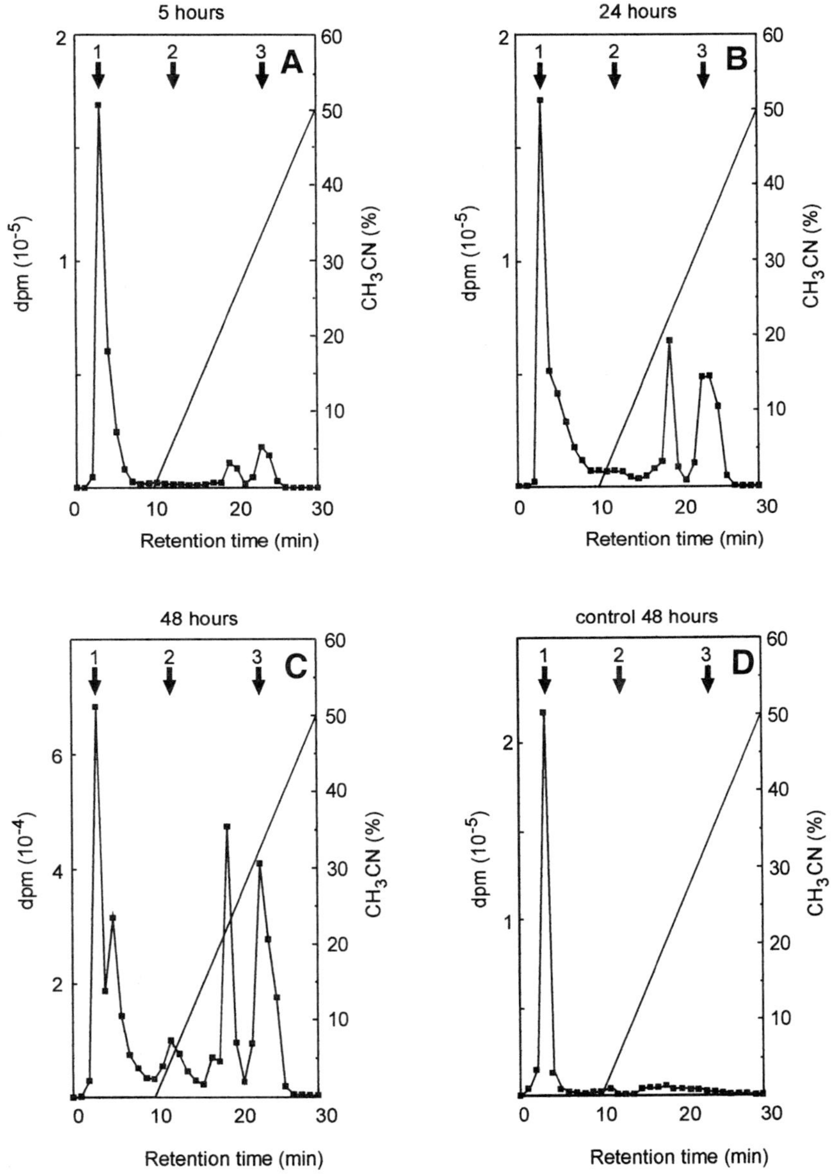

Fig. 2. Example of separation by reversed-phase HPLC of cell extracts from control and GalNAcα-*O*-benzyl–treated cells after metabolic labeling with [^{3}H]Gal *(12)*. Incorporation of [^{3}H]Galactose was examined after 5 h **(A)**, 24 h **(B)**, and 48 h **(C)** of exposure to 5 m*M* GalNAcα-*O*-benzyl. The HPLC profile of control cells **(D)** is shown for 48 h of incubation with [^{3}H]Gal. The retention times of [^{14}C]labeled standards are indicated: 1, [^{14}C]Gal; 2, [^{14}C] NeuAcα2-3Galβ1-3GalNAcα-*O*-benzyl; 3, [^{14}C]Galβ1-3GalNAcα-*O*-benzyl.

containing 10 mL of 50 m*M* citrate buffer, pH 6.0, 0.9% NaCl, 0.1% $CaCl_2$ supplemented with 50 mU mL of sialidase from *Clostridium perfringens*. After a 16-h of incubation at 37°C, go back to the current protocol at **step 2**.

4. Development of the coloration must be performed without shaking. The revelation of mucins should appear within the first 2 to 3 min. Increasing the times is not advised and may enhance the nonspecific staining.

References

1. Roussel, P., Lamblin, G., Lhermitte, M., Houdret, N., Lafitte, J. J., Perini, J. M., Klein, A., and Scharfman, A. (1988) The carbohydrate diversity of human respiratory mucins : a protection of the underlying mucosa? *Biochimie* **70,** 1471–1482.
2. Strous, G. J. and Dekker, J. (1992) Mucin-type glycoproteins *Crit. Rev. Biochem. Mol. Biol.* **27,** 57–92.
3. Verma, M. and Davidson, E. A. (1994) Mucin genes : structure, expression and regulation *Glycoconjugate J.* **11,** 172–179.
4. Van Klinken, B. J. W., Dekker, J., Büller, H. A., and Einerhand, A. W. C. (1995) Mucin gene structure and expression : protection vs adhesion. *Am. J. Physiol.* **269,** *(Gastrointest. Liver Physiol.* **32),** G613–G627.
5. Lesuffleur, T., Zweibaum, A., and Real, F. X. (1994) Mucins in normal and neoplastic human gastrointestinal tissues *Crit. Rev. Oncol. Hematol.* **17,** 153–180.
6. Kim, Y. S., Gum, J. R., and Brockhausen, I. (1996) Mucin glycoproteins in neoplasia *Glycoconjugate J.* **13,** 693–707.
7. Zweibaum, A., Laburthe, M. Grasset, E., and Louvard, D. (1991) Use of cultured cells lines in studies of intestinal cell differentiation and function, in "Intestinal Absorption and Secretion. " *Handbook of Physiology*. vol. 4 (Field, M. and Frizzel, R. A., eds.), American Physiological Society, Bethesda, MD, pp. 223–255.
8. Kuan, S. F., Byrd, J. C., Basbaum, C., and Kim, Y.S. (1989) Inhibition of mucin glycosylation by aryl-N-acetyl-α-galactosaminides in human colon cancer cells *J. Biol. Chem.* **264,** 19271–19277.
9. Brockhausen, I., Rachaman, E. S., Matta, K., and Schachter, H. (1983) The separation by liquid chromatography (under elevated pressure) of phenyl, benzyl, and O-nitrophenyl glycosides of oligosaccharides. Analysis of substrates and products for four N-acetyl-D-glucosaminyltransferases involved in mucin synthesis. *Carbohydr. Res.* **120,** 3–16.
10. Huang, S., Byrd, J. C., Yoor, W. H., and Kim, Y. S. (1992) Effect of benzyl-α-GalNAc, an inhibitor of mucin glycosylation, on cancer-associated antigens, in human colon cancer cells *Oncol. Res.* **4,** 507–515.
11. Huet, G., Kim, I., de Bolos, C., Lo-Guidice, J. M., Moreau, O., Hemon, B., Richet, C., Delannoy, P., Real, F. X., and Degand, P. (1995) Characterization of mucins and proteoglycans synthesized by a mucin-secreting HT-29 cell subpopulation. *J. Cell Sci.* **108,** 1275–1285.
12. Delannoy, P., Kim, I., Emery, N., de Bolos, C., Verbert, A., Degand, P., and Huet, G. (1996) Benzyl-N-acetyl-α-D-galactosaminide inhibits the sialylation and the secretion of mucins by a mucin secreting HT-29 cell subpopulation *Glycoconj. J.* **13,** 717–726.
13. Houdret, N., Perini, J. M., Galabert, C., Scharfman, A., Humbert, P., Lamblin, G., and Roussel, P. (1986) The high lipid content of respiratory mucins in cystic fibrosis is related to infection. *Biochem. Biophys. Acta* **880,** 54–61.

14. Sheehan, J. K., Oates, K., and Carlstedt, I. (1986) Electron microscopy of cervical, gastric and bronchial mucus glycoproteins. Biochem. J 239, 147–153.
15. Lesuffleur, T., Barbat, A., Dussaulx, E., and Zweibaum, A. (1990) Growth adaptation to methotrexate of HT-29 human colon carcinoma cells is associated with their ability to differentiate into colummar absorptive and mucus-secreting cells. *Cancer Res.* **50,** 6334–6343.

23

O-Linked Chain Glycosyltransferases

Inka Brockhausen

1. Introduction

The complex *O*-linked oligosaccharide chains (*O*-glycans) attached to the polypeptide backbone of mucins are assembled by glycosyltransferases. These enzymes act in the Golgi apparatus in a controlled sequence that is determined by their substrate specificities, their localization in Golgi compartments, and their relative catalytic activities *(1)*. Activities are controlled by many factors, including the membrane environment, metal ions, concentrations of donor and acceptor substrates, cofactors, and, in some cases, posttranslational modifications of enzymes. Cloning of glycosyltransferases has revealed the existence of families of homologous glycosyltransferases with similar actions but encoded by different genes. Thus, many steps in the pathways of *O*-glycosylation appear to be catalyzed by several related glycosyltransferases that may show slight differences in properties and substrate specificities. The relative expression levels of these enzymes is cell typespecific and appears to be regulated during the growth and differentiation of cells and, during tissue development, and is altered in many disease states *(2,3)*.

Figure 1 shows the biosynthetic pathways of *O*-glycans with the common mucin *O*-glycans core structures 1–4. The biosynthesis of other less common core structures *(1)* has not been studied in detail. Core structures can be elongated by repeating GlcNAcβ1-3Galβ1-4 or GlcNAc1-3Galβ1-3 structures (poly-*N*-acetyllactosamine chains, i antigens). Poly-*N*-acetyllactosamine chains may contain branches of GlcNAcβ1-6 residues linked to Gal- (I antigen), and may be terminated by blood group or tissue antigens (blood group ABO and Lewis antigens) or by sialic acid and sulfate. Many of the enzymes involved in these elongation and termination reactions also act on *N*-linked oligosaccharides of glycoproteins and on glycolipids.

From: *Methods in Molecular Biology, Vol. 125: Glycoprotein Methods and Protocols: The Mucins*
Edited by: A. Corfield © Humana Press Inc., Totowa, NJ

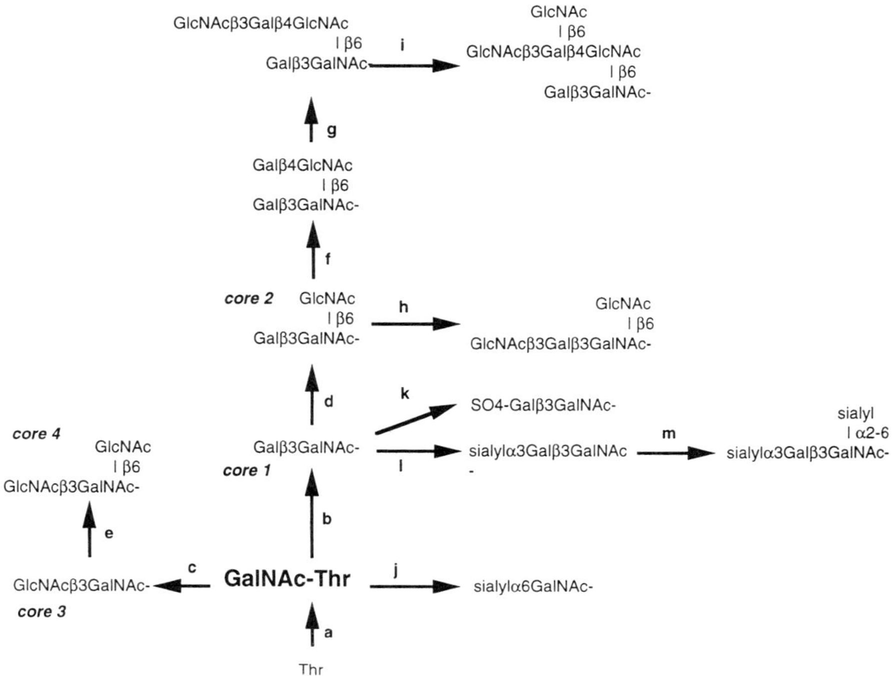

Fig. 1. Pathways of O-glycan biosynthesis. The first step of *O*-glycosylation is catalyzed by polypeptide α-GalNAc-transferase (path a), which acts preferably on Thr in vitro. The occurrence of various *O*-glycan structures is cell type specific and varies with cellular activation and differentiation, and in disease states. Core 1 and 2 structures are the most common core structures in mucins, and are synthesized by core 1 β3-Gal-transferase (path b) and core 2 β6-GlcNAc-transferase (path d). Core 3 is synthesized by core 3 β3-GlcNAc-transferase (path c) and core 4 by core 4 β6-GlcNAc-transferase (path e). GalNAc- may be sialylated by α6-sialyltransferases (path j) to form sialylα2-6GalNAc-, which cannot be converted to any of the core structures. After the synthesis of cores, chains may be elongated, sulfated, fucosylated, or sialylated, and blood group and other antigenic determinants may be added. Core 1 is sialylated by α3-sialyltransferase (path l), and this reaction blocks core 1 branching and elongation with the exception of α6-sialylation by α6-sialyltransferase (path m), which may differ from those catalyzing path j. The α3-sialyltransferase can also act on the Gal residue of core 2. The Gal residue of core 1 may be sulfated by Gal 3-sulfotransferase (path k). Sulfation will also block core 1 elongation and branching. Cores 1 and 2 are elongated by elongation β3-GlcNAc-transferase (path h). On galactosylation of the GlcNAc residue of core 2 by β4-Gal-transferase (path f), the poly-N-acetyllactosamine chains can be assembled by the repeated actions of β4-Gal-transferase and i β3-GlcNAc-transferase (paths f and g, respectively). GlcNAcβ1-6 Gal branches (I antigen) may be introduced into these chains by I β6-GlcNAc-transferase (path i).

Golgi glycosyltransferases have a domain structure characteristic of type II membrane proteins; the amino terminus extends into the cytoplasm, followed by a membrane anchor domain, and a catalytic domain at the carboxy terminus, which extends into the lumen of the Golgi. Mainly the membrane anchor and adjacent amino acid sequences, but also other protein determinants of these enzymes, as well as the membrane structure, determine the localization of transferases in various Golgi compartments *(4)*. The donor substrates for mucin glycosyltransferases are nucleotide sugars:

$$\text{nucleotide-sugar} + \text{acceptor} \rightarrow \text{product} + \text{nucleotide} \quad (1)$$

Mucin sulfotransferases transfer sulfate from 3'-phosphoadenosine 5'-phosphosulfate (PAPS) to the hydroxyls of sugars:

$$\text{PAPS} + \text{acceptor} \rightarrow \text{product} + \text{PAP} \quad (2)$$

These donor substrates are synthesized in the cytosol, with the exception of CMP-sialic acid, which is made in the nuclear compartment, and are transported into the Golgi by specific transporter systems *(5)*.

In the sequences of glycosylation reactions, glycosyltransferases often compete for a common substrate. For example, the enzymes that synthesize cores 1 and 3 (**Fig. 1**, paths b and c) compete for GalNAc-R substrates. Conversely, certain products formed may block further reactions; for example, no glycosyl transferase acts on sialylα2-6GalNAc (**Fig. 1**, path j), which therefore blocks extension of chains. Alternatively, certain reactions may be required prior to further conversions. For example, core 1 has to be formed before a GlcNAcβ1-6 residue can be added to GalNAc in the synthesis of core 2 (**Fig. 1**, path d). The distinct specificities of glycosyltransferases therefore regulate the pathways, and thus the relative amounts of final *O*-glycan structures found in mucins. The peptide backbones as well as existing glycosylation of substrates near the *O*-glycosylation sites also have an important function in regulating *O*-glycosylation *(6)*. Thus, primary *O*-glycosylation as well as the synthesis of various *O*-glycan core structures appear to be sitedirected by peptide sequences and their glycosylation patterns.

Based on many different studies, *O*-glycosylation appears to be initiated mainly in early Golgi compartments. The first enzyme in the *O*-glycosylation pathways, polypeptide α-GalNAc-transferase (**Fig. 1**, path a), has been localized to the *cis* Golgi compartment in porcine submaxillary gland *(7)* but can be more broadly distributed throughout the Golgi in other cell types *(8)*. The various members of this glycosyltransferase family have slightly different specificities toward their peptide substrates, have different cell type–specific expression patterns, and may be localized to different subcellular compartments *(9,10)*. Polypeptide α-GalNAc-transferase does not require a specific peptide sequence in the substrate; however, particular charged amino acids *(11)* as well as existing glycosylation *(12)* influence the activity.

Most mucins and other glycoproteins contain *O*-glycans with the core 1 structure, and the enzyme synthesizing core 1, core 1 β3-Gal-transferase (**Fig. 1**, path b), is a ubiquitous enzyme *(1,13)*. The activity is a prerequisite for the synthesis of T-antigens and sialylated core 1 structures, as well as core 2 structures *(14)*. The peptide sequences of glycopeptide substrates and the existing glycosylation determine the activity of core 1 β3-Gal-transferase *(15)*. Erythrocytes from patients with permanent mixed-field polyagglutinability *(16)*, human T-lymphoblastoid Jurkat cells *(17)*, and human colon cancer cells LSC *(18)* lack the enzyme and therefore cannot make cores 1 and 2 structures. A similar effect can be introduced by the use of GalNAcα-benzyl, which is an alternative substrate for enzymes extending *O*-glycan chains, and which can penetrate cell membranes to compete with endogenous substrates of mucin. Thus, cells treated with this *O*-glycosylation inhibitor exhibit truncated *O*-glycans, terminating mainly in GalNAc *(19)*. These truncated chains no longer carry ligands for cell-cell interactions, and the binding of colon cancer cells to the endothelium via E-selectin is significantly reduced *(20)*.

The synthesis of core 2 (**Fig. 1**, path d) is catalyzed by core 2 β6-GlcNAc-transferase *(21)*. Several apparently related β6-GlcNAc-transferases exist that synthesize GlcNAcβ1-6 branches on Gal or GalNAc *(1,22,23)*. The L-type core 2 β6-GlcNAc-transferase occurs in leukocytes and other cells and only synthesizes core 2. The M-type enzyme is found in most mucin-secreting cell types and can synthesize the GlcNAcβ1-6 branch of core 2, core 4 (**Fig. 1**, path e), and the I antigen (**Fig. 1**, path i) *(24)*. The L-enzyme activity increases during cellular activation and differentiation *(25,26)*. The M enzyme may be affected in cancer cells *(3,27)*. Core 2 β6-GlcNAc-transferase appears to be localized to *cis* and medial Golgi compartments *(27a,28)*, which is in agreement with its role in synthesizing a central *O*-glycan core structure.

The enzymes synthesizing *O*-glycan cores 3 and 4 (**Fig. 1**, paths c and e, respectively) appear to occur exclusively in mucin-secreting tissues since these cores have not been found in nonmucin molecules *(1)*. Core 3 is synthesized by core 3 β3-GlcNAc-transferase *(29)*. The enzyme is enriched in colonic tissues but reduced in colon cancer tissue *(30,31)* and is lacking in many other tissues. The activity apparently is lacking in colon cancer cell lines *(27)*. The enzyme activity synthesizing core 4, core 4 β6-GlcNAc-transferase, resides in the M-type core 2 β6-GlcNAc-transferase *(24,29)*.

Poly-*N*-acetyllactosamine chains of mucins are assembled by the repeating actions of β4-Gal-transferase (**Fig. 1**, path f) *(32)* and i β3-GlcNAc-transferase (**Fig. 1**, path g) *(33)*. These enzymes are ubiquitous, and may be considered as housekeeping enzymes. However, their expression is often up- or downregulated in healthy tissues as well as in a number of disease states *(2,3,34)*. The reaction catalyzed by β4-Gal-transferase occurs mainly in the *trans*-Golgi *(35)*. Yet another elongation β3-GlcNAc-transferase elongates core 1 and 2 structures, also by a GlcNAcβ1-3Gal linkage **Fig. 1**, path h) *(36)*.

Poly-*N*-acetyllactosamine chains may acquire GlcNAcβ1-6 (GlcNAcβ1-3) Gal branches in a developmentally regulated fashion, which leads to a change from the i to the I antigenicity. Some of the I β6-GlcNAc-transferases (**Fig. 1**, path i) synthesizing the I branch act on terminal Gal residues whereas others recognize internal

Gal residues *(1,37,38)*. Most of the enzymes synthesizing blood group ABO, Lewis, and other antigenic determinants act on *O*- and *N*-glycans as well as glycolipids. By contrast, sialyltransferases often prefer one type of glycoconjugate *(39)*. Two sialyltransferase families, α3- and α6-sialyltransferase, act preferably on mucin-type *O*-glycans.

α3-Sialyltransferase acts on Gal residues of cores 1 and 2 (**Fig. 1**, path l) *(24,40,41)*. The enzyme is developmentally regulated in thymocytes *(42)* and increased in leukemia cells *(43)* and several cancer models *(3,30)*. The α3-sialyltransferase has been localized to medial and *trans*-Golgi compartments *(44)*. The sialylation reaction catalyzed by this enzyme has an important role in keeping *O*-glycan chains short and sialylated. Since the enzyme acts relatively early in the *O*-glycan extension pathways (**Fig. 1**, path l), it has the ability to compete with branching and elongation reactions. Once core 1 is α3-sialylated, it is no longer a substrate for extension reactions although it can still be converted to the disialylated core 1 by α6-sialyltransferase (**Fig. 1**, path m).

The α6-sialyltransferase (**Fig. 1**, path j) that acts on GalNAc-R to form the sialyl-Tn antigen, sialylα2-6GalNAc-Th/Ser *(45)*, requires glycoproteins as substrate and cannot act on GalNAc-benzyl or nitrophenyl substrates *(46,47)*. However, another type of α6-sialyltransferase (α6-sialyltransferase III) does not have a peptide requirement, but is specific for the α3-sialylated core 1 structure *(48)*. The disialylated core structure can probably be synthesized by α6-sialyltransferase III and other α6-sialyltransferases (**Fig. 1**, path m). Modifications of the sialic acid residues of mucins include *O*-acetylation, catalyzed by specific *O*-acetyltransferases acting in the Golgi *(49)*.

The common sulfate ester linkages in mucins are SO_4-6-GlcNAc and SO_4-3-Gal. Several types of sulfotransferases have been described that act on the 6-position of GlcNAc *(50)* or the 3-position of Gal of core 1 (**Fig. 1**, path k) and *N*-acetyllactosamine structures *(51,52)*. Sulfated oligosaccharides appear to play an important role in cell adhesion through binding to selectins and in the control of bacterial binding *(53,54)*. Sulfation also functions in directing the biosynthetic pathways of complex *O*-glycans by blocking certain reactions. For example, sulfation of core 1 prevents the branching reaction to form core 2 *(51)*.

The enzymes catalyzing the reactions depicted in **Fig. 1** assemble mucin-type *O*-linked carbohydrate chains and are listed in **Table 1**, together with their substrates, enzyme products, and the high performance liquid chromatography (HPLC) conditions of product separation. Probably none of these enzymes are specific for mucins, but also act on other glycoproteins that carry *O*-glycans, and can act on various glycopeptides with *O*-linkages. In vitro, many of these enzymes utilize synthetic compounds as substrates in which the peptide chain is replaced by a hydrophobic group. The substrate should be clean, specific, and easy to isolate in order to determine the enzyme activity and specificity accurately. For a few enzymes, purified mucins with defined glycosylation are available as substrates. However, mucins are usually too heterogeneous in their carbohydrate structures, and therefore the use of synthetic compounds with defined structure is preferred. In addition, it is much easier to determine the product structure of synthetic substrates as a proof of the assayed activity. When a compound is a potential substrate for several glycosyltransferases present in the enzyme preparation, or several reactions occur in sequence, the various products have to be separated and identified. This can usually be achieved by HPLC. With the exception of β1,6-GlcNAc-transferases, UDP-sugar binding enzymes require the presence of divalent metal ion for optimal activity. Thus, measuring a β6-GlcNAc-transferase activity in the presence of EDTA will eliminate the activity of other GlcNAc-transferases potentially acting on the same substrate. Enzymes utilizing CMP-sialic acid may be stimulated by metal ions but usually can act in their absence. Unless they are released and secreted, or are produced as soluble recombinant enzymes, glycosyltransferases are membrane-bound enzymes, and their activities are stimulated by detergents.

A convenient way of identifying and quantifying glycosyltransferase products is by the use of nucleotide-sugar donors that contain ^{14}C or ^{3}H-labeled radioactive sugar. Similarly, the sulfate moiety of PAPS can be labeled with ^{35}S. Calculations of sulfotransferase activities must take into account the relatively short half-life of ^{35}S (about 87 d).

2. Materials

2.1. Preparation of Enzymes

1. 0.25 M Sucrose.
2. 0.9% NaCl.
3. Potter-Elvehjem hand homogenizer.
4. Low-speed centrifuge (10,000*g*).
5. Ultracentrifuge (100,000*g*).
6. Small pieces of tissue, or cells.

2.2. Preparation of Substrates and Standard Compounds

1. Commercially available oligosaccharides: GlcNAc, GalNAcα-benzyl, Galβ1-3 GalNAcα-benzyl, GlcNAcβ1-3 GalNAcα-p-nitrophenyl [pnp], Galβ1-4 GlcNAc, GlcNAcβ1-3 Galβ-methyl (Sigma, St. Louis MO); Galβ1-3 GalNAcα-pnp (Toronto Research Chemicals, Toronto, Canada).
2. Thr-peptides, synthesized by Hans Paulsen, University of Hamburg, Germany *(15,55)*.
3. Frozen sheep submaxillary glands (Pel-Freez, Rogers, AR) to isolate ovine submaxillary mucin (OSM), 0.1 *N* H_2SO_4, bovine testicular β-galactosidase (Boehringer, Laval, Canada), 0.1 *M* Na-citrate buffer, Sephadex G25 column.
4. Components of enzyme assays to prepare product standards enzymatically GlcNAcβ1-3 GalNAcα-benzyl, GlcNAcβ1-6 (GlcNAcβ1-3) GalNAcα-pnp, GlcNAcβ1-3 Galβ1-4 GlcNAc, GlcNAcβ1-6 (GlcNAcβ1-3 Galβ1-3) GalNAcα-benzyl, GlcNAcβ1-6 (GlcNAcβ1-3) Galβ-methyl, sialylα2-3 Galβ1-3 GalNAcα-pnp, SO_4-3 Galβ1-3 GalNAcα-benzyl, SO_4-3 Galβ1-4 GlcNAc.
5. Enzymatically prepared substrates: GlcNAcβ1-6 (Galβ1-3) GalNAcα-benzyl, sialylα2-3 Galβ1-3 GalNAcα-pnp

6. Nuclear magnetic resonance (NMR) and mass spectrometers, reagents for methylation analysis.

2.3. Separation and Identification of Glycosyltransferase and Sulfotransferase Products

1. HPLC apparatus.
2. HPLC columns C18, NH_2 (amine), PAC (cyano-amine).
3. Acetonitrile/water mixtures.
4. Dionex system for high-performance anion-exchange chromatography (HPAEC).
5. Bio-Gel P4 or P2 column (80 × 1.6 cm) (Bio-Rad, Hercules, CA).
6. Ion-exchange columns (AG1 × 8, 100–200 mesh, Bio-Rad).
7. High-voltage electrophoresis apparatus, 1% Na-tetraborate, Whatman No. 1 paper.
8. C18 Sep-Pak columns, methanol.
9. 0.05 *M* KOH/1 *M* $NaBH_4$ for β-elimination.
10. Scintillation fluid, scintillation counter.

2.4. Polypeptide α-GalNAc-Transferase Assays

1. 5% Triton X-100.
2. 0.5 *M* N-morpholino ethanesufonate (MES) buffer, pH 7.
3. 0.05 *M* Adenosine 5'-monophosphate (AMP) to inhibit pyrophosphatases.
4. 0.5 *M* $MnCl_2$.
5. 10 m*M* UDP-GalNAc (2000 dpm/nmol) donor substrate.
6. 5 m*M* Acceptor substrate solution: Thr-containing peptide.
7. Enzyme homogenate or solution.

2.5. β3- and β6-GlcNAc-Transferase Assays

1. 5% Triton X-100.
2. 0.5 *M* $MnCl_2$ (for β3-GlcNAc-transferases only).
3. 0.5 *M* MES buffer, pH 7.
4. 0.05 *M* AMP.
5. 0.5 *M* GlcNAc to inhibit *N*-acetylglucosaminidases.
6. 50 m*M* γ-galactonolactone (if substrate with terminal Gal is used) to inhibit galactosidases.
7. 10 m*M* UDP-GlcNAc (2000 dpm/nmol).
8. 5 m*M* Acceptor substrate solution: GalNAcα-benzyl, Galβ1-3 GalNAcα-benzyl, Galβ1-4 GlcNAc, GlcNAcβ1-3Galβ-methyl, or GlcNAcβ1-6 (Galβ1-3) GalNAcα-benzyl.
9. Enzyme homogenate or solution.

2.6. Core 1 β3-Gal- and β4-Gal-Transferase Assays

1. 5% Triton X-100.
2. 0.5 *M* $MnCl_2$.
3. 0.5 *M* MES buffer, pH 7.
4. 0.05 *M* AMP.
5. 0.05 *M* γ-galactonolactone.
6. 10 m*M* UDP-Gal (2000 dpm/nmol).

7. 5 m*M* Acceptor substrate solution: GalNAcα-benzyl or GlcNAc.
8. Enzyme homogenate or solution.

2.7. α3- and α6-Sialyltransferase Assays

1. 5% Triton X-100.
2. 0.5 *M* Tri-HCl buffer, pH 7.
3. 0.05 *M* AMP.
4. 10 m*M* CMP-sialic acid (2000 dpm/nmol).
5. 5 m*M* Acceptor substrate solution: DS-OSM. with 3 m*M* GalNAc concentration, Galβ1-3 GalNAcα-pnp, or sialylα2-3 Galβ1-3 GalNAcα-pnp.
6. Enzyme homogenate or solution.
7. High-voltage electrophoresis apparatus.
8. 20 m*M* EDTA/2 % Na-tetraborate.
9. 1% Na-tetraborate.
10. Whatman No. 1 paper.
11. HPLC apparatus.

2.8. Sulfotransferase Assays

1. 5% Triton X-100.
2. 0.1 *M* magnesium-acetate.
3. 0.1 *M* NaF to inhibit sulfatases.
4. 0.5 *M* Tris-HCl buffer, pH 7
5. 0.05 *M* adenosine triphosphate (ATP).
6. 0.1 *M* 2,3-Mercaptopropanol to inhibit PAPS degradation.
7. 0.3 m*M* PAPS (2000 dpm/nmol).
8. 5 m*M* Acceptor substrate solution: Galβ1-3GalNAcα-benzyl, Galβ1-4 GlcNAc, or GlcNAcβ1-3Galβ-methyl.
9. Enzyme-homogenate or solution.
10. High-voltage electrophoresis apparatus.
11. 20 m*M* EDTA/2% Na-tetraborate.
12. 1% Na-tetraborate.
13. Whatman No. 1 paper.
14. HPLC apparatus.
15. Dionex system for HPAEC.

3. Methods

3.1. Preparation of Enzymes

Ideally, enzymes are present in the highly purified state, and soluble in the assay mixture. A number of enzymes have been purified. However, these procedures depend on the specific enzyme and tissue and may take several months or years. Therefore, purification protocols are not described here. Purified enzymes may be stable at 4°C for months but are usually more stable at lower temperatures. With tissue homogenates or microsomes, however, this is rarely the case. The enzyme preparations inevitably contain interfering substances and degradative enzymes. For example pyrophosphatases and phosphatases that degrade nucleotide sugar donors, glycosidases that degrade substrates and products, and proteases that degrade the peptide moiety of substrates and products may be present, and deactivate the enzyme to be assayed. These unwanted reactions can be suppressed with specific inhibitors.

1. To prepare crude homogenate, hand homogenize tissue in 10 times the volume of 0.25 *M* sucrose. For most studies of crude enzymes, this preparation is sufficient. The homogenate can be stored at –20°C for a few months, or at –70°C for several years. If sufficient material is available, a more enriched enzyme fraction can be prepared as microsomes. Microsomes may be prepared from homogenates by first removing a low-speed pellet by centrifugation at 10,000*g*, followed by the precipitation of microsomes from the supernatant at 100,000*g*. The microsomal pellet is hand homogenized in 10 times the volume of 0.25 *M* sucrose.
2. Enzymes from cultured cells are prepared similarly. Harvest cells from the culture plate, and wash three times with 0.9% NaCl by gently stirring and centrifuging cells. After washing, hand homogenize cells in 0.25 *M* sucrose (1 mL/10^8 cells) and store as described in **step 1**.

3.2. Preparation of Substrates and Standard Compounds

Substrates may be purchased, prepared by chemical synthesis or combined chemical-enzymatic synthesis, or prepared by enzymatic synthesis or degradation from natural glycoproteins.

1. GalNAc-OSM is prepared from purified sheep submaxillary mucin, ovine submaxillary mucin (OSM) ***(29)***. Treat OSM with 0.1 *N* H_2SO_4 for 1 h at 80°C to remove sialic acid and fucose. For high purity, follow by digestion with bovine testicular β-galactosidase ***(56)***, which removes the small amount of β1-3–linked Gal residues present in OSM ***(30)***.
2. Substrate and product compounds that are not commercially available are synthesized with a known source of the desired enzyme under the conditions described for the standard transferase assay.
3. Low molecular weight compounds are isolated by gel filtration on Bio-Gel P4 or P2 columns, followed by HPLC (**Table 1**). The purity and linkages of all compounds should be verified by mass spectrometry (MS) and ^{1}H-NMR. The concentrations of individual sugars can be determined after acid hydrolysis (1 h at 80°C with 33% trifluoroacetic acid for sialic acid–containing compounds, 1 h at 100°C with 6 *N* HCl for neutral sugars) by HPAEC (Dionex system) as described in **Subheading 3.4.5.**

3.3. Separation and Identification of Glycosyltransferase and Sulfotransferase Products

To demonstrate that an enzyme activity is synthesizing a certain sugar linkage, the product has to be isolated and its structure determined. This is especially important when a new enzyme activity is to be assayed or when a novel variant of a known activity is expected.

1. Produce large amounts of glycosyltransferase product in a standard assay, possibly after incubation for 8–24 h, and pass through an AG1 × 8 column to remove nucleotide sugar

and negatively charged molecules. For sulfotransferase products, run high-voltage electrophoresis after the standard assay.

2. Purify low molecular weight compounds by HPLC or using the Dionex system, as described in **Subheading 3.4.5.** Low molecular weight compounds separated by high-voltage electrophoresis, can be eluted off the paper with water, by placing the paper into syringes and centrifuging at low speed. Borate is removed by repeated flash evaporation with methanol.
3. Purify mucin substrates by passing incubation mixtures through AG1 × 8 columns, followed by gel filtration on Bio-Gel P4 or P2. *O*-glycans are released from mucins by β-elimination (0.05 *N* KOH/1 *M* Na BH_4 at 45°C for 16 h). After neutralization, purify reduced *O*-glycan-alditols by gel filtration on Bio-Gel P4 or P2 columns, and by HPLC.
4. Carry out structural analysis of all low molecular compounds and oligosaccharide-alditols by NMR, MS (fast atom bombardment, electrospray, or matrix-assisted laser desorption ionization), and methylation analysis. If small amounts of product are available, chromatographic methods, including HPLC and the Dionex system, with the use of standard compounds, and sequential glycosidase digestion are useful ***(29,57–60)***.

*3.4. Glycosyltransferase Assays (see **Notes 1 and 2**)*

3.4.1. Ion-Exchange Assay

The ion exchange assay is simple, quick, and inexpensive, and can be applied to all transferase assays using neutral acceptor substrates.

1. After the incubation, stop the reaction with 100 mL of ice-cold water. Apply mixture to a column (a Pasteur pipet) of 0.4 mL of AG1 × 8, which removes excess radioactive nucleotide sugar. Wash the column three times with 0.6 mL of water and collect the eluate.
2. Add 5 mL of scintillation fluid and estimate radioactivity with a scintillation counter. Since the radioactivity in the eluate includes free radioactive sugar (originating from nucleotide sugar breakdown) and radioactive products from various endogenous substrates, the radioactivity of assays lacking exogenous substrates has to be substracted from the disintegrations per minute obtained. The specific enzyme activity is calculated as nanomoles/(hour·milligrams of protein).
3. Regenerate AG1 × 8 columns with 5 *M* NaCl, followed by thorough washing with water.

3.4.2. C18 Column Assay

The C18 column (Sep-Pak) assay can be applied when substrates contain a hydrophobic group. This method is often not reliable when charged (sialylated or sulfated) products are formed, unless very large hydrophobic groups are present in the enzyme product. A methyl aglycone group does not provide sufficient hydrophobicity to bind to Sep-Pak C18 columns.

1. After the incubation, apply the mixture onto a Sep-Pak C18 column, previously washed in water. Wash columns with 5 mL of water to elute excess nucleotide sugar and free radioactive sugar.

2. Elute radioactive product with 5 mL of methanol and count in 5 mL of scintillation fluid.
3. Regenerate Sep-Pak columns with 10 mL methanol followed by 20 mL of water.

3.4.3. HPLC Assay

All assays using low molecular weight substrates can be carried out by HPLC. Depending on the structure of the enzyme product, various conditions are used (**Table 1**). This method usually allows the separation of enzyme product from all other components of the assay; it also can separate multiple products. The structure of the product can be identified by HPLC if standard compounds are available. HPLC can be carried out after the incubation mixtures have been passed through AG1 × 8 or Sep-Pak columns.

1. Adjust the HPLC conditions, using substrate and product standards so that product elutes in 20–40 min. The flow rate for analytical columns is usually 1 mL/min but can be reduced for better resolution. Acetonitrile/water mixtures are usually successful. When compounds have charged groups (sialic acid and sulfate), water is replaced by buffer at low pH, e.g., 15 m*M* KH_2PO_4, pH 5.4. For glycopeptides, water may be replaced by 0.1% trifluoroacetic acid.
2. Inject an aliquot of the assay mixture into the HPLC, collect fractions, measure absorbance at about 200 nm and radioactivity of fractions, and compare elution times to those of standard compounds. This will assess the identity and quantity of the enzyme products.

3.4.4. High-Voltage Electrophoresis

Mucin products and sialylated and sulfated products can be separated on paper by high-voltage electrophoresis.

1. Stop enzyme reactions with 10 μL of 20 m*M* EDTA/2% borate.
2. After the incubation, apply samples and radioactive standards on paper, separated by 2.5 cm. Wet paper with 1% borate buffer, let the buffer evenly soak in, and place paper so that samples are just above the surface of the borate buffer in the electrophoresis tank. Run electrophoresis at 1500 V for about 1 h. The current is about 150 mA.
3. Dry paper, cut into 2-cm strips, add 7 mL of scintillation fluid, and measure radioactivity with a scintillation counter.

3.4.5. Dionex System

The Dionex system HPAEC allows the quantification of small amounts of free monosaccharides, or oligosaccharides, for which standards are available. It can also achieve excellent separation of mono- and oligosaccharides and, in particular, sialylated and sulfated compounds that would otherwise be difficult to obtain by HPLC.

1. For free sugars, use 15 m*M* NaOH as the mobile phase; for sialylated and sulfated compounds, use 15 m*M* NaOH/0.1 *M* Na-acetate, or higher concentrations of Na-acetate.
2. Sugars can be detected amperometrically. Once an elution time has been established, collect a fraction of eluting samples and detect radioactivity by scintillation counting.

Table 1
Mucin Glycosyltransferases, Their Substrates and Products, and HPLC Conditions for Product Isolation

Path	Enzyme	Substrate	Product	HPLC column	% AN
a	Polypeptide α-GalNAc-T	Thr-peptide	GalNAc-Thr-peptide	C18	0–20
b	Core 1 β3-Gal-T	1. GalNAcα-Bn	Galβ3GalNAcα-Bn	C18	10
		2. GalNAcα-OSM	Galβ3GalNAcα-OSM	—	—
c	Core 3 β3-GlcNAc-T	1. GalNAcα-Bn	GlcNAcβ3GalNAcα-Bn	C18	10
		2. GalNAcα-OSM	GlcNAcβ3GalNAcα-OSM	—	—
d	Core 2 β6-GlcNAc-T	Galβ3GalNAcα-pnp	GlcNAcβ6(Galβ3)GalNAcα-pnp	C18	8
e	Core 4 β6-GlcNAc-T	GlcNAcβ3GalNAcα-pnp	GlcNAcβ6(GlcNAcβ3)GalNAcα-pnp	C18	8
f	β4Gal-T	GlcNAc	Galβ4GlcNAc	NH2	85
g	i β3-GlcNAc-T	Galβ4GlcNAc	GlcNAcβ3Galβ4GlcNAc	NH2	82
h	Elongation β3-GlcNAc-T	GlcNAcβ6(Galβ3)-GalNAcα-Bn	GlcNAcβ6(GlcNAcβ3Galβ3)-GalNAcα-Bn	C18	8
				PAC	82
i	I β6-GlcNAc-T	GlcNAcβ3Galβ-CH_3	GlcNAcβ6(GlcNAcβ3)Galβ-CH_3	NH2	85
j	α6-sialyl-T	GalNAcα-OSM	Sialylα6GalNAcα-OSM	—	—

k	Gal 3-sulfo-T	Galβ3GalNAcα-Bn	SO_4-3-Galβ3GalNAcα-Bn	NH2	80[a]
l	α3-sialyl-T	Galβ3GalNAcα-pnp	Sialylα3Galβ3GalNAcα-pnp	NH2	80[a]
m	α6-sialyl-T III	Sialylα3Galβ3GalNAcα-pnp	Sialylα3Galβ3 (sialylα2-6) GalNAcα-pnp	NH2	80[a]

[a]AN, acetonitrile; Bn, benzyl; Gal, galactose; GalNAc, *N*-acetylgalactosamine; GlcNAc, *N*-acetylglucosamine; OSM, ovine submaxillary mucin; pnp, paranitrophenyl; T, transferase; C18, reversed phase C18 column; NH2, amine column; PAC, mixed cyano-amine column. Compounds containing large peptides (OSM) and charged groups (sialic acid, sulfate) are separated by high-voltage electrophoresis. Sulfated compounds also separate well in the Dionex system. Low molecular weight compounds are separated by HPLC, and eluted in acetonitrile/water mixtures. The paths (a–l) catalyzed by transferases refer to those indicated in **Fig. 1.**

[b]HPLC separation is carried out at 80% acetonitrile and 20% 15 m*M* KH_2PO_4 or $KHCO_3$, pH 5.4.

If more than 10 μL of sample are collected, neutralize the solution before adding scintillation fluid.

3.5. Assays for Specific Enzymes

Nucleotide donor substrates should be radioactively labeled with ^{14}C or ^{3}H (sugar) for sensitivity and unequivocal identification of enzyme product.

3.5.1. Assay for Polypeptide GalNAc-transferase (path a) (Note 3)

1. Add the following to the incubation mixture to a total volume of 40 μL: 8 μL of 5 m*M* peptide substrate, dried in the assay tube, 4 μL of 50 m*M* AMP, 1 μL of 0.5 *M* $MnCl_2$, 10 mL of 0.5 *M* MES, pH 7.0, 1 μL of 5% Triton X-100, 4 μL of 10 m*M* UDP-*GalNAc, 10 μL of enzyme source.
2. Incubate 1 h at 37°C, and then proceed as described for ion-exchange or HPLC assay.

3.5.2. Assay for Core 3 β6-GlcNAc-Transferase (path c)

1. Add the following to the incubation mixture to a total volume of 40 μL: 16 μL of 5 m*M* GalNAcα-benzyl substrate, dried in the assay tube, 4 μL of 50 m*M* AMP, 1 μL of 0.5 *M* $MnCl_2$, 10 μL of 0.5 *M* GlcNAc, 10 μL of 0.5 *M* MES, pH 7.0,1 μL of 5% Triton X-100, 4 μL of 10 m*M* UDP-*GlcNAc, 10 μL enzyme source.
2. Incubate for 1 h at 37°C, and then proceed as described for ion-exchange assay, C18 column, or HPLC assay.

3.5.3. Assay for Core 2 β6-GlcNAc-Transferase (path d)

1. Add the following to the incubation mixture to a total volume of 40 μL: 16 μL of 5 m*M* Galβ1-3GalNAcα-benzyl or Galβ1-3 GalNAcα-pnp substrate, and 10 μL of 0.5 *M* GlcNAc, dried in the assay tube, 4 μL of 50 m*M* AMP, 10 μL of 0.5 *M* MES, pH 7.0,1 μL of 5% Triton X-100, 8 μL of 50 m*M* γ-galactonolactone, 4 μL of 10 m*M* UDP-*GlcNAc, 10 μL of enzyme source.
2. Incubate for 1 h at 37°C, and then proceed as described for ion-exchange, C18 column, or HPLC assay.

3.5.4. Assay for Core 4 β6-GlcNAc-Transferase (path e)

1. Add the following to the incubation mixture to a total volume of 40 μL: 16 μL of 5 m*M* GlcNAcβ1-3 GalNAcα-pnp substrate, dried in the assay tube, 4 μL of 50 m*M* AMP, 10 μL of 0.5 *M* GlcNAc, 10 μL of 0.5 *M* MES, pH 7.0, 1 μL of 5% Triton X-100, 4 μL of 10 m*M* UDP-*GlcNAc, 10 μL of enzyme source.
2. Incubate for 1 h at 37°C, and then proceed as described for ion-exchange, C18 column, or HPLC assay.

3.5.5. Assay for i β3-GlcNAc-Transferase (path g)

1. Add the following to the incubation mixture to a total volume of 40 μL: 16 μL of 5 m*M* Galβ1-4GlcNAc substrate, dried in the assay tube, 4 μL of 50 m*M* AMP, 1 μL of 0.5 *M* $MnCl_2$, 10 μL of 0.5 *M* GlcNAc, 10 μL of 0.5 *M* MES, pH 7.0, 1 μL of 5% Triton X-100, 4 μL of 10 m*M* UDP-*GlcNAc, 10 μL of enzyme source.
2. Incubate for 1 h at 37°C, and then proceed as described for ion-exchange or HPLC assay.

3.5.6. Assay for Elongation β3-GlcNAc-Transferase (path h)

1. Add the following to the incubation mixture to a total volume of 40 μL: 16 μL of 5 m*M* GlcNAcβ1-6 (Galβ1-3) GalNAcα-benzyl substrate, dried in the assay tube, 4 μL of 50 m*M* AMP, 1 μL of 0.5 *M* $MnCl_2$, 10 μL of 0.5 *M* GlcNAc, 10 μL of 0.5 *M* MES, pH 7.0, 8 μL of γ-galactonolactone, 1 μL of 5% Triton X-100, 4 μL of 10 m*M* UDP-*GlcNAc, 10 μL of enzyme source.
2. Incubate for 1 h at 37°C, and then proceed as described for ion-exchange, C18 column or HPLC assay.

3.5.7. Assay for I β6-GlcNAc-Transferase (path i)

1. Add the following to the incubation mixture to a total volume of 40 μL: 16 μL of 5 m*M* GlcNAcβ1-3 Galβ-methyl substrate, dried in the assay tube, 4 μL of 50 m*M* AMP, 10 μL of 0.5 *M* GlcNAc, 10 μL of 0.5 *M* MES, pH 7.0, 1 μL of 5% Triton X-100, 4 μL of 10 m*M* UDP-*GlcNAc, 10 μL of enzyme source.
2. Incubate for 1 h at 37°C, and then proceed as described for ion-exchange or HPLC assay.
3. For some of the I β6-GlcNAc-transferases, substrates with terminal Gal residues may be required such as Galβ1-4 GlcNAcβ1-3 Galβ-R ***(38)***.

3.5.8. Assay for Core 1 β3-Gal-Transferase (path b)

1. Add the following to the incubation mixture to a total volume of 40 μL: 16 μL of 5 m*M* GalNAcα-benzyl substrate, dried in the assay tube, 4 μL of 50 m*M* AMP, 1 μL of 0.5 *M* $MnCl_2$, 10 μL of 0.5 *M* MES, pH 7.0, 1 μL of 5 % Triton X-100, 8 μL of 50 m*M* γ-galactonolactone, 4 μL of 10 m*M* UDP-*Gal, 10 μL of enzyme source.
2. Incubate for 1 h at 37°C, and then proceed as described for ion-exchange, C18 column, or HPLC assay.
3. If DS-OSM is used as a substrate, add 50 μL of DS-OSM (3 m*M* GalNAc) and lyophilize, instead of GalNAc-benzyl. Product is isolated by high-voltage electrophoresis. Alternatively, mucin product can be isolated by precipitation with ethanol.

3.5.9. Assay for β4-Gal-Transferase (path f)

1. Add the following to the incubation mixture to a total volume of 40 μL: 16 μL of 5 m*M* GlcNAc substrate, dried in the assay tube, 4 μL of 50 m*M* AMP, 1 μL of 0.5 *M* $MnCl_2$, 10 μL of 0.5 *M* MES, pH 7.0, 1 μL of 5% Triton X-100, 8 μL of 50 m*M* γ-galactonolactone, 4 μL of 10 m*M* UDP-*Gal, 10 μL of enzyme source.
2. Incubate 1 h at 37°C, and then proceed as described for ion-exchange or HPLC assay.

3.5.10. Assay for α6-Sialyltransferase (paths *j* and *m*)

1. Add the following to the incubation mixture to a total volume of 40 μL: 50 μL of DS-OSM (3 m*M* GalNAc) substrate, dried in the assay tube, 4 μL of 50 m*M* AMP, 10 μL of 0.5 *M* Tris-HCl, pH 7.0, 1 μL of 5% Triton X-100, 4 μL of 10 m*M* CMP-*sialic acid, 10 μL of enzyme source.
2. Incubate for 1 h at 37°C, and then proceed as described for high-voltage electrophoresis assay.
3. Use sialylα2-3 Galβ1-3 GalNAcα-pnp as a substrate for α6-sialyltransferase III. After the incubation, add 1 mL of water to the incubation mixture and boil for 3 min to destroy CMP-sialic acid ***(61)***. Then perform HPLC separation of product.

4. Certain α6-sialyltransferases may utilize Galβ1-3 GalNAc-peptide as a substrate. Antifreeze glycoprotein with only Galβ1-3 GalNAc chains may be a good substrate, or porcine submaxillary mucin that has been processed by acid and glycosidases and possesses GalNAc and Galβ1-3 GalNAc chains. Perform high-voltage electrophoresis assay.

3.5.11. Assay for α3-Sialyltransferase (path l)

1. Add the following to the incubation mixture to a total volume of 40 μL: 16 μL of 5 m*M* Galβ1-3 GalNAcα-pnp substrate, dried in the assay tube, 4 μL of 50 m*M* AMP, 10 μL of 0.5 *M* Tris-HCl, pH 7.0, 1 μL of 5% Triton X-100, 4 μL of 10 m*M* CMP-*sialic acid, 10 μL of enzyme source.
2. Incubate 1 h at 37°C, and then proceed as described for high-voltage electrophoresis assay. Product can be eluted off the paper and isolated by HPLC. Alternatively, 1 mL of water is added after the incubation, and samples are boiled for 3 min, filtered, and then subjected to HPLC analysis.

3.5.12. Assay for Gal 3-Sulfotransferase (path k)

1. Add the following to the incubation mixture to a total volume of 40 μL: 16 μL 5 of m*M* Galβ1-3 GalNAcα-benzyl substrate, dried in the assay tube, 4 μL of 50 m*M* ATP, 2 μL of 0.1 *M* Mg-acetate, 4 μL of 0.1 *M* NaF, 4 μL of 0.1 *M* 2,3-mercaptopropanol, 10 μL of 0.5 *M* Tris-HCl, pH 7.0, 1 μL of 5% Triton X-100, 4 μL of 0.3 m*M* $PAP^{35}S$, 10 μL of enzyme source.
2. Incubate for 1 h at 37°C, then proceed as described for high-voltage electrophoresis assay. Product can be eluted off the paper and isolated by HPLC or by the Dionex system.
3. Other sulfotransferases can be assayed similarly, using Galβ1-4GlcNAc, GlcNAcβ1-3Galβ-CH_3 substrate or other low molecular weight substrates with mucin-type *O*-glycan structures.

4. Notes

1. When comparisons between tissues are made (e.g., between cancer and normal tissue), the maximal velocities of enzyme activities should be measured. This is done using high concentrations (at least twice the affinity constant [K_M]) of donor and acceptor substrates.
2. For the isolation of large amounts of products, the assay should first be optimized for the chosen tissue, e.g., by testing various concentrations of metal ions, detergents, incubation times, and enzyme concentrations. In addition, some preliminary kinetic studies should be carried out.
3. For reasons that are not yet clear, most polypeptide α-GalNAc-transferases act in vitro only on Thr-containing substrates although in vivo Ser is also glycosylated. However, it may be possible that activities exist in certain tissues that accept Ser-containing substrates with significant activity.

References

1. Brockhausen, I. (1995) *Glycoproteins*, New Comprehensive Biochemistry Vol 29A, (Montreuil, J., Vliegenthart, J., and Schachter, H., eds.), Elsevier, New York, pp. 201–259.

2. Brockhausen, I. (1993) Clinical aspects of glycoprotein biosynthesis. *Crit. Rev. Clin. Lab. Sci.* **30,** 65–151.
3. Brockhausen, I. and Kuhns, W. (1997) *Glycoproteins and human disease.* Medical Intelligence Unit, CRC Press and Mosby-Yearbook, Chapman & Hall, New York.
4. Gleeson, P. A., Teasdale, R. D., and Burke, J. (1994) Targeting of proteins to the Golgi apparatus. *Glycoconj. J.* **11,** 381–394.
5. Hirschberg, C. B. (1997) Transporters of nucleotide sugars, nucoleotide sulfate and ATP on the Golgi apparatus: where next? *Glycobiology* **7,** 169–171.
6. Brockhausen, I., Möller, G., Merz, G., Adermann, K., and Paulsen, H. (1990) Control of glycoprotein synthesis: the peptide portion of synthetic O-glycopeptide substrates influences the activity of O-glycan core 1 uridine 5'-diphospho-galactose: N-acetylgalactosamineα-R β3-galactosyl-transferase. *Biochemistry* **29,** 10,206–10,212.
7. Roth, J., Wang, Y., Eckhardt, A. E., and Hill, R. L. (1994) Subcellular localization of the UDP-N-acetyl-D-galactosamine: polypeptide N-acetylgalactosaminyltransferase-mediated O-glycosylation reaction in the submaxillary gland. *Proc. Natl. Acad. Sci. USA* **91,** 8935–8939.
8. Röttger, S., White, J., Wandall, H. H., Olivo, J. C., Stark, A., Bennett, E. P., Whitehouse, C., Berger, E. G., Clausen, H., and Nilsson, T. (1998) Localization of three human polypeptide GalNAc-transferases in HeLa cells suggests initiation of O-linked glycosylation throughout the Golgi apparatus. *J. Cell Sci.* **111,** 45–60.
9. Clausen, H. and Bennett, E. P. (1996) A family of UDP-GalNAc: polypeptide N-acetylgalactosaminyl-transferases control the initiation of mucin-type O-linked glycosylation. *Glycobiology* **6,** 635–646.
10. Sutherlin, M., Nishimori, I., Caffrey, T., Bennett, E., Hassan, H., Mandel, U., Mack, D., Iwamura, T., Clausen, H., and Hollingsworth, M. (1997) Expression of three UDP-N-acetyl-α-D-galactosamine: polypeptide GalNAc N-acetylgalactosaminyltransferases in adenocarcinoma cell lines. *Cancer Res.* **57,** 4744–4748.
11. O'Connell, B. C., Hagen, F. K., and Tabak, L. A. (1992) The influence of flanking sequence on the O-glycosylation of threonine in vitro. *J. Biol. Chem.* **267,** 25,010–25,018.
12. Brockhausen, I., Toki, D., Brockhausen, J., Peters, S., Bielfeldt, T., Kleen, A., Paulsen, H., Meldal, M., Hagen, F., and Tabak, L. (1996) Specificity of O-glycosylation by bovine colostrum UDP-GalNAc: polypeptide α-N-acetylgalactosaminyltransferase using synthetic glycopeptide substrates. *Glycoconj. J.* **13,** 849–856.
13. Schachter, H. and Brockhausen, I. (1992) *Glycoconjugates: Composition, Structure and Function.* (Allen, H. J. and Kisailus, E. C., eds.), Marcel Dekker, New York, pp. 263–332.
14. Brockhausen, I., Möller, G., Pollex-Krüger, A., Rutz, V., Paulsen, H., and Matta, K. L. (1992) Control of O-glycan synthesis: specificity and inhibition of O-glycan core 1 UDP-galactose:N-acetylgalactosamine-α-R β3-galactosyltransferase from rat liver. *Biochem. Cell Biol.* **70,** 99–108.
15. Granovsky, M., Bielfeldt, T., Peters, S., Paulsen, H., Meldal, M., Brockhausen, J., and Brockhausen, I. (1994) O-glycan core 1 UDP-Gal: GalNAc β3-galactosyltransferase is controlled by the amino acid sequence and glycosylation of glycopeptide substrates. *Eur. J. Biochem.* **221,** 1039–1046.

16. Cartron, J., Andrev, J., Cartron, J., Bird, G. W. G., Salmon, C., and Gerbal, A. (1978) Demonstration of T–transferase deficiency in Tn–polyagglutinated blood samples. *Eur. J. Biochem.* **92,** 111–119.
17. Thurnher, M., Rusconi, S., and Berger, E. G. (1993) Persistent repression of a functional allele can be responsible for galactosyltransferase deficiency in Tn syndrome. *J. Clin. Invest.* **91,** 2103–2110.
18. Brockhausen, I., Yang, J., Dickinson, N., Ogata, S., and Itzkowitz, S. (1998) Mechanism leading to the expression of the cancer-associated sialyl-Tn and Tn antigens in human cancer cells. *Glycoconj. J.* **15,** 595–603.
19. Kuan, S. F., Byrd, J. C., Basbaum, C., and Kim, Y. S. (1989) Inhibition of mucin glycosylation by aryl-N-acetyl-α-galactosaminides in human colon cancer cells. *J. Biol. Chem.* **264,** 19271–19277.
20. Kojima, N., Handa, K., Newman, W., and Hakomori, S.–I. (1992) Inhibition of selectin–dependent tumor cell adhesion to endothelial cells and platelets by blocking O–glycosylation of these cells. *Biochem. Biophys. Res. Commun.* **182,** 1288–1295.
21. Williams, D. and Schachter, H. (1980) Mucin synthesis. I. Detection in canine submaxillary glands of an N-acetylglucosaminyltransferase which acts on mucin substrates. *J. Biol. Chem.* **255,** 11,247–11,252.
22. Bierhuizen, M. F. A. and Fukuda, M. (1992) Expression cloning of a cDNA encoding UDP-GlcNAc:Galbeta1-3-GalNAc-R (GlcNAc to GalNAc) beta1-6GlcNAc transferase by gene transfer into CHO cells expressing polyoma large tumor antigen. *Proc. Natl. Acad. Sci. USA* **89,** 9326–9330.
23. Bierhuizen, M., Mattei, M. G., and Fukuda, M. (1993) Expression of the developmental I antigen by a cloned human cDNA encoding a member of a beta-1,6-N-acetylglucosaminyltransferase gene family. *Genes Dev.* **7,** 468–478.
24. Kuhns, W., Rutz, V., Paulsen, H., Matta, K. L., Baker, M. A., Barner, M., Granovsky, M., and Brockhausen, I. (1993) Processing O-glycan core 1, Galβ1-3GalNAcα-R. Specificities of core 2 UDP-GlcNAc:Galβ1-3GalNAc-R β6-N-acetylglucosaminyltransferase and CMP-SA: Galβ1-3GalNAc-R α3-sialyltransferase. *Glycoconj. J.* **10,** 381–394.
25. Piller, F., Piller, V., Fox, R., and Fukuda, M. (1988) Human T-lymphocyte activation is associated with changes in O-glycan biosynthesis. *J. Biol. Chem.* **263,** 15,146–15,150.
26. Heffernan, M., Lotan, R., Amos, B., Palcic, M., Takano, R., and Dennis, J. W. (1993) Branching beta 1-6N-acetylglucosaminyltransferase and polylactosamine expression in mouse F9 teratocarcinoma cells and differentiated counterparts. *J. Biol. Chem.* **268,** 1242–1251.
27. Vavasseur, F., Yang, J., Dole, K., Paulsen, H., and Brockhausen, I. (1995) Synthesis of core 3: characterization of UDP-GlcNAc: GalNAc β3–N-acetylglucosaminyl-transferase activity from colonic tissues. Loss of the activity in human cancer cell lines. *Glycobiology* **5,** 351–357.
27a. Whitehouse, C., (1998) PhD thesis. University of London, London, UK.
28. Skrincosky, D., Kain, R., El-Battari, A., Exner, M., Kerjaschki, D., and Fukuda, M. (1997) Altered Golgi localization of core 2 β-1,6-N-acetylglucosaminyltransferase leads to decreased synthesis of branched O-glycans. *J. Biol. Chem.* **272,** 22,695–22,702.

29. Brockhausen, I., Matta, K. L., Orr, J., and Schachter, H. (1985) Mucin synthesis. VI. UDP-GlcNAc:GalNAc-R β3-N-acetylglucosaminyltransferase and UDP-GlcNAc: GlcNAcβ1-3GalNAc-R (GlcNAc to GalNAc) β6-N-acetylglucosaminyltransferase from pig and rat colon mucosa. *Biochemistry* **24,** 1866–1874.
30. Yang, J., Byrd, J., Siddiki, B., Chung, Y., Okuno, M., Sowa, M., Kim, Y., Matta, K., and Brockhausen, I. (1994) Alterations of O-glycan biosynthesis in human colon cancer tissue. *Glycobiology* **4,** 873–884.
31. King, M. J., Chan, A., Roe, R., Warren, B. F., Dell, A., Morris, H. R., Bartolo, C. C., Durdey, P., and Corfield, A. P. (1994) Two different glycosyltransferase defects that result in GalNAcα-O-peptide (Tn) expression. *Glycobiology* **4,** 267–269.
32. Shaper, J. H. and Shaper, N. L. (1992) Enzymes associated with glycosylation. *Curr. Opin. Struct. Biol.* **2,** 701–709.
33. Koenderman, A. H., Koppen, P. L., and van den Eijnden, D. H. (1987) Biosynthesis of polylactosaminoglycans. *Eur. J. Biochem.* **166,** 199–208.
34. Harduin-Lepers, A., Shaper, J. H., and Shaper, N. L. (1993) Characterization of two cis-regulatory regions in the murine β1,4-galactosyltransferase gene: evidence for a negative regulatory element that controls initiation at the proximal site. *J. Biol. Chem.* **268,** 14,348–14,359.
35. Roth, J. (1995) Compartmentation of glycoprotein biosynthesis, *Glycoproteins* (Montreuil, J., Vliegenthart, J. F. G., and Schachter, H., eds.), Elsevier, Amsterdam, pp 287–312.
36. Brockhausen, I., Orr, J., and Schachter, H. (1984) Mucin synthesis. the action of pig gastric mucosal UDP-GlcNAc:Galβ1-3(R_1)GalNAc-R_2 (GlcNAc to Gal) β3-N-acetylglucosaminyltransferase on high molecular weight substrates. *Can. J. Biochem. Cell Biol.* **62,** 1081–1090.
37. Piller, F., Cartron, J.-P., Maranduba, A., Veyrieres, A., Leroy, Y., and Fournet, B. (1984) Biosynthesis of blood group I antigens: identification of a UDP-GlcNAc: GlcNAcβ1-3Gal (-R) β1-6(GlcNAc to Gal) N-acetylglucosaminyltransferase in hog gastric mucosa. *J. Biol. Chem.* **259,** 13,385–13,390.
38. Leppänen, A., Penttilä, L., Niemelä, R., Helin, J., Seppo, A., Lusa, S., and Renkonen, O. (1991) Human serum contains a novel beta1,6-N-acetylglucosaminyltransferase activity that is involved in midchain branching of oligo(N-acetyllactosaminoglycans). *Biochemistry* **30,** 9287–9296.
39. Harduin-Lepers, A., Recchi, M.-A., and Delannoy, P. (1995) 1994, the year of sialyltransferases. *Glycobiology* **5,** 741–758.
40. Lee, Y. C., Kurosawa, N., Hamamoto, T., Nakoaka, T., and Tsuji, S. (1993) Molecular cloning and expression of Gal-beta-1,3GalNAc-alpha-2,3-sialyltransferase from mouse brain. *Eur. J. Biochem.* **216,** 377–385.
41. Kurosawa, N., Hamamoto, T., Inoue, M., and Tsuji, S. (1995) Molecular cloning and expression of chick Gal beta 1,3GalNAc alpha 2,3-sialyltransferase. *Biochim. Biophys. Acta* **1244,** 216–222.
42. Gillespie, W., Paulson, J., Kelm, S., Pang, M., and Baum, L. (1993) Regulation of α2,3-sialyltransferase expression correlates with conversion of peanut agglutinin $(PNA)^+$ to PNA^- phenotype in developing thymocytes. *J. Biol. Chem.* **268,** 3801–3804.
43. Baker, M., Taub, R., Kanani, A., Brockhausen, I., and Hindenburg, A. (1985) Increased activity of a specific sialyltransferase in chronic myelogenous leukemia. *Blood* **66,** 1068–1071.

44. Whitehouse, C., Burchell, J., Gschmeissner, S., Brockhausen, I., Lloyd, K., and Taylor-Papadimitriou, J. (1997) A transfected sialyltransferase that is elevated in breast cancer and localizes to the medial/trans-Golgi apparatus inhibits the development of core-2-based O-glycans. *J. Cell Biol.* **137,** 1229–1241.
45. Beyer, T. A., Sadler, J. E., Rearick, J. I., Paulson, J. C., and Hill, R. L. (1981) *Advances in Enzymology* Vol 52, (Meister, A. ed.), John Wiley & Sons, New York, pp. 23–175.
46. Kurosawa, N., Hamamoto, T., Lee, Y. C., Nakaoka, T., Kojima, N., and Tsuji, S. (1994) Molecular cloning and expression of GalNAc alpha 2,6-sialyltransferase. *J. Biol. Chem.* **269,** 1402–1409.
47. Kurosawa, N., Kojima, N., Inoue, M., Hamamoto, T., and Tsuji, S. (1994) Cloning and expression of Gal beta 1,3GalNAc-specific GalNAc alpha 2,6-sialyltransferase. *J. Biol. Chem.* **269,** 19,048–19,053.
48. Bergh, M. L. E. and van den Eijnden, D. H. (1983) A glycan specificity of fetal calf liver and ovine and porcine submaxillary gland α-N-acetylgalactosaminide α2-6 sialyltransferase. *Eur. J. Biochem.* **136,** 113–118.
49. Varki, A. (1992) Diversity in the sialic acids. *Glycobiology* **2,** 25–40.
50. Carter, S., Slomiany, A., Gwozdzinski, K., Liau, Y., and Slomiany, B. (1988) Enzymatic sulfation of mucus glycoprotein in gastric mucosa. *J. Biol. Chem.* **263,** 11,977–11,984.
51. Kuhns, W., Jain, R., Matta, K., Paulsen, H., Baker, M., Geyer, R., and Brockhausen, I. (1995) Characterization of a novel mucin sulfotransferase activity synthesizing sulfated O-glycan core 1, 3-sulfate-Gal-beta1-3GalNAc-alpha-R. *Glycobiology* **5,** 689–697.
52. Lo-Guidice, J., Perini, J., Lafitte, J., Ducourouble, M., Roussel, P., and Lamblin, G. (1995) Characterization of a sulfotransferase from human airways responsible for the 3-O-sulfation of terminal galactose in N-acetyllactosamine-containing mucin carbohydrate chains. *J. Biol. Chem.* **270,** 27,544–27,550.
53. Brockhausen, I. and Kuhns, W. (1997) Role and metabolism of glycoconjugate sulfation. *Trends Glycosci. Glycotechnol.* **9,** 379–398.
54. Tsuboi, S., Isogai, Y., Hada, N., King, J. K., Hindsgaul, O., and Fukuda, M. (1996) 6'-Sulfo sialyl Le^x but not 6-sulfo sialyl Le^x expressed on the cell surface supports L-selectin-mediated adhesion. *J. Biol. Chem.* **271,** 27,213–27,216.
55. Paulsen, H. and Aderman, K. (1991) Synthese von O-Glycopeptid-Sequenzen des N-terminus von Interleukin-2. *Liebigs Ann Chemie* 751–769, 771–780.
56. Distler, J. J. and Jourdian, G. W. (1973) The purification and properties of β-galactosidase from bovine testis. *J. Biol. Chem.* **248,** 6772–6780.
57. Hounsell, E. F., Davies, M. J., and Renouf, D. V. (1996) O-linked protein glycosylation structure and function. *Glycoconj. J.* **13,** 19–26.
58. Brockhausen, I., Matta, K. L., Orr, J., Schachter, H., Koenderman, A. H. L., and van den Eijnden, D. H. (1986) Mucin synthesis: conversion of R_1-β1-3Gal-R_2 to R_1-β1-3(GlcNAcβ1-6)Gal-R_2 and of R_1-β1-3GalNAc-R_2 to R_1-β1-3(GlcNAcβ1-6)GalNAc-R_2 by a β6-N-acetylglucosaminyltransferase in pig gastric mucosa. *Eur. J. Biochem.* **157,** 463–474.
59. Lo-Guidice, J.-M., Herz, H., Lamblin, G., Plancke, Y., Roussel, P., and Lhermitte, M. (1997) Structures of sulfated oligosaccharides isolated from the respiratory mucins of a non-secretor (O, Le^{a+b-}) patient suffering from chronic bronchitis. *Glycoconj. J.* **14(1),** 113–125.

60. Capon, C., Leroy, Y., Wieruszeski, J., Ricart, G., Strecker, G., Montreuil, J., and Fournet, B. (1989) Structures of O-glycosidically linked oligosaccharides isolated from human meconium glycoproteins. *Eur. J. Biochem.* **182,** 139–152.
61. Dall'Olio, F., Malagolini, N., Guerrini, S., Lau, J., and Serafini-Cessi, F. (1996) Differentiation-dependent expression of human β-galactoside α2,6-sialyltransferase mRNA in colon carcinoma CaCo-2 cells. *Glycoconj. J.* **13,** 115–121.

VIII

Mucin Gene Detection

24

Mucin cDNA Cloning

Jean-Pierre Aubert and Nicole Porchet

1. Introduction

Much information on the structure, organization, and expression of many genes has been acquired by using the powerful technique of cDNA cloning ***(1–5)***. cDNA clones differ from genomic DNA clones in that they represent a permanent copy of an mRNA and are representative of the parts of a gene that are expressed as mature RNA. This comparison of cDNA clones with their genomic homologs has resulted in the discovery of introns in most eukaryotic genes. Amino acid sequence of gene products can only be obtained from full-length cDNA sequences. cDNA clones can also be used to express the protein products of genes in prokaryotes.

A cDNA library must be created by a series of enzymatic reactions. The first one consists in generating a first DNA strand complementary to RNA followed by a removal of the RNA strand and replacement by a "second strand" of DNA. The new double-stranded DNA is ligated into an appropriate vector. Two kinds of vectors can be used: plasmid or bacteriophage cloning vectors. Most of the cDNA cloning experiments of human mucins have been done in bacteriophage λgt11 and/or Lambda ZAP® Vector. Both are protein expression cDNA vectors. They are designated to generate cDNA libraries from small amounts of DNA (between 5000 and 10,000 clones per nanogram of cDNA). cDNA inserts are cloned into the 3' end (λgt11) or the 5' end (λZAP®) of the β-galactosidase gene-coding region, allowing blue/white color screening of clones with inserts in the presence of X-Gal and isopropyl-1-thiol-β-D-galactopyranoside (IPTG).

Due to the large size of the mRNA of mucins (superior to 20 kb) described, it is nearly impossible to obtain a full-length cDNA by screening cDNA libraries. Therefore, methods have been developed to amplify DNA sequences from an mRNA template between a known internal site and the unknown sequences of either the 3' or the 5' end of the mRNA. These methods were first described by Frohmann et al. ***(6)*** and are referred to as RACE (rapid amplification of cDNA ends).

From: *Methods in Molecular Biology, Vol. 125: Glycoprotein Methods and Protocols: The Mucins*
Edited by: A. Corfield © Humana Press Inc., Totowa, NJ

This chapter describes the protocols that have been successfully used in our laboratory to clone three cDNAs corresponding to MUC4, MUC5AC, and MUC5B *(7)*.

2. Materials

2.1. Biological Materials

1. Mucus-secreting cultured cells (e.g., LS174T, HT29 MTX) or mucosae from human or animals (e.g., tracheobronchial, gastric, intestinal) are snap-frozen in liquid nitrogen and stored at –80°C until used.

2.2. Monoclonal or Polyclonal Antibodies

1. Antibodies must be directed against the peptide moiety of mucins. A chemical deglycosylation of whole mucins or of mucin glycopeptides is needed. The quality of the antibodies must be checked by immunohistochemistry before screening ***(7–10)***. *See* Chapters 29 and 30.

2.3. Buffers

1. Phenol/chloroform: Add a volume of chloroform equal to that of TE (*see* **item 3**) and saturated phenol and mix. Allow the phases to separate. Store in dark bottles at 4°C.
2. Chloroform: Prepare isoamylalcohol by mixing 24 vol of chloroform with 1 vol of isoamyl alcohol. Store at room temperature in a dark bottle.
3. TE buffer: Dissolve 1.21 g of Tris base and 0.37 g of EDTA-disodium salt per litre. Adjust to pH 8.0 with HCl. Sterilize by autoclaving.
4. STE buffer: Dissolve 5.84 g of NaCl, 1.21 g of Tris base and 0.37 g of EDTA-disodium salt per liter. Adjust pH to 8.0 with HCl. Sterilize by autoclaving.
5. SM buffer: Dissolve 5.8 g of NaCl, 2 g of $MgSO_4$-$7H_2O$, 6.05 g of Tris base, 5 mL of 20 g gelatine per liter. pH to 7.5 with HCl. Sterilize by autoclaving.
6. TBS buffer: Dissolve 6.05 g of Tris base and 8.76 g of NaCl per liter. Adjust pH to 8.0 with HCl. Sterilize by autoclaving.
7. TBST buffer: Add 0.5 mL of Tween-20 to 1 L of TBS. Sterilize by autoclaving.
8. 3 *M* sodium acetate: Dissolve 408.1 g of sodium acetate 3 H_2O in 800 mL of water. Adjust pH to 5.2 with glacial acetic acid. Adjust volume to 1 L. Dispense in aliquots and sterilize by autoclaving.
9. 0.25 *M* EDTA: add 9.3 g of disodium EDTA-$2H_2O$ to 80 mL water. Adjust to pH 9.0 with 10 *N* NaOH and adjust the volume to 100 mL. Sterilize by autoclaving.
10. 10 m*M* magnesium sulfate ($MgSO_4$-$7H_2O$): Dissolve 7.4 g of magnesium sulfate/100 mL of distilled water. Sterilize by autoclaving.
11. 0.5 *M* calcium chloride ($CaCl_2$-$2H_2O$): Dissolve 7.4 g of calcium chloride/100 mL of distilled water. Sterilize by autoclaving.
12. 1 *M* magnesium chloride ($MgCl_2$-$6H_2O$): Dissolve 20.3 g of magnesium chloride/100 mL of distilled water. Sterilize by autoclaving.
13. PEG/NaCl solution: Dissolve 20 g of polyethylene glycol 6000 and 11.7 g of NaCl/100 mL of SM buffer pH 7.5. Sterilize by autoclaving.
14. 20% maltose: Dissolve 10 g of maltose/50 mL of distilled water. Filter sterilize. Store at 2–8°C.
15. Ampicillin stock solution: Dissolve 0.25 g of ampicillin/10 mL of distilled water. Filter sterilize. Store at –20°C.
16. Luria-broth (LB): Dissolve in 800 mL of H_2O, 10 g of tryptone, 5 g of yeast extract, and 10 g of NaCl. Adjust to pH 7.0 with 5 *M* NaOH and increase volume to 1 L. Sterilize by autoclaving immediately.

17. M-broth + ampicillin : just before use, add 20% maltose, to a final concentration of 0.4% and ampicillin, to a final concentration of 50 μg/mL in L-broth.
18. L-amp plates: Add 15 g of agar per litre to L-broth, prior to autoclaving. Cool to approx 50°C, add 2 mL of ampicillin stock solution, and pour 25 mL into each 90 mm sterile Petri dish (*see* **Note 1**).
19. L-top agar: Add 0.8 g of agar per 100 mL of LB before autoclaving (*see* **Note 2**).
20. M-top agar + ampicillin: Melt L-top agar, cool to 45°C and add 20% maltose to a final concentration of 0.4 *M* and ampicillin (25 mg/mL stock) to a final concentration of 50 μg/mL.
21. Color-selection reagents: 100 m*M* IPTG; dissolve 23.8 mg isopropyl β-D-thiogalactopyranoside/mL in water. Filter sterilise. 2% X-gal: Dissolve 2% (w/v) 5-bromo-4-chloro-3-indoly-β-D-galactopyranoside in dimethyl formamide. Add 10 μL of each solution/mL of top agar just before plating at 45°C.
22. Denaturing solution: Dissolve 87.66 g of NaCl and 20 g of NaOH in 800 mL H_2O slowly with care. Adjust volume to 1 L.
23. Neutralizing solution: Dissolve 175.32 g of NaCl in 500 mL of 1 *M* Tris-HCl, pH 7.0. Adjust to 1 L with water.

2.4. Kits and Modules

1. Poly A tract® mRNA isolation system (Promega, Charbonnieres, France).
2. mRNA purification kit (Pharmacia Biotech, Orsey, France).
3. cDNA synthesis module (Amersham,Les Ulis, France).
4. cDNA rapid adaptor ligation module (Amersham).
5. cDNA rapid cloning module (Amersham).
6. 5'/3' RACE kit (Boehringer Mannheim, Roche Diagnostics, Meylan, France).

2.5. Equipment

1. Several water baths at a range of temperatures from 12 to 90°C and incubators at 32, 37, and 43°C.
2. Microcentrifuge.
3. Horizontal gel electrophoresis apparatus.
4. Thermocycler.

3. Methods

3.1. Preparation of mRNA

1. Total RNA is isolated from cultured cells or human mucosae as described in Chapter 25. In spite of the fact that the purification yield of poly A^+ mucin fraction is very poor, the purification step of polyA^+ RNA from total RNA cannot be omitted prior to cloning (careful elimination of ribosomal RNA is needed).
2. Different technologies are available for the purification of mRNA. We have successfully used the poly A tract mRNA isolation system from Promega or the mRNA purification kit from Pharmacia Biotech.

3.2. cDNA Synthesis (see Note 3)

Many kits are commercially available. We have used the cDNA synthesis module from Amersham and strictly followed the manufacturer's instructions.

1. Synthesize the first strand cDNA at 42°C for 1 h by the AMV reverse transcriptase.
2. Synthesize the second-strand cDNA by the T4 DNA polymerase after hydrolysis of the remaining RNA by the *Escherichia coli* Ribonuclease H.

3. Purify the double-stranded cDNA by phenol/chloroform extraction and ethanol precipitation.
4. If a radiolabeled nucleotide has been added during the synthesis of either the first- or the second-strand cDNA, the yield of cDNA synthesized can be calculated and the quality of the cDNA evaluated by alkaline gel electrophoresis in agarose.

3.3. cDNA Cloning (see Notes 4 and 5)

We have used the cDNA rapid adaptor ligation module and the cDNA rapid cloning module from Amersham and followed the manufacturer's instructions.

1. Ligation of adaptors to cDNA: Each adaptor contains a blunt- end for ligation with cDNA and an *Eco*RI cohesive end to permit ligation with any *Eco*RI-digested vector.
2. Purification of "adapted" cDNA: Use the spun columns of the kit.
3. Phosphorylation of the *Eco*RI overhangs carrying 5'-hydroxyl groups with T4 polynucleotide kinase.
4. Ligations into λgt11 vector arms: The λgt11 vector arms and the cDNA are ligated to generate recombinant DNA molecules, which also form linear concatemers by ligation of their cos ends.
5. In vitro packaging of ligation mixtures: The recombinant DNA molecules are packaged into infectious phage particles. The necessary cell extracts are supplied with Amersham λ-DNA in vitro packaging module.
6. Phage plating cells: Use Y1090 *E. coli* bacteria.
7. Titration of λgt11 recombinants: As predicted by the manufacturer's instructions, we have regularly obtained approx 10^6–10^7 recombinants per microgram of cDNA.
8. Quality of λgt11 cDNA library immunoscreening: The quality of immunoscreening depends on the specificity of the antibodies (monoclonal or polyclonal).
   a. Plate out the λgt11 library at the required plaque density. The best result is obtained when each individual plaque lysis is still well separated from the surrounding plaques.
   b. Induce the expression by overlaying plates with IPTG-impregnated filters. Incubate at 37°C overnight.
   c. Mark the filters and then block the nonspecific binding by incubation of the filters in 20% fetal calf serum (FCS) for 1h at room temperature.
   d. Bind protein-specific primary antibodies diluted in TBS + 20% FCS for at least 12 h at 4°C.
   e. Wash the filters three times with TBS and then TBST.
   f. Bind peroxidase-labeled second antibody in TBS + 20% FCS for 1 h at room temperature.
   g. Wash the filters as described in **step e**.
   h. Incubate filters with DAB substrate + H_2O_2 diluted in TBS.
   i. Align filter and original plate. Identify and pick positive plaque(s). Rescreen to check and isolate clone(s).
9. Characterize the selected clones by purification of phage DNA either by polymerase chain reaction (PCR) using forward and reverse λgt11 primers or by small-scale liquid culture and phage DNA purification.
10. The cDNA inserts can now be subcloned into plasmids or M13 phagemids for sequencing.

3.4. 5'/3' RACE PCR

Since most human mucin mRNAs are usually very large (up to 24 kb), it is not judicious to try to obtain the full-length cDNA by screening libraries (*see* **Note 6** and **7**). We have successfully used the 5'/3' RACE kit from Boehringer Mannheim by scrupulously following the manufacturer's instruction.

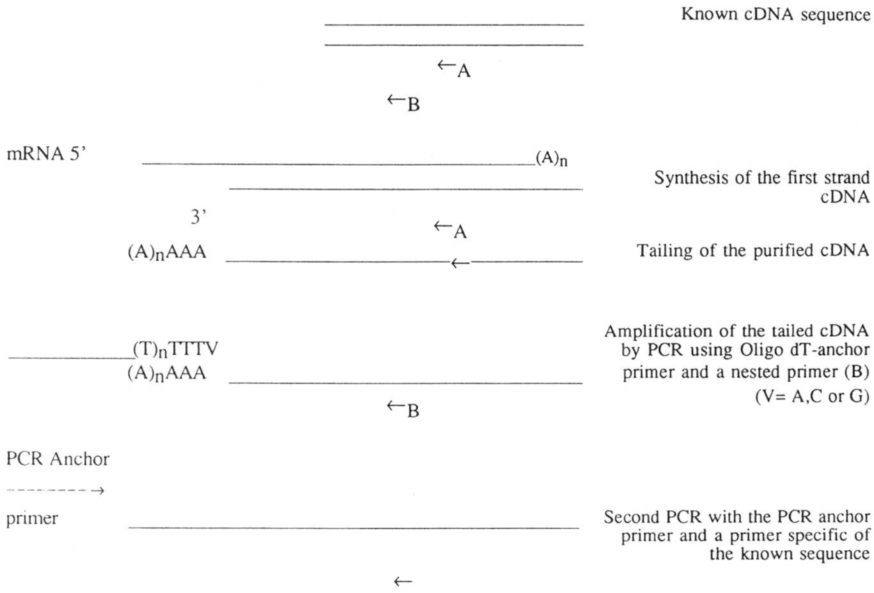

Fig. 1.

3.4.1. 5' RACE PCR (*Fig. 1*)

1. Synthesize first-strand cDNA from poly (A^+) mRNA using a cDNA-specific primer (designated from the first cDNA sequences obtained) and avian myeloblastosis virus (AMV) reverse transcriptase.
2. Used terminal transferase to add a homopolymeric A-tail to the 3'-end of the first strand.
3. Then amplify tailed cDNA by PCR using an oligo dT-anchor primer and a second specific primer of the cDNA of interest (nested primer).
4. Clone the PCR product in a vector such as pCR2.1 (Invitrogen) and sequence.

3.4.2. 3' RACE PCR (*Fig. 2*)

1. Initiate first-strand cDNA synthesis at the natural poly(A^+) tail of mRNA using an oligo dT-anchor primer.
2. Then directly perform amplification using the PCR anchor primer and a specific primer designated from the first cDNA sequence obtained.

4. Notes

1. L-amp plates may be stored inverted in plastic bags at 4°C.
2. Sterile L-top agar may be stored solid and melted in a microwave oven before use.
3. To optimize the yield of cDNA synthesis, the enzymatic steps must be initiated using as templates freshly purified poly(A^+)RNAs.
4. Perform at least three separate cDNA-containing ligation reactions to find the optimal cloning efficiency. We use generally 20-, 50-, 100-, and 150-ng linkered cDNA for 1 µg of λgt11 arms.

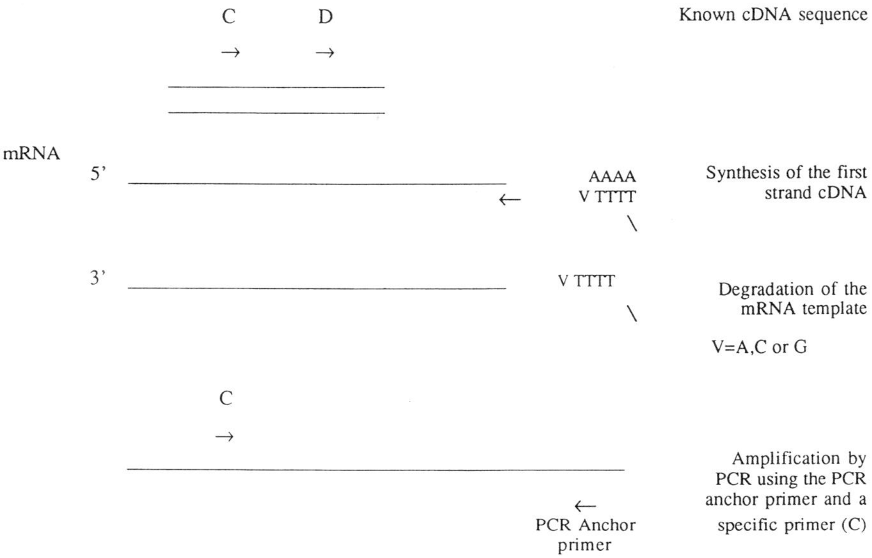

Fig. 2.

5. To obtain the best yield of recombinants, a crucial step is the speed with which the extracts are mixed.
6. Such cDNA libraries can be screened with oligonucleotide or DNA radiolabeled probes.
7. In the case of large human mucin cDNAs, it will probably not be possible to obtain the full-length 5'-end cDNA in one attempt. From the new sequence determined, the 5'-end RACE procedure can be repeated.

Acknowledgments

The authors wish to thank Dr. M. Crépin for his collaboration and D. Petitprez for her excellent technical assistance.

To the memory of G. Vergnes who contributed to selection of the first recombinant clones of human tracheobronchial apomucins.

References

1. Davis, R. V., Botstein, D., and Roth, J. R. (1980) *Advanced Bacterial Genetics*, Cold Spring Harbor Laboratory, Cold Spring Harbor, NY.
2. Williams, J. G. (1981) *Genetic Engineering*, vol. 1, (Williamson, R., ed.), Academic Press, London, pp. 1.
3. Maniatis, T., Fritsch, E. F., and Sambrook, J. (1982) *Molecular cloning: A Laboratory Manual*, Cold Spring Harbor Laboratory Press , Cold Spring Harbor, NY.
4. Huynh, T. V., Young, R. A., and Davis, R. W. (1985) *DNA Cloning : A Practical Approach,* vol. 1 (Glover, D., ed.), IRL, Oxford, UK, pp. 49.
5. Watson, C. J. and Jackson, J. F. (1985) *DNA Cloning : A Practical Approach* vol. 1 (Glover, D., ed.), IRL, Oxford, UK, pp. 79.

6. Frohmann, M. (1994) Beyond classic RACE. *PCR Methods Appl.* **4,** 540–558.
7. Aubert, J. P., Porchet, N., Crépin, M., Duterque-Coquillaud, M., Vergnes, G., Mazzuca, M., Debuire, B., Petitprez, D., and Degand, P. (1991) Evidence for different human tracheobronchial mucin peptides deduced from nucleotide cDNA sequences. *Am. J. Respir. Cell Mol. Biol.* **5,** 178–185.
8. Crépin, M., Porchet, N., Aubert, J. P., and Degand, P. (1990) Diversity of the peptide moiety of human airway mucins. *Biorheology* **27,** 471–484.
9. Gum, J. R., Byrd, J. C., Hicks, J. W., Toribara, N. W., Lamport, D. T. A., and Kim, Y. S. (1989) Molecular cloning of human intestinal mucin cDNAs. Sequence analysis and evidence for genetic polymorphism. *J. Biol. Chem.* **264,** 6480–6487.
10. Toribara, N. W., Robertson, A. M., Ho, S. B., Kuo, W. L., Gum, E., Hicks, J. W., Gum, J. R., Byrd, J. C., Siddiki, B., and Kim, Y. S. (1993) Human gastric mucin. Identification of a unique sequence by expression cloning. *J. Biol. Chem.* **268,** 5879–5885.

25

Northern Blot Analysis of Large mRNAs

Nicole Porchet and Jean-Pierre Aubert

1. Introduction

Northern blot analysis has historically been one of the most common methods used to provide information on the number, length, and relative abundance of mRNAs expressed by a single gene. This technique also generates a record of the total mRNA content expressed by a cell culture or by a tissue, which can be analyzed and compared on the same specimens by successive hybridizations with specific probes.

There are two main difficulties often associated with this technique. The fisrt is that Northern blotting is generally considered to be rather insensitive, requiring large amounts of starting material and consuming large amounts of tissue. The second problem stems from the fact that the RNA isolated from cells or tissues must be of high purity and high quality, and nondegraded; maintaining these qualities can be difficult, specifically in the case of large mRNAs, even for experienced workers.

Messenger RNAs, larger than 10 kb, encoding human titin (23 kb), nebulin (20.7 kb), apolipoprotein B-100 (14.1 kb), dystrophin (14 kb), and secreted mucins MUC2, MUC3, MUC4, MUC5B, MUC5AC, MUC6 (14–24 kb) usually show more or less polydisperse patterns on Northern blots, which are attributable to artifactual causes. These patterns are in the form of a smear or very wide bands and result mainly from two technical problems: (1) the high sensitivity of large mRNAs to mechanical damage that occurs during extraction and purification steps, and (2) the lack of efficiency of the transfer of large RNAs onto membranes, and thus the poor detection of the large intact mRNA species.

Moreover, efficiency of large poly(A^+)RNA selection is very poor with preferential loss of the largest transcripts. Hence, there is a risk of misinterpretation of the data in the case of mucin genes that express allelic transcripts of different size owing to variable numbers of tandem repeat polymorphism.

This chapter describes in detail the protocols for carrying out Northern blots that have successfully been used in our laboratory to examine mucin gene expression both in cell cultures (HT-29 MTX) and in tissues from various mucosae (trachea, bronchus,

From: *Methods in Molecular Biology, Vol. 125: Glycoprotein Methods and Protocols: The Mucins*
Edited by: A. Corfield © Humana Press Inc., Totowa, NJ

stomach, colon, small intestine). These protocols can also be adapted to analyze other large mRNAs such as ApoB transcripts ***(1,2)***.

2. Materials

2.1. Preparation of mRNA (see Note 1)

1. Cultured cells or tissues: snap-freeze in liquid nitrogen and store in liquid nitrogen until used.
2. Homogenization buffer: 4 *M* guanidinium isothiocyanate buffer is prepared by dissolving 23.6 g of guanidinium isothiocyanate, 73.5 mg of sodium citrate, and 250 mg of sodium *N*-lauroylsarcosine in 50 mL of water, treated with 0.1% diethylpyrocarbonate (DEPC), and then autoclave. Add 2-mercaptoethanol to 100 m*M* just before use.
3. Cesium chloride 5.7 *M* EDTA cushion: Dissolve 19.2 g of cesium chloride at 60°C in 0.1 *M* EDTA, pH 7.5, treated with 0.1% DEPC to a final volume of 20 mL and then autoclaved. Keep at 4°C in dark bottles.
4. TE buffer: Dissolve 1.21g of Tris base and 0.37g of EDTA-disodium salt per liter. Adjust to pH 8.0 with HCl. Sterilize by autoclaving.
5. Chloroform:*n*-butyl alcohol (4:1).
6. 3 *M* Sodium acetate, pH 5.5: Dissolve 408.1 g of sodium acetate $3H_2O$ in 800 mL of water. Adjust pH to 5.5 with glacial acetic acid. Adjust volume to 1 L. Dispense in aliquots and sterilize by autoclaving.
7. Ethanol: 100 and 95%.

2.2. Electrophoresis

1. Gel box, castinng tray, and combs: carefully clean with 0.1 *N* NaOH overnight or for at least 3 h, rinse with distilled water (verify the pH), rinse with ethanol, and give a final rinse with sterile distilled water just before use. Use gloves in all steps. Set up the casting tray and comb in a fume hood because of toxic vapors given off during the pouring and setting of the gel (hot formaldehyde).
2. DEPC-treated water.
3. 10X Morpholino-propane-sulfonic acid (MOPS) stock solution (0.2M MOPS): Dissolve MOPS (10 g) in DEPC water (200 mL) containing 3 *M* sodium acetate (4.2 mL), and 0.5 *M* EDTA, pH 8.0 (5 mL). Adjust the pH to 7.0 with 3 *M* NaOH and the final volume to 250 mL with DEPC-treated water.
4. Gel running buffer: 0.02 *M* MOPS, pH 7.0 (1X MOPS stock solution).
5. Denaturing buffer: 50% deionized formamide, 18% deionized formaldehyde, 0.02 *M* MOPS, pH 7.0.
6. Denaturing gel: 0.9% agarose gel (13 × 18 × 0.3 cm) containing 18% formaldehyde and MOPS stock solution, pH 7.0 (final concentration : 1X).
7. Loading buffer: 0.1% xylene cyanol, 0.1% bromophenol blue, 1X MOPS, pH 7.0, solution, 50% glycerol. Make with DEPC-treated water and autoclaved glycerol.
8. Molecular weight markers (Roche Diagnostics, Meylan, France).
9. Fluorescent indicator F 254 (Merck, Darmstadt, Germany).

2.3. Transfer and Crosslinking

1. 20X Sodium chloride sodium citrate (SSC) buffer: Dissolve 175 g of NaCl and 88.2 g of trisodium citrate dihydrate per liter. Adjust to pH 7.0.
2. Hybond™ N+ membrane (Amersham).
3. Ultraviolet (UV) light source, 254 nm.

2.4. Filter Hybridization

1. Random-primed labeling kit (Boehringer Mannheim).
2. 20X SSPE buffer: Dissolve 174 g of NaCl, 27.6 g of sodium dihydrogen phosphate monohydrate, and 7.4 g of EDTA per liter. Adjust to pH 7.4 with 10 *N* NaOH.
3. 50X Denhardt's solution: 1 g of Ficoll 400, 1 g of polyvinylpyrrolidone, and 1 g of bovine serum albumin are dissolved in 100 mL of H_2O. Sterilize by filtration.
4. Salmon sperm DNA (Boehringer Mannheim).
5. Hybridization solution: 50% formamide, 5X SSPE, 10X Denhardt's solution, 2% (w/v) sodium dodecyl sulfate (SDS), and 100 mg/mL of sheared salmon sperm DNA.
6. Kodak X-Omat film (Kodak, Rochester, NY).

3. Methods

3.1. Preparation of mRNA (see Notes 1–3)

The original guanidinium isothiocyanate method is recognized for the purity and quality of the RNA obtained. Guanidinium isothiocyanate combines the strong denaturing characteristic of guanidine with the chaotropic action of isothiocyanate and efficiently solubilizes tissue homogenates. Effective disruption of cells can be obtained without the use of a homogenizer, which has otherwise been used routinely, especially when the starting material comes from tissues. In the specific case of large mRNAs, great care must be taken to prevent all risks of mechanical degradation. For comparison, RNA was also isolated in our laboratory by other methods using guanidinium isothiocyanate-phenol/chloroform *(3)*, lithium chloride-urea *(4)*, or different optimized commercial total RNA preparation kits from Bioprobe (Montreuilsous Bois, France) or Clontech Inc. (Palo Alto, CA). The best protocol to prepare intact large RNAs was the following improved method that we developed, derived from the guanidinium isothiocyanate protocol *(5)*:

1. Grind cells (1.5×10^6) or tissues (optimal weight of 1 g) to a fine powder in a mortar and pestle in liquid nitrogen and mix with 10 mL of homogenization buffer (**Subheading 2.1., item 2**) still in liquid nitrogen.
2. Then allow the homogenate to thaw gradually at room temperature, during which time the guanidinium isothiocyanate and 2-mercaptoethanol efficiently solubilize the cell or tissue mixture.
3. Gently transfer the homogenate obtained onto 3.2 mL of 5.7 *M* cesium chloride cushion (*see* **Subheading 2.1., item 3**). Ultracentrifugation is performed for 16 h at 29,500 rpm in a Beckman SW41 rotor. Remove the supernatant, cut off the bottom of the tube, and carefully resuspend the white pellet of total RNA in 2 × 1 mL TE buffer, pH 8.0 (*see* **Subheading 2.1., item 4**), 0.1% SDS by using wide-mouth pipets, carefully avoiding all shear forces.
4. Remove chloride cesium salts from the pellet by two washings with 2 vol of chloroform/*n*-butyl alcohol (4:1) mixture. This purification of the pellet of RNA makes it easier to dissolve. During the washing steps, the tubes must be mixed only by gentle inversion, and vortexing is strictly avoided.
5. Carefully remove the top aqueous phase, which contains the RNA, with wide mouth pipettes and transfer it to a fresh tube. Precipitated the RNA by adding 0.1 vol of 3 *M* sodium acetate, pH 5.5, and 2.5 vol of ethanol, at –80°C for 15 h. Centrifuge the precipitate of RNA at 10,000*g* for 30 min at 4°C, wash it with ice-cold 95% ethanol, and then 100% ethanol, centrifuge it again at 10,000*g* for 30 min at 4°C, and let it air-dry.

6. Redissolve the RNA pellet carefully in DEPC-treated water and quantify by measuring the A_{260nm} of an aliquot.
7. Store the final preparation at –80°C until it is needed.

3.2. Electrophoresis

Studying the isolation of poly(A+)RNA by using standard protocols or more recent systems (Poly [A] tract®RNA Isolation System from Promega, Charbonnieres, France), we concluded that selection of poly(A+) is not recommended in the case of large mRNAs because of a very poor yield and additional risks of mechanical damage *(1)*. Thus, electrophoresis must be performed starting from total RNA. Moreover, in the case of large mRNAs, no risk of misinterpretation of the data can be expected from the presence of ribosomal bands.

The methods for electrophoresis of RNA have been described in many books. The protocol below represents a modified version of the standard technique described in **ref.** ***6***, and only the modifications introduced to fractionate mucin RNAs are described in detail.

1. Prepare the RNA samples: The optimal quantity is 10 µg (2 µL or less). Adjust to a final volume of 10 µL with deionized formamide (5 µL), deionized formaldehyde (1.78 µL), 10x MOPS stock solution pH 7.0 (1 µL), and DEPC-treated water.
2. Heat shock the samples to denature the RNA at 68°C for 10 min in a water bath and cool on ice. Add 3 µL of loading buffer (*see* **Subheading 2.2., item 7**) and load the gel immediately.
3. Run the gel (*see* **Subheading 2.2., item 6**) for 16 h at 30 V.
4. Stop electrophoresis and cut off the molecular weight markers, and RNA control lanes. The different bands (markers) or ribosomal bands (control) appear as shadows when put onto a silica gel plate containing a fluorescent indicator F 254 (Merck) when exposed to UV illumination.

3.3. Transfer and Crosslinking (see Note 4)

1. Prior to transfer, soak the gel in 0.05 *N* NaOH with gentle shaking. Obtain the optimal signal after treatment for 20 min (for a 3-mm thick gel).
2. Then rinse the gel in DEPC-treated water and soak for 45 min in 20X SSC (*see* **Subheading 2.3., item 1**).
3. Use capillary transfer in 20X SSC to transfer the RNA from the gel in a standard manner *(7)* or via vacuum blotting for 1 h.
4. The UV-crosslinking method proposed is based on tests designed to optimize the permanent binding of RNA to Hybond N+ membrane: bake at 80°C in a vacuum for 30 min and then expose to 254 nm of UV light for 4 min. The filter is now ready for hybridization.

3.4. Filter Hybridization

A large variety of hybridization buffers are available and can be used with equal success in the filter hybridization. In this method, all the probes used MUC1 *(8)*, MUC2 *(9)*, MUC3 *(10)*, MUC4 *(11)*, MUC5AC *(12)*, MUC5B *(13)*, and MUC6 *(14)*, and the apoB-100 probes *(15–17)* are labeled with [^{32}P] dCTP using a commercial random-primed labeling kit according to the manufacturer's protocol (*see* **Subheading 2.4., item 1**). These probes are used at 1×10^6 cpm/mL and 10^6 cpm per lane.

1. Preform prehybridization and hybridization in 10 mL of hybridization solution (*see* **Subheading 2.4., item 5**) for 2 and 16 h, respectively, at 42°C in a hybridization oven.

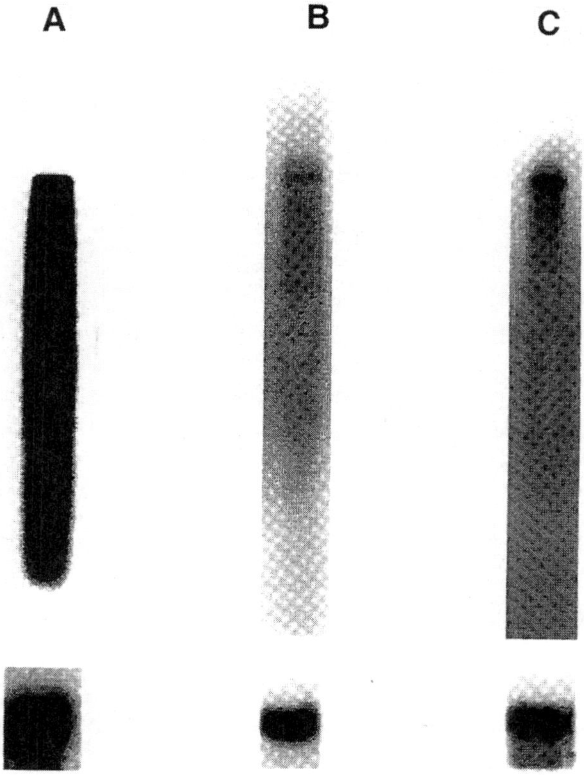

Fig. 1. Efficiency of this improved protocol for large RNA isolation. Total RNA from the same human colon mucosa specimen was isolated by the original guanidinium isothiocyanate-ultracentrifugation protocol **(A)** or by this improved method (**B** and **C**). In (A) a large smear from up to 20 kb to about 0.5 kb is detected by MUC2 probe while in (B) and (C) a discrete unique band is obtained. In (C) (compared to B) the efficiency of the transfer was increased at least ten fold by using a pre-treatment of the gel with 0.05 *N* NaOH.

2. Remove the filter and rinse with 50 mL of 2X SSPE (*see* **Subheading 2.4., item 2**) at room temperature to remove most of the nonhybridized probe.
3. Wash the membranes twice in 0.1X SSPE and 0.1% SDS buffer for 15 min at 65°C.
4. After a final wash with 6X SSPE, at room temperature, wrap the membrane in plastic film while moist. Expose to autoradiography at –80°C with an intensifying screen and Kodak X-Omat film (**Fig. 1**).
5. After analysis of the results, strip the filters by two washings in a 0.1% SDS boiling solution for 15 min (this can be repeated if necessary, testing for remaining label by autoradiography).

3.5. Estimation of the Sizes of Large Mucin mRNAs

The use of standard RNA molecular weight markers or total RNA controls (28S and 18S ribosomal bands) is useful to evaluate the quality of electrophoretic migration but is not adapted to determine large sizes (nonlinear curves). Moreover the demonstration of the integrity of the RNA preparation using a probe such as β-actin, GAPDH, or 28S rRNA is misleading in the case of large RNAs because these probes hybridize to short messages, for which the degradation problem is not encountered using usual protocols of RNA preparation. Using the improved protocols presented here, we found that each of the MUC2-6 genes expresses mRNAs of much larger size than usually found in eukaryotes (14–24 kb). Moreover, allelic variations in length of these mucin transcripts were observed, directly related to the variable number of tandem repeat polymorphisms seen at the DNA level ***(1)***. So it is of great interest, since mucin gene expression and polymorphism is implicated in the increase of susceptibility to any pathology, to estimate the size of the mucin transcripts. Because the largest transcripts found in standard RNA molecular size markers are not larger than 10 kb, we use apoB-100 ***(15–17)*** and MUC5B (Laine, A., unpublished results) probes, which detect large unique transcripts of 14.1 kb (small intestine-colon) and 17.6 kb (bronchus, trachea, gallbladder, submaxillary glands), respectively ***(1,18)***.

The standard curve is derived by using β-actin (*see* **Fig. 2**) (2 kb, point A), 28S rRNA (5 kb, point B) apoB-100 (14.1 kb, point C) ,and MUC5B (17.6 kb, point D) as standards. The curve between points A and C is approximately a straight line. The curve between points A and D is a nonlinear curve of the following formula: $Y = 31.05 \exp(-0.273\ X)$, where Y represents size in kb and X represents distance migration in centimeters. A simple method of drawing this nonlinear curve consists in joining points A, B, and C (first straight line) and points C and D (second straight line). Using this method, we found that the largest transcrpts of human mucin were from MUC4 (24 kb). Their size is deduced from the second straight line (point E). In the future, once it has been precisely sized after complete sequencing, it will also be possible to use the largest MUC4 allele, which is common, as an additional size standard for a better size estimation of mRNAs encoding mucins and other large RNAs.

4. Notes

1. Methods for isolation and analysis of large RNAs require the same precautions as for all other RNAs and involve using of pure high-grade analytical reagents and taking care to avoid accidental introduction of RNAses. Standard precautions can be used to avoid problems with RNAses such as wearing gloves at all times, autoclaving buffers and deionized or distilled water and decontaminating general laboratory glassware or plasticware with 0,1% DEPC *(6,7)*.
2. Extreme care must be taken at each step during the experiments to prevent any risk of mechanical degradation. Homogenization, the use of syringes, and vortexing are strictly avoided.
3. After centrifugation, if the RNA pellet is difficult to dissolve, warm it to 45°C for 20 min.
4. Soaking the gel in NaOH before transfer is an important step that must not be omitted and must be optimized according to the gel thickness. This treatment partially hydrolyzes the RNA and improves (at least 10-fold) the efficiency of transfer of very large RNA species.

Acknowledgments

Special thanks to Dr. Virginie Debailleul for her contribution to the development of the method presented herein. The authors also thank Dr. Dallas Swallow for her con-

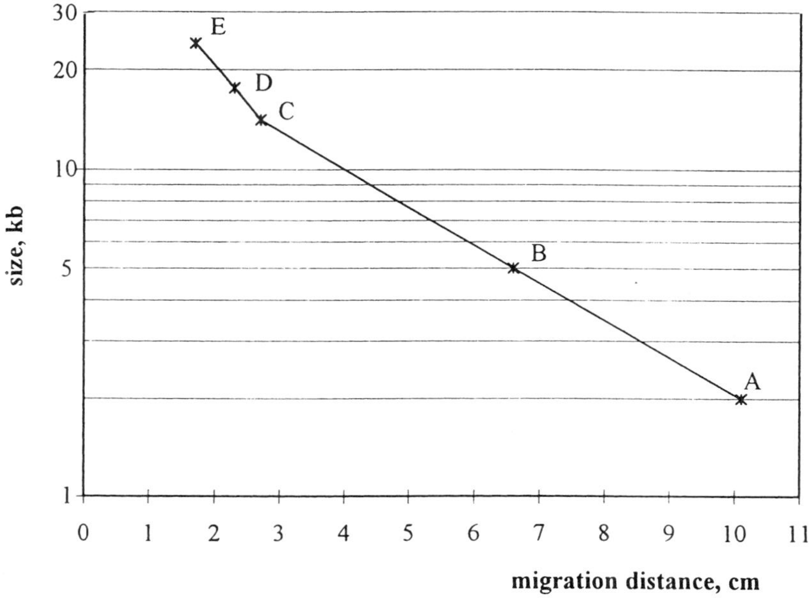

Fig. 2. Estimation of the size of large mRNA species. The use of standard RNA molecular size markers is not adapted to determine large sizes. We therefore recommend use of apoB-100 and MUC5B for analysis of mucin messages. The standard curve is derived by using 4 points: β-actin (2 kb **point A**), 285 rRNA (5 kb, **point B**), apo-100 (14.1 kb, **point C**) and MUC5B (17.6 kb, **point D**). The best formula corresponding to the nonlinear curve joining these four points is: Y = 31.05 exp (–0.273x) where *Y* represents size in kb and X represents distance migration in centimeter. A simple method of drawing this curve is shown in this graph: a first straight lane is obtained joining **points A**, **B**, and **C**. A second straight is obtained by joining **points C** and **D**. An additional **point E** represents the size of the largest MUC4 mRNA message which is deduced from this second straight lane.

stant encouragement and assistance in preparing the manuscript. Support was received from l'Association de Recherche sur le Cancer, the Comité du Nord de la Ligue Nationale sur le Cancer and the CH et U de Lille (n°96/29/9595).

References

1. Debailleul, V., Laine, A., Huet, G., Mathon, P., Collyn d'Hooghe, M., Aubert, J.P. and Porchet, N. (1998) Human mucin genes MUC2, MUC3, MUC4, MUC5AC, MUC5B and MUC6 express stable and extremely large mRNAs and exhibit a variable length polymorphism: an improved method to analyse large mRNAs. *J. Biol. Chem.* **273**, 881–890.
2. Debailleul, V. (1997) Expression des gènes de mucines humaines: étude de la stabilité, de l'hétérogénéité et du polymorphisme des ARNm. Thèse de Doctorat d'Université des Sciences de la Vie et de la Santé, Lille, France.
3. Chomczynski, P. and Sacchi, N. (1987) Single-step method of RNA isolation by acid guanidinium thiocyanate-phenol-chloroform extraction. *Anal. Biochem.* **162**, 156–159.
4 Auffray, C. and Rougeon, F. (1980) Purification of mouse immunoglobulin heavy chain RNAs from myeloma tumor RNA. *Eur. J. Biochem.* **107**, 303–314.

5. Chirgwin, J. M., Przybyla, A. E., McDonald, R. J., and Rutter, W. J. (1979) Isolation of biologically active ribonucleic acid from sources enriched in ribonuclease. *Biochemistry* **18**, 5294–5299.
6. Sambrook, J., Fritsch, E. F., and Maniatis, T. (1989) *Molecular cloning: A Laboratory Manual* (Cold Spring Harbor Laboratory Press, Cold Spring Harbor, New York) pp. 7.3–7.52.
7. Krumlauf, R. (1991) Northern blot analysis of gene expression, in *Methods in Molecular Biology*, vol. 7: *Gene Transfer and Expression Protocols* (Murray, E. J., ed.), Humana, Clifton, NJ.
8. Gendler, S. J., Lancaster, C. A., Taylor-Papadimitriou, J., Duhig, T., Peat, N., Burchell, J., Pemberton, L., Lalani, E., and Wilson, D. (1990) Molecular cloning and expression of a human tumor-associated polymorphic epithelial mucin. *J. Biol. Chem.* **265**, 15,286–15,293.
9. Gum, J. R., Byrd, J. C., Hicks, J. W., Toribara, N. W., Lamport, D. T. A., and Kim, Y. S. (1989) Molecular cloning of human intestinal mucin cDNA. Sequence analysis and evidence for genetic polymorphism. *J. Biol. Chem.* **264**, 6480–6487.
10. Gum, J. R., Hicks, J. W., Swallow, D. M., Lagace, R. L., Byrd, J. C., Lamport, D. T. A., Siddiki, B. and Kim, Y. S. (1990) Molecular cloning of cDNA derived from a novel human intestinal mucin gene. *Biochem. Biophys. Res. Commun.* **171**, 407–415.
11. Porchet, N., Nguyen, V.C., Dufossé, J., Audié, J. P., Guyonnet-Dupérat, V., Gross, M. S., Denis, C., Degand, P., Beinheim, A., and Aubert J. P. (1991) Molecular cloning and chromosomal localization of a novel human tracheo-bronchial mucin cDNA containing tandemly repeated sequences of 48 base pairs. *Biochem. Biophys. Res. Commun.* **175,** 414–422.
12. Guyonnet-Dupérat,V., Audié, J. P., Debailleul, V., Laine, A., Buisine, M. P., Galiègue-Zouitina, S., Pigny, P. Degand, P., Aubert, J. P., and Porchet, N. (1995) Characterization of the human mucin gene MUC5AC: a consensus cysteine-rich domain for 11p15 mucin genes? *Biochem. J.* **305**, 211–219.
13. Dufossé, J., Porchet, N., Audié, J. P., Guyonnet-Dupérat, V., Laine, A., Van Seuningen, I., Marrakchi, S., Degand, P., and Aubert, J. P. (1993) Degenerate 87 base pair tandem repeats create hydrophilic/hydrophobic alternating domains in human mucin peptides mapped to 11p15. *Biochem. J.* **293**, 329–337.
14. Toribara, N. W., Robertson, A. M., Ho, S. B., Kuo, W. L., Gum, E., Hicks, J. W., Gum, J. R., Byrd, J. C., Siddiki, B., and Kim, Y. S. (1993) Human gastric mucin. Identification of a unique species by expression cloning *J. Biol. Chem.* **268**, 5879–5885.
15. Chen, S. H., Yang, C. Y., Chen, P. F., Setzer, D., Tanimura, M., Li, W. H., Gotto, A. M., and Chan, L. (1986) The complete cDNA and amino acid sequence of human apoliprotein B-100. *J. Biol. Chem.* **261**, 12,918–12,921.
16. Law, S. W., Grant, S. M., Higuhi, K., Hospattankar, A., Lackner, K., Lee, N., and Brewer H. B., Jr (1986) Human liver apolipoprotein B-100 cDNA complete nucleic acid and derived amino acid sequence. *Proc. Natl. Acad. Sci. USA* **83**, 8142–8146.
17. Cladaras, C., Hadzopuolou-Cladaras, M., Nolte, R. T., Atkinson, D., and Zannis V.J. (1986) The complete sequence and structural analysis of human apolipoprotein B-100 relationship between apo B-100 and apo B-48 forms. *EMBO J.* **5**, 3495–3507.
18. Vinall, L. E., Hill, A. S., Pigny, P., Gum, J. R., Kim, Y. S., Porchet, N., Aubert, J. P., and Swallow, D. M. (1998) Variable number tandem repeat polymorphism of the mucin genes located in the complex on 11p15.5. *Hum. Genet.* **102,** 357–366.

26

Southern Blot Analysis of Large DNA Fragments

Nicole Porchet and Jean-Pierre Aubert

1. Introduction

Pulsed-field gel electrophoresis (PFGE) has been used successfully to generate physical maps of a large region from many genomes. In addition, PFGE is useful for determining the order of genes or markers more precisely than is possible with genetic linkage analysis. Since a book in the Methods in Molecular Biology series *(1)* has already been devoted to this subject, the aim of this chapter is to give protocols that were successfully used in our laboratory for the human mucin genes. Whenever possible, we refer to the relevant chapters of this book or other references in which strategies or techniques are discussed in detail.

Some types of PFGE, including contour-clamped homogeneous electric field (CHEF) give excellent separations of a wide range of DNA fragments in straight lanes *(2)*. The CHEF technique utilizes a hexagonal electrode array surrounding the gel. The array comprises two sets of driving electrodes oriented 120° apart. An electric potential is periodically applied across each set for equal time intervals (the pulse time). The DNA fragments reorient with each change in the electric field and zigzag through the gel, but the net direction is perfectly straight. This technique allows precise comparison of the sizes of several fragments analyzed on the same gel. The results obtained depend on several factors including the electric field strength, the temperature, the agarose composition and concentration, the pulse time and the angle between alternating electric fields. The results obtained with a given set of conditions can also be affected by the particular apparatus used *(3)*. We used a noncommercially built hexagonal CHEF device *(4)*.

We used two windows of resolution in CHEF electrophoresis. Molecular size resolution optimal between 50 and 800 kb allowed us to study the organization of *MUC2*, *MUC5AC*, *MUC5B*, and *MUC6* genes within a complex of genes mapped to 11p15.5 *(5)* and to construct a detailed physical map of the *MUC* cluster. Molecular size resolution optimal between 400 and 3000 kb was useful to integrate and orientate this map in the general physical map of 11p15.5 including HRAS, D11S150, and IFG2 reference markers.

From: *Methods in Molecular Biology, Vol. 125: Glycoprotein Methods and Protocols: The Mucins*
Edited by: A. Corfield © Humana Press Inc., Totowa, NJ

Even when the chromosomal localization of a gene of interest is already known, it is not a simple matter to locate it precisely. Fortunately, in mammals at least, evolution has selected unusual (G+C)-rich sequences at the 5' ends of most genes. Usually these sequences, named CpG islands, are nonmethylated whereas the remainder of the genome is heavily methylated at CpG. Thus, these CpG sequences, which are gene markers, can be located using certain types of rare cutting restriction endonucleases, thereby facilitating the mapping of genes. These enzymes have two important properties: first, they recognize one or two CpGs in their restriction sites and second, cleavage is blocked by methylation.

When the source of DNA is cultured cell lines that have been established for a long time, a variable degree of methylation may be expected even at CpG islands of nonessential genes *(6)*. It is therefore generally useful to choose multiple sources of genomic DNA for each study. In each DNA sample, a variable pattern of methylation of genes occurs, and different and complex sets of fragments can be hybridized to the probes resulting from incomplete cleavages by rare cutters. The study of these partially digested fragments is very useful for construction of long-range maps because they may allow some islands to be bridged, but they may also fail to detect other CpG islands. The choice of biological starting material is also dictated by the availability of suitable sources of DNA: fresh blood (circulating lymphocytes), lymphoblastoid or fibroblast cell lines.

2. Materials

In our study we used three lymphoblastoid cell lines, including the Karpas 422 cell line, one erythroblastoid cell line K 562 (CCL243), one breast epithelial cell line HBL 100 (HTB124) and normal human circulating lymphocytes from one individual. Karyotype and characteristics of each source of DNA are described in **ref. *5***. Some cell lines were cultured either in the absence or in the presence of 5-azacytidine as methylation inhibitor.

1. RPMI 1640 medium (Gibco-BRL).
2. J Prep medium (J. Bio, Les Ullis, France).
3. Low melting point agarose (Bio-Rad).
4. Lysis buffer: 0.5 *M* EDTA, pH 8.0, 1% sarcosyl, and 100 µg /mL proteinase K.
5. TE buffer: 10 m*M* Tris-HCl, pH 7.0, 1 m*M* EDTA.
6. Phenylmethylsulfonylfluoride (PMSF) (Sigma).
7. Restriction enzymes: *Asc*I, *Sac*I, *Ksp*I, *Pac*I (Biolabs); *Not*I, *Bss*HII, *Nar*I, *Mlu*I, *Nru*I, *Swa*I, *Spe*I, *Ssp*I, *Cla*I, *Pvu*II (Boehringer Mannheim).
8. Spermidine (Sigma).
9. 10X TBE stock solution: 890 m*M* Tris-HCl, pH 8.3, 890 m*M* boric acid, and 2.5 m*M* EDTA.
10. CHEF apparatus (noncommercial apparatus *[4]*).
11. Size markers: λ4-phage concatemers and chromosomes of *S. cerevisiae* (225-2200 kb and chromosomes from *Saccharomyces pombe* (3.5, 4.6, and 5.7 Mb) (Bio-Rad).
12. Ethidium bromide (Sigma).
13. Neutralizing buffer; 0.5 *M* Tris-HCl, pH 7.5, 3 *M* NaCl.
14. 20X sodium dodecyl sulfate (SSC) buffer: Dissolve 175 g of NaCl and 82.2 g of trisodiumcitrate dihydrate per liter. Adjust to pH 7.0.

15. Hybond™ N+ membrane (Amersham).
16. 5X Denhardt's solution (Appligene).
17. Dextran sulfate (Pharmacia).
18. Sheared herring sperm DNA (Sigma).

3. Methods

3.1. Preparation of DNA

To generate intact restriction fragments of up to several megabases from mammalian genomes, the DNA must be protected from shearing forces during its preparation. Whole intact cells are thus embedded in a solid matrix of agarose gel prior to DNA extraction and enzyme cleavage *(7,8)*.

3.1.1. Human Blood

1. Dilute 20 mL of fresh blood collected in citrate tubes with RPMI 1640 medium (v:v) and separate lymphocytes on J Prep medium and wash with phosphate-buffered saline (PBS) buffer.

3.1.2. Cultured Cells

1. Harvest cells from culture.
2. Suspend cells three times in 1X PBS at 37°C to be washed.
3. Pellet the cells by centrifugation at 3000 rpm for 3 min, and resuspend at a concentration of 3.5×10^7/mL in 1X PBS at 4°C.

3.1.3. Embedding Cells in Agarose Blocks

1. Dilute the cell suspension with an equal volume of 1% low melting point agarose dissolved in PBS and held at 50°C.
2. Mix by gentle inversion: do not allow bubbles to form.
3. Dispense into plastic molds that have the same dimensions as the gel comb (80 µL, approx 10^6 cells or 10 µg of DNA).
4. Leave the agarose blocks to set on ice for 30 min.
5. Incubate 20 agarose blocks in two changes of 5 mL of lysis buffer in a sterile plastic tube at 50°C for 24 h.
6. Decant the agarose plugs into sterile tubes and wash once with 5 mL of 1X TE buffer (10 m*M* Tris-HCl, pH 7.0, 1 m*M* EDTA) and twice with 5 mL of 1X TE buffercontaining 0.04 mg/mL of PMSF dissolved in isopropanol, at 50°C for 30 min. The DNA is ready to be digested with restriction enzymes (*see* **Note 1**).

3.2. Restriction Enzyme Digestion (see Notes 2 and 3)

3.2.1. Restriction Enzyme Digestion of Plugs

1. For restriction enzymes, choose enzymes according to the sites they recognize:
   a. (G+C) rich sites included in CpG islands: *Not*I, *Asc*I (Group I); *Bss*HII, *Sac*I, *Ksp*I (Group II);
   b. those independant of the presence of CpG islands: *Nar*I (Group III), *Mlu*I, *Nru*I (Group IV), or
   c. (A+T)-rich sites: *Swa*I, *Spe*I, *Ssp*I, *Pac*I, *Cla*I, *Pwe*I.
2. For enzyme digestion, use 25 U of enzyme at 37°C except for *Bss*HII (50°C); add BSA to enzyme buffers (10 µL pf a 1 mg/mL solution) when *Mlu*I, *Not*I, or *Nru*I is used; add

spermidine (5 μL of a 100 m*M* spermidine solution) when digestion is performed with *Ksp*I, *Not*I, *Nru*I.

3. Incubate each agarose block just before digestion in 5 mL 1X TE buffer, pH 7.6, for 20 min. Repeat this procedure twice to eliminate EDTA.
4. Perform each digestion on one half block (40 μL) containing 5 μg of DNA. Transfer the plugs to individual Eppendorf tubes and add 400 μL of appropriate 1X restriction buffer for 30 min at 4°C.
5. Replace the buffer with fresh buffer including BSA or spermidine as specified, the final volume being 100μl.
6. Add enzyme and incubate for 4 h.
7. For complete digestion, add again 25 U of enzyme and incubate overnight.
8. Obtain partial enzyme digests by including variable $MgCl_2$ concentrations (from 0.3 to 10 m*M*) in the digestion buffer and/or using variable amounts of restriction enzyme (from 0.5 to 25 U).
9. Stop digestion by washing three times in agarose half-blocks in 5 mL of cold 1X TE buffer, pH7.6, for 1 h at 4°C prior to loading into the gel. Test each DNA preparation for absence of nuclease activity by incubating a block in standard conditions without restriction enzyme.

3.2.2. Pulsed-Field Gel Electrophoresis

1. Prepare an agarose gel at the required concentration (typically low melting point agarose 1%) in 0.25X TBE buffer (starting from 10X TBE stock solution).
2. Equilibrate blocks in running buffer and push the blocks into the slots in the gel. Seal each slot with low melting point agarose at an agarose concentration equivalent to that of the running gel.
3. Run the gel in a CHEF apparatus at a constant temperature of 12°C (use a recirculating pump) ***(10)***.
4. Program the switching device and constant voltage power supply.
   a. Molecular size resolution optimal between 50 and 800 kb: pulse time of 50 s for 16 h, 30 s for 8 h, and then 80 s for 16 h, constant voltage 190 V.
   b. Molecular size resolution optimal between 700 and 3000 kb: pulse time of 30 s for 8 d and then 5 min for 2 d, constant voltage 80 V.
5. Size markers: λ4 phage concatemers and chromosomes of *S. cerevisiae* (225–2200 kb) for conditions in **step 4a** (50–800 kb); these and chromosomes from *S. pombe* (3.5, 4.6, and 5.7 Mb) for conditions in **step 4b** (700–3000 kb) ***(10)***.

3.2.3. Southern Blotting of PFGE DNA (see **Note 4**)

1. After electrophoresis, stain the gel for 15 min in 4 μg/mL of ethidium bromide (*see* **Note 4**) with constant shaking, destain in water for 40 min, and photograph using a transilluminator with fluorescent rulers.
2. Because large DNA fragments are not efficiently transferred onto membranes, DNA fragments separated by PFGE must be cleaved by depurination before Southern blotting ***(10)***. Perform depurination by putting the gel in 500 mL of 0.25 *N* HCl for 15 min at room temperature. Denature the DNA by putting the gel twice in 1.5 *M* of NaCl and 0.5 *N* of NaOH for 30 min.
3. Then, neutralize the gel by two 30-min treatments with 0.5 *M* Tris-HCl, pH 7.5 and 3 *M* NaCl at room temperature for 30 min.
4. Transfer the DNA by capillary blotting using 20 X SSC as transfer solution for at least 24 h, or alternately, for 4 h under vacuum. In our study, Hybond N^+ membrane (Amersham) was used.

5. Carefully remove the blotting papers. Mark the location of the wells and the orientation of the membrane. Rinse briefly in 2X SSC. Dry the membrane on an absorbent paper (Whatman 3MM). Fix the DNA onto the membrane by baking in an oven at 80°C for 10 min under vacuum and then ultraviolet (UV) crosslinking with UV light for 124 mJ in a UV oven.

3.2.4. Radioactive Probing of PFGE Blots

Radiolabeling is the preferred method of probing because the detection sensitivity of PFGE blots is lower than that of conventional Southern blots. In our study, human mucin probes used corresponded to cDNA probes from tandem repeats ***(11–17)***.

1. Perform prehybridization at 65°C for at least 2 h in 6X SSC, 5X Denhardt's, and 0.5% sodium dodecyl sulfate (SDS).
2. Then hybridize the membranes with the same buffer in which 10% dextran sulfate and 500 µg/mL of sheared herring sperm DNA are added, at 65°C overnight. Use probes at 3×10^6 cpm/mL and 2.5×10^6 cpm/lane.
3. Wash membranes twice at 65°C in 0.1 X SSC, 0.1% SDS for 15 min.
4. Perform autoradiography at –80°C for 24 h to 2 wk (several days in the case of repetitive mucin probes, 1 or 2 wk in the case of other probes corresponding to unique sequences).
5. To remove the probe, wash the filter twice in a 0.1% SDS boiling solution, rinse in water at room temperature, and check for probe removal by autoradiography.

3.2.5. Data Interpretation

Construction of long-range restriction maps involves sequential hybridization of each blot with probes from different genes or markers to assess whether any of the probes recognize the same DNA fragments ***(18)***. To do this accurately, all the autoradiographs must be perfectly aligned with each other, and therefore the precise position of each lane must be marked starting from the corresponding well.

The fact that certain probes cohybridize with one DNA fragment suggests that the genes or markers they recognize are physically linked. However, it is necessary to establish that a physical linkage exists between markers, and that the markers do not recognize distinct DNA fragments of similar size that comigrate. Confirmation can be easily made by the fact that a great number of different pieces of information are available from several PGFE blots:

1. estimation of the size of the hybridizing fragments,
2. existence of fragments corecognized by several markers,
3. similarities or differences observed comparing complete/partial or single/combined digestions,
4. analysis of several sources of DNA,
5. identification of CpG islands. (CpG islands surrounding a marker are identified when, whatever the combination of [G+C]-rich site rare cutting enzyme is used, the same fragments are constantly detected.)

The results obtained from PFGE blots are generally more difficult to interpret than those obtained from conventional Southern blots. The hybridization patterns are usually complex because the majority of enzymes used are methylation sensitive, giving rise to incomplete digestion. Thus, the bands are rarely seen as sharp and discrete bands, and, in most cases, the most informative fragments are not seen as the major bands.

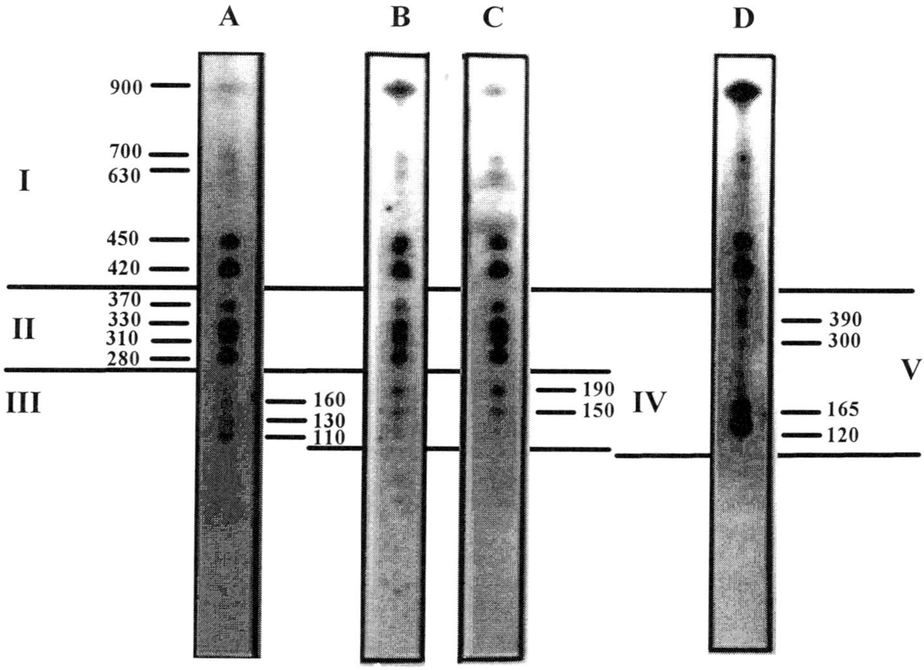

Fig. 1. Organization of the MUC cluster located on 11p15.5. CHEF analysis of DNA fragments cleaved by *Mlu*I from this source of DNA (170 lymphoblastoid cell line) was useful to determine the restriction map of the *MUC* cluster and the relative location of the four genes: (**probe A**) *MUC6*, (**probe B**) *MUC2*, (**probe C**) *MUC5AC*, and (**probe D**) *MUC5B*. *Mlu*I recognizes variably methylated (G+C)-rich sites which are independent of the presence of CpG islands. Enzymes belonging to this group of rare cutting enzymes generate numerous fragments useful to join the markers. In this figure, fragments from 900 kb to 420 kb (I) are common to the four *MUC* genes, indicating that the *MUC* cluster spans between 420 and 370 kb. Fragments II are common to *MUC6*, *MUC2*, and *MUC5AC*, but not to *MUC5B* (specific fragments V). This indicates that *MUC5B* is situated at one end of the *MUC* cluster. The presence of fragments specific to *MUC6* (III) is helpful in locating *MUC6* to the other end of the cluster. The fact that similar sets of short fragments (IV) were obtained with *MUC2* and *MUC5AC* probes indicates that these two genes are adjacent and situated in the central part of the *MUC* cluster.

From the combined data, a long-range restriction map should be constructed based on the most informative fragments. In our work, studying the organization of *MUC* genes in a cluster of mucin genes mapped to 11p15.5, analysis of digestions performed with (A+T)-rich site rare cutting enzymes (*Swa*I, *Pac*I, *Cla*I) was very useful to initiate the construction of the map.

4. Notes

1. The washes once with 5 mL of 1X TE buffer (10 m*M* Tris-HCl, pH 7.0, 1 m*M* of EDTA) and twice with 5 mL of 1X TE buffer containing 0.04 mg/mL of PMSF serve to remove sarcosyl and to inhibit proteinase K. The DNA is ready to be digested with restriction enzymes or transferred to 0.5 *M* EDTA, pH 8.0, for storage at 4°C (several years).

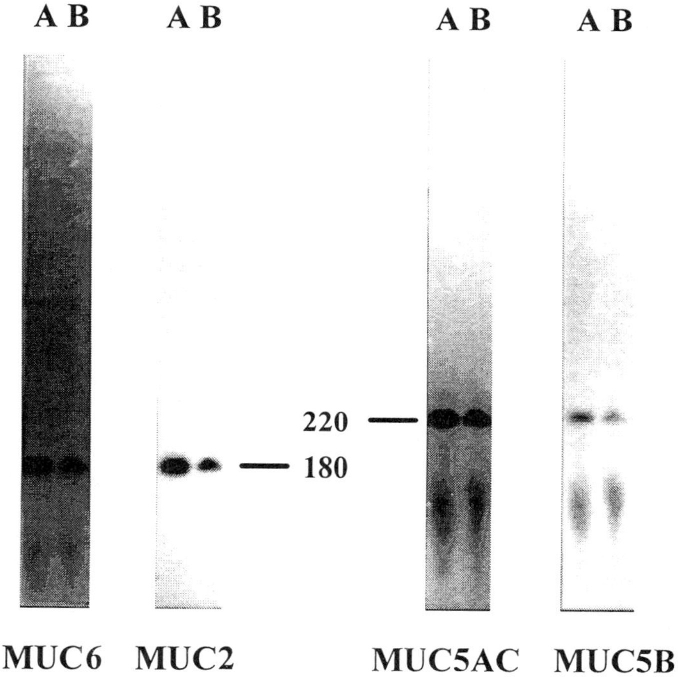

Fig. 2. Organization of the *MUC* cluster located on 11p15.5. CHEF analysis of DNA fragments cleaved by *Swa*I from two cell lines, one erythroblastoid cell line (K 562, lane A) and one lymphoblastoid cell line (170, lane B). *Swa*I recognizes (A+T)-rich sites, and cleavage by this enzyme is not affected by methylation of DNA. Simple PFGE patterns were obtained that clearly indicate that *MUC6* and *MUC2* are adjacent and situated on a common 180-kb fragment whereas *MUC5AC* and *MUC5B* are separated from these two genes and are situated on another common fragment that is 220 kb in length.

2. All the restriction enzymes used for PFGE analysis produce large DNA fragments because they cut at recognition sites that occur rarely in mammalian genome. Two classes of enzyme are available: (1) enzymes recognizing (G+C)-rich sequences found in CpG islands and enzymes recognizing rare sites (eight or more nucleotides), independent of the presence of CpG islands, and (2) those recognizing (A+T)-rich sites ***(19)***. The first class of enzymes usually generates complex patterns of fragments revealed by the probes (**Fig. 1**) whereas the second class of enzymes are not affected by methylation and produces simpler patterns (**Fig. 2**). Enzymes must be chosen from both of these two classes. The first class of enzymes is useful to construct the map around each marker or gene studied by the probe. The second class of enzymes allows production of continuous fragments that join several genes. The first class of enzymes is divided into several subclasses depending on the length and G+C content of the site, and the specificity to CpG islands ***(19)***. In our study, we used the following panel of enzymes:
   a. *Nru*I was useful to obtain very large DNA fragments.
   b. *Not*I was useful for obtaining many large or medium partial fragments.
   c. *Cla*I, *Mlu*I, and *Bss*HII allowed us to obtain medium-size complex patterns of bands.

   d. *Pac*I and *Swa*I gave simple patterns of medium-size fragments useful for determining bridges in parts of the map, when used alone or in combined digestions.
3. A double restriction enzyme digestion procedure may be done as two consecutive steps of complete digestion, each step in the appropriate buffer.
4. Ethidium bromide is not added to DNA before electrophoresis because it is thought to modify the migration of DNA.

Acknowledgments

The methods described here were developed with the help of Veronique Guyonnet-Duperat *(20)* and Pascal Pigny *(21)*. The authors thank Dr. Alexander S. Hill, Dr. Wendy S. Pratt, and Dr. Dallas Swallow (The Galton Laboratory MRC London) for further discussions, encouragement, and support and Dr. J. P. Kerckaert and S. Galiègue-Zouitina (U.124 INSERM, Lille) for the help and use of their PFGE apparatus. Support was received from l'Association de Recherche sur le Cancer, the Comité du Nord de la Ligue Nationale Contre le Cancer.

References

1. Burmeister, M. (1992) Strategies for mapping large regions of mammalian genomes, in *Pulsed-Field Gel Electrophoresis* : *Protocols, Methods and Theories, Methods in Molecular Biology*, vol. 12 (Burmeister, M., and Vlanovsky, L., eds.), Humana, Totowa, NJ pp. 259–284.
2. Chu, G., Vollrath, D., and Davis, R. W. (1986) Separation of large DNA molecules by coutour clamped homogeneous electric fields. *Science* **234**, 1582–1585.
3. Vollrath, D. (1992) Resolving multimegabase DNA molecules using contour-clamped homogeneous electric fields (CHEF), in *Pulsed-Field Gel Electrophoresis, Methods in Molecular Biology*, vol. 12 (Burmeister, M., and Vlanovsky, L., eds.), Humana , Totowa, NJ, pp. 19–30.
4. Galiègue-Zouitina, S., Collyn-d'Hooghe, M., Denis, C., Mainardi, À., Hildebrand, M. P., Tilly, H., Bastard, C., and Kerckaert, J. P. (1994) Molecular cloning of a t(11;14)(q13;q32) translocation breakpoint centromeric to the *BCL1*-MTC. *Genes, Chromosomes Cancer* **11**, 246–255.
5. Pigny, P., Guyonnet-Dupérat, V., Hill, A. S., Pratt, W. S., Galiègue-Zouitina, S., Collyn-d'Hooghe, M., Laine, A., Van Seuningen, I., Degand, P., Gum, J. R., Kim, Y. S., Swallow, D. M., Aubert, J. P., and Porchet, N. (1996) Human mucin genes assigned to 11p15.5: identification and organization of a cluster of genes. *Genomics* **38**, 340–352.
6. Antequera, F., Boyes, J., and Byrd, A. (1990) High levels of de novo methylation and altered chromatin structure at CpG islands. *Cell* **62**, 503–514.
7. Barlow, D. P. (1992) Preparation, restriction and hybridization analysis of mammalian genomic DNA for pulsed-field gel electrophoresis, in *Pulsed-Field Gel Electrophoresis* , *Methods in Molecular Biology*, vol. 12 (Burmeister , M. and Vlanovsky, L., eds.), Humana, Totowa, NJ, pp. 107–128.
8. Overhauser, J. (1992) Encapsulation of cells in agarose beads, in *Pulsed-Field Gel Electrophoresis, Methods in Molecular Biology*, vol. 12: (Burmeister, M. and Vlanovsky, L., eds.), Humana, Totowa, NJ, pp. 129–134.
9. Boultwood, J. (1994) Physical mapping of the human genome by pulsed field gel electrophoresis, in *Protocols for Gene Analysis, Methods in Molecular Biology*, vol. 31: (Harwood, A. J., ed.), Humana, Totowa, NJ, pp. 121–134.

10. Birren, B. and Lai, E. (1993) *Pulsed Field Gel Electrophoresis. A Practical Guide.* Academic, San Diego, CA.
11. Gendler, S. J., Lancaster, C. A., Taylor-Papadimitriou, J., Duhig, T., Peat, N., Burchell, J., Pemberton, L., Lalani, E., and Wilson, D. (1990) Molecular cloning and expression of a human tumor-associated polymorphic epithelial mucin. *J. Biol. Chem.* **265**, 15,286–15,293.
12. Gum, J. R., Byrd, J. C., Hicks, J. W., Toribara, N. W., Lamport, D. T. A., and Kim, Y. S. (1989) Molecular cloning of human intestinal mucin cDNA. Sequence analysis and evidence for genetic polymorphism. *J. Biol. Chem.* **264**, 6480–6487.
13. Gum, J. R., Hicks, J. W., Swallow, D. M., Lagace, R. L., Byrd, J. C., Lamport, D. T. A., Siddiki, B. and Kim, Y. S. (1990) Molecular cloning of cDNA derived from a novel human intestinal mucin gene. *Biochem. Biophys. Res. Commun.* **171**, 407–415.
14. Porchet, N., Nguyen, V.C., Dufossé, J., Audié, J. P., Guyonnet-Dupérat, V., Gross, M. S., Denis, C., Degand, P., Beinheim, A., and Aubert J. P. (1991) Molecular cloning and chromosomal localization of a novel human tracheo-bronchial mucin cDNA containing tandemly repeated sequences of 48 base pairs. *Biochem. Biophys. Res. Commun.* **175,** 414–422.
15. Guyonnet-Dupérat,V., Audié, J. P., Debailleul, V., Laine, A., Buisine, M. P., Galiègue-Zouitina, S., Pigny, P. Degand, P., Aubert, J. P., and Porchet, N. (1995) Characterization of the human mucin gene MUC5AC : a consensus cysteine-rich domain for 11p15 mucin genes? *Biochem. J.* **305,** 211–219.
16. Dufossé, J., Porchet, N., Audié, J. P., Guyonnet-Dupérat, V., Laine, A., Van Seuningen, I., Marrakchi, S., Degand, P., and Aubert, J. P. (1993) Degenerate 87 base pair tandem repeats create hydrophilic/hydrophobic alternating domains in human mucin peptides mapped to 11p15. *Biochem. J.* **293,** 329–337.
17. Toribara, N. W., Robertson, A. M., Ho, S. B., Kuo, W. L., Gum, E., Hicks, J. W., Gum, J. R., Byrd, J. C., Siddiki, B., and Kim, Y. S. (1993) Human gastric mucin. Identification of a unique species by expression cloning. *J. Biol. Chem.* **268,** 5879–5885.
18. Burmeister, M. (1992) Strategies for mapping large regions of mammalian genomes, in *Pulsed-Field Gel Electrophoresis, Methods in Molecular Biology,* vol. 12 (Burmeister, M. and Vlanovsky, L., eds.), Humana, Totowa, NJ, pp. 259–284.
19. Bickmore, W. A. and Byrd, A. P. (1992) Use of restriction endonucleases to detect and isolate genes from mammalian cells, in *Recombinant DNA, part G, Methods in Enzymology,* vol. 216 (Wu, R., ed.), Academic, San Diego, CA.
20. Guyonnet-Dupérat, V. (1993) Etude des gènes de mucines humaines localisées en 11p15: mégacartographie et approche de l'organisation génomique de MUC5B et MUC5AC. Thèse de Doctorat d'Université en Sciences de la Vie et de la Santé, Lille, France.
21. Pigny, P. (1997) Les gènes de mucines humaines localisées en 11p15—Polymorphisme, cartographie physique et approches de régulation. Thèse de Doctorat d'Université des Sciences de la Vie et de la Santé.

27

In Situ Hybridization Techniques for Localizing Mucin mRNA

Ilene K. Gipson

1. Introduction

Progress in understanding how mucosal surfaces are protected is closely related to the development of morphologic techniques to study the structure and secretory function of the mucosal epithelia. Morphologic methods have allowed characterization of mucus-secreting cells of the epithelia of the eye, and the respiratory, gastrointestinal (GI), and reproductive tracts. Characteristics of the mucus-secreting cells of these tissues vary, and many questions remain regarding special characteristics of mucus present over the differing mucosal surfaces. Recent progress in cloning and characterization of mucin genes has facilitated the use of *in situ* hybridization (ISH) to begin to characterize the mucin gene repertoires and specific functions of mucins expressed by the various epithelia, either those covering mucosal surfaces or glandular epithelia contributing to the mucous layer on the surface of the tissue. ISH has been a particularly valuable method in this regard, since antibodies to specific mucin proteins are often difficult to use on tissues or secretions without heroic methods to deglycosylate in order to make protein epitopes available.

Mucins, because of their heavy glycosylation and size, have presented major technical difficulties to biochemists and molecular biologists struggling to characterize them *(1–3)*. The use of molecular techniques to sequence the mucin gene has identified a characteristic common to all mucin genes, that of tandemly repeated sequences in their amino acid/nucleotide sequence. (For review *see* **refs. *4*** and ***5***). This character greatly facilitates application of ISH methods to localize specific mucin mRNAs in tissues and cells. Probes to the tandemly repeated nucleotide sequences bind at multiple sites along cellular mRNA, providing an amplified signal and excellent visualization of the presence of specific mucin mRNAs. For once, there is something about mucin character that facilitates ease of application of a method! While this enhanced signal is useful, it is an impediment to quantitative assays. One cannot rely on the use of tandem repeat (TR) probes to quantitate mRNA levels, especially with those mucin genes that exhibit polymorphisms.

From: *Methods in Molecular Biology, Vol. 125: Glycoprotein Methods and Protocols: The Mucins*
Edited by: A. Corfield © Humana Press Inc., Totowa, NJ

Currently, the two probes of choice for ISH are riboprobes of usually 100–300 bp (RNA transcribed from cDNA probes) or oligonucleotide probes of 18–100 bp (which match the cDNA sequence), the latter being less sensitive. Because of the enhanced signal obtained with TR probes, straightforward simple *in situ* methods can be applied. One can thus employ less sensitive, labeled oligonucleotide probes with radioisotope detection or with nonradioisotope immunodetection methods; the latter disclosure method is also less sensitive. Special efforts to preserve all RNA in the tissues, usually a requirement for tissues with low levels of message, is not always necessary; thus, archived, less stringently fixed and processed tissue sometimes can be used. Because of its relative simplicity, our probe of choice for ISH of mucin mRNA is, therefore, the oligonucleotide probe, but we usually use radioisotope labeling at least in initial experiments until we determine signal levels.

ISH methods have been applied to the study of mucin genes in two ways: (1) for chromosomal localization of specific mucin genes, and (2) for tissue or cellular localization of specific mucin mRNAs. This chapter describes protocols for tissue localization only; for chromosome localization methodologies, readers are referred to **ref. *6***. The methods described in this chapter are those that have been successfully applied in our laboratory to determine specific mucin mRNA localization in epithelia covering the eye, reproductive tract, and GI tract ***(7–11)***. Since the signal for mucin message is usually easily detected in tissues, one does not have to be as concerned with loss of low-level message and access to message. Thus, one can use paraformaldehyde-fixed, paraffin-embedded tissue rather than frozen tissue and benefit from the better preservation of tissue architecture.

Both radioisotope (^{35}S) and immunodetection (digoxygenin [DIG]) methods of ISH (colorimetric and fluorescence disclosure) are described, and both methods work well with routine mucin mRNA localization. Of the methods described, the most sensitive is that of the radioisotope labeling of probes. The colorimetric DIG protocol is useful if one chooses not to use radioisotope methods, is not equipped for the work, or does not have access to dark-field microscopy. It has the disadvantage that with colorimetric disclosure methods, interpretation can be difficult to distinguish in counterstained tissues with low expression levels. The fluorescent DIG ISH method gives the best resolution of message within the cytoplasm of cells, especially when viewed with confocal microscopy. In our hands, however, this method is the most capricious of the three disclosure methods and does not provide a permanent record. **Figures 1** and **2** show examples of several methods of ISH as applied to mucin mRNA localization.

The protocols that follow are described in a rather practical and simple fashion. For complete descriptions of the theory and practice of ISH, readers are referred to **refs. *12–17***.

2. Materials

All materials and solutions are prepared RNase free. Baked glassware is used and all materials and equipment are handled with latex gloves. All water and buffers are made RNase free by diethylpyrocarbonate (DEPC) treatment (*see* **Note 1**).

2.1. Equipment

1. Microtome.
2. Water bath: 30–60°C.

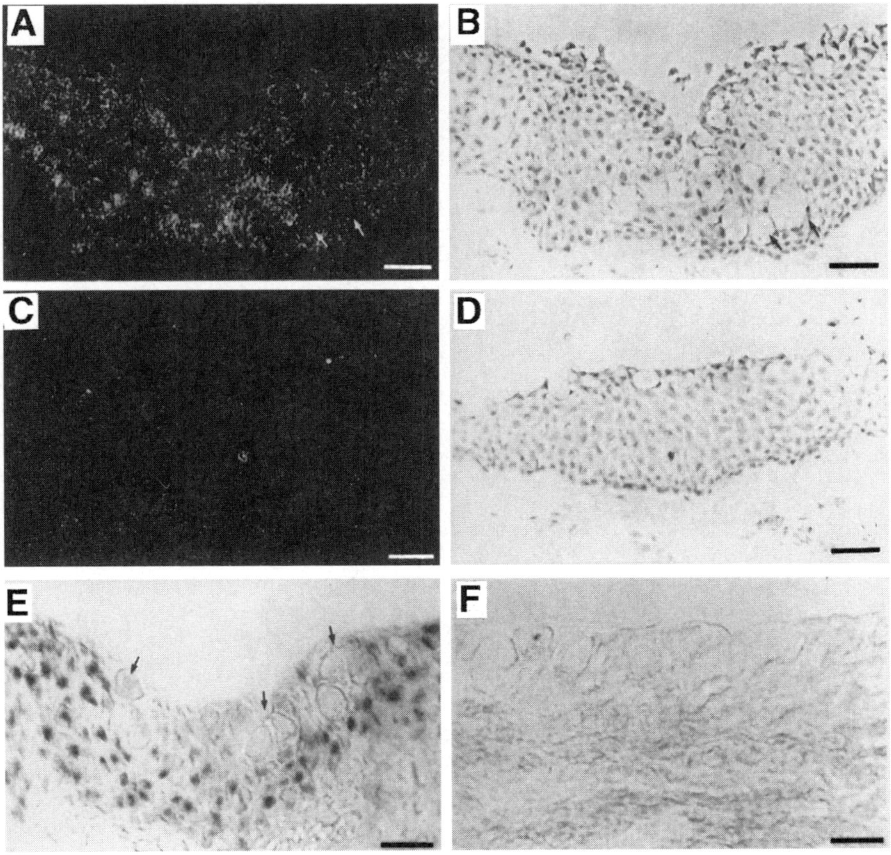

Fig. 1. Micrographs demonstrating two methods of disclosure of oligonucleotide probes binding to mucin mRNA. (**A**, **C**) Dark field; (**B**) and (**D**) H&E of the same field of conjunctival epithelium, respectively. In (A), ^{35}S-labeled oligonucleotide (48 mer) to MUC4 TR sequence is localized in all cell layers of the stratified epithelium. (B) is the sense control of (A); (E) is antisense, and (**F**) is sense control of DIG-labeled MUC4 oligonucleotide probe disclosed with alkaline phosphatase/NBT. Bars = 50 μm. (Reproduced by permission from **ref.** ***8***.)

3. Microfuge.
4. Vortex mixer.
5. Oven/incubator: 30–60°C.
6. Heat block/water bath adjustable to 80°C.
7. Water bath (42°C) for autoradiography.
8. Light-tight darkroom.

2.2. Fixation and Embedding of Tissue in Paraffin

1. 4% Paraformaldehyde in 0.1 *M* phosphate buffer, pH 7.4.
2. 0.1 *M* phosphate buffer, pH 7.4.
3. 100, 95, 70, and 50% ETOH.

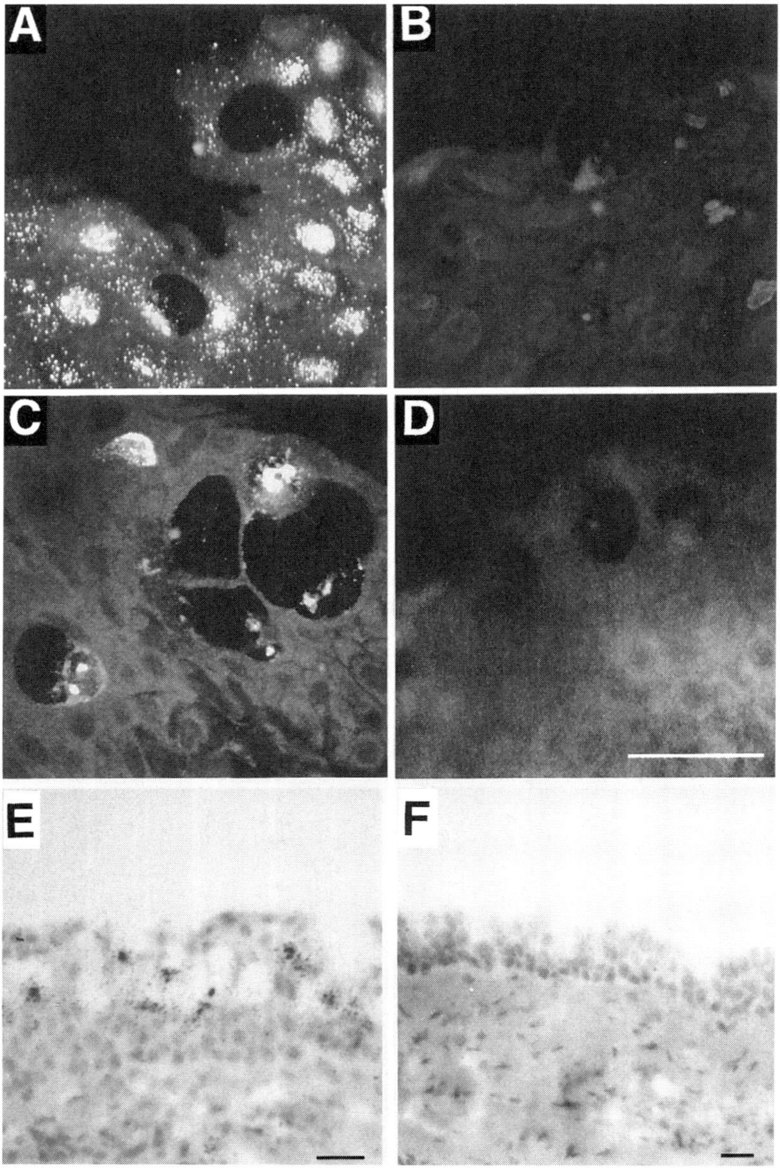

Fig. 2. Example of three methods of ISH using mucin mRNA probes on sections of human conjunctiva. **(A, B)** Localization of MUC4 mRNA using antisense (A) and sense (B) oligonucleotide probes labeled with DIG and disclosed with fluorescently labeled anti-DIG. Note MUC4 message surrounds the nuclei of all the cells in the epithelium (A). **(C–E)** Use of riboprobes to localize MUC5AC in goblet cells of human conjunctival epithelium. In (C) the riboprobe was labeled with DIG-labeled UTP and the DIG was disclosed with fluorescently labeled anti-DIG. In (E), the riboprobe was labeled with ^{35}S UTP and disclosed by autoradiography. Note that 5AC mRNA is restricted to goblet cells. (D) and (F) are sense controls for (C) and (E), respectively. Bars = 20 μm. (Reproduced from **ref. *9*.**)

4. Xylene.
5. Paraffin (e.g., Paraplast).
6. Embedding molds.

2.3. Preparation of Slides and Sectioning of Tissue

1. Microscope slides.
2. Gelatin.
3. Sodium potassium chromate.
4. 4% Paraformaldehyde in 0.1 *M* phosphate buffer, pH 7.4.
5. Routine paraffin-sectioning supplies.
6. Coplin jars or glass staining dishes.

2.4. Preparation and Labeling of Oligonucleotide Probes

Synthesized oligonucleotides, both antisense and sense (>18 mer), appropriately purified ***(16)*** are available from a variety of manufacturers. (For discussion of design and synthesis, *see* **refs. *17–19***.)

Commercially available 3'-labeling (tailing) kits are available for labeling oligoprobes with either radionucleotides or DIG. The kits are convenient and can be an economical method. Companies providing kits include Boehringer Mannheim (Mannheim, Germany), Promega (Madison, WI), and Stratagene (La Jolla, CA).

2.4.1. Labeling of Oligoprobes with ^{35}S

Kits containing **items 1** and **2** can be purchased; they usually also contain 5 m*M* $CoCl_2$ included in the buffer.

1. 5X buffer: 1 *M* potassium cacodylate, 0.125 *M* Tris-HCl, 1.25 mg/mL bovine serum albumin (BSA), pH 6.6.
2. Terminal transferase.
3. 0.2 *M* EDTA, pH 5.2.
4. 3 *M* Na acetate, pH 5.2.
5. tRNA.
6. 75% ETOH.

2.4.2. Labeling with DIG

1. Kits for labeling oligonucleotides with DIG that contain the following:
   a. 5X reaction buffer: 1 *M* potassium cacodylate, 0.125 *M* Tris-HCl, 1.25 mg/mL of BSA, pH 6.6.
   b. 25 m*M* $CoCl_2$ solution.
   c. 1 m*M* DIG-deoxy uridine triphosphate (dUTP).
   d. 10 m*M* deoxyadenosine triphosphate (dATP) in Tris buffer.
   e. Terminal transferase: 50 U/µL in 0.2 *M* potassium cacodylate, 1 m*M* EDTA, 200 m*M* KCl, 0.2 mg/mL of BSA, pH 6.5, 50% (v/v) glycerol.
   f. Control oligonucleotide: unlabeled, 20 pmol/µL.
   g. Control oligonucleotide: DIG-dUTP/dATP, tailed 2.5 pmol/µL.
   h. 0.25 mg/mL of supercoiled pUC18 control DNA in 10 m*M* Tris-HCl, pH 7.6, 1 m*M* EDTA.
   i. 20 mg/mL of glycogen solution.
   j. DNA dilution buffer: 50 µg/mL of herring sperm DNA in 10 m*M* Tris-HCl, 1 m*M* EDTA, pH 8.0.
   k. 10 mg/mL of poly (A) solution.

2. 0.2 *M* EDTA, pH 5.2.
3. 3 *M* Na acetate, pH 5.2.
4. tRNA.
5. 75% ETOH.

2.5. Prehybridization Solutions

1. Phosphate-buffered saline (PBS), pH 7.4.
2. Proteinase K.
3. 100 m*M* Tris-HCl, pH 7.6.
4. 0.5 *M* EDTA, pH 7.5.
5. 0.2% Glycine in PBS.
6. 4% Paraformaldehyde in PBS.
7. 1 *M* Triethanolamine, pH 8.0.
8. Acetic anhydride.
9. 20X Sodium chloride/sodium citrate (SSC) buffer: 3 *M* NaCl, 0.3 *M* sodium citrate; adjust pH to 7.0 with 1 *M* HCl.

2.6. Hybridization Solutions and Supplies

1. Formamide (Sigma, St. Louis, MO).
2. 10X salt buffer: 3 *M* NaCl, 0.1 *M* Tris-HCl, pH 7.6, 50 m*M* EDTA, 0.2% Ficoll 400, 0.2% polyvinylpyrrolidone, 0.2% BSA.
3. 1 *M* Dithiothreitol (DTT) (not necessary for immunodetection method).
4. ^{35}S or DIG-labeled sense and antisense oligonucleotide probes.
5. 50% Dextran sulfate.
6. tRNA.
7. Cover Wells™, or Probe Clips®, which are cover slips with sealing gaskets that provide moist, well-sealed chambers for the hybridization step (available from GBL, Pontiac, MI, or PGC Scientific, Frederick, MD).
8. Slide holder (Sigma Humid Chamber, cat. no. 6644).
9. Sealable moist plastic box.

2.7. Posthybridization Solutions

1. 20X SSC.
2. Formamide.
3. 14 *M* β-mercaptoethanol.
4. Ribonuclease (i.e., RNase) (Boehringer Mannheim).

2.8. Autoradiography/Counterstaining for Disclosure of ^{35}S Oligonucleotide Binding

1. Kodak NTB2 Autoradiography Emulsion (cat. no. 165 4433, Kodak, Rochester, NY).
2. Light-tight black box.
3. Kodak D19.
4. Kodak fixer.
5. Hematoxylin and eosin (H&E) stain.

2.9. Disclosure of DIG-Labeled Oligonucleotide Probe

2.9.1. Colorimetric-Alkaline Phosphatase

Kits are available from Boehringer Mannheim.

1. Buffer 1: 0.1 *M* Tris-HCl, 0.15 *M* NaCl, pH 7.5.
2. 1% Dry milk in buffer 1; alternate 1% BSA in buffer 1.
3. Anti-DIG-alkaline phosphatase conjugate: sheep anti-DIG, Fab fragments, conjugated with alkaline phosphatase, 750 U/mL.
4. Alkaline reaction buffer: 0.1 *M* Tris, 0.1 *M* NaCl, 50 m*M* $MgCl_2$, 0.1% Tween-20, pH 9.5.
5. Nitroblue tetrazolium (NBT) salt.
6. 5-bromo-4-chloro-3-indolyl phosphate toluidinium (BCIP) salt.

2.9.2. Fluorescent ISH (FISH)

1. Buffer 1: 0.15 *M* NaCl, 0.1 *M* Tris-HCl, pH 7.5.
2. 1% Dry milk in buffer 1.
3. Anti-DIG-rhodamine-Fab fragments (Boehringer Mannheim) with a final concentration of 20 µg/mL.

3. Methods

3.1. Preparation of Tissue

1. To prepare fixative (4% paraformaldehyde in RNase-free 0.1 *M* phosphate buffer, pH 7.4), heat buffer to 60°C and add 4 g of paraformaldehyde/100 mL of buffer. Add 1–3 drops of 1 *N* NaOH. Store at 4°C for up to 1 mo.
2. Fix tissue for 1–24 h at room temperature. Ideally, tissues are fixed immediately or within 1 h of biopsy or death (*see* **Note 2**).
3. Rinse tissue in RNase-free 0.1 *M* of phosphate buffer, pH 7.4 (three times for 15 min), dehydrate in ETOH series followed by xylene, and embed in paraffin, using standard but RNase-free techniques. Store blocks at 4°C.

3.2. Preparation of Slides

Prepare gelatin-coated slides. In our experience, gelatin is superior to other "subbing" compounds in that loss of tissue sections during the hybridization procedure is minimal *(14)*. For "subbing slides":

1. Transfer slides to metal carrier and soak in 100% ETOH overnight in glass dishes.
2. Discard ETOH and bake slides in glass dishes for 2 h at 180°C.
3. Cool at room temperature.
4. Dip in 1% gelatin/0.1% chromium potassium sulfate solution for 10 min, and then allow to dry. The gelatin solution is made by dissolving 3 g of gelatin in 200 mL: of DEPC-treated water, which is warmed to 60°C until gelatin is completely dissolved. Separately, 0.3 g of chromium potassium sulfate is added to 100 mL of DEPC-treated water and mixed at room temperature until dissolved. The two solutions are combined.
5. Fix slides in 4% paraformaldehyde in 0.1 *M* phosphate buffer for 15 min.
6. Wash in DEPC-treated distilled water two times for 5 min each.
7. Dry and store at room temperature in a clean box.

3.3. Sectioning of Tissue

1. Section paraffin-embedded tissue at 6 µm, mount on subbed slides, and store slides at 4°C until use.
2. Heat slides to 40°C overnight just prior to use.

3.4. Preparation and Labeling of Oligonucleotide Probes

Oligonucleotide probes and riboprobes (*see* **Note 3**) have been used for radioisotope or immunodetection of mucin mRNA in tissues. **Table 1** lists examples of probes used in ISH studies of mucin gene expression. Oligonucleotides to mucin TR sequence work quite well and provide the simplest method for labeling. Riboprobes provide a more sensitive method of message detection and are useful for quantitative assays when using non-TR probes. For examples of ^{35}S-labeled oligonucleotide and riboprobe as well as DIG-alkaline phosphatase or FISH, *see* **Figs. 1** and **2**.

3.4.1. Labeling of Oligoprobes with ^{35}S

1. Mix the following in a microfuge tube on ice:
   a. 4 µL of 5X reaction buffer.
   b. 10 pmol of oligonucleotide probe.
   c. 2.5 µL of ^{35}S-dATP.
   d. 1 µL of terminal deoxynucleotidyl transferase (TdT).
   e. Water to a final volume of 20 µL.
2. Incubate at 37°C for 1 h and microfuge for 1–2 min.
3. Add 4 µL of 0.2 *M* EDTA to terminate the reaction on probe purification (*see* **Note 4**).
4. To precipitate the probe, add 0.1 vol of 3 *M* Na acetate, pH 5.2, 2.5 vol of alcohol, and 0.2 vol of 1 mg/mL tRNA in DEPC-treated water and store at –80°C for 2 to 3 h or overnight.
5. Centrifuge the probe at 12,000*g* (15–20 min), wash the pellet with 50 µL of cold ethanol (75%, v/v), and air- or vacuum-dry the pellet.
6. Dissolve the pellet in 10 µL of DEPC-treated water.
7. Use 1 µL of the probe to check counts per minute in scintillation counter.
8. Store probe at –80°C.

3.4.2. Labeling of Oligonucleotide Probe with DIG-UTP (see **Note 5**)

1. Mix well the following in an RNase-free microfuge tube on ice:
   a. 10 pmol of probe.
   b. 4 µL of 5X reaction buffer.
   c. 4 µL of 25 mM $CoCl_2$.
   d. 1 µL of 1 m*M* DIG-dUTP solution.
   e. 1 µL of 10 m*M* dATP.
   f. 1 µL of TdT.
   g. Water to final volume of 20 µL.
2. Incubate at 37°C for 1 h.
3. Add 2 µL of stop solution (mixture of 1 µL of glycogen and 200 µL of 0.2 *M* EDTA) to stop the reaction. (*See* **Note 4** on NucTrap Column.)
4. Precipitate the labeled oligonucleotide probe with 0.1 vol of 3 *M* Na acetate, pH 5.2, 2.5 vol of ETOH, and 0.2 vol of 1 mg/mL tRNA at –80°C for 2 h or overnight.
5. Centrifuge the probe at 12,000*g* (15–20 min), wash the pellet with 50 µL of cold ethanol (75%, v/v), and air- or vacuum-dry the pellet.
6. Dissolve the pellet in 10 µL of DEPC-treated water.
7. Store labeled oligonucleotide at –80°C for up to 1 yr.

3.5. Prehybridization, Proteinase K Treatment, and Acetylation

1. Select enough slides so that both antisense and sense probes may be used (see **Note 6**). You may put in an extra slide for quick X-ray film assay to determine the success of labeling (*see* **Note 7**). Incubate at 40°C overnight.

Table 1
Examples of Probes and Disclosure Methods for *In Situ* Hybridization to Localize Mucin mRNAs

Mucin gene	Probe designation/type	Probe length (bp)	Disclosure used	Refs.
MUC1	MUC1-1/2	450	^{35}S	*(7,10,20)*
	Oligo	48	^{35}S	*(21)*
MUC2	HAM-1	92	^{35}S	*(20–22)*
	SMUC41	836	DIG	*(22,23)*
	Oligo	48	^{35}S	*(24)*
Muc3	Riboprobe	473	^{35}S	*(11)*
	Oligo	48	^{35}S	*(24)*
MUC4	Oligo	48	^{35}S	*(24)*
	Oligo	48	DIG/FISH	*(9)*
	Oligo	48	DIG/colorimetic	*(8)*
MUC5AC	Oligo	48	^{35}S	*(24)*
	4F	494	^{35}S	*(8)*
	4F	494	DIG/FISH	*(9)*
	PM5	1300	DIG/colorimetric	*(25)*
MUC5B	ngBM4-1	984	^{35}S	*(10,26)*
	Oligo	48	^{35}S	*(21,24)*
MUC6	Oligo	30	DIG/colorimetric	*(27)*
	Oligo	39	^{35}S	*(10)*
	Oligo	74	^{35}S	*(28)*
	Oligo	48	DIG/colorimetric	*(29)*
	PM6	840	DIG/colorimetric	*(25)*
MUC7	Oligo	48	^{33}P	*(30)*
	Oligo	48	DIG/colorimetric	*(27)*

2. Deparaffinize slides in the following:
   a. Xylene for 10 min (two times).
   b. 100% ETOH for 4 min.
   c. 90% ETOH for 4 min.
   d. 75% ETOH for 4 min.
3. Fix in 4% paraformaldehyde in PBS for 10 min.
4. Rinse in PBS for 3 min.
5. Treat with proteinase K to increase accessibility of probe to mRNA in fixed tissue. Stock: 2.5 mg/mL in 10 m*M* Tris-HCl, pH 7.6. Warm proteinase solution to 37°C before adding slides.
   a. 10 mM Tris-HCl/1 mM EDTA (TE buffer) for 5 min.
   b. 1 µg/mL ofproteinase K in TE buffer for 20 min at 37°C.
   c. 0.2% Glycine in PBS for 5 min.
   d. PBS for 3 min.
   e. 4% Paraformaldehyde in PBS for 20 min.
   f. PBS for 5 min.
6. Treat with acetic anhydride to block nonspecific binding. Important: Add acetic anhydride to triethanolamine just prior to treating slides.

   a. 0.1 *M* Triethanolamine, pH 8.0, containing 1/200 vol/vol of acetic anhydride for 10 min.
7. Store slides in 2X SSC until hybridization solution is ready.

3.6. Hybridization

1. Make hybridization buffer (need 200 μl/slide), i.e., add:
   a. 1200 μL of formamide.
   b. 480 μL of 50% dextran sulfate.
   c. 24 μL (leave out for DIG method) of 1 *M* DTT.
   d. 240 μL of 10X salt buffer.
   e. 50 μL of 1 mg/mL tRNA.
   f. DEPC-treated water to a final volume of 2400 μL.
2. Heat hybridization buffer at 80°C for 5 min and mix well.
3. To make hybridization solution, add probe to buffer to a final concentration of 5×10^3 cpm/μL for ^{35}S-labeled oligoprobe or to an amount of 0.5–1 μg/mL of DIG-labeled probe.
4. Take slides out of SSC and air-dry.
5. Add 200 μL of hybridization solution to each slide and cover with a 200 μL-Probe Clip or cover well to seal.
6. Place flat on to a slide holder and then into a sealed moist chamber (plastic box), and place in a 37°C-oven overnight.

3.7. Posthybridization Washes (see Note 8)

1. Dip slides once in 2X SSC to remove Probe Clip or cover well. Then wash as follows:
   a. 2X SSC at room temperature for 30 min.
   b. 1X SSC at room temperature for 30 min.
   c. 0.5X SSC at 37°C for 30 min.
   d. 0.5X SSC at room temperature for 30 min.
2. For disclosure by autoradiography, dry slides overnight; for disclosure by the immunodetection method, enter the slides into the detection protocol after the last SSC wash.

3.8. Autoradiographic Detection of Probe

1. Warm water bath to 43°C to melt emulsion.
2. Use Kodak NTB2 Autoradiography Emulsion diluted 1:1 with warm water. Use emulsion under safelight or in complete darkness.
3. Dip slides, one at a time, in the emulsion and stand to dry for 2 h. This should be done under a dim safelight or in complete darkness. Always dip a completely blank slide as a negative control for the quality of the autoradiography for each condition/development time.
4. Store slides in a light-tight black box with desiccant, taped shut in a black plastic envelope at 4°C for 1–4 wk.
5. Develop slides as follows:
   a. Allow slides to come to room temperature before developing (~30 min).
   b. Develop in Kodak D-19 1:1 dilution with water for 5 min room temperature.
   c. Stop in distilled water and briefly rinse.
   d. Fix in full-strength Kodak Fixer for 15 min at room temperature.
   e. Check slides under safelight to make sure fixation is complete (slides are clear, not opaque).
6. Wash in gently flowing water for 30–60 min.
7. Lightly counterstain in H&E.
8. Dehydrate through ethanols to xylene and cover slip with Permount (Fisher Scientific, Pittsburgh, PA).

3.9. Immunodetection of Probe

3.9.1. Colorimetric–Alkaline Phosphatase Detection of Binding of DIG-Labeled Oligonucleotide Probe to Tissue Sections

1. Wash slides in buffer 1 for 10 min.
2. Block in 1% dry milk in buffer at room temperature, 30 min.
3. Incubate with 1/500 anti-DIG-alkaline phosphatase conjugate at room temperature for 2 h or overnight at 4°C.
4. Wash with buffer 1 two times for 15 min each.
5. Equilibrate the slides with alkaline reaction buffer for 2 min.
6. Develop the color with color-substrate solution: 1:50 dilution of NBT/BCIP stock from Boehringer Mannheim (45 µL of NBT solution and 35 µL of BCIP solution in 10 mL of buffer 1 at room temperature).
7. Stop the reaction with distilled water and wash twice.
8. Mount with Vectashield and a cover slip.

3.9.2. Fluorescent Antibody Detection of Binding of DIG-Labeled Oligonucleotide Probes to Tissue Sections

1. Wash slides in buffer 1 for 5 min.
2. Block in 1% dry milk in buffer 1 for 30 min.
3. Incubate with 20 µg/mL of anti-DIG-rhodamine at 37°C for 1 h in a sealed moist chamber.
4. Wash in buffer 1 three times for 10 min each.
5. Mount sections with mounting media.

4. Notes

1. For general methods, glassware baking, and DEPC treatment of water, bake all glassware slides and heat-resistant equipment for 2 h at 180°C to make RNase free. To make RNase-free water and buffers, add 0.1% DEPC, stir for 10 min, let sit overnight, and autoclave the following day.
2. Fixation time may depend on the size of the excised tissue. Reports vary on overfixation ***(14–16)***: Some investigators suggest that overfixation causes excessive crosslinking that prevents probe access, whereas others suggest that it induces nonspecific binding. Underfixation in the center of tissues of a large block is cited as another potential problem ***(14)***. All agree that fixation of tissue as soon as possible after death or biopsy is key (<30 min).
3. Riboprobes (RNA probes) are currently considered the gold standard of ISH because they provide a more sensitive assay. This is owing in part to the fact that the riboprobe is labeled with the disclosing agent, be it radioisotope or DIG, as it is transcribed from the cDNA template. Labeled nucleotide is incorporated all along the RNA sequence; thus, the probe is heavily labeled, allowing an enhanced signal. Riboprobes are especially suitable for tissues in which there are low copy numbers of the mRNA in question or if mucin probes to be used are not TR probes (suitable for quantitative assay). The preparation of the riboprobe is labor-intensive. One must derive the riboprobe from appropriate template cDNA. The template cDNA of ideally 100–300 bp may be obtained in several ways: (1) elution from an agarose gel of a polymerase chain reaction product obtained by designing primers for a known sequence of a mucin molecule from RNA or DNA of tissue known to express the mucin; (2) excision of a smaller fragment from a large cDNA by endonucleases; (3) in a plasmid vector, from a colleague. After the cDNA template is obtained, the following steps are required for riboprobe production and labeling. The

template cDNA is ligated into an appropriate plasmid vector (i.e., Bluescript), the plasmid vector is transfected into *Escherichia coli*, and positive colonies isolated. Positive colonies are amplified, plasmid DNA is purified from *E. coli* cultures, and the presence and direction (to determine sense, antisense sequence) of inserted template cDNA is verified and determined, respectively. The plasmid is then linearized and reverse transcribed with polymerases, with the choice depending on vector; the common vector Bluescript uses SP6, T7, or T3 with simultaneous labeling with either ^{35}S-UTP or DIG-UTP, resulting in a labeled riboprobe ready for ISH. For methods of template cDNA, riboprobe production, and labeling, the reader is referred to ***(6,16)***.

4. One alternate method of removing unincorporated nucleotides is the use of a Sephadex G-25 column (5 × 0.5 cm). Equilibrate the column with buffer (10 m*M* Tris-HCl, pH 7.5, 1 m*M* EDTA, 0.1 M NaCl), add labeled probe mixture, incubate, and elute. Unbound free nucleotides remain in the column. A second alternate and convenient method is the use of commercially available "push column" devices (i.e., NucTrap Push Columns, Stratagene Cloning Systems, cat. no. 400701, La Jolla, CA).
5. For direct/indirect FISH, one can also use fluorescently labeled DIG to label the oligoprobe. In our hands, we obtained enhanced signal using the indirect detection method, that of fluorescently labeled anti-DIG antibody localization.
6. For slides, selections, and numbers, select a minimum of three slides of each tissue tested—one for sense, two for antisense. Add extra slides of control tissues (e.g., epithelia known to express the mucin gene in question), and several sets of slides for autoradiography, so that several exposure times may be tested (e.g., 1 and 2 wk).
7. For X-ray film preexposure, one can obtain some indication of strength of signal and gross distribution by exposing slides to fast X-ray film. For ^{35}S, 2–5 d of exposure may be required. For complete protocol, *see* **ref. *16***. With mucin mRNA localization, 1 wk is often sufficient exposure for autoradiographic slides. Thus, the intermediate X-ray film method may not be labor saving.
8. The posthybridization conditions of stringency (highest stringency equals lowest SSC concentration) and temperature can be manipulated to remove nonspecific, less specifically associated oligoprobes so as to improve specific signal. Conditions must be empirically determined for each probe.

References

1. Carlstedt, I., Sheehan, J. K., Corfield, A. P., and Gallagher, J. T. (1995) Mucous glycoproteins: A gel of a problem. *Essay Biochem.* **20,** 40–76.
2. Strous, G. J. and Dekker, J. (1992) Mucin-type glycoproteins. *Crit. Rev. Biochem. Mol. Biol.* **27,** 57–92.
3. Gum, J. R., Jr. (1992) Mucin genes and the proteins they encode: Structure diversity and regulation. *Am. J. Respir. Cell. Mol. Biol.* **7,** 557–564.
4. Gendler, S. J. and Spicer, A. P. (1995) Epithelial mucin genes. *Annu. Rev. Physiol.* **57,** 607–634.
5. Gum, J. R., Jr. (1995) Mucins: Their structure and biology. *Biochem. Soc. Trans.* **23,** 795–799.
6. Choo, K. H. A., ed. (1994) *In situ Hybridization Protocols*, Humana, Inc., Totowa, NJ.
7. Inatomi, T., Spurr-Michaud, S., Tisdale, A. S., and Gipson, I. K. (1995) Human corneal and conjunctival epithelia express MUC1 mucin. *Invest. Ophthalmol. Vis. Sci.* **36,** 1818–1827.
8. Inatomi, T., Spurr-Michaud, S., Tisdale, A. S., Zhan, Q., Feldman, S. T., and Gipson, I. K. (1996) Expression of secretory mucin genes by human conjunctival epithelia. *Invest. Ophthalmol. Vis. Sci.* **37,** 1684–1692.

9. Gipson, I. K. and Inatomi, T. (1999) Cellular origin of mucins of the ocular surface. in *Advances in Experimental Medicine and Biology* (Sullivan, D. A., ed.), Plenum, New York, pp. 221–228.
10. Gipson, I. K., Ho, S. B., Spurr-Michaud, S. J., Tisdale, A. S., Zhan, Q., Torlakovic, E., Pudney, J., Anderson, D. J., Toribara, N. W., and Hill, J. A., III. (1997) Mucin genes expressed by human female reproductive tract epithelia. *Biol. Reprod.* **56,** 999–1011.
11. Shekels, L. L., Hunninghake, D. A., Tisdale, A. S., Gipson, I. K., Kieliszewski, M., Kozak, C. A., and Ho, S. B. (1998) Cloning and characterization of mouse intestinal MUC3 mucin: 3' sequence contains EGF-like domains. *Biochem. J.* **300,** 1301–1308.
12. Polak, J. M. and McGee, J. O. D., eds. (1990) *In Situ Hybridization: Principles and Practice.* Oxford University Press, New York.
13. Young, W. S., III (1990) In situ hybridization histochemistry, in *Handbook of Chemical Neuroanatomy,* (Björklund, A., Hökfelt, R., Wouierlood, F. G., and van den Pol, A. N., eds.), Elsevier Science Publishers, UK, pp. 481–512.
14. Ausubel, F. M., Brent, R., Kingston, R. E., Moore, D. D., Seidman, J. G., Smith, J. A., and Struhl, K. (1995) In situ hybridization and immunohistochemistry. *Curr. Protocols Mol. Biol.* **2,** 14.0.1–14.8.24.
15. Millen, K. J. and Hui, C.-C. (1996) Radioactive in situ hybridization of tissue sections, in *A Laboratory Guide to RNA: Isolation, Analysis, and Synthesis* (Krieg, P. A., ed.), Wiley-Liss, New York, pp. 339–355.
16. Penschow, J. D. and Coghlan, J. P. (1994) Hybridization histochemistry using radiolabeled oligodeoxyribonucleotide probes, in *In Situ Hybridization Protocols,* (Andy Choo, K. H., ed.), Humana, Totowa, NJ, pp. 277–292.
17. Beck, F., Tucci, J., and Senior, P. V. (1994) Tissue distribution of gene expression in mammalian development, in *In Situ Hybridization Protocols* (Choo, K. H. A., ed.), Humana, Totowa, NJ, pp. 313–344.
18. Gait, M. J., ed. (1984) *Oligonucleotide Synthesis: A Practical Approach,* IRL, Oxford, UK.
19. Cowell, I. G. (1997) Cloning sequence-specific DNA-binding factors from cDNA expression libraries using oligonucleotide binding site probes, in *cDNA Library Protocols,* (Cowell, I. G. and Austin, C. A., eds.), Humana, Totowa, NJ, pp. 161–170.
20. Chambers, J. A., Hollingsworth, M. A., Trezise, A. E. O., and Harris, A. (1994) Developmental expression of mucin genes MUC1 and MUC2. *J. Cell. Sci.* **107,** 413–424.
21. Balagué, C., Audié, J.-P., Porchet, N., and Real, F. X. (1995) In situ hybridization shows distinct patterns of mucin gene expression in normal, benign, and malignant pancreas tissues. *Gastroenterology* **109,** 953–964.
22. Jany, B. H., Gallup, M. W., Yan, P.-S., Gum, J. R., Kim, Y. S., and Basbaum, C. B. (1991) Human bronchus and intestine express the same mucin gene. *J. Clin. Invest.* **88,** 77–82.
23. Li, D., Wang, D., Majumdar, S., Jany, B., Durham, S. R., Cottrell, J., Caplen, N., Geddes, D. M., Alton, E. W., and Jeffery, P. K. (1997) Localization and up-regulation of mucin (MUC2) gene expression in human nasal biopsies of patients with cystic fibrosis. *J. Pathol.* **181,** 305–310.
24. Audie, J. P., Janin, A., Porchet, N., Copin, M. C., Gosselin, B., and Aubert, J. P. (1993) Expression of human mucin genes in respiratory, digestive, and reproductive tracts ascertained by in situ hybridization. *J. Histochem. Cytochem.* **41,** 1479–1485.
25. Byrd, J. C., Yan, P., Sternberg, L., Yunker, C. K., Scheiman, J. M., and Bresalier, R. S. (1997) Aberrant expression of gland-type gastric mucin in the surface epithelium of *Helicobacter pylori*-infected patients. *Gastroenterology* **113,** 455–464.
26. Keates, A. C., Nunes, D. P., Afdhal, N. H., Troxler, R. F., and Offner, G. D. (1997) Molecular cloning of a major human gall bladder mucin: complete C-terminal sequence and genomic organization of MUC5B. *Biochem. J.* **324,** 295–303.

27. Nielsen, P. A., Bennett, E. P., Wandall, H. H., Therkildsen, M. H., Hannibal, J., and Clausen, H. (1997) Identification of a major human high molecular weight salivary mucin (MG1) as tracheobronchial mucin MUC5B. *Glycobiology* **7,** 413–419.
28. Reid, C. J., Hyde, K., Ho, S. B., and Harris, A. (1997) Cystic fibrosis of the pancreas: Involvement of MUC6 mucin in obstruction of pancreatic ducts. *Mol. Med.* **3,** 403–411.
29. Bolos, C. D., Garrido, M., and Real, F. X. (1995) MUC6 apomucin shows a distinct normal tissue distribution that corrrelates with Lewis antigen expression in the human stomach. *Gastroenterology* **109,** 723–734.
30. Aust, M. R., Madsen, C. S., Jennings, A., Kasperbauer, J. L., and Gendler, S. J. (1997) Mucin mRNA expression in normal and vasomotor inferior turbinates. *Am. J. Rhinol.* **11,** 293–302.

28

Detection of Mucin Gene Polymorphism

Lynne E. Vinall, Wendy S. Pratt, and Dallas M. Swallow

1. Introduction

1.1. The MUC Genes

The polypeptide backbones of mucins and mucin-type glycoproteins are each encoded by one of multiple genes . At least nine distinct genes (*MUC1, MUC2, MUC3, MUC4, MUC5AC, MUC5B, MUC6, MUC7*, and *MUC8*) that encode mucin-type proteins expressed in epithelial cells have been reported in humans ***(1,2)***. The genes encoding mucins are dispersed in the human genome, although a family of four related genes—*MUC2*, *MUC5AC*, *MUC5B*, and *MUC6*—each of which encodes an apomucin expressed in specialized secretory cells, is found on chromosome 11p15.5 ***(1)***. The other genes appear to be rather different. *MUC1,* the first epithelial mucin gene to be identified, is located on chromosome 1q21, and encodes a relatively small molecule with a transmembrane anchor, which is widely expressed in epithelia and can be detected at low levels in certain other cells ***(3)***. *MUC3* (7q22) and *MUC4* (3q29) are extremely large and also have transmembrane anchors ***(5–7,7a–c)***. *MUC3* and *MUC4*, like the 11p15.5 mucin genes, show a restricted tissue distribution, but are expressed in columnar cells as well as in specialized secretory cells ***(8,9)***. *MUC7* (4q) encodes a very small secreted glycoprotein (MG2) expressed primarily in salivary glands ***(10,11)***, but there is little information about *MUC8* (12q24.3) ***(2)***.

A common feature of the *MUC* genes is that they contain tandem repeats (TR) of DNA sequence that lead to tandem repetition of amino acid motifs. These repeated regions may comprise 50% or more of the polypeptide. The repeat units vary in sequence and in length, from 24 nucleotides in *MUC5AC* to 507 nucleotides in *MUC6,* and also in the extent to which they are conserved within each array ***(7,12–17)***.

1.2. MUC Gene Polymorphism (see *Notes 1–9*)

It has been known for some time that the human mucin genes show a high level of polymorphism ***(7,16–22)***. The occurrence of polymorphism owing to variable numbers of the tandem repeats (VNTRs) in mucin genes was first shown for *MUC1*.

From: *Methods in Molecular Biology, Vol. 125: Glycoprotein Methods and Protocols: The Mucins*
Edited by: A. Corfield © Humana Press Inc., Totowa, NJ

The same restriction fragment polymorphism was observed with a number of different restriction enzymes, each of which cuts outside the TR region *(22)*, and the polymorphism was readily detectable in the protein product by sodium dodecyl sulfate (SDS) gel electrophoresis *(22,23)* and also in the messenger RNA *(24,25)*. To date, the extent and nature of polymorphism has been assessed in seven of the nine *MUC* genes on a large number of unrelated individuals, mainly of European extraction, in our laboratory and elsewhere. *MUC1* (*see* **Note 1**), *MUC2* (*see* **Note 2**), *MUC3* (*see* **Note 3**), *MUC4* (*see* **Note 4**), *MUC5AC* (*see* **Note 5**), *and MUC6* (*see* **Note 6**) were all found to be highly polymorphic at least partly owing to VNTR, whereas *MUC5B* showed no evidence of VNTR variation (*see* **Note 7**) *(26)*. The relevant published work is referenced in **Table 1**. We have not studied polymorphism of *MUC7* and *MUC8,* but it has been reported that there is a variation in the number (five or six) of the 69-bp TRs of the small *MUC7* molecule (*see* **Note 8**) (**Table 1**).

It is not clear at this stage what impact variation in the length of the TR regions of the *MUC* genes is likely to have, but it should be noted that the predicted differences in polypeptide length of *MUC1*, *MUC2, MUC4*, and *MUC6* are substantial, since the allele length differences are attributable to coding sequence and do not apparently contain introns. Although it has been known for a long time that the VNTR polymorphism of *MUC1* is detectable in the protein, direct evidence for this in the case of the other genes is only now becoming available. For example, recent data reveal evidence of the same VNTR polymorphism in the mRNAs encoding *MUC2, MUC4*, and *MUC6* *(27)* and corresponding size differences in MUC2 glycoprotein subunits *(27a)*. In the case of *MUC2* the smallest allele that our group has observed is approx 3.5 kb and the very largest allele ever observed, by our collaborators, is 14 kb *(26)*, a difference of more than 150 23 amino acid repeat units. These sizes indicate that the alleles encode full-length MUC2 polypeptides (prior to glycosylation) of about *Mr* 350,000 and 680,000, respectively, a twofold difference in size *(28)*. *MUC4* shows a dramatic 20-kb difference in size between the smallest and largest alleles so far observed. If these alleles are transcribed and translated in their entirety, this difference corresponds to about M*r* 700,000. It seems probable that such substantial differences will be of functional importance, as, e.g., appears to be the case for apolipoprotein(a), which shows similar variation in polypeptide length *(29)*. Variation in length of mucins is likely to have an impact on the properties of the mucous gel; thus, studies to investigate possible disease susceptibility associated with extreme allele lengths are worthwhile.

With the exception of *MUC7,* these polymorphisms involve gene length differences that are kilobases in size, and thus have been analyzed by electrophoresis of restriction enzyme-digested DNA and hybridization with gene-specific cDNA probes after transfer of the DNA onto nylon membranes (Southern blotting). We have devized a procedure whereby six of the genes can be analyzed using only two restriction enzyme digests (*Hin*fI and *Pvu*II). **Table 1** also lists other restriction enzymes that can be used to detect VNTR variation in these genes. In each case, it is important to select an enzyme which cuts close to the repeats and to avoid enzymes that cut within the TRs such as *Taq*1 in *MUC2 (16)*, *MUC5AC*, and *MUC6 (26)*. Note, however, that rare allelic variation involving nucleotide substitutions that involve the restriction sites

Table 1
Size and Distribution of the TR Domains and Enzymes Used for Their Detection

Gene	Chromosomal location		Main TR	Recommended enzyme	VNTR range	Other possible enzymes	Refs.
MUC1	1q21	60 bp	20 amino acids:	*Hin*fI	2.8–8.0 kb	*Eco*RI/*Pst*I, *Alu*I, etc[a]	***22,24,33, 35,36***
MUC2	11p15.5	69 bp	23 amino acids:	*Hin*fI	3.3–11.4 kb	*Pst*I, *Bam*HI/*Hin*dIII	***16,20,26***
MUC3	7q22	51 bp	17 amino acids (two zones):	*Pvu*II	7.0–15.0 kb	*Pst*I	***4,18***
					20–50 kb	[a]	
MUC4	3q29	48 bp	16 amino acids:	*Pvu*II	6.5–27 kb	*Pst*I/*Eco*RI	***7,19***
MUC5AC	11p15.5	24 bp	8 amino acids (interrupted):	*Hin*fI	6.6kb/7.4kb	*Pst*I	***26***
						*Pvu*II[a]	
MUC5B	11p15.5	87 bp	29 amino acids (interrupted):	*Bgl*II	16 kb	[a]	***26***
MUC6	11p15.5	507 bp	169 amino acids:	*Pvu*II	8–13.5kb[a]	[a]	***17,26***
MUC7	4q13-21	69 bp	23 amino acids:	PCR[a]	5/6 repeats	[a]	***10,11***
MUC8	12q24.3	41 bp			Unknown	No information	***2***

[a]*See* text for comments.

themselves may sometimes complicate the picture. The detailed protocols are given in **Subheading 3.** Full protocols for *MUC7* are not given, but the appropriate literature is cited (*see* **Note 8**).

Although at present the only way of analyzing this variation is by Southern blotting, as outlined in **Subheading 3.1.**, it may eventually be possible to find polymerase chain reaction (PCR) formattable polymorphisms that are in linkage disequilibrium with the VNTR alleles, which would allow analysis of more samples and would use less DNA. A protocol for such a polymorphism within *MUC1* is presented here (*see* **Note 9**).

2. Materials (*see* Notes 10 and 11)

1. Puregene kit for genomic DNA preparation (Flowgen, Sittingbourne, UK).
2. Restriction enzymes: *Hin*fI and *Pvu*II (Gibco-BRL, Life Technologies, Paisley, Scotland).
3. TBE buffer (1X = 0.89 *M* Tris-HCl, 0.1 *M* borate, 0.002 *M* EDTA buffer, pH 8.3): prepared as a 10X or 5X stock (*see* **Note 10**).
4. For agarose electrophoresis: Horizon 20:25 apparatus, a 30-sample comb (Gibco-BRL), and a small gel tank (minihorizontal unit, Anachem, Luton, UK) or equivalents.
5. Agarose (Sigma, Poole, UK).
6. Loading buffer for agarose gels: 0.25% bromophenol blue, 0.25% xylene cyanol, 40% sucrose in water.
7. Transilluminator (U.V.P. International, Ultra-Violet Products, Cambridge, UK).
8. Hybond N+ membranes (Amersham Pharmacia Biotech, Buckinghamshire, UK).
9. Vacuum blotter (Vacugene XL, Amersham Pharmacia Biotech).
10. Multiprime DNA labeling kit (Amersham Pharmacia Biotech).
11. Sodium chloride/sodium citrate (SSC)-containing solutions: prepare from a stock of 20X SSC (3 *M* NaCl, 0.3 *M* trisodium citrate) (*see* **Note 11**).
12. Denhardt's solution: make as a 100X stock (2% [w/v] Ficoll 2% [w/v] polyvinylpyrrolidone, 2% [w/v] bovine serum albumin, pH 7.2) and filter sterilized.
13. Sonicated Herring sperm DNA (Promega, Southampton, UK).
14. Molecular weight markers for agarose electrophoresis: Raoul markers (Appligene, Durham, UK), 1-kb ladder (Gibco-BRL), λ*Hin*dIII (Gibco-BRL), and control genomic DNA samples containing alleles of known length.
15. Shaking water bath at 65°C.
16. Cling film (e.g., Clingorap, Terinex, Bedford, UK).
17. Luminescent marking solution (Glo-bug X-ray marking solution, Radleys, Saffron Walden, UK).
18. Oligonucleotide primers AGAGAGTTTAGTTTTCTTGCTCC (CAS) and TTCTTGGCT CTAATCAGCCC (CAA) (nucleotides 5915–5937 and 6092–6073 in Genbank/EMBL M61170), one of which is labeled with fluorescein.
19. PCR machine (e.g., Perkin Elmer, Beaconsfield, UK).
20. *Taq* polymerase in storage buffer A (Promega).
21. Deoxynucleotides: "DNA polymerization mix" (Amersham Pharmacia Biotech). Make a 2 m*M* stock (1/10 of solution supplied).
22. Automated sequencing machine (ALF, Amersham Pharmacia Biotech).
23. Polyacrylamide gels prepared using 6% acrylamide (19:1 acrylamide to bis, Bio-Rad, Herts, UK) in the molds supplied with the automated Sequencing machine.
24. Marker for the ALF acrylamide gels: 250-bp Sizer (Amersham Pharmacia Biotech).
25. Loading buffer: 5 µg/mL of Dextran Blue in 100% formamide.

3. Methods

3.1. Southern Blot Analysis (see Notes 5 and 12–16)

1. Prepare genomic DNA samples from whole blood or other convenient source, using the appropriate Puregene kit, or other standard protocol or kit.
2. Quantify the DNA by measurement of OD_{259}. Dilute sample approx 1/100 and then multiply by the dilution factor and the conversion factor of 50 to convert OD to micrograms per milliliter.
3. Check the integrity of the DNA by agarose electrophoresis of 1 µL of each sample plus 2 µL of loading buffer on small gels (0.8% in 1X TBE) in the presence of 50 ng/µL ethidium bromide, and inspection under ultraviolet (UV) light using a transilluminator.
4. Treat 5–7 µg of DNA with restriction enzymes *Hin*fI or *Pvu*II, in a final volume of 25 µL (with the buffer provided and as recommended by the manufacturers).
5. Check digestion of the DNA by electrophoresis of 3 µL of each sample plus 2 µL of loading buffer on small gels (0.8% in 1X TBE) in the presence of 50 ng/µL of ethidium bromide, and inspection under UV light .
6. For analysis of *MUC1*, *MUC2*, and *MUC5AC*, separate the *Hin*fI fragments (22 µL digest plus 7 µL of loading buffer) by electrophoresis using 0.8% 20 × 25cm agarose gels in 1X TBE, for 24 h at 2 V/cm.
7. For analysis of *MUC3*, *MUC4*, and *MUC6*, separate the *Pvu*II fragments (22 µL digest plus 7 µL of loading buffer) by electrophoresis using 0.5% 20 × 25cm agarose gels in 1X TBE, at 2 V/cm for 24 h, followed by a complete change of the tank buffer, and continued electrophoresis at 1.2 V/cm for a further 19 h.
8. Apply four kinds of markers to each gel: Raoul markers, 1-kb ladder, λ*Hin*dIII, and DNA samples with alleles of known size.
9. Following electrophoresis, visualize the markers by poststaining with 0.4 mg/mL of ethidium bromide in distilled water for 20 min (*see* **Note 12**).
10. Record the migration of the marker bands by making a photographic record including a clear ruler aligned to the leading edge of the wells.
11. Depurinate the DNA with 0.25 *M* HCl for 30 min, with occasional gentle agitation.
12. Denature with 1.5 *M* NaCl and 0.5 *M* NaOH for 30 min, with occasional gentle agitation.
13. Neutralize with 0.5 *M* Tris-HCl, 1.5 *M* NaCl, and 0.001 M EDTA, pH 7.2 for 30 min, with occasional gentle agitation (*see* **Note 13**).
14. Transfer the digested DNA onto Hybond N+ membranes by capillary blotting overnight or vacuum blotting for 2 h, both as recommended by the manufacturers, again aligning the top of the membrane accurately.
15. Fix the DNA on to the filters by baking at 80°C for 2 h.
16. Detect the *MUC* genes using TR cDNA probes: PUM24P for *MUC1* ***(30)***, SMUC41 for *MUC2* ***(13)***, SIB124 for *MUC3* ***(31)***, JER64 for MUC4 ***(32)***, JER58 for *MUC5AC* ***(15)***, and the cDNA reported in ***(17)*** for *MUC6*, and, when used, JER57 for *MUC5B* ***(14)***. Label 25 ng by random primed labeling utilizing the Multiprime DNA labeling kit using the solutions and protocol provided.
17. Prehybridize the filters in a plastic box in 200 mL of 6xSCC, 5X Denhardt's and 0.5% (w/v) SDS in a shaking water bath at 65°C (*see* **Note 14**).
18. After approx 4 h, prepare the hybridization solution. Add 500 µg of sonicated Herring sperm DNA (Promega) to the labeled probe and boil for 5 min.
19. Add to the prehybridization solution and agitate the box to ensure that the probe is dispersed evenly.
20. Hybridize the filters overnight in the shaking water bath.

21. Wash the filters down in several changes of SSC, with a final stringent wash of 0.1X SSC and 0.1% SDS at 65°C for 10 min.
22. Cover the wet filters with cling film, place luminescent Glo-bug marks on pieces of tape near the filter, and conduct autoradiography using Fuji X-ray film.
23. Determine the relative sizes of the fragments by plotting a standard curve using the control *MUC* alleles (detected after transfer by autoradiography) as well as the commercial size markers (**Note 15**). Carefully transfer the position of the top of the filter onto the autoradiograph after development by using luminescent Glo-bug marks to reposition the autoradiograph in the cassette. Measure all distances from this start line.
24. For the allele length distribution studies, plot results in histogram form grouping the fragment size in 500-bp steps (*see* **Note 16**). Analyze *MUC5AC* as two size classes as indicated (*see* **Note 5**).

3.2. MUC1 CA Microsatellite PCR (see Notes 9 and 17)

1. Add approx 100 ng of genomic DNA to a 50-µL reaction mix containing a final concentration of 200 m*M* dNTPs in 1X Promega *Taq* polymerase buffer and then denature for 5 min at 95°C (*see* **Note 17**).
2. Add 1.25 U of *Taq* polymerase, and run the PCR machine for 30 cycles as follows: denaturation for 20 s, at 94°C, annealing for 20 s at 45°C, and elongation for 20 s at 70°C.
3. Mix 0.5–1 µL of PCR product with 0.5 µL of 250-bp Sizer (Pharmacia) and 4 µL of loading buffer (5 µg/mL of Dextran Blue in 100% formamide).
4. Denature the samples for 5 min at 95°C and snap cool on ice before loading onto the gel, which is prewarmed to 42°C. Set the gel conditions such that the gel runs at 42°C, limiting at 1900 V, 55 mA, and 38 W, for 240 min. Use PCR product amplified from a clone or DNA from a homozygous individual that has been sequenced as a size standard loaded twice on each gel (*see* **Note 9**).

4. Notes

1. The TRs in *MUC1* seem to be rather conserved, such that several enzymes (e.g., *Sma*I) cut almost every repeat unit whereas many others (e.g., *Hin*fI, *Eco*RI, *Alu*I, *Pst*I, *Pvu*II, *Taq*I) do not cut at all within the array ***(12,22,24,25)***. Thus, many different restriction enzymes detect the *MUC1* VNTR polymorphism. Here we recommend the use of HinfI (**Fig. 1**), which reveals allelic band sizes of 2.8–8.0 kb with a bimodal distribution (**Fig. 2**) and heterozygosity of 0.78 in the U.K. population. Larger sizes are seen with *Eco*RI, but this enzyme has also been effectively used for disease association studies with the *MUC1* VNTR alleles ***(33)***.
2. *MUC2* shows two TR domains, the larger one containing conserved 69-bp repeats and upstream from that a smaller one with poorly conserved 48-bp repeats. Although polymorphism in *MUC2* can be detected with a large number of restriction enzymes ***(16,20)*** several of these cut one or more times within the 69-bp TR array. *Hin*fI, *Pst*I, and *Bam*HI/*Hin*dIII detect VNTR polymorphism, but of these only *Hin*fI cuts immediately either side of the 69-bp TR domain. The *Hin*dIII site is located downstream of the 69-bp repeat domain, whereas the *Bam*HI site is located upstream of the 48-bp TR domain ***(16)***. The observation that the *Bam*HI/*Hin*dIII fragments show the same relative mobility as the *Hin*fI fragments suggests that the poorly conserved TR region does not show common variation. Electrophoresis of *Hin*fI-digested DNA under the conditions described (**Fig. 1**) reveals more than 12 distinct alleles (size range: 3.3–11.4 kb in the U.K. population tested; heterozygosity: 0.59). Our studies have shown the distribution of allele lengths to be

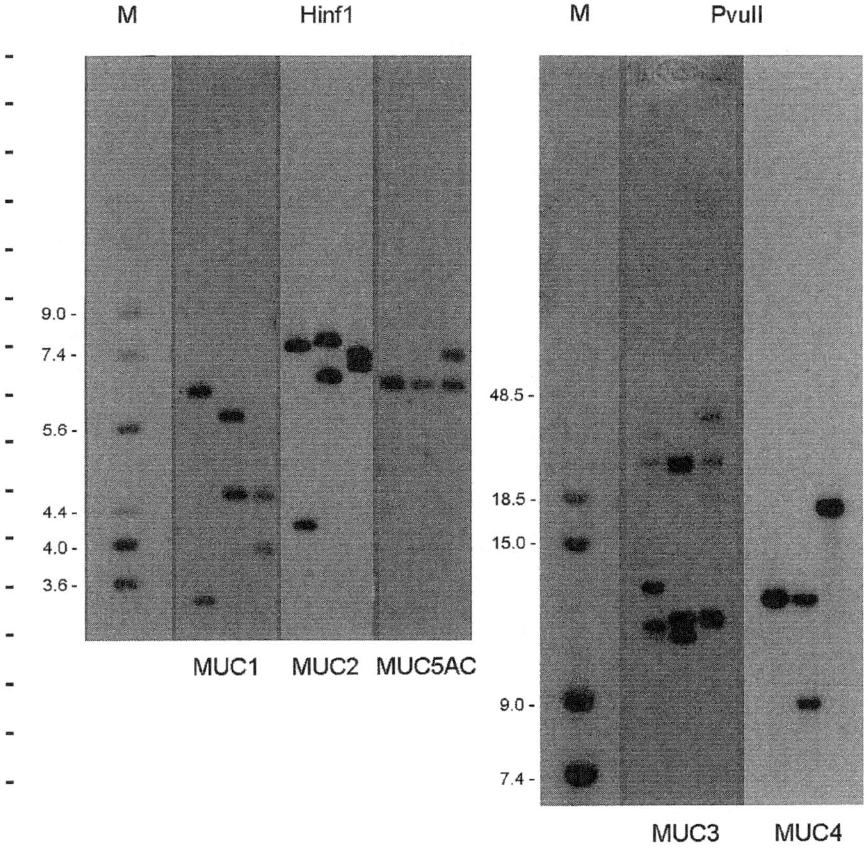

Fig. 1. Southern blot analysis of *MUC1*, *MUC2*, *MUC5AC*, *MUC3*, and *MUC4*. Examples of the *Hin*fI and *Pvu*II systems to show typical mobilities under the conditions described. *MUC6* is also run on the *Pvu*II system (not shown). The Raoul size markers (M) that are visible with the *MUC* probes are shown or their position is indicated (48.5 kb), and the sizes are given in kilobases. The dashes show the scale in centimeters.

bimodal, with the majority of the alleles approx 6.5–7.0-kb in size and a second very small peak comprising alleles of mean size of 3.5–4.0 kb. The distribution of allele lengths in unrelated individuals from the United Kingdom and including only those of northern European extraction is shown in **Fig. 2** and shows that in the U.K. population the small alleles are very rare.

3. The smallest single DNA fragment that can be detected with the *MUC3* TR probe SIB124 is an *Swa*I fragment of approx 200 kb (by pulsed field gel electrophoresis). When digested with *Pst*I or *Pvu*II, SIB124 recognises two distinct sets of very large polymorphic bands. Each set shows independent allelic variation, and there is no apparent association between the two polymorphic regions in either case, that is, the variation seen in the upper set of fragments is not dependent on that seen in the lower set. Two hundred and twenty-six unrelated northern Europeans have been tested with *Pvu*II using the protocols described

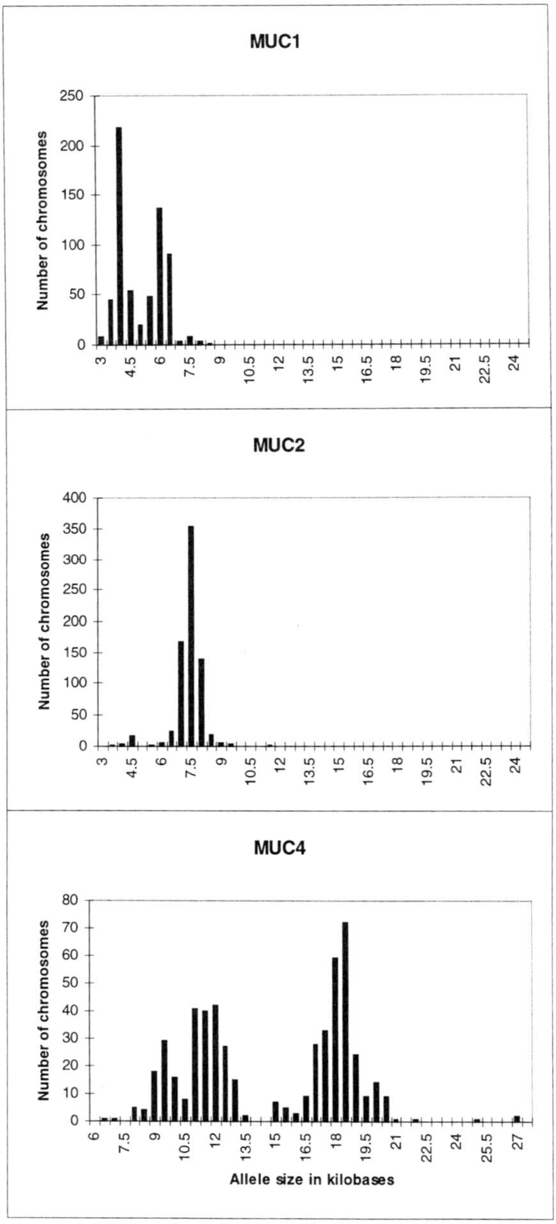

Fig. 2. Histograms showing the distribution of different *MUC 1, MUC2*, and *MUC4* size alleles in the U.K. population. The fragment sizes are grouped in 500-bp intervals to reflect the approximate accuracy of the size determinations, and the size groups are labeled such that 7 kb, e.g., contains all alleles between 7.0 and 7.4 kb inclusive; however, note that some bars on the histogram correspond to several alleles of slightly different size. Samples were taken from unrelated volunteers and include healthy persons and members of our chest and intestinal disease surveys.

here. The apparent size range of the upper set varies from 20 kb to greater than the 48.5-kb marker, with a multimodal distribution, and the most frequent allele at 24 kb, and a heterozygosity of 0.67 in the U.K. population. The lower set of polymorphic fragments detected vary in size from 7 to 15 kb, with a unimodal distribution with a peak at about 12 kb and a heterozygosity of 0.51 in the U.K. population. Examples are shown in **Fig. 1**. The *Pst*I bands have not been sized accurately but they are very similar in size to the *Pvu*II bands. Initially, the broad similarity of the patterns observed with both *Pvu*II and *Pst*I indicated that the polymorphism was simply owing to variation in the number of 51-bp TRs in the two zones. However, it was later noted that the relative mobilities of the bands detected with *Pvu*II and *Pst*I are not always consistent. The simplest interpretation of these observations is that there is some VNTR variation with additional polymorphism of a *Pst*I site, although polymorphism at *Pvu*II sites cannot be excluded. The explanation for the very large size of the zones that contain 51-bp TR, detected with *Pvu*II and *Pst*I, is unclear. The total size of two haploid sets of polymorphic *Pvu*II fragments can vastly exceed 50 kb, and this does not even cover the whole transcript since the recently described ***(4)*** large TR zone contains regular *Pvu*II sites. If, in fact, most of this sequence was expressed, a very large message would be produced. Indeed, the largest and smallest alleles of the upper set of fragments detected with *Pvu*II differ in size by approx 30,000 bp. This is not compatible with the size of the mRNA transcripts detected by Northern blotting, which has recently been estimated as 16 or 17.5 kb (two distinct size alleles in three individuals ***[27]***). These observations may indicate that either one or both of the VNTR zones contains intronic sequences and that the total length of intronic sequence may differ in different alleles, perhaps also owing to VNTR polymorphism. Alternatively, one region may represent a pseudogene. If this were true, it would be tempting to speculate that the smaller set of bands represents the expressed gene.

4. Polymorphism of *MUC4* is detected with all restriction enzymes tested (*Bam*HI, *Hin*dII, *Pst*I, *Eco*RI, *Taq*I, *Pvu*II, *Hin*fI, and *Rsa*I ***[19]***; Vinall et al., unpublished data) using the TR probe JER64. Of these, *Rsa*I gives a complex pattern of bands, and *Hin*fI a pattern of one, two, or three bands. *MUC4* shows more allele length diversity than any of the other *MUC* genes. *Pvu*II digestion and electrophoresis under the conditions described (**Fig. 1**) reveals a range of allele sizes from 6.5–27 kb with a trimodal frequency distribution (**Fig. 2**), and heterozygosity of 0.78 in the U.K. population. A similar pattern (with slightly smaller bands) is revealed by double digestion with *Pst*I and *Eco*RI, which cut closer to the tandem repeats.
5. *MUC5AC* is also highly polymorphic and polymorphism can be readily detected with a variety of enzymes ***(21,26)***. Evidence of VNTR variation comes from the correspondance of patterns observed with several restriction enzymes. With *Hin*fI and *Pst*I band sizes largely fall into two major classes (a: *Hin*fI 6.6 kb and *Pst*I 8.4 kb; b: *Hin*fI 7.4 kb and *Pst*I 9.0 kb), but these clearly represent more than two alleles since there are additional fine variations that are not correlated in the two enzyme digests. Several other enzymes (e.g., *Pvu*II, *Taq*I, and *Msp*I) reveal more than one set of bands. The large bands detected with *Pvu*II correlate well in relative mobility with the *Pst*I and *Hin*fI bands, but the large bands observed with *Taq*I and *Msp*I are different from these. However, a correspondence in relative mobility is evident between the small additional bands detected with *Pvu*II, *Taq*I, and *Msp*I. The 24-bp TR array of *MUC5AC* is interrupted by cysteine-rich sequences ***(15)***. Our results suggest a length variation involving two zones within this domain and that *Hin*fI cuts outside one of these zones but several times within the second variable region, whereas *Pst*I (and also *Hin*dIII) cut outside the whole TR region, as discussed in

(26). The polymorphism may simply involve differences in numbers of 24-bp TRs but may also involve duplication of larger stretches of sequence. Since it is more likely that the larger differences in size have an impact on function than the very small variations, and also the small size differences observed with *Pst*I are hard to evaluate, we have chosen to use *Hin*fI to assign the two major alleles, a and b. The allele frequencies found in the U.K. population are a = 0.79 and b = 0.21, with two rare alleles in 334 individuals.

6. Several restriction enzymes (e.g., *Hae*III, *Msp*I, and *Pst*I) reveal a relatively complex pattern of multiple bands that show person to person variation while *Hin*dIII and *Eco*RI each give a "single" very large band (>>30 kb) but also show hints of person-to-person variation. However, with *Pvu*II, a very clear length polymorphism is detected. A quite similar pattern is seen with *Taq*I, but *Taq*I also cuts once or twice (in different alleles) within the TRs, making it unsuitable for VNTR analysis. *Pvu*II is the only enzyme identified so far that cuts outside the TRs but close enough to reveal the polymorphism clearly. A simple pattern of bands is observed with this enzyme, composed of one or two large bands in each individual, owing to 11 or more distinct alleles, ranging in size from 8 to 13.5 kb. The frequency distribution of these alleles is approximately unimodal, with a peak at about 10 kb. A heterozygosity of 0.70 was obtained in our previous studies for the unrelated chromosomes from the CEPH families ***(26)***. *MUC6* has not yet been analyzed in our U.K. population on precisely the same gel system as described here, so no size distribution data are yet available and no examples are shown here.
7. *MUC5B* contrasts with the other mucins in showing little variation ***(26)***. Multiple bands are detected in DNA digested with several enzymes (e.g., *Msp*I, *Pst*I, and *Taq*I). Relatively infrequent variant patterns involving the presence or absence of one or more small bands were detected with *Pst*I and *Taq*I. A single large band is detected with *Eco*RI (27 kb) and with *Hin*dIII (25 kb). In most individuals (52/54), a single large band (16.5 kb) is detected in DNA digested with *Bgl*II, which cuts immediately outside the TR domain, but two individuals showed an additional band (19.5 or 15.5 kb). With *Eco*RI these two individuals both showed the common phenotype of a single 27-kb band, suggesting that the variant phenotypes are owing to nucleotide changes within *Bgl*II sites rather than numbers of TR. *MUC5B* is therefore not included in our main protocol.
8. Limited VNTR variation has been reported in the small *MUC7* gene. Analysis of 14 individuals by PCR amplification across a region containing 69-bp TRs revealed that the most common allele contains six repeats whereas a less common allele contains five repeats ***(10)***. PCR was conducted using the primers CTGGACTGCTAGCTCACCAGA AGCCG and TTCAGAAGTGTCAGGTGCAAG located at nucleotides 242–267 and 1068–1048 in Genbank/EMBL L13283.
9. Two other polymorphisms have been identified within the *MUC1* gene ***(34,35)***, one in exon 2 and one in intron 6. Both are in linkage disequilibrium with the VNTR alleles and can be detected by PCR-based techniques ***(35)***. Here we describe the protocol for detection of the CA microsatellite polymorphism in intron 6—the easier of the two sites to assess. Three common alleles of 176, 178, and 180 bp corresponding to CA_{11}, CA_{12}, and CA_{13} are detected in Europeans.
10. 10X TBE stock comes out of solution when cold.
11. SSC for hybridization is autoclaved.
12. Ethidium bromide should not be included in the gel because it distorts the electrophoretic separation and mobilities, particularly of *MUC2*.
13. **Steps 11–13** are only for passive blotting. For vacuum blotting, the same solutions are used, but as recommended by the manufacturer of the vacuum blotter.
14. Several filters can be probed together, with a blank filter layered on top.

15. The Raoul markers are also visible after transfer, since they are usually revealed with the *MUC* probes.
16. Despite all the size markers, slight gel-to-gel variations mean that it is not possible to size the bands more accurately. Analysis of individual gels makes it apparent that several alleles exist within each size range, with the exception of *MUC6* (with a repeat unit of 507 bp), but it is not practicable to rerun large numbers of samples in different combinations to assign the alleles more precisely, and, in any case, alleles that differ by a single repeat unit are unlikely to be separated on these gels.
17. The reaction mix is covered with mineral oil (Sigma) unless a PCR machine with a heated cover is used.

Acknowledgments

The authors are grateful for the support of the British Lung Foundation. They also wish to thank Drs. J. P. Aubert, N. Porchet, P. Pigny, D. Mitchell, J. R. Gum, N. Toribara, M. Sarner, and J. Lennard-Jones for their collaboration and contributions to this work.

References

1. Pigny, P., Guyonnet-Dupérat, V., Hill, A. S., Pratt, W. S., Galiègue-Zouitina, S., Collyn d'Hooge, M., Laine, A., Van-Seuningen, I., Degand, P., Gum, J. R., Kim, Y. S., Swallow, D. M., Aubert, J.-P., and Porchet, N. (1996) Human mucin genes assigned to 11p15. 5: identification and organisation of a cluster of genes. *Genomics* **38,** 340–352.
2. Shankar, V., Pichan, P., Eddy, R. L., Tonk, V., Nowak, N., Sait, S. N. J., Shows, T. B., Schultz, R. E., Gotway, G., Elkins, R. C., Gilmore, M. S., and Sachdev, G. P. (1997) Chromosomal localisation of a human mucin gene (MUC8) and cloning of the cDNA corresponding to the carboxy terminus. *Am. J. Respir. Physiol. Cell. Biol.* **16,** 232–241.
3. Gendler, S. J. and Spicer, A. P. (1995) Epithelial mucin genes. *Annu. Rev. Physiol.* **57,** 607–634.
4. Gum, J. R., Ho, J. L., Pratt, W. S., Hicks, J. W., Hill, A. S., Vinall, L. E., Roberton, A. M., Swallow, D. M., and Kim, Y. S. (1997) MUC3 Human intestinal mucin. *J. Biol. Chem.* **272,** 26,678–26,686.
5. Khatri, I. A., Forstner, G. G., and Forstner, J. F. (1997) The carboxyl-terminal sequence of rat intestinal mucin RMuc3 contains a putative transmembrane region and two EGF-like motifs. *Biochim. Biophys. Acta* **1326,** 7–11.
6. Shekels, L. L., Hunninghake, D. A., Tisdale, A. S., Gipson, I. K., Kieliszewski, M., Kozak, C. A., and Ho, S. B. (1998) Cloning and characterisation of mouse intestinal MUC3: 3' sequence contains EGF-like domains. *Genbank/EMBL AF027137.*
7. Nollet, S., Moniaux, N., Maury, J., Petitprez, D., Degand, P., Laine, A., Porchet, N., and Aubert, J.-P. (1998) Human mucin gene MUC4: organisation of its 5'-region and polymorphism of its central tandem repeat array. *Biochem. J.* **332,** 739–748.

7a. Crawley, S. C., Gum, J. R. Jr., Hicks, J. W., Pratt, W. S., Aubert, J. P., Swallow, D. M., and Kim, Y. S. (1999) Genomic organization and structure of the 3' region of human *MUC3*: alternative splicing predicts membrane-bound and soluble forms of the mucin. *Biochem. Biophys. Res. Commun.* **263,** 728–736.

7b. Williams, S. J., Munster, D. J., Quin, R. J., Gotley, D. C., and McGuckin, M. A. (1999) The *MUC3* gene encodes a transmembrane mucin and is alternatively spliced. *Biochem. Biophys. Res. Commun.* **261,** 83–89.

7c. Moniaux, N., Nollet, S., Porchet, N., Degand, P., Laine, A., and Aubert, J. P. (1999) Complete sequence of the human mucin *MUC4*: a putative cell membrane-associated mucin. *Biochem. J.* **338,** 325–333.
8. Chang, S.-K., Dohrman, A. F., Basbaum, C. B., Ho, S. B., Tsuda, T., Toribara, N. W., Gum, J. R., and Kim, Y. S. (1994) Localization of mucin (MUC2 and MUC3) messenger RNA and peptide expression in human normal intstine and colon cancer. *Gastroenterology* **107,** 28–36.
9. Audie, J. P., Janin, A., Porchet, N., Copin, C., Gosselin, B., and Aubert, J. P. (1993) Expression of human mucin genes in respiratory, digestive, and reproductive tracts. *J. Histochem. Cytochem* **41,** 1479–1485.
10. Biesbrock, A. R., Bobek, L. A., and Levine, M. J. (1997) MUC7 gene expression and genetic polymorphism. *Glycoconjugate J.* **14,** 415–422.
11. Bobek, L. A., Liu, J., Saint, S. N. J., Shows, T. B., Bobek, Y. A., and Levine, M. J. (1996) Structure and chromosomal localization of human salivary mucin gene, MUC7. *Genomics* **31,** 277–282.
12. Gendler, S., Taylor-Papadimitriou, J., Duhig, T., Rothbard, J., and Burchell, J. (1988) A highly immunogenic region of a human polymorphic epithelial mucin expressed by carcinomas is made up of tandem repeats. *J. Biol. Chem.* **263,** 12,820–12,823.
13. Gum, J. R., Byrd, J. C., Hicks, J. W., Toribara, N. W., Lamport, D., and Kim, Y. S. (1989) Molecular cloning of human intestinal mucin cDNAs. Sequence analysis and evidence for genetic polymorphism. *J. Biol. Chem.* **264,** 6480–6487.
14. Dufossé, J., Porchet, N., Audié, J. P., Guyonnet, D. V., Laine, A., VanSeuningen, I., Marrakchi, S., Degand, P., and Aubert, J. P. (1993) Degenerate 87-base-pair tandem repeats create hydrophilic/hydrophobic alternating domains in human mucin peptides mapped to 11p15. *Biochem. J.* **293,** 329–337.
15. Guyonnet-Dupérat, V., Audié, J.-P., Debailleul, V., Laine, A., Buisine, M.-P., Galiègue-Zouitina, S., Pigny, P., Degand, P., Aubert, J.-P., and Porchet, N. (1995) Characterization of the human mucin gene MUC5AC: a consensus cysteine-rich domain for 11p15 mucin genes? *Biochem. J.* **305,** 211–219.
16. Toribara, N. W., Gum, J. J., Culhane, P. J., Lagace, R. E., Hicks, J. W., Petersen, G. M., and Kim, Y. S. (1991) MUC-2 human small intestinal mucin gene structure. Repeated arrays and polymorphism. *J. Clin. Invest.* **88,** 1005–1013.
17. Toribara, N., Roberton, A., Ho, S., Kuo, W., Gum, E., Hicks, J., Gum, J. J., Byrd, J., Siddiki, B., and Kim, Y. (1993) Human gastric mucin. Identification of a unique species by expression cloning, *J. Biol. Chem.* **268,** 5879–5885.
18. Fox, M. F., Lahbib, F., Pratt, W., Attwood, J., Gum, J. R., Kim, Y., and Swallow, D. M. (1992) Regional localisation of the intestinal mucin gene MUC3 to chromosome 7q22. *Ann. Hum. Genet.* **56,** 281–287.
19. Gross, M.-S., Guyonnet-Dupérat, V., Porchet, N., Bernheim, A., Aubert, J. P., and Nguyen, V., C, (1992) Mucin 4 (MUC4) gene: regional assignment (3q29) and RFLP analysis. *Ann. Genet.* **35,** 21–26.
20. Griffiths, B., Mathews, D. J., West, L., Attwood, J., Povey, S., Swallow, D. M., Gum, J. R., and Kim, Y. S. (1990) Assignment of the polymorphic intestinal mucin gene (MUC2) to chromosome 11p15. *Ann. Hum. Genet.* **54,** 277–285.
21. Pigny, P., Pratt, W. S., Laine, A., Leclercq, A., Swallow, D. M., Nguyen, V. C., Porchet, N., and Aubert, J.-P. (1995) The MUC5AC gene: RFLP analysis with the Jer 58 probe. *Hum. Genet.* **96,** 367–368.

22. Swallow, D. M., Gendler, S., Griffiths, B., Corney, G., Taylor-Papadimitriou, J., and Bramwell, M. E. (1987) The human tumour-associated epithelial mucins are coded by an expressed hypervariable gene locus PUM. *Nature* **328,** 82–84.
23. Karlsson, S., Swallow, D. M., Griffiths, B., Corney, G., and Hopkinson, D. A. (1983) A genetic polymorphism of a human urinary mucin. *Ann. Hum. Genet.* **47,** 263–269.
24. Hareuveni, M., Tsarfaty, I., Zaretsky, J., Kotkes, P., Horev, J., Zrihan, S., Weii, M., Green, S., Lathe, R., Keydar, I., and Wreschner, D. N. (1990) A transcribed gene, containing a variable number of tandem repeats, codes for a human epithelial tumor antigen. cDNA cloning, expression of the transfected gene and over-expression in breast cancer tissue. *Eur. J. Biochem.* **189,** 475–486.
25. Ligtenberg, M., Vos, H. L., Gennissen, A., and Hilkens, J. (1990) Episialin, a carcinoma-associated mucin, is generated by a polymorphic gene encoding splice variants with alternative amino termini. *J. Biol. Chem.* **265,** 5573–5578.
26. Vinall, L. E., Hill, A. S., Pigny, P., Pratt, W. S., Toribara, N., Gum, J. R., Young, K. I., Porchet, N., Aubert, J. P., and Swallow, D. M. (1998) Variable number tandem repeat polymorphism of the mucin genes located in the complex on 11p15. 5. *Hum. Genet.* **102,** 357–366.
27. Debailleul, V., Laine, A., Huet, G., Mathon, P., Collyn d'Hooghe, M., Aubert, J. P., and Porchet, N. (1998) Human mucin genes MUC2, MUC3, MUC4, MUC5AC, MUC5B and MUC6 express stable and extremely large mRNAs and exhibit a variable length polymorphism. An improved method to analyse large size RNAs. *J. Biol. Chem.* **273,** 881–890.
27a. Hermann, A., Davies, J. R., Lindell, G., Martensson, S., Packer, N. H., Swallow, D. M., and Corstedt, I. (1999) Studies on the insoluable glycoprotein complex from human colon: identification of reduction-insensitive MUC2 oligomers and C-terminal cleavage. *J. Biol. Chem.* **274,** 15,828–15,836.
28. Gum, J. R., Hicks, J. W., Toribara, N. W., Siddiki, B., and Kim, Y. S. (1994) Molecular cloning of human intestinal mucin (MUC2) cDNA-identification of the amino-terminus and overall sequence similarity to pre-pro von-Willibrand. *J. Biol. Chem.* **269,** 2440–2446.
29. Brunner, C., Lobentanz, E.-M., Petho-Schramm, A., Ernst, A., Kang, C., Dieplinger, H., Muller, H. J., and Uterman, G. (1996) The number of identical Kringle IV repeats in Apolipoprotein (a) affects its processing and secretion by HepG2 cells. *J. Biol. Chem.* **271,** 32,403–32,410.
30. Yonezawa, S., Byrd, J. C., Dahiya, R., Ho, J., Gum, J. R., Griffiths, B., Swallow, D. M., and Kim, Y. S. (1991) Differential mucin gene expression in human pancreatic and colon cancer cells. *Biochemical J.* **276,** 599–605.
31. Gum, J. R., Hicks, J. W., Swallow, D. M., Lagace, R. L., Byrd, J. C., Lamport, D., Siddiki, B., and Kim, Y. S. (1990) Molecular cloning of cDNAs derived from a novel human intestinal mucin gene. *Biochem. Biophys. Res. Commun.* **171,** 407–415.
32. Porchet, N., Cong, N. V., Dufosse, J., Audie, J. P., Guyonnet-Dupérat, V., Gross, M. S., Denis, C., Degand, P., Bernheim, A., and Aubert, J. P. (1991) Molecular cloning and chromosomal localization of a novel human tracheo-bronchial mucin cDNA containing tandemly repeated sequences of 48 base pairs. *Biochem. Biophys. Res. Commun.* **175,** 414–422.
33. Carvalho, F., Seruca, R., David, L., Amorim, A., Seixas, M., Bennett, E., Clausen, H., and Sobrinho-Simoes, M. (1997) MUC1 gene polymorphism and gastric cancer- an epidemiological study. *Glycoconjugate J.* **14,** 107–111.
33a. Moniaux, N., Nollet, S., Porchet, N., Degand, P., Laine, A., and Aubert, J.-P. (1999) Complete sequence of the human mucin MUC4: a putative cell membrane-associated mucin. *Biochem. J.* **338,** 325–333.

34. Ligtenberg, M., Gennissen, A., Vos, H. L., and Hilkens, J. (1991) A single nucleotide polymorphism in an exon dictates allele dependent differential splicing of episialin mRNA. *Nucleic Acids Res.* **19,** 297–301.
35. Pratt, W. S., Islam, I., and Swallow, D. M. (1996) Two additional polymorphisms within the hypervariable MUC1gene: association of alleles either side of the VNTR region. *Ann. Hum. Genet.* **60,** 21–28.
36. Gendler, S. J., Lancaster, C. A., Taylor-Papadimitriou, J., Duhig, T., Peat, N., Burchell, J., Pemberton, L., Lalani, E. N., and Wilson, D. (1990) Molecular cloning and expression of human tumor-associated polymorphic epithelial mucin. *J. Biol. Chem.* **265,** 15,286–15,293.

IX

Preparation of Antimucin Antibodies

29

Polyclonal and Monoclonal Techniques

Francisco X. Real, Carme de Bolós, and Egbert Oosterwijk

1. Introduction

In the past twenty years, much progress in the study of mucins has resulted from the development of antibodies recognizing carbohydrate and peptide epitopes. Although antibodies are extremely useful reagents in biology, antibodies of ambiguous specificity, or antibodies whose specificity has been poorly characterized, can also add considerable confusion to any field of study. Whereas it is relatively straightforward to develop antibodies recognizing a known molecule or epitope, it is often much more difficult and time-consuming to establish the specificity of antibodies raised against complex molecular mixtures. Furthermore, because the universe of antigens cannot be systematically tested, it is extremely important to bear in mind that antibodies, whether polyclonal or monoclonal, raised against a known molecule may crossreact with unrelated molecules. The latter may share with cognate antigen chemical characteristics that are easy to identify (i.e., primary amino acid sequence), or lack apparent structural relatedness. Therefore, when using antibodies, a word of caution is always necessary.

Because antibodies are tools, the most important question at the outset is, What do we need the antibodies for? The answer to this question will impose constraints on the experimental strategies used to obtain them. Several questions, that are applicable in general to any antibody techniques, and specifically to mucins because of their biochemical complexity, should be taken into account:

1. Are we searching antibodies of well-known, defined specificity or antibodies of novel specificity?
2. What kind of immunogen is available?
3. What application are the antibodies to be used for?
4. How do we plan to establish antibody specificity?
5. Is it necessary to prepare polyclonal or monoclonal antibodies (MAbs)?
6. How much antibody do we foresee we will need?
7. Is batch-to-batch variation an important issue?
8. What are the financial constraints?

From: *Methods in Molecular Biology, Vol. 125: Glycoprotein Methods and Protocols: The Mucins*
Edited by: A. Corfield © Humana Press Inc., Totowa, NJ

1.1. Antigen Source

1.1.1. Purified Mucins

When dealing with mucins as antigens, no matter how carefully the purification is carried out, the concept "purified antigen" is not operationally useful (*see* Chapters 7, 8, 15, and 16). First, mucins are extremely large macromolecules consisting of a large peptide backbone (e.g., the predicted molecular mass for an MUC2 apomucin monomer is >550 kDa) and a large number of carbohydrate chains. Second, mucin gels may consist in vivo of hetero-oligomers of multiple apomucin species since more than one mucin gene is expressed in a given tissue and even at the single-cell level. Third, there is extensive variability in the structure of the carbohydrate chains linked to a given apomucin monomer. Fourth, detailed molecular characterization of a complex biochemical mix is simply difficult, and often impossible, beyond a certain level (i.e., it is difficult and time-consuming to establish whether specific carbohydrate chains are selectively linked to Ser or Thr in the context of specific amino acid motifs). Finally, if the antigen preparation to be used for immunization is also to be used for the analysis of antibody specificity, the preceding difficulties will apply.

These caveats do not imply that antisera raised against native mucins are not useful; for example, affinity purification on a solid matrix may be used to obtain highly specific antibodies. It is even possible that certain specificities may only be attained by using such preparations.

The potential difficulties associated with the use of preparations of purified mucin would suggest that peptides, glycopeptides, or carbohydrates obtained either by synthesis or biochemical purification would be more appropriate immunogens. However, the main advantage of such molecules, a simpler biochemical structure, may become a disadvantage because complex and perhaps highly specific epitopes resulting from apomucin folding (i.e., in the poorly glycosylated domains), oligomerization, or glycosylation may be missed.

1.1.2. Deglycosylated Mucins

Procedures regularly used for deglycolsylation of mucin lead both to the removal of carbohydrate moieties and to peptide proteolysis. This treatment may thus affect the immunogenicity of the resulting molecules. Preparations of deglycosylated mucin have proven useful to generate antibodies for the screening of cDNA libraries ***(1–3)*** as well as for the production of MAbs ***(4)***, and their use remains an interesting possibility.

1.1.3. Fresh Tissue and Cultured Cells

The main advantage of using whole-tissue extracts or cells is that one may aim at isolating antibodies, primary MAbs, recognizing yet unknown molecules. Polyclonal antibodies raised against colonic or gastric tissue extracts have been quite useful both for the cell biological characterization of mucus-producing cells and for the isolation of mucin-encoding cDNAs ***(5,6)***.

Despite the fact that many investigators have raised MAbs against fresh tissues and cultured epithelial cells, very few of these antibodies have turned out to recognize apomucin epitopes, either in the repetitive or nonrepetitive regions, except for MUC1 ***(7)***. This observation points to a relative immunodominance of carbohydrate epitopes.

1.1.4. Synthetic Peptides

Synthetic peptides have been extensively used for the production of both polyclonal antibodies and MAbs; the production of mAbs recognizing the repetitive domains of mucins is reviewed elsewhere (*see* Chapter 30). In vitro glycosylated peptides have been successfully used to produce MAbs specific for cabohydrate-peptide epitopes ***(8)***.

1.1.5. Recombinant Proteins

Few investigators have reported the use of proteins produced from recombinant cDNAs in *Escherichia coli* or baculovirus to make antibodies. Based on the scanty information available in the literature and the experience of our own laboratory, the isolation of recombinant proteins often presents solubility and/or purification problems, thus complicating their use. The Ser^{-}/Thr-rich regions are normally glycosylated in nature and they may not acquire stable conformations when unglycosylated. In addition, the Ser^{-}/Thr-poor domains are Cys rich and, intra- and intermolecular disulfide bonds may be formed in vitro that do not represent the actual conformation of these domains of mucin in the natural molecules. Despite these difficulties, more attention should be given to this strategy since it is based on relatively less complex antigen preparations.

1.1.6. Other Strategies

Several additional strategies have been used to obtain antibodies of novel specificity when using complex antigen preparations. Among them are the induction of tolerance in newborn mice ***(9)*** or the preincubation of antigen preparations with known antibodies in order to decrease the likelihood of raising antibodies with the same specificity ***(10)***. Although these possibilities are intelectually stimulating and have been used with some success, there is limited information in the literature regarding their usefulness.

1.2. Polyclonal vs MAbs

Because the development of MAbs is relatively expensive and time-consuming, if the availability of highly purified immunogen is not limiting, it is reasonable to first produce polyclonal antibodies, analyze their specificity, and then assess the interest of making MAbs. The major reason to prepare the latter is to have an indefinite source of reagents of identical specificity. When considering the preparation of polyclonal antibodies, several points should be taken into account:

1. Natural antibodies may contribute to substantial background; this is particularly true for antibodies recognizing carbohydrate epitopes and is, therefore, a relevant issue when dealing with mucins.
2. There may be considerable variation in the immune response obtained in animals from different species. In general, stronger immune responses are obtained across greater phylogenetic differences. Detailed information on the homology of proteins across species may facilitate decision making. Usually, greater homology is associated with lower immunogenicity.
3. It is possible to obtain very large amounts of antibodies from certain species, e.g., horse. Some inexpensive and easy-to-handle laboratory animals, such as chicken, can provide substantial antibody amounts, e.g., in the egg yolk.

4. Even among animals sharing a genetic background, significant interindividual variation can be observed in the immune response. Therefore, more than one animal should be immunized.
5. The immune response is a dynamic process and subtle differences in the specificity of serum antibodies can be demonstrated when comparing different bleedings from the same animal.

1.3. Screening Strategies

The general principle that "one gets what one looks for" should not be forgotten. Antibodies, whether monoclonal or polyclonal, often work best when using certain, (but not all), techniques. Therefore, when the application is known at the outset, it is most useful to devise screening strategies accordingly. For example, if one wishes to use antibodies for studies on paraffin-embedded tissue sections, it is best to include immunohistochemical analysis on paraffin-embedded sections in the screening procedure. This is not always feasible when preparing hybridomas because some methods are too tedious to be used for routine screening (e.g., immunoprecipitation). In such a case, it may be better to use a simple screening strategy that will allow the identification of a small number of antigen-reactive clones and then apply a second round of screening with the specific technique desired.

1.4. Financial and Other Considerations

Making polyclonal antibodies is less time-consuming and cheaper than making hybridomas. The commercial cost of polyclonal antipeptide antibodies is in the range of $3,000 (U.S.). By contrast, the production of hybridomas will cost approx $8,000 (U.S.). Among the considerations to be taken into account is the experience of the group. For an unexperienced laboratory, setting up the hybridoma technology will require some effort whereas it is easy to prepare polyclonal antibodies. Another consideration is time. Provided that the antigen preparation is reasonably immunogenic, it takes less time to prepare affinity-purified polyclonal antibodies (3–4 mo) than to make hybridomas (6–12 mo until a purified antibody preparation is available).

2. Materials

2.1. Polyclonal Sera

1. Complete Freund's adjuvant (Sigma, St. Louis, MO).
2. Incomplete Freund's adjuvant (Sigma).

2.2. Monoclonal Sera

1. RPMI (Gibco-BRL, Edinburgh, UK).
2. Neubauer chamber.
3. Trypan blue (Sigma).
4. 96-well plates (Nunc, Life Technologies, Edinburgh, UK).
5. Complete medium: RPMI with 10% fetal bovine serum (FBS), 1% glutamine, 1% nonessential amino acids, 1% penicillin/streptomycin.
6. Avertin (Aldrich, Gillingham, UK): 2,2,2-tribromoethylalcohol (10 g) in 10 mL of tert-amyl alcohol (stock solution).
7. Lysis buffer: 150 m*M* NH_4Cl, 0.01 *M* $KHCO_3$ 0.1 m*M* EDTA, pH 7.0.

8. Mouse myeloma cell lines (American Type Culture Collection, Rockville, MD or European Collection of Cell Cultures, Salisbury, UK).
9. Polyethylene glycol (PEG) 1500 (Boehringer Mannheim, Germany).
10. HAT medium: RPMI supplemented with 1 m*M* sodium pyruvate, 2 m*M* L-glutamine, 1% nonessential amino acids, 100 U/mL of penicillin, 100 mg/mL of streptomycin, 10% FBS, 2% HAT stock (Sigma) (hypoxanthine, aminopterin, thymidine).
11. Phosphate buffered saline (PBS): 1.9 m*M* NaH_2PO_4, 8.1 m*M* Na_2HPO_4, 154 m*M* NaCl, pH 7.2.
12. Dimethyl sulfoxide (DMSO) (Sigma).
13. Tissue culture flasks, 25 and 175 cm^2 (Nunc, Life Technologies).
14. Pristane (Sigma).
15. Agarose type VII (low melting temperature) (Sigma).
16. Nitrocellulose filters (0.45 mm, 6 cm diameter) (Schleicher and Schuell, Dassell, Germany).
17. Sintered glass funnel (Millipore, Bedford, MA).
18. Ultraviolet (UV) light (30 W, Philips TUV).
19. Mycoplasma polymerase chain reaction (PCR) detection kit (Stratagene, La Jolla, CA).

3. Methods

3.1. Polyclonal Antisera

Several species can be used to obtain polyclonal antibodies. Rabbits are most commonly used, but chickens provide certain advantages that should be considered since it is relatively simple to house them. Up to 5 mg of total IgY can be isolated from the yolk of one egg. The procedure for making rabbit polyclonal sera is described next.

3.1.1. Preimmune Serum

Prior to the first immunization, obtain a preimmune serum sample from the auricular artery using a 20-gage needle and a 20-cm^3 syringe. To dilate the artery, it is advisable to rub the ear with alcohol and warm it under a spotlight. If mice are used, obtain the preimmune serum sample from the retroauricular plexus by means of a Pasteur pipet.

3.1.2. Screening of the Preimmune Rabbit Sera

Screen the preimmune rabbit sera for lack of reactivity using the assay that will be applied for testing the postimmune serum. This allows the selection of rabbits whose serum yields low background. Therefore, it is advisable to analyze preimmune serum from several rabbits.

3.1.3. Preparation of the Immunogen for Inoculation

Prepare the immunogen for inoculation by mixing 200 µg of the immunogen (in 500 µL) with 500 µL of complete Freund's adjuvant until a good emulsion is obtained (amount per rabbit). In general, 1 mL of mixture yields 0.8 mL of emulsion. For mice, 10–20 µg of antigen are used.

It is best to mix both components using a 2-mL syringe with a needle until the mixture is hard. Alternatively, join two syringes (without needle) with a rubber tube and transfer the mixture from one syringe to the other until it is hard. Be careful, because as the mix hardens the needle or the rubber can slip off the syringe and the immunizing material may be lost. The emulsion is ready when, on applying a drop of it on water, it is not dispersed.

The use of Freund's adjuvant has been contested recently because of the extensive inflammation it induces. Detoxified endotoxin represents a reasonable alternative as an adjuvant.

3.1.4. First Immunization

Perform the first immunization in the leg muscle with the mixture prepared as described in **Subheading 3.1.3.** It is better to split the immunogen in two injection sites. Perform subsequent inoculations in the same way except for prepraring the emulsion with incomplete Freund's adjuvant.

3.1.5. Bleeding of Rabbits

A few days after the fourth immunization, bleed rabbits (or mice) as described in **Subheading 3.1.1.** At this time, perform a screening with the preimmune and post-fourth immunization serum samples to determine serum titer and specificity. In general, after the fourth immunization, a good antibody titer should be obtained. Otherwise, start thinking of other immunization strategies.

3.1.6. Immunization and Final Bleeding

Immunization and bleeding proceed on a monthly basis following the same protocol. If the serum titer is good, it is advisable to bleed the animals regularly (every 2 wk) to store quickly a large amount of serum because antibody specificity may change with time. Then, proceed to a final bleeding prior to killing the animal.

3.2. Monoclonal Antibodies

3.2.1. Liquid Culture Technique

3.2.1.1. Immunization

Immunize Balb/c mice three to five times subcutaneously with 20 μg of the corresponding immunogen in complete Freund's adjuvant/incomplete Freund's adjuvant as described for polyclonal antibodies. Three days before the cell fusion boost the animals with the immunogen/incomplete Freund's adjuvant.

Balb/c mice are normally used, but there have been few studies addressing the importance of mouse strain use in hybridoma production. The optimal schedule and route of immunization have seldom been analyzed in a systematic way. Therefore, the final decision is left to the imagination of the investigator. It is recommended to check the serum antibody titer prior to performing the fusion: in the absence of high antibody titers, the likelihood of getting a high yield of specific hybrids is low.

3.2.1.2. Feeder Layer Cells

Feeder layer cells are used to produce soluble factors that increase hybrid growth. Alternative sources of feeder cells are mouse thymocytes, mouse spleen cells, and certain fibroblast cell lines.

1. To obtain peritoneal macrophages, kill a normal nonimmunized mouse of the same strain as the immunized mice the day prior to the cell fusion. It is recommended to plate the cells 24 h before the cell fusion and check for contamination at the fusion time.

2. Using a 25-gage needle, inject 5 mL of cold serum-free RPMI intraperitoneally. Palpate the abdomen gently and withdraw the medium using a 21-gage needle (*see* **Note 1**).
3. Centrifuge at 700*g* for 5 min and wash the cells twice with RPMI.
4. Count the cells in a Neubauer chamber and check for viability by Trypan blue exclusion. Usually one mouse will yield 2.5×10^6 peritoneal macrophages with >95% viability.
5. Plate peritoneal macrophages in 96-well plates at $2–5 \times 10^3$/100 µL/well in complete medium (RPMI supplemented with 10% FBS, 1% glutamine, 1% nonessential amino acids, 1% penicillin/streptomycin). In general, peritoneal macrophages from one mouse will suffice for fusion of one spleen.

3.2.1.3. Spleen Cells

1. On the day of the fusion, anesthesize the mouse by intraperitoneal injection of 0.2 mL of a 2.5% avertin solution (prepared from stock). Collect blood from the retro-orbital plexus by means of a Pasteur pipet. Store serum to test for specific polyclonal antibodies.
2. Kill the mouse, remove aseptically the spleen, and transfer to a 50-mm-diameter Petri dish filled with serum-free RPMI medium. Wash the spleen twice with medium under sterile conditions.
3. Obtain a single-cell suspension of splenocytes by perforating the spleen tip with a 20-gage needle and squeezing the splenocytes out with a forceps. Disperse cells gently with a Pasteur pipet or by passage through a fine-mesh metal screen.
4. Transfer spleen cells into a 15-mL conical tube and fill up with serum-free RPMI.
5. Centrifuge at 400–700*g* for 5 min and discard supernatant.
6. Lyse red blood cells by adding 10 mL of lysis buffer and incubating for 5 min at 37°C.
7. Centrifuge and wash the cells twice with serum-free medium.
8. Resuspend splenocytes in 10 mL of RPMI, count, and assess viability with Trypan blue. Approx 100–200 million cells/spleen should be obtained.

3.2.1.4. Myeloma Cells

Several mouse myeloma cell lines are available. The choice may be determined by the fact that some lines (NS1) produce, but do not secrete, Ig light chains. Since these can associate with Ig heavy chains produced by a hybridoma, and be secreted, the preferred myeloma cells are the nonproducing lines P3X63-Ag8 or Sp2/O-Ag14.

It is essential to make sure that the myeloma cells are free of mycoplasma contamination since this results in a dramatic drop in hybrid outgrowth. We recommend to test myeloma cells regularly using PCR. If fusions are unproductive, one should first suspect that myeloma cells are not in good shape.

1. Culture mouse myeloma cells in complete medium. It is best to use cells that have been in culture for a few days. The day before the fusion, change the medium and seed cells at 10^5 cells/mL to ensure that they will be growing exponentially.
2. Harvest myeloma cells in a 25-mL conical tube and wash three times with serum-free RPMI.
3. Count the cells and assess viability by Trypan blue exclusion. It is important that viability be >95%.

3.2.1.5. Fusion

1. Mix in a 15-mL conical tube the splenocytes and myeloma cells at a ratio of 5:1. Fill up the tube with RPMI and centrifuge at 200*g* for 5 min. Usually, ratios from 1:1 to 1:10 yield high fusion efficiency.

2. Discard the media and mix gently to soften the cell pellet.
3. Add 0.5 ml (dropwise) of prewarmed (37°C) PEG 1500 over 1 min. Mix gently as the PEG solution is added to the cells.
4. Let the cells rest for 1 min and add 10 mL of prewarmed RPMI dropwise mixing gently over 10 min (1 mL/min).
5. Centrifuge at 200*g* for 5 min.
6. Wash twice with RPMI.
7. Resuspend the cells in HAT medium (*see* **Subheading 2.2., item 10**) in the appropriate volume to yield a suspension of 1×10^6 splenocytes/mL (*see* **Note 2**).
8. Seed the cells in 96-well plates at 10^5 splenocytes/well. Incubate at 37°C in a 5% CO_2 atmosphere (*see* **Note 3**).
9. Check cells, by visualization with an inverted microscope, twice a week for contamination and cell growth.
10. Change 100 μL of medium weekly (optional if cells are kept in a wet incubator). After the second week, fusion medium may be replaced by complete medium supplemented with 2% hypoxanthine, thymidine (HT) (*see* **Note 4**).
11. When clones are ready, remove 100 μL of supernatant for screening and replace with fresh medium (*see* **Note 5**).

3.2.1.6. Subcloning by Limiting Dilution

1. Gently disperse clones producing antigen-reactive antibodies of the desired specificity with an automatic pipet, and count cells.
2. Cells are cloned by limiting dilution. Seed in 96-well plates at densities of 1–1000 cells/well (approximately half-plate per dilution) in complete medium supplemented with 2% HT (*see* **Note 6**).
3. Perform several rounds of subcloning until >95% of the clones seeded at 0.1–1 cells/well are reactive with the immunogen.
4. First expand cloned hybridoma cells in wells of 24-well plates, and then freeze in liquid N_2 (in 90% FBS–10% DMSO). It is recommended to confirm the persistence of antibody-producing clones after each step of expansion. Make sure to freeze cell stocks at each step of subcloning and expansion.

3.2.1.7. Scaling Up Antibody Production

To obtain large quantities of specific MAb, the following two methods are generally used.

3.2.1.7.1. Production of Hybridoma Culture Supernatant

1. Seed the hybridoma cells in a 25-cm² tissue culture flask at a density of 10^5 cells/ml in complete medium (there are commercially available synthetic media that support hybridoma growth in the absence of FBS) (*see* **Note 7**).
2. When cells reach a density of $1–2 \times 10^6$ or the medium becomes acidic, centrifuge cell suspension, collect the supernatant, and seed the cells in a 175-cm² tissue culture flask with the appropriate amount of medium. Replace 80% of the medium regularly and substitute with fresh medium. A fraction of the cells will also have to be replaced, or the culture expanded, as indicated in the section above. Collect the spent medium, store at 4°C until use, or freeze at –40°C for long-term storage. If FBS-free medium is used, hybridoma culture medium can be concentrated in order to obtain high titer antibody preparations.

*3.2.1.7.2. Ascites Fluid Production (see **Note 8**)*

1. Inject pristane (2-, 6-, 10-, 12-tetramethylpentadecane) intraperitoneally to mice of the same strain as those used for immunization (0.5 mL/mouse) 1 wk before inoculation with hybridoma cells (*see* **Note 9**).
2. Culture hybridoma cells as described in **Subheding 3.2.7., step 1.** Centrifuge the cells, and wash twice with PBS or medium without FBS. Count the cells and determine viability.
3. Inoculate 5×10^5 cells/mouse.
4. Wait 1 to 2 wk for ascites formation.
5. To harvest ascites, wash mouse belly with 70% ethanol, insert an 18-gage needle, and let ascites drop into a conical polypropylene tube.
6. To remove the cells, centrifuge ascites at 1100–1400*g* for 10 min and discard the pellet.
7. Add 0.1% sodium azide (0.1% final concentration), make aliquots, and freeze at –80°C.
8. Continue to tap the animals at 1- to 3-d intervals (*see* **Note 10**).

*3.2.2. Filter Plaque Technique (see **Notes 11–20**)*

The following methodology is adapted from **ref. *11***. **Steps 1**, **3**, **4**, **5**, and **7** in **Subheading 3.2.1.** are to be followed.

1. To prepare 20 agarose plates, sterilize 45 mL of a 3% agarose solution in 0.9% NaCl and allowed to cool to approx 50°C. Add 40 mL to 160 mL of complete HAT medium at 37°C (final concentration 0.6% low-melting agarose).
2. Add 10 mL of fluid to each dish and allow the agarose to solidify at room temperature. Avoid bubbles on top of the agarose layer. Do not solidify the agarose at 4°C, because this will result in entrapment of air bubbles.
3. Fuse splenocytes and myeloma cells as described in **Subheading 3.2.1.5.** (*see* **Notes 21 and 22**).
4. Resuspend the primary fusion products in 30 mL complete HAT medium at 37°C, add 4 mL 3% agarose solution at 40°C and gently mix, and layer the cell suspension over the 20 dishes. Place the dishes in the refrigerator at 4°C for 20 min and subsequently place in a regular CO_2 incubator.
5. Examine outgrowth of cells at d 5 and every other day subsequently (*see* **Note 23**).
6. Prepare nitrocellulose filters as follows: cut filters from a larger nitrocellulose sheet, including a tab to facilitate filter handling (*see* **Fig. 1**). Alternatively, filters of 6 cm diameter are available from Schleicher and Schuell. Wash filters in 0.9% NaCl. This step removes toxic reagents from the nitrocellulose and is absolutely necessary. Homemade filters can be prepared with a 6-cm-diameter stainless steel punch, with a sharpened edge. By leaving a 5-mm opening in the sharpened edge, a small triangle can be cut, resulting in filters shaped as shown in **Fig. 1**.
7. Coat filters with purified antigen or tissue cell homogenates. Place filters on a sintered glass funnel and suck 10 mL of the antigen solution through each filter. For purified molecules, 10–50 µg per filter is sufficient; lower amounts can be used when the amount of material available for the assay is a limiting factor. For cell or tissue homogenates, mince 1 g of tissue and homogenize in 20 mL of PBS. Remove larger debris by low-speed centrifugation at 3000 rpm for 10 min. For coating purposes, prepare a 10-fold dilution, and use 10 mL of this solution per filter.
8. Punch three small holes at the sides of filters coated with irrelevant antigen to shape a triangle to facilitate positioning (**Fig. 1**).
9. Air-dry filters and sterilize by UV light (20 min per side, twice) in a laminar flow hood.

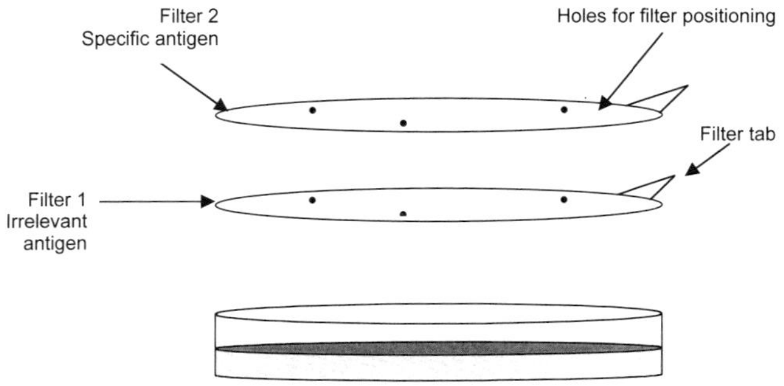

Fig. 1. Screening using the filter plaque technique. Placement of filters on top of the agarose layer and array of holes for filter positioning.

Wash several times in sterile water and store filters overnight in sterile water. The next day, soak filters in complete medium supplemented with 2% HAT to block remaining protein binding sites. Filters can be stored at –20°C or used immediately after preparation. Immediate use is preferable to reduce contamination problems.

10. To place the filters on top of the agarose layer, first completely remove excess fluid from the agarose plates with a Pasteur pipet. Unwanted excess fluid will result in flotation of the filters; tilt the plate to remove excess fluid.
11. Place two filters on every dish: Briefly dry one filter, coated with irrelevant protein, and place directly, coated side up, on top of the agarose; on a stack of sterile filter paper; briefly dry the second filter on sterile filter paper and placed on top of filter 1, coated side up (**Fig. 1**). Placement of both filter tabs in exactly the same position will help subsequent filter removal. Because contact between both filters is crucial, check for air bubbles between the filters. These show up as light spots.
12. Gently place the filters on the brim of the dish and slide the filter onto the agarose. Punch through three small holes in the upper filter with an injection needle, corresponding to the larger holes in the lower filter. Using an inverted microscope, precisely label in the bottom of the plate the position of the three holes and the position of the filter tabs with a marker, and incubate overnight in a regular CO_2 incubator at 37°C.
13. The next day, place the dishes in the refrigerator at 4°C for 1 h to facilitate further handling, and then remove excess fluid again. Grab both filter tips with a sterile forceps, tilt the Petri dish gently, and gently remove both filters. Number the filters and dishes and place in PBS.
14. Incubate the filters for 1 h at room temperature with horseradish peroxidase-labeled rabbit-antimouse Ig at the appropriate dilution. After extensive washing with PBS or any other appropriate buffer, reactions are developed with 0.05% 3-3' diaminobenzidine and 0.03% H_2O_2 in 50 m*M* Tris-HCl, pH 7.4. A direct comparison of corresponding filters, using the holes as landmarks, allows the identification of spots specific for the protein of interest, i.e., those present in the antigen-coated filters and absent from the control filters. These spots are punched out, and the filters accurately repositioned underneath the corresponding Petri dish, using the markings on the dish.
15. Pasteur pipets drawn in a flame to produce small bent tips and sterilized are used to retrieve by gentle suction the colonies that should be located in the position indicated by the punched out hole. Colonies retrieved are grown in suspension in HAT medium in 96-

wells. After 5–10 d undiluted culture media can now be tested from these monoclonal cultures. If more than one colony is identified at the target position, retrieve each of them separately. When accurate repositioning is difficult, spots of colonies producing irrelevant antibody can be used as additional landmarks. Take up the colonies as neatly and cleanly as possible, without contaminating cells in the neighborhood.

16. Cultures of interest are subcloned two to three times by agarose seeding following the same basic strategy described above: two dishes are prepared per clone and approx 1000 and 2500 cells/dish are plated.
17. After approx 10 d, 10–20 colonies are retrieved from the plates, grown in culture and retested. Because the initial screening generally results in monoclonal populations, testing of more colonies is not necessary.

*3.2.3. Analysis of Antibody Specificity (see **Note 24**)*

The analysis of the antibodies produced by the techniques described here may be carried out in many ways, as described in Chapters 12, 30, 37, and 41. However there are a number of points to consider with regard to the determination of specificity.

4. Notes

1. Cold medium causes the gut to retract and reduces the risk of contamination.
2. HAT medium will kill the non fused myeloma cells. Unfused spleen cells will die in a few days.
3. It is best to keep cells in a wet incubator although this requires very careful control of sterility throughout the procedure. To avoid evaporation, the outermost row of wells in the plate may be filled with PBS instead of cells.
4. In general, clones containing a few cells should be seen 1 wk after the fusion. Clones should be ready for screening (visible by the naked eye, approx 1 mm diameter) at the end of the second week.
5. It is essential that the screening method be optimized before the fusion is initiated because once clones are ready for screening, cells will grow very rapidly. It is also essential to perform the screening as rapidly as possible (maximum 2 d). To decrease the workload, it is important to devise a screening strategy that allows the identification of uninteresting clones (i.e., weak reactivity, wide reactivity, low avidity, false positive) early on.
6. If there are not enough cells, seed first in a well of a 24-well plate with 2 mL of medium/well. It is very important to perform the subcloning rapidly because nonproducing or low-producing cells tend to overgrow the cultures. Cloning efficiency generally increases when mouse peritoneal macrophages are used as feeder layers.
7. Many hybridomas grow well at this stage in the absence of feeder layers since better growing cells are selected throughout the procedure. Nevertheless, some hybridomas will require the continuous use of a feeder layer. Most hybridomas grow well in a flask placed upright whereas others may prefer to grow in a flask placed flat.
8. Ascites production is not allowed by Animal Ethics and Experimentation Committees in certain countries.
9. Normal mice can be used. In some cases, a greater yield is obtained when using immunocompromised nude mice.
10. It is useful to freeze stocks of cells adapted to grow as mouse ascites.
11. The filter screening technique was specifically adapted for the isolation of antibodies used in diagnostic pathology. When using conventional hybridoma techniques (*see* **Subheading 3.2.1.**) several fused cells may grow out in a particular well. Therefore, immedi-

ate immunohistochemical screening cannot be performed: the presence of cell-specific antibodies produced by one clone will be masked by widely reactive antibodies produced by another hybridoma.

12. Another limiting factor is the number of supernatants that can be tested using immunohistochemical assays. After a successful fusion, screening of 1500--2000 wells is needed, obviously well beyond a screenable number. Although initial screening by ELISA may result in the isolation of useful hybridoma cultures, test conditions in ELISA differ considerably from immunohistochemical requirements. Coating conditions of protein on plates will impose conformational constraints. Therefore, it is not uncommon that antibodies that show appropriate reactivity appropriately by ELISA fail to react in immunohistochemical assays.
13. By choosing the filter placed directly on the dish, one can actually loosely screen for hybridomas producing high avidity antibody/high amounts of antibody. In general we prefer to place filters coated with irrelevant proteins below filters coated with relevant antigen. Thus, clones producing low amounts of irrelevant antibody will be less likely to show up as colonies of interest. In theory, coating with pure antigen should result in the identification of colonies producing antigen-specific antibody only. Nonetheless, we always find high numbers of spots on control filters as well as specific filters. We reason that this can be explained by the production of "sticky," crossreactive antibody by many fused cells. Therefore, we invariably use two filters for screening purposes. Obviously, two filters are necessary when crude tissue homogenates are used for coating purposes to distinguish hybridoma colonies producing tissue-specific antibody from hybridomas producing crossreactive antibody.
14. One of the main advantages of this technique is that monoclonal cultures are immediately available for screening purposes. Only two to three additional subclonings are necessary. The latter are performed to stabilize the antibody producing phenotype, and are not really necessary to obtain a monoclonal culture. Instead of the regular limiting dilution method for subcloning purposes, we always use agarose seeding.
15. Clonal stability. One should be aware that a small percentage of colonies will fail to grow in suspension. This may be due to growth factors available in the vicinity of the clones when growing in agar. Alternatively, cells die, for example, because of instability, and the secreted antibody produced before death was detected by this filter technique, regardless of the cells' status. Likewise, in conventional screening, wells may appear to contain cells excreting antibody with the desired specificity, which, however, is lost upon further expansion. This is possibly due to similar events, i.e., the cells secreting the antibody were not viable. This technique was also successfully used to rescue extremely unstable hybridoma cultures. These are seeded at two or three different densities (1000–3000 cells/dish) and subsequently screened with antimouse Ig-coated filters. Antibody producing colonies can be detected by incubation with antimouse Ig conjugates. This alleviates cumbersome subcloning by limiting dilution, which may not be successful for extremely instable hybridomas with a small percentage of cells still producing antibody. Additionally, filter screening can be performed within 2 weeks, much quicker than with serial dilution subcloning.
16. We have used synthetic antigens, purified antigens ***(12)***, as well as whole tissue homogenates for screening purposes ***(13–15)***. Obviously, the rate of success heavily depends upon the stipulations of the investigator and the immunogenicity of the antigen of interest. However, this holds true for the conventional hybridoma technology as well. This method was successfully used for example, for the isolation of tumor-specific mAbs,

tissue-specific mAbs, carbohydrate specific mAbs, and single cytokeratin specific mAbs *(12–15)*. For example, for the isolation of carbohydrate-specific MAbs after immunization with a synthetic carbohydrate-carrier conjugate, a total of 10^2 colonies were retrieved. Additional ELISA screening showed carbohydrate specificity in $43/10^2$ wells, $5/10^2$ showed cross-reactivity with the protein carrier, and $54/10^2$ did not show reactivity with any tested substance. The number of nonreactive colonies may seem high, but one has to bear in mind that in general more than one colony is picked at each position identified.

17. It is noteworthy that in many cases tissue or tumor-specificMAbs were obtained that failed to react in Western blot analysis. This is most likely a reflection of the recognized determinants that might be nonlinear. As mentioned earlier, this method may allow for a more native conformation of the antigens.
18. As with all sterile techniques, vigorous care should be taken to avoid any contamination during the whole procedure. In particular sterilization of the filters should be carried out carefully. We found it convenient to place the filters in a laminar flow in rows at fair distances, enabling us to flip the sterilized side of the filters on the sterilized laminar flow bench as if reading a book. One should try not to reach over the filters during the sterilization process, to prevent contamination. Although repositioning of the filters appears cumbersome and difficult, this is not the case, provided care was taken when the holes in the filters at the bottom of the petri dish are marked.
19. A major concern is the filter removal. Obviously, the top agarose layer should not be disturbed, despite the intimate contact between the lower filter and the agarose. The position of the filter should never be changed: it will result in complete destruction of the integrity of the agarose. In rare cases agarose will stick to the filter upon removal. Although we do not know why this happens, additional cooling of the remaining plates usually alleviates the problem.
20. The temperature of the agarose to be added to the fused cells should not exceed 40°C. Do not add too cold agarose to the cells, the agarose will lump, resulting in uneven distribution of the top layer and fluid top layers, disallowing any screening. The concentration of the top agarose should never be below 0.35%, otherwise the agarose becomes too fragile and fluid. Never look at the plates before d 5, it is detrimental to the cells. Remove as much fluid from the Petri dishes before screening, it facilitates screening, because the filters will adhere well to the agarose. If the top agarose is removed during the screening, it may be helpful to place the dishes at 4°C again. This will solidify the agarose, making it less fragile. Also, it may be helpful to remove the filters one by one.
21. In selected cases, e.g., when the yield of fused cells may be crucial, hybridoma formulated media can be used. However, this is not really required: fusion efficiency is highly investigator dependent, as well as dependent on the quality of the fusion partners. Outgrowth of fused cells does not seem to be heavily affected by the media choice.
22. Myeloma cells to be used for fusion have previously been adapted to agarose growth using standard techniques. In the Nijmegen laboratory, we have mainly used Sp2/O cells.
23. Generally, colonies of 50–200 cells are present 9 to 10 d after seeding, a colony size sufficient for filter screening. Usually, 100–150 colonies are present per dish, accounting for 2000–3000 colonies per fusion. This is comparable to the fusion efficiency of conventional hybridoma technology.
24. Several points should be made regarding the analysis of antibody specificity:
   a. antibodies may bind to other molecules through their Fc region, and the use of isotype-matched Ig of irrelevant specificity is always the best control (i.e., omission of antibody is not an appropriate control),

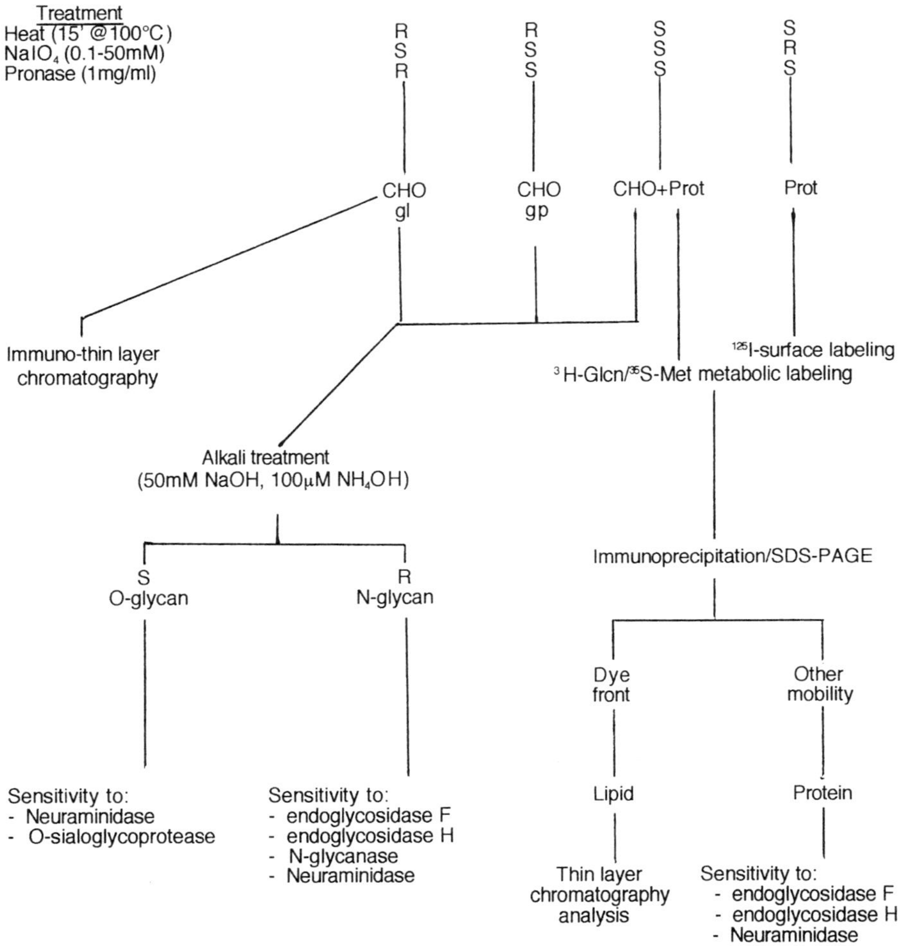

Figure 2. R, resistant; S, sensitive; gl, glycolipid; gp, glycoprotein; Prot, protein.

b. inhibition studies are useful at confirming the specificity of antibody binding but crossreactions cannot be evidenced using this type of assay,
c. antibody "specificity" (in the sense of tissue-specificity) is highly concentration dependent and it is necessary to be cautious when drawing conclusions,
d. it is always best to use multiple antibodies to different epitopes of the same molecule in order to establish specificity, and
e. when appropriate, alternative confirmatory techniques should be used (i.e., *in situ* hybridization can confirm results obtained using immunohistochemistry).

When defined molecular species are used for immunization, specificity analysis is relatively straightforward. In that case, the major point to be made is that the antibodies raised may cross-react with related epitopes on other molecules. For example, we have raised antibodies against the repetitive sequence of MUC5AC that show cross-reactivity with the repetitive sequence of MUC2. This is not so surprising considering that both structures are rich in Ser/Thr tracts. Furthermore, mucin cDNA cloning has also revealed that the poorly glycosylated domains of the different apomucins also displays significant homology (i.e., 110 amino acid Cys-rich repeats, von Willebrand-like D domains). In view of this homology, it is mandatory to be cautious. The analysis of the specificity of antimucin antibodies is much more difficult when complex antigenic mixtures are used for immunization *(16)*. In that case, it may first be necessary to establish, for the sake of simplicity, if the antibody recognizes a carbohydrate or a peptide epitope. Of course, complex epitopes dependent on both carbohydrate and peptide exist and their precise identification is very complicated. If a carbohydrate specificity is suspected, it will be useful to determine if the epitope is present not only on glycoproteins but also on glycolipids. In the latter case, it is preferable to pursue the specificity analysis on glycolipids as these can be purified to homogeneity. A protocol for epitope characterization is suggested in **Fig. 2**.

Acknowledgments

Work at the Institut Municipal d'Investigació Mèdica was supported by grants from Fondo de Investigación Sanitaria (FIS 94/1228) and Comisión Interministerial de Ciencia y Tecnología (SAF97-0085). The authors thank Dr. Dallas M. Swallow for critical reading and L. Español for help in the preparation of the manuscript.

References

1. Lan, M. S., Batra, S. K., Qi, W. N., Metzgar, R. S., and Hollingsworth, M. A. (1990) Cloning and sequencing of a human pancreatic tumor mucin cDNA. *J. Biol. Chem.* **265,** 15294–15299.
2. Gum, J. R., Byrd, J. C., Hicks, J. W., Toribara, N. W., Lamport, D. T. A., and Kim, Y. S. (1989) Molecular cloning of human intestinal mucin cDNAs. Sequence analysis and evidence for genetic polymorphism. *J. Biol. Chem.* **264,** 6480–6487.
3. Aubert, J. P., Porchet, N., Crepin, M., Duterque-Coquillaud, M., Vergnes, G., Mazzuca, M., Debuire, B., Petitprez, D., and Degand, P. (1991) Evidence for different human tracheobronchial mucin peptides deduced from nucleotide cDNA sequences. *Am. J. Respir. Cell. Mol. Biol.* **5,** 178–185.
4. Gambús, G., de Bolós, C., Andreu, D., Francí, C., Egea, G., and Real, F. X. (1993) Detection of the MUC2 apomucin tandem repeat with a mouse monoclonal antibody. *Gastroenterology* **104,** 93–102.
5. Lesuffleur, T., Porchet, N., Aubert, J. P., Swallow, D., Gum, J. R., Kim, Y., Real, F. X., and Zweibaum, A. (1993) Differential expression of the human mucin genes MUC1 to MUC5 in relation to growth and differentiation of different mucus-secreting HT-29 cell subpopulations. *J. Cell Sci.* **106,** 771–783.
6. Lesuffleur, T., Roche, F., Hill, A. S., Lacasa, M., Fox, M., Swallow, D. M., Zweibaum, A., and Real, F. X. (1995) Characterization of a mucin cDNA clone isolated from HT-29 mucus-secreting cells. The 3' end of MUC5AC? *J. Biol. Chem.* **270,** 13665–13673.
7. Abe, M. and Kufe, D. W. (1987) Identification of a family of high molecular weight tumor-associated glycoproteins. *J. Immunol.* **139,** 257–261.
8. Reis, C. A., Sorensen, T., Mandel, U., David, L., Mirgorodskaya, E., Roesptorff, P., Kihlberg, J., Hansen, J. E. S., and Clausen, H. (1998) Development and characterization of an antibody directed to an α-N-acetyl-D-galactosamine glycosylated MUC2 peptide. *Glycoconj. J.* **15,** 51–62.

9. Fradet, Y., LaRue, H., Parent-Vaugeois, C., Bergeron, A., Dufour, C., Boucher, L., and Bernier, L. (1990) Monoclonal antibody against a tumor-associated sialoglycoprotein of superficial papillary bladder tumors and cervical condylomas. *Int. J. Cancer* **46,** 990–997.
10. Fradet, Y., Islam, N., Boucher, L., Parent-Vaugeois, C., and Tardif, M. (1987) Polymorphic expression of a human superficial bladder tumor antigen defined by mouse monoclonal antibodies. *Proc. Natl. Acad. Sci. USA* **84,** 7227–7231.
11. Sharon, J., Morisson, S. L., and Kabat, E. A. (1979) Detection of specific hybridoma cells by replica immunoadsorption of their secreted antibodies. *Proc. Natl. Acad. Sci. USA* **76,** 1420–1424.
12. Oosterwijk, E., Kalisiak, A., Wakka, J. C., Scheinberg, D. A., and Old, L. J. (1991) Monoclonal antibodies against Gal alpha 1-4Gal beta 1-4Glc (Pk, CD77) produced with a synthetic glycoconjugate as immunogen: reactivity with carbohydrates, with fresh frozen human tissues and hematopoietic tumors. *Int. J. Cancer* **48,** 848–854.
13. Oosterwijk, E., Ruiter, D. J., Hoedemaeker, P. J., Pauwels, E. K., Jonas, U., and Zwartendijk, J. (1986) Warnaar, S. O.: Monoclonal antibody G 250 recognizes a determinant present in renal-cell carcinoma and absent from normal kidney. *Int. J. Cancer* **38,** 489–494.
14. Oosterwijk, E., Ruiter, D. J., Wakka, J. C., Huiskens-van-der-Meij, J. W., Jonas, U., Fleuren, G. J., Zwartendijk, J., Hoedemaeker, P., and Warnaar, S. O. (1986) Immunohistochemical analysis of monoclonal antibodies to renal antigens. Application in the diagnosis of renal cell carcinoma. *Am. J. Pathol.* **123,** 301–309.
15. van-Muijen, G. N., Ruiter, D. J., Franke, W. W., Achtstatter, T., Haasnoot, W. H., Ponec, M., and Warnaar, S. O. (1986) Cell type heterogeneity of cytokeratin expression in complex epithelia and carcinomas as demonstrated by monoclonal antibodies specific for cytokeratins nos. 4 and 13. *Exp. Cell Res.* **162,** 97–113.
16. Bara, J., Chastre, E., Mahiou, J., Singh, R. L., Forgue-Lafitte, M. E., Hollande, E., and Godeau, F. (1998) Gastric M1 mucin, an early oncofetal marker of colon carcinogenesis, is encoded by the MUC5AC gene. *Int. J. Cancer* **75,** 767–773.

30

Monoclonal Antibodies to Mucin VNTR Peptides

Pei Xiang Xing, Vasso Apostolopoulos, Jim Karkaloutsos, and Ian F. C. McKenzie

1. Introduction

One of the interesting technical aspects of working with mucins is that it is relatively easy to make antibodies to different mucin glycoproteins—mainly because the repeat sequences in the variable numbers of tandem repeat (VNTR) region are highly immunogenic. Indeed, all the mucin genes (MUC1–MUC8) *(1,2)* were originally cloned using polyclonal antisera and *Escherichia coli* DNA expressions systems, in which, because of the repeated sequences, the expressed cDNAs could be detected and cloned. We found this of particular interest because we had tried very hard in the early days of cloning to isolate lymphocyte surface antigens with monoclonal antibodies (MAbs)—all these efforts failed. Because the VNTRs are so highly immunogenic, immunization of mice with human tumors, mucin-containing materials such as the human milk fat globule membrane (HMFGM) (isolated from human milk), cell membranes or synthetic peptides, all lead to the production of MAbs. We have made numerous MAbs to human mucin 1, 2, 3, and 4 VNTRs; to variants, and to mouse muc1 *(3–8)*. As will be described herein it is not difficult to make these antibodies, and, for the most part, these can be easily characterized and the antibodies recognize linear amino acids of peptides—whether the peptides are present in tissues or as native molecules (immunohistological detection), or the examination of synthetic peptides, whether they are bound to a solid support, in solution, on pins with one end tethered, or conjugated to other proteins, e.g., keyhole limpet hemocyanin (KLH). In almost all circumstances, the reactions obtained are clear-cut, which contrasts with many other antipeptide antibodies that react with nonlinear structures (requiring appropriate secondary or tertiary folding for detection), which makes detection erratic. We describe here the methods used to make the antibodies and the principles of their characterization. In addition, we summarize the properties of MAbs to human MUC1 VNTR peptide; to MUC1 variant peptides; to MUC2, MUC3, MUC4 VNTR peptides; and to mouse muc1.

From: *Methods in Molecular Biology, Vol. 125: Glycoprotein Methods and Protocols: The Mucins*
Edited by: A. Corfield © Humana Press Inc., Totowa, NJ

2. Materials

1. Cell culture medium: Dulbecco's modified Eagle's medium (DMEM) containing 2 m*M* glutamine, 100 µL/mL of penicillin, 100 µg/mL of streptomycin, and 10% fetal calf serum (FCS).
2. Tissue culture flasks, canted neck (Falcon, Becton Dickinson Labware, NJ).
3. Microtest tissue culture plate, 96-well flat-bottomed with low-evaporation lid (Falcon, Becton Dickinson Labware, NJ).
4. Applied Biosystems Model 430A automated peptide synthesizer (Foster City, CA).
5. Reagents for peptide synthesis were purchased from Applied Biosystems, except for the amino acid derivatives, which were purchased from Auspep (South Melbourne, Australia).
6. Anhydrous trifluoromethanesulfonic acid (TFMSA).
7. Liquid chromatography reversal-phase high-performance liquid chromatography (HPLC) (Waters, Milford, MA).
8. Brownlee C8-Aquapore RP-300 column (Applied Biosystems, Foster City, CA).
9. Polyethylene pins (Pepscan) (Cambridge Research Biochemicals, Cambridge, UK) ***(9,10)***.
10. Human milk was obtained from nursing mothers, and mouse milk from lactating mammary glands of nursing mice ***(11)***.
11. Buffers for preparation of HMFGM (**Subheading 3.3.**)
    a. Buffered saline solution: 0.15 *M* NaCl, 2 m*M* $MgCl_2$, 10 m*M* Tris-HCl, pH 7.4.
    b. Medium buffer: 75 m*M* MaCl, 1 m*M* $MgCl_2$, 5 m*M* Tris-HCl.
    c. Sucrose buffer: 0.3 *M* sucrose, 70 m*M* KCl, 2 m*M* $MgCl_2$, 10 m*M* Tris-HCl, pH 7.4.
12. Glutathione, insolublized on cross linked beaded agarose (Sigma, St Louis, MO).
13. Thrombin, Thrombostat, 5000 U/5 mL (Parke Davis).
14. pGEX2T vector (Pharmacia, Uppsala, Sweden).
15. KLH, Slurry (Calibiochem, La Jolla, CA).
16. Phosphate-bufferd saline (PBS), pH 7.4.
17. Complete Freund's adjuvant (Sigma).
18. Female BALB/c mice.
19. Female Lewis rats.
20. Mouse myeloma cell line NS1.
21. Coating buffer: 0.05 *M* carbonate-bicarbonate buffer, pH 9.6.
22. Polyvinyl chloride (PVC) U-bottomed microtiter plates (Costar, Cambridge, CA).
23. 50X HAT solution: containing 0.8 m*M* thymidine (Sigma), 5 m*M* hypoxanthine (6-hydroxypurine, Sigma). 20 µ*M* aminopterin (Sigma).
24. Sheep antimouse immunoglobulin (Ig) conjugated with horseradish peroxidase (HRP, Amersham, Buckinghamshire, UK).
25. Sheep antirat Ig labeled with HRP (Amersham).
26. Rabbit antimouse Ig linked to HRP (Dakopatts, Copenhagen, Denmark).
27. Antimouse subclass antibodies (Serotec, Oxford, UK).
28. Substrate buffer for enzyme-linked immunosorbent assay (ELISA):
    a. 0.03% 2, 2-azino-bis-(3-ehtylbenzathiazoline 6-sulfonate (ABTS), in 0.1 *M* citrate buffer, pH 4.0, containing 0.02% H_2O_2.
    b. 0.05% ABTS, in 0.1 *M* citrate buffer, pH 4.0, containing 0.12% H_2O_2.
29. ELISA plate reader (Kinetic Reader, Model EL312E, BIO-TEK Instruments, Inc., Winsooki, VT).
30. Buffers for ELISA to test peptide on pins:
    a. Blocking buffer: of ELISA to test peptide on pins: 1% ovalbumin, 1% bovine serum albumin (BSA), 0.1% Tween-20 in PBS, pH 7.2, and 0.05% sodium azide.

b. Disruption buffer: 1% sodium dodecyl sulfate (SDS), 0.1% 2-mercaptoethanol, 0.1 *M* sodium dihydrogen othophosphate.
31. Sonicator (Unisonics, Sydney, Australia).
32. O.C.T. Compound (Tissue-Tek, Torrence, CA).
33. Aminoalkylsilane coated slides ***(12)***.
34. Microtome Cryostat HM 500 OM (Microm Laborgerate, Waldorf, Germany)
35. 3,3 Diaminobenzidine (DAB) (Sigma), 1.5 mg/mL in PBS containing 0.1% H_2O_2.
36. Electrophoresis power supply, EPS 500/100 (Phamacia, Uppsala, Sweden).
37. Flow cytometer (Becton Dickson).
38. 50% polyethyelene glycol 4000 (Merck, Darmstadt, Germany) in DMEM.
39. 37°C, 10% CO_2 in a humidified incubator.
40. BIAcore™ 2000 biosensor (Pharmacia) ***(13)***. CM5 sensor chip and the amine coupling kit ***(14)***.

3. Methods

3.1. Solid-Phase Peptide Synthesis

The peptides were produced using an Applied Biosystems Model 430A automated peptide synthesizer, based on the standard Merrifield solid-phase synthesis method ***(15,16)***. All reagents for synthesis were purchased from Applied Biosystems, except for the amino acid derivatives, which were purchased from Auspep.

3.1.1. Peptide Synthesis

Solid-phase peptide synthesis (SPPS) was formulated by Merrifield ***(15)***. The concept has undergone many improvements and is now a widely established technique.

1. Synthesis occurs from the carboxyl to the amino terminal of the peptide. The α-carboxyl group of the C-terminal amino acid is covalently bonded to an insoluble polystyrene resin bead via an organic linker.
2. The α-amino group of this amino acid and all subsequent amino acids used in the synthesis are protected by an organic moiety. There are two fundamental organic moieties that serve as protecting groups of the α-amino group: tertiary butyloxycarbonyl (tBOC), which is acid labile; and fluorenylmethyloxycarbonyl chloride (Fmoc), which is base labile. In our study, tBoc chemistry was employed to synthesize the MUC1, MUC2, MUC3, and MUC4 peptides. There are three sites on an amino acid that are potentially reactive: the α-amino group (NH_2), the carboxyl group (COOH), and at certain side-chain functional groups (R). The carboxyl group is not chemically protected because it is the site that forms an amide bond with the α-amino group of the amino acid that was previously coupled to the growing peptide chain.
3. The synthesis cycle consists of three chemical reactions repeated for each amino acid:
   a. Deprotection: This is carried out by using trifluoroacetic acid (TFA) to effectively remove (deprotect) the tBoc protecting group. This procedure allows the next amino acid to react at that site to form an amide (peptide) bond.
   b. Activation: This involves the formation of symmetric anhydrides that are very effective, activated carboxyl forms of amino acids. Dicyclohexylcarbodimide (DCC) was used to generate symmetric anhydrides. There are three amino acids that do not form stable symmetric anhydrides, and they can begin to degrade within 4 min of formation. The three amino acids asparagine, glutamine, and arginine are coupled as 1-hydroxybenzotriazole (HOBt) esters. When DCC/HOBt activation is utilized, significantly improved coupling is achieved.

c. Coupling: This occurs when the activated amino acid (symmetric anhydride) forms an amide bond (CO-NH) with the growing peptide chain.

3.1.2. Postsynthesis: Cleavage

When a peptide has been synthesized, it is then "cleaved." Cleavage is the process that chemically "cuts" (cleaves) the peptide from the resin and any side chain protecting groups that are present. The chemical linkers and protecting groups used in tBoc peptide synthesis normally require very harsh conditions for effective removal. Powerful acids, such as hydrofluoric acid (HF) or TFMSA are needed in conjunction with scavengers. Scavengers are chemical moieties that have the ability to bind irreversibly to amino acid protecting groups. These trapped cations are thus prevented from undergoing further reactions. There are numerous varieties of scavengers. Ethanedithiol (EDT) has proved to be a most efficient scavenger for tertiary-butyl protecting groups (a widely used protecting group). However, using a combination of scavengers is usually necessary. Thioanisole, water, phenol, *p*-cresol, phenol, and dimethylsulfide, to name a few, have also been shown to be efficient scavengers in trapping protecting groups and suppressing certain reactions from taking place (such as alkylation when tryptophan or methionine are present in the peptide sequence) that would otherwise proceed under normal cleavage conditions.

1. The cleavage mixture incorporated for the MUC1, MUC2, MUC3, and MUC4 peptides is 80% TFA, 8% TFMSA, 8% thioanisole, and 4% EDT. The mixture is allowed to react with the peptide-resin for 30 min at room temperature.
2. After this time, the entire contents are filtered.
3. Thirty milliliters of chilled diethyl ether is used to precipitate the peptide, which is then centrifuged and the supernatant discarded.
4. The pellet (crude peptide) is then dissolved with 6 *M* of guanidine hydrochloride, pH 7.5.

Care must be taken when selecting for appropriate scavengers, for instance, water is an essential scavenger when using Fmoc synthesis. The combination of scavengers implemented is determined by that protecting groups present on the peptide. The type of acid needed is determined by the binding strength of the organic linker and side-chain protecting groups. As an example, the organic linkers and protecting groups for Fmoc synthesis can be readily cleaved with TFA. tBoc synthesis requires much stronger acids such as HF or TFMSA.

3.1.3. Purification

The method of choice for peptide purification is reversed-phase HPLC. This technique separates compounds based on the principles of hydrophobicity.

1. The peptides are purified using a Waters Model 441 HPLC, on a C8-Aquapore RP-300 column (Brownlee) using a gradient solvent system of 0.1% aqueous TFA 0.1% TFA, 39.9% H_2O, and 60% CH_3CN.

The purity of synthetic peptides was approx 90% as judged by HPLC and mass spectometry.

3.1.4. The Peptides Synthesized

The peptides synthesized were derived from the MUC1, MUC2, MUC3, MUC4 peptides, which include (**Table 1**):

Table 1
Synthetic Peptides Used in Our Study

Peptide	Amino acid sequence[a]
MUC1	
VNTR	
p1-40	PDTRPAPGSTAPPAHGVTSA PDTRPAPGSTAPPAHGVTSA
p1-24	PDTRPAPGSTAPPAHGVTSAPDTR
p5-20	PAPGSTAPPAHGVTSA
p13-32	PAHGVTSAPDTRPAPGSTAP
C-p13-32	(C)PAHGVTSAPDTRPAPGSTAP
p1-15	PDTRPAPGSTAPPAH
A-p1-15	APDTRPAPGSTAPPAH
N-terminal to VNTR	
p31-55	TGSGHASSTPGGEKETSATQRSSVP
p51-70	RSSVPSSTEKNAVSMTSSVL
C-terminal to VNTR	
p344-364	NSSLEDPSTDYYQELQRDISE
p408-423	TQFNQYKTEAASRYNL
Cytoplasmic tail of MUC1	
p471-493	AVCQCRRKNYGQLDIFPARDTYH
p507-526	(C)YVPPSSTDRSPYEKVSAGNG
CT18	(C)SSLSYTNPAVVTTSANL
Variants of MUC1	
SP11 (splicing peptide)	(CY)TEKNAFNSS
sMUC1 (secreting peptide)	VSIGLSFPMLP
MUC2	
MI-29	(KY)PTTTPISTTTMVTPTPTPTGTQTPTTT
MUC3	
SIB-35	(C)HSTPSFTSSITTTETTSHSTPSFTSSITTTETTS
MUC4	
M4.22	(C)TSSASTGHATPLPVTDTSSAS
MUC5	
M5	(C)HRPHPTPTTVGPTTVGSTTVGPTTVGSC
Mouse muc-1	
MP26	(C)TSSPATRAPEDSTSTAVLSGTSSPA
Mouse CD4	
T4NI	KTLVLGKEQESAELPCECY

[a](C), (CY), (KY): These extra amino acids were added to the peptide.

1. MUC1 VNTR peptide: the peptide Cp13-32, derived from MUC1 VNTR region (containing an N-terminal cysteine to form dimers); peptides from N- and C-terminal regions to the VNTR, and cytoplasmic tail peptides of MUC1. The peptides were named by either position number in the protein sequence (e.g., p344–364, **Table 1**) or in a two continuous 20-amino acid repeats (e.g., p1–40, **Table 1**), or by individual amino acid name combined with the following peptide name, e.g., A-p1-15 (**Table 1**).

2. Two variant peptides of MUC1 *(17,18)*:
   a. Splicing peptide SP11, consisting of the amino acids 58–62 of the MUC1 (TEKNA) and amino acids of 343–346 (FNSS), lacking VNTRs (C and Y are added to N-terminus for conjugation and dimer formation).
   b. Secreted form of MUC1, sMUC1, consisting of the 14 amino acids derived from a secreted cDNA isoform; cytoplasmic tail peptide CT18, which is derived from the last 17 amino acids of the MUC1 cytoplasmic tail region (15 of 17 amino acids are identical with the mouse muc1 cytoplasmic tail) (**Table 1**).
3. MUC2, MUC3, and MUC4 VNTR peptides: peptides MI29 derived from the MUC2 VNTR gene, consisting of one repeat unit of 23 amino acids and part of the next repeat of four amino acids PTTT *(19)*; SIB35, derived from the MUC3 VNTR gene, containing two repeat units of 17 amino acids *(20)*, M4.22, derived from MUC4 VNTR gene *(21)*, corresponding to the thirty-first and thirty-eighth repeat (16 amino acids) and part of the next repeat (5 amino acids, TSSAS).
4. Mouse muc1 peptide Mp26, derived from mouse tandem repeats (TRs), containing 20 amino acids of the seventh repeat and 5 amino acids of the eighth repeat *(22)*. Cysteine was added at the N-termini of the mucin VNTR peptides to aid disulfide bond dimer formation as indicated in **Table 1**.
5. T4N1 representing the N-terminal of mouse CD4 was used as a negative control.

Hydrophilicity and antigenicity of the peptides were analyzed as described elsewhere *(23–25)*.

3.2. Peptide Synthesis Using Polyethylene Pins

1. Peptides are synthesized on polyethylene pins (Pepscan) (Cambridge Research Biochemicals, Cambridge, UK) *(9,10)*, and in our studies consisted of 20 overlapping 6-mer peptides of MUC1 VNTR, e.g., PDTRPA, DTRPAP, TRPAPG, ······ APDTRP, that were made to map the MUC1 epitopes reacting with MAbs (**Table 2**).
2. To map the epitopes of other mucin MAbs, overlapping peptides of MUC2, MUC3, MUC4 and mouse muc1 are synthesized by the Pepscan method *(4,5,7,8)* (also commercial available from Chiron, Australia).

3.3. Production of HMFGMS (25,26)

1. Human milk (50 mL) is obtained from nursing mothers.
2. Dilute the milk with 50 mL buffered saline solution (*see* **Subheading 2., item 11a**).
3. Centrifuge the diluted milk at 2500*g* for 15 min.
4. Collect the floating cream and wash it three times with buffered saline.
5. Resuspend in cold medium buffer (*see* **Subheading 2., item 11b**) and homogenize it using a homogenizer (T8.01, IKA Labortechnik, Stauffen, Germany).
6. Centrifuge crude membranes at 10,000*g* for 1.5 h at 4°C. Resuspend the pellet in sucrose buffer (*see* **Subheading 2., item 11c**) and store at –70°C.

3.4. Production of Human and Mouse MUC1 Glutathione-Fusion Protein

1. A human fusion protein (hFP) containing a glutathione-*S*-transferase (GST) and five VNTR repeats of MUC1 is produced in *E. coli* using methods described elsewhere *(6,27)*.
2. The 5 VNTR repeats are cleaved from FP using the site-specific protease thrombin.
3. Using the same method, a mouse fusion protein (mFP) containing 550 bp of TR region (total 1065 bp, 16 repeats) is also produced *(11,22)*. mFP consists of GST and 184 amino acids of the mouse muc1 TR region (repeats 7—16).

Table 2
MAbs to Mucin Peptides

Name	Immunogen	Host	Ig Class	Minimum epitope[a]
MUC1				
BCP7	C-P13-32	Mouse	IgG2a	VTSA
BCP8	C-P13-32	Mouse	IgG2b	DTR
VA1	hFP[b]	Mouse	IgG1	APG
VA2	hFP[b]	Mouse	IgG1	DTRPA
CT1.53	CT18	muc1 deficient mouse	IgG1	NT
CT91	CT18	Rat	IgG1	NT[a]
SEC1	sMUC1	Mouse	IgG2b	NT
SEC2	sMUC1	Mouse	IgG1	NT
SEC3	sMUC1	Mouse	IgM	NT
SP3.9	Sp11	Mouse	IgG1	NT
MUC2				
CCP31	MI29	Mouse	IgA	STTT
CCP37	MI29	Mouse	IgG1	PTT
CCP58	MI29	Mouse	IgG1	GTQTP
MUC3				
M3.1	SIB35	Mouse	IgG2a	SITTIE
M3.2	SIB35	Mouse	IgG2a	NA
M3.3	SIB35	Mouse	IgG1	PFSTSS
MUC4				
M4.171	M4.22	Mouse	IgG2a	TPL
M4.275	M4.22	Mouse	IgG1	PLPV
Mouse muc1				
M30	Mp26	Rat	IgM	TSS
MFP25	mFP[c]	Rat	IgM	LSGTSSP
MFP32	mFP[c]	Rat	IgM	NA

[a]NT, not tested; NA, not available.
[b]Human mucin 1 fusion protein.
[c]Mouse mucin 1 fusion protein.

4. Both hFP and mFP are prepared from transformed *E. coli* DH5α, induced with 0.1 m*M* isopropyl-β-D-thiogulactopyranoside, lysed by sonication and 1% Triton X-100 buffer, and purified from the lysate using a GST-agarose column and eluted with 10 m*M* reduced glutathione ***(6,27)***.
5. GST is prepared using pGEX2T vector, without any insert, as a negative control.

3.5. Production of MAbs to Mucin VNTR Peptides

1. To produce antipeptide MAbs, two groups of antigens were used: peptide and fusion proteins (hFP and mFP).
2. To prepare peptide as immunogen, mix 1 mL of peptide (2 mg/mL) and 1 mL of KLH (2 mg/mL) with 1 mL of 0.25% glutaraldehyde for 8 h at room temperature. Dialyze the mixture in a dialysis tube against PBS, pH 7.4.
3. To immunize mice, emulsify conjugated peptides or fusion protein with equal volume of complete Freund's adjuvant, and inject 0.2 mL of the antigen-adjuvant mixture intraperitoneally into female Balb/c mice.

4. Inject the mice with 60–100 μg in 100–200 μL PBS after 4 and 6 wk of first injection.
5. Collect blood samples from immunized mice 1 wk after the third immunization.
6. Test the serum by ELISA using peptide-coated plates (*see* **Subheading 3.6.1.**).
7. Perform a fusion 3 d after the fourth injection of the conjugated peptides ***(4)***.
8. To produce B-cell hybridomas to human MUC1, fuse the mouse myeloma cell line NS1 (2×10^7 cells) with the spleen cells (10^8) of immunized Balb/c mice as described elsewhere ***(4)***.
9. To produce B-cell hybridomas to mouse muc1, immunize Lewis female rats with mouse muc1 fusion protein or conjugated peptides (100 μg) as described under **steps 2** and **3**).
10. Screen the hybridoma supernatants on the immunizing peptide and a negative peptide by ELISA, and tested further by immunoperoxidase staining on tissues, by flow cytometry, or by any other method.
11. Determine the isotypes of MAbs by using antimouse or antirat subclass antibodies by 1% agarose gel immunodiffusion.

3.6. ELISA Tests

3.6.1. Direct Binding

1. Coat PVC U-bottomed microtiter plates with 50 μL of 20 μg/mL of peptides in 0.05 *M* carbonate-bicarbonate buffer, pH 9.6, at 37°C for 2 h or overnight at 4°C.
2. Wash plate twice with PBS-0.05% Tween-20.
3. Block nonspecific binding sites with 100 μL of 2% BSA for 1 h at room temperature.
4. Wash the plate with PBS-0.05% Tween-20, and add 50 μL of tissue culture supernatants of hybridomas or purified antibody to the peptide-coated plate at room temperature for 1 h ***(3)***.
5. Thoroughly wash the plate 10 times with PBS-0.05% Tween-20, add 50 μL of sheep antimouse Ig conjugated with HRP (Amersham) at 1:500 dilution in PBS, incubate 1 h at room temperature.
6. Wash the plate 10 times with PBS-0.05% Tween-20, add the substrate, ABTS (*see* **Subheading 2., item 28a**) and incubate the plate at room temperature for 10–30 min until the positive well becoming the blue color.
7. Measure the absorbance at 405 nm using an ELISA plate reader.

3.6.2. Inhibition ELISA

1. Preincubate the MAb at a constant concentration (which was determined as 50% binding in the direct binding ELISA) with peptides or relevant antigens (inhibitors) in a series dilution for 2 h at room temperature.
2. Add the mixtures to the plates coated with antigens under **Subheading 3.6.1.**, and further incubated overnight at 4°C.
3. Detect the binding of residual MAb as described under **Subheading 3.6.1.**
4. Calculate the percentage of inhibition by comparing the binding of MAbs preincubated with and without antigen (inhibitor): % of inhibition = [1 – (binding of MAb with inhibitor/binding without inhibitor)] × 100%.

3.6.3. ELISA Tests of Peptides on Pins

Using small peptides (5–9-mer) on pins gives the ability to screen rapidly many small peptides to find the reactive epitope (Pepscan). This procedure was first described by Geysen et al. ***(9)*** and has been used as the standard method to define linear epitopes.

1. Synthesize peptides on the pins following the standard method or purchase them from Chiron.
2. Block the pins for 1 h in microtiter plates using blocking buffer (*see* **Subheading 2., item 30a**) at room temperature with agitation.
3. Add 150 µL of antibody to each well and incubate the antibody with the pins in the plate overnight at 4°C.
4. Wash the pins four times for 10 min each in a tub containing 50 mL of PBS-0.05% Tween-20 at room temperature with agitation.
5. Wash the microtiter plate four times 10 min each with PBS-0.05% Tween-20 at room temperature with agitation.
6. Add 150 µL of sheep antimouse or antirat Ig labeled with HRP to each well in PBS-0.05% Tween-20 (1:500), and incubate the pins with conjugate for 1 h at room temperature in the microtiter plates.
7. Wash the pins with vigorous agitation (four times for 10 min each) in PBS-0.05% Tween-20 and incubated at room temperature in the dark in microtiter plates containing 150 µL/well of substrate 0.05% ABTS (*see* **Subheading 2., item 28a**), containing 0.04 mL H_2O_2 /100 mL buffer. Stop the incubation when the plates appear to have sufficient color by removing the pins, after which 50 µL from each well is transferred to another microtiter plate and the absorbance read immediately at 405 nm using a plate reader.
8. The pins and their irreversibly bound peptides can be used many times if bound antibodies are efficiently removed after each assay. To remove the bound antibodies, prewarm the disruption buffer (*see* **Subheading 2., item 30b**) to 60°C. Place the pins in a sonication bath with sufficient disruption buffer to ensure that the pins are well covered. Sonicate the pins for 60 min.
9. Wash the pins four times with hot distilled water (60°C).
10. Wash the pins in boiling methanol for 2 min in a tub, and then air-dry. The pins may be stored at room temperature in plastic bags containing silica gel.

3.7. Immunoperoxidase Staining

The reactivity and specificity of the MAbs can be determined by immunoperoxidase staining on snap-frozen fresh human tissues or formalin-fixed tissues ***(4,28)***.

1. To prepare frozen sections, embed the tissues in OCT, snap-frozen in liquid nitrogen, and stored at –20°C.
2. Cut the tissues at 6–8 µm using Microtom cryostat, attach the section to aminoalkylsilane-coated slides ***(12)***, air-dry, and keep at –20°C until tested.
3. Fix slides with cold acetone for 10 min and air-dry.
4. Block endogenous peroxidase activity for 40 min at room temperature using 0.5% of H_2O_2, and wash once with PBS.
5. Cover tissue sections with at least 100 µL of diluted tissue culture supernatant (1/5–1/10), ascites (1/1000–1/10,000), or purified antibody (1–5 µg). Make dilutions with 0.5% BSA/plain DMEM (without FCS).
6. Incubate slides with antibody in a humidified container for 40 min at room temperature.
7. Remove excess antibody by immersion of slides in PBS for 5 min, and repeat it three times.
8. Add 50–100 µL of rabbit antimouse Igs linked to HRP (1/50 in 0.5% BSA/DME) to cover the sections in the slides for 40 min at room temperature.
9. Wash slides in PBS for 5 min, three times.
10. Cover the sections with 1.5 mg/mL of DAB (*see* **Subheading 2., item 35**) in PBS containing 0.1% H_2O_2 for 5 min or until brown color is visible on sections.

11. Remove excess DAB by immersion of slides in running tap water.
12. Staining and mounting slides: Place slides in hemotoxylin for 2–10 min, wash with tap water and Scotts water, and rinse in running tap water.
13. Pass tray of slides through 75, 95, and 100% alcohol and three times of shellex (1 min each).
14. Mount slide with cover-slip.
15. Grade the staining by microscope according to the following:
    a. The percentage of cells stained: –, < 5%; +, 5–25%; 2+, 25–50%; 3+, 50–75%; 4+, 75–100%.
    b. The density of staining: –, no staining; +, weak staining; 2+, moderate staining (dark brown color); 3+, strong staining (dark brown color); 4+, very strong staining (condensed brown color).
16. For inhibition immunoperoxidase staining, preincubate MAbs with antigens (inhibitors) as under **Subheading 3.6.2.**

3.8. Other Tests

Cells or cell lines can be tested by flow cytometry using standard methods. In addition, further characterization can be done as for any MAbs, e.g., Western blotting.

3.9. Affinity Measurements by Biosensor

Affinity of the MAbs can be routinely and easily performed–provided that there is access to a biosensor machine. This adds a further level to selection at the screening stage, at which MAbs of high or low affinity/avidity can be measured. In our studies, we used MAbs at 100 μg/mL and a BIAcore 2000 biosensor (Pharmacia) ***(14)***.

1. The antigens peptide C-p13-32, FP, and 5 repeats of MUC1 VNTR peptide are immobilized on a CM5 sensor chip using the amine coupling kit ***(14)***.
2. Sensor chips are regenerated with 10 m*M* glycine-HCl (pH 2.4).
3. K_a (association kinetic constants) and k_d (dissociation kinetic constants) are calculated from sensorgram plots of MAb amount bound to immobilized antigens vs time. The affinity of the MAbs (K_A)are obtained by dividing K_a by $K_{d.}$ The data is analysed to their relative affinity and recorded: low, $K_A < 2.5 \times 10^7$; medium, $1 \times 10^8 > K_A \geq 2.5 \times 10^7$; high, $K_A \geq 1 \times 10^8$.

4. Notes

1. Generally, MAbs to mucins have been produced by immunizing with crude or purified mucins—particularly HMFGM for MUC1, or whole tumors or their extracts. In this way, many antibodies to MUC1 have been made which are reactive with breast cancer cells and on further analysis these have clearly fallen into several groups, wherein the antibody reacts with either predominantly carbohydrate, or peptide epitopes, or both carbohydrate and peptide epitope ***(4,10,28)***. However it is sometimes difficult to make antibodies to particular mucin sequences using crude antigens. With recent advances in both synthetic peptide chemistry (*see* Chapters 11–12) and in the cloning of the cDNA encoding the protein core of the mucins (*see* Chapters 24), comes the ability to make antimucin antibodies by using synthetic moieties—as described herein. This represents a substantial advance in being able to use clearly defined antigens rather than a mixture of materials obtained by extracts from tissues or secretions. On the basis of mucin cDNA sequences, and synthetic peptides, we and others have successfully made MAbs to synthetic peptides

derived from the cDNA sequences of VNTR regions of MUC1, MUC2, MUC3, and MUC4 *(4,5,7,8)*, and Reis et al. ***(30)*** have made MAb to MUC5AC mucins.

2. Our approach to produce the antibodies was to synthesize a peptide representing the whole sequence of a VNTR, with an extension into the next VNTR so that all potential amino acid epitopes were represented and to make a dimer peptide by adding a cysteine at the N-terminal (cysteine can also be added to C- or both N- or C-terminals to form a dimer or a ring form) in favor of forming a native construct ***(31)***.
3. It has been possible with some peptides—indeed, to some mucins—to predict the immunogenic regions based on hydrophilicity analysis ***(23–25)***; for example, with MUC1, the most hydrophilic region contains the amino acid APDTR, and indeed most antibodies made to MUC1 native mucin react with this sequence. But this is not necessarily the case when synthetic peptides are used for immunization, as some of the antibodies obtained are to hydrophobic, and not to hydrophilic regions, indicating that with peptides there is no absolute correlation between antigenicity and hydrophilicity. Similar results were also reported by Geyson et al. ***(32)***. Nevertheless, selection of the hydrophilic region as a peptide to synthesize is still the first choice to make MAbs for an interesting region, but it should be noted that the hydrophobic region could also be immunogenic.
4. For immunization, the peptides should be conjugated to KLH or BSA, rabbit serum albumin (RSA) using glutaraldehyde, since free peptides (<15 mer) are poorly immunogenic ***(13)***.
5. Producing FP is an alternative way to make a longer peptide, and it can satisfactorily be used as an antigen to make MAbs.
6. The hybridoma supernatants should be screened on the immunizing peptide and a negative peptide by ELISA, and further tested by immunoperoxidase staining on tissues, by flow cytometry or by any other method. It was of interest to us that screening on tissue sections can be done at ~400 sections being rapidly read for a +/– reaction, or indeed any specific staining patterns selected.
7. BSA- or RSA-conjugated peptide can also be used rather than unconjugated peptide in ELISA; the concentration of coating antigen, such as a fusion protein or HMFGM can vary from 0.1 to 20 μg/mL.
8. Snap-frozen tissues for immunoperoxidase staining should be used since most peptide epitopes are damaged or masked by formalin fixation.
9. Inhibition tests in both ELISA and immunoperoxidase staining should be used to demonstrate specific reactions.
10. The MAbs to peptide are useful reagents in the study of the mucin glycosylation and recognition of T- and B-cells.
11. Producing FP is an alternative way to make a longer peptide and can be satisfactorily used to make MAbs.

References

1. Gendler, J. D. and Spicer, A. P. (1995) Epithelial mucin genes. *Annu. Rev. Physiol.* **265,** 607–34.
2. Shankar, V., Glimore, M. S., Elkins, R. C., and Sachdev, G. P. (1994) A novel human airway mucin cDNA encodes a protein with unique tandem-repeat organization. *Biochem. J.* **300,** 295–298.
3. Xing, P. X., Reynolds, K., Tjandra, J. J.,. Tang, X. L., and McKenzie, I. F. C. (1990) Synthetic peptides reactive with anti-human milk fat globule membrane. *Cancer Res.* **50,** 89–96.

4. Xing, P. X., Prenzoska, J., Quelch, K., and McKenzie, I. F. C. (1992) Second generation anti-MUC1 peptide monoclonal antibodies. *Cancer Res.* **52,** 2310–2317.
5. Xing, P.-X., Prenzoska, J., Layton, G.T., Devine, P., and McKenzie, I.F.C. (1992) Second-generation monoclonal antibodies to intestinal MUC2 peptides reactive with colon cancer. *J. Natl. Cancer Inst.* **84,** 699–703.
6. Apostolopoulos, V., Xing, P.-X., Trapani, J. A., and McKenzie, I. F. C. (1993) Production of anti-breast cancer monoclonal antibodies using a glutathione-S-transferase-MUC1 bacterial fusion protein. *Br. J. Cancer.* **67,** 713–720.
7. Apostolopoulos, V., Xing, P. X., and McKenzie, I.F.C. (1995) Anti-peptide monoclonal antibodies reactive to intestinal mucin 3. *J. Gastroenterol. Hepatol.* **10,** 555–561.
8. Xing, P. X., Prenzoska, P., Apostolopoulos, V., and McKenzie, I. F C. (1997) Monoclonal antibodies to a MUC 4 synthetic peptide reactive with lung cancer. *Int. J. Oncol.* **11,** 289–295.
9. Geysen, H. M., Rodda, S. J., Mason, T. J., Tribbick, G., and Schoofs, P. G. (1987) Strategies for epitope analysis using peptide synthesis. *J. Immunol. Meth.* **102,** 259–274 .
10. Xing, P.-X., Prenzoska, J., and McKenzie, I.F.C. (1991) Epitope mapping of anti-breast and anti-ovarian mucin monoclonal antibodies. *Molecular Immunol.* **29,** 641–650.
11. Xing, P. X., Lees, C., Lodding, J., Prenzoska, J., Poulos, G., Sandrin, M., Gendler, S., and McKenzie, I. F.C. (1988) Mouse mucin 1 (muc1) defined by monoclonal antibodies. *Int. J. Cancer*, in press.
12. Rentrop, M., Knapp, B., Winter, H., and Schweizer, J. (1986) Aminoalkylsilane-treated slides as support for in situ hybridization of keratin cDNAs to frozen tissue sections under varying fixation and pretreatment condition. *Histochem. J.* **18,** 271–276.
13. Briand, J. P., Muller, S., and Van Regenmorted, M. H. V. (1985) Synthetic peptides as antigens: pitfalls of conjugation methods. *J. Immunol. Meth.* **78,** 59–69 .
14. Karlsson, R, Michaelsson, A., and Mattsson, L. (1991) Kinetic analysis of monoclonal antibody-antigen interactions with a new biosensor based analytical system. *J. Immunol. Meth.* **145,** 229–240.
15. Merrifield, R. B. (1965) Automated synthesis of peptides. *Science* **150,** 178–185.
16. Kent, S. B. H. and Hood, L. E. A. (1985) A novel approach to automated peptide synthesis based on new insights into solid phase chemistry, in *Peptides chemistry 1984* (Izumiya, N., ed.), Protein Research Foundation, Osaka, Japan, pp. 217–222.
17. Lightenberg, M. J. L., Vos, H. L. , Gennissen, A. C. M., and Hilkens, J. (1990) Episialin, a carcinoma-associated mucin, is generated by a polymorphic gene encoding splice variants with alternative amino-termini. *J. Biol. Chem.* **265,** 5573–5578.
18. Wreschner, D. H., Hareuveni, M., Tsarfaty, I., Smorodinsky, N., Horev, J., Zaretsky, J., Kotkes, P., Weiss, M., Lathe, R., Dion, A., and Keydar, I. (1990) Human epithelial tumor antigen cDNA sequences - differential splicing may generate multiple protein forms. *Eur. J. Biochem.* **189,** 463–474.
19. Gum, J. R., Byrd, J. C., Hicks, J. W., Toribara, N. W., Lamport, D. T. A., and Kind, Y. S. (1989) Molecular cloning of human intestinal mucin cDNAs: Sequence analysis and evidence for genetic polymorphism. *J. Biol. Chem.* **264,** 6480–6487.
20. Gum, J. R., Hicks, J. W., Swallow, D. M., Lagace, R. L., Byrd, J. E., Lamport, D. T. A., Siddick, B., and Kim, Y. (1990) Molecular cloning of cDNAs derived from a novel human intestinal mucin gene. *Biochem. Biophy. Res. Comm.* **171,** 407–415.
21. Porchet, N., Van Cong, N., Dufosse, J., Audie, J.P., Guyonnet-Duperat, V., Gross, M.S., Denis, C., Degand, P., Bernhelum, A., and Aubert, J.P. (1991) Molecular cloning and chromosomal localization of a novel human tracheo-bronchial mucin cDNA containing tandemly repeated sequences of 48 base pairs. *Biochem. Biophys. Res. Commun.* **175,** 414–422.

22. Spicer, A. P., Parry, G., Patton, S., and Gendler, S. J. (1991) Molecular cloning and analysis of the mouse homologue of the tumor-associated mucin, MUC1, reveals conservation of potential O-glycosylation sites, transmembrane, and cytoplasmic domains and a loss of minisatellite-like polymorphism. *J. Bio. Chem.* **266,** 15,099–15,109.
23. Hopp, T. P.and Woods, K. R. (1983) A computer program for predicting protein antigenic determinants. *Mol Immunol.* **20,** 483–489.
24. Hopp, T. (1985) Computer prediction of protein surface features and antigenic determinants, in *Molecular Basis of Cancer, Part B.* Alan R. Liss. New York, p. 367.
25. Welling, G. W., Weijer, W.J., van der Zee, R., and Weeling-Wester, S. (1985) Prediction of sequential antigenic regions in proteins. *FEBS Lett.* **188,** 215–218.
26. Jarasch, E., Bruder, G., Keenan, T. W., and Franke, W. W. (1977) Redox constituents in milk fat globule membranes and rough endoplasmic reticulum from lactating mammary gland. *J. Cell Biol.* **73,** 223–241.
27. Smith, D. B. and Jonson, K. S. (1988) Single-step purification of polypeptides expressed in Escherichia coli as fusions with glutathione S-transferase. *Gene* **67,** 31–40.
28. Xing, P. X., Tjandra, J. J., Stacker, S. A., Teh, J. G., Thompson, C. H., McLaughlin, P. J., and McKenzie, I. F. C. (1989) Monoclonal antibodies reactive with mucin expressed in breast cancer. *Immunol. Cel. Biol.* **67,** 183–185.
29. Devine, P, L,. Clark, B. A., Birrell, G. W., Layton, G. T., Ward, B. G., Alewood. P. F., and McKenzie, I. F. C. (1991) The breast tumor-associated epitope defined by monoclonal antibody 3E1.2 is an O-linked mucin carbohydrate containing N-Glycosylneuraminic acid. *Cancer Res.* **51,** 5820–5836.
30. Reis, C. A., David, L, Nielsen, P. A, Clausen, H., Mirgorodskaya, K, Roepstorff, P., and Sobrinho-Simoes, M. (1997) Immunohistochemical study of MUC5AC expression in human gastric carcinomas using a novel monoclonal antibody. *Int. J. Cancer.* **74,** 112–121.
31. Dorow, D.S., Shi, P-T., Carone, F.R., Minasian, E., Todd, P. E. E., and Leach, S. J. (1985) Two large immunogenic and antigenic myoglobin peptides and the effects of cyclisation. *Mol. Immunol.* **22,** 1255–1264.
32. Geyson, H. M., Tainer, J.A., Rodda, S. J., Mason, T, Alxander, J. H., Getzoff, E. D., and Lerner, R. J. (1987) Chemistry of antibody binding to a protein. *Science* **235,** 1184–1190.

X

Enzymatic Degradation of Mucins

31

Mucinase Activity

Roger M. Stark, Rebecca Wiggins, Elizabeth Walley, Sally J. Hicks, Gulnaz A. Gill, Stephen D. Carrington, and Anthony P. Corfield

1. Introduction

Turnover of the mucous "barrier gel" overlying mucosal surfaces is essential for hydration, mechanical protection, the physical removal of contaminants and toxins, the generation of sacrificial binding ligands that prevent microbial penetration, and the provision of a suitable environment to renew other defensive molecules that are incorporated into mucus. Mucinase activity is crucial to this turnover process in locations such as the gut and the reproductive tract. Similar activity may also be of relevance at other mucosal surfaces that are not normally colonised by significant microbial populations, such as the eye and the respiratory tract.

Mucinase activity is owing to a mixture of enzymes that are expected to include proteinases, peptidases, glycosidases, and sulphatases of prokaryotic or eukaryotic origin. In addition, the presence and action of other hydrolases, including phosphatases esterases and lipases should also be considered.

The assay of individual enzyme activity resulting in the release of single components from mucin substrates does not give a complete analysis of the total mucin-degrading potential of the sample under study. The examination of mucinase activity from any source must be correlated with the nature and origin of the mucin used as substrate. Mucins are fragmented differently as a result of structural variability between apomucin peptides and their individual patterns of glycosylation. Equally significant is the composition of the mucinase activity that they encounter. A careful examination of the fragments derived from incubations of specific mucins with mucinase activity from different sources can yield information on the succession of degradative steps, as well as the enzymes that carry them out.

The development of methods for detecting mucinase activity and resolving it into its component stages is in its infancy. However, with the increasing knowledge of the structure and organization of mucin *(1,2)* it is now becoming possible to match frag-

From: *Methods in Molecular Biology, Vol. 125: Glycoprotein Methods and Protocols: The Mucins*
Edited by: A. Corfield © Humana Press Inc., Totowa, NJ

mentation patterns with known sequence information for both peptides and oligosaccharides. Clearly, this is best assessed using a whole mucin substrate and analyzing the progressive degradation to small molecular weight fragments. However, the analysis of size-fractionated products for carbohydrate and/or amino acid composition requires large amounts of mucin substrate, is time-consuming, and is therefore not suitable for the screening of large numbers of samples. Larger-scale screening for mucinase activity is possible using radiolabeled mucins prepared from organ or cell cultures. Mucins can be labeled with suitable precursors; for example, cultures of colonic mucosa can be labeled with [^{35}S]sulfate, [^{3}H]glucosamine, and [^{3}H]threonine. The purified mucins are subsequently incubated with a source of mucinase activity, and the degradation is assessed by gel filtration *(3–5)*.

An alternative to the radioactive methods is the use of biotinylated mucin substrates tagged through either the peptide or carbohydrate moieties of the molecule *(6,7)*. Such assays have been described for protein substrates and adapted for use with microtiter plates *(8)*. They require small amounts of pure substrate, have high sensitivity, are rapid, and can be applied to large numbers of samples. Such assays are suitable for the detection of "total" mucinase activity. Thus, their main use is for the rapid screening of sources of mucinase activity that may be enzymatically heterogeneous. The known structure and properties of mucins, together with the ability to biotinylate either their protein or carbohydrate domains, has important implications for the interpretation of microtiter plate assays. It is important to be aware that the release of adherent mucin from a plastic surface may be owing to several factors. First, the degradation of mucin substrates and their release from the plates may be entirely the result of mucinase action over the whole domain structure of the mucin. Second, it may result from the cleavage of regions of the mucin molecules which specifically bind to microtiter plates. A third caveat is the possibility that paradoxical results may be obtained when digesting mucins labeled through their peptide vs their carbohydrate residues. In extreme cases, e.g., where adhesion of mucins to the plastic of microtiter plates is mainly related to *hydrophobic* binding to peptide sequences, mucins labeled through their carbohydrate may *appear* to be fully digested when exposed only to peptidases. In such cases, however, any released glycopeptides would be protected from digestion by their carbohydrate chains. Where adhesion is mainly related to *ionic* interactions between charged carbohydrates and the microtiter plate, sialidase or sulfatase activity alone may release the *majority* of bound mucin while causing only *limited* degradation of the whole molecule. Therefore, the microtitre plate assay may not appear to be specific for enzyme type with respect to the nature of the mucin biotin label. In reality, however, binding is probably the result of multiple binding interactions, which tends to reduce the significance of such extremes of binding interaction.

Nevertheless, the composition of mixed enzyme preparations cannot be deduced directly from the plate assays. To obtain this information, an examination of fragmentation profiles by size fractionation and electrophoresis would define the pattern and sequence of degradation of mucin for both methods of biotinylation. Further information can be obtained by comparing such profiles with those obtained using commercially available peptidases and glycosides, and by subsequent enzyme inhibition experiments.

2. Materials

2.1. Enzyme Sources

The nature of possible samples to be screened for mucinase activity is diverse and the sources given here serve only as examples.

1. Commercially prepared enzymes: proteases, e.g., trypsin (Sigma, Poole, UK), pronase E (Boehringer Mannheim, Lewes, UK), pepsin (Calbiochem-Novabiochem, Nottingham, UK); glycosidases, e.g., α-sialidase, α- and β-galactosidase (Oxford Glycosciences, Abingdon, UK), α- and β-*N*-acetylhexosaminidase (Boehringer Mannehim), α-fucosidase (Sigma), *O*-glycanase (Calbiochem -Novabiochem).
2. Bacterial culture supernatants and cell suspensions or cell extracts.
3. Animal/human secretions and excretions that contain enzymes can also be used, e.g., plasma, urine, tears, fecal extracts, sputum, mucosal washings.
4. Animal and insect cell culture supernatants, cell suspensions, or cell extracts.

2.2. Preparation of Radiolabeled Mucin Substrates

1. Radiolabeled mucin singly or dual labeled with [^{3}H]glucosamine or [^{3}H]threonine combined with either [^{35}S]sulfate or [^{14}C]threonine. This mucin must be purified and should elute as a single high molecular weight peak at the V_o of a Sepharose CL-2B column. The preparation of these substrates is detailed in Chapter 19 (*see* **Note 1**).
2. Gel filtration buffer: 10 m*M* Tris-HCl, pH 8.0.
3. Sepharose CL-2B (Pharmacia, Uppsala, Sweden).

2.3. Preparation of Biotinylated Mucin Substrates: Biotinylation of Mucins

1. Prepare mucin using density gradient centrifugation and gel filtration and characterize for the presence of noncovalently associated contaminants as described in Chapters 1, 7, and 8 (*see* **Note 1**).
2. Sephadex G25 (Pharmacia). Sephadex column preparation: hydrate Sephadex G25 in phosphate-buffered saline (PBS) (*see* **Subheading 2.2.1., item 8**) and pack in an all-glass column approx 1 × 15 cm. Thoroughly equilibrated the packed column in PBS.
3. Protein biotinylation reagent: δ-biotinyl-ε-aminocaproic-*N*-hydroxysuccinimide ester (BNHS) (Sigma).
4. Carbohydrate biotinylation reagent: δ-biotinyl-ε-aminocaproic-acid hydrazide (BACH) (Boehringer Mannheim).
5. Dimethylformamide (DMF) (Sigma).
6. Dimethylsulfoxide (DMSO) (Sigma).
7. Sodium periodate (Sigma).
8. PBS: 0.375 g of sodium dihydrogen phosphate dihydrate, 1.155 g of disodium hydrogen phosphate, and 8.765 g of sodium chloride in 1000 mL water.

2.4. Mucinase Assay with Biotinylated Mucin

*2.4.1. Coating of Plates with Biotinylated Mucin (*see ***Notes 4–6*)**

1. Microtiter plates, 96-well, Nunc-Immuno™ plates, MaxiSorp Surface™ (Nalge Nunc, Life Technologies, Glasgow, Scotland) (*see* **Note 2**).
2. Coating buffers: 0.1 *M* sodium acetate buffer, pH 5.0, for carbohydrate labelled mucin and 0.1 *M* sodium phosphate buffer, pH 7.0, for protein label (*see* **Notes 3** and **4**).

2.4.2. Detection of Biotin-Labeled Mucin

1. PBS (*see* **Subheading 2.3.1., item 8**).
2. PBST: Add Tween-20 (Sigma) to PBS to give a final concentration of 0.2%.
3. Blocking buffer: 1% bovine serum albumin in PBST. Use enough to fill the well, 200–300 µL.
4. Streptavidin-horseradish peroxidase (HRP) solution: Streptavidin-HRP conjugate (Vector, Peterborough, UK) at 1 mg/mL is diluted to 1:1500 in blocking solution (75 µL/well)
5. OPD Solution: 1,2-phenylenediamine dihydrochloride (Dako, High Wycombe, Bucks, UK). Dissolve four (2 mg) tablets in 12 mL of distilled water and add 5 µL of 30% hydrogen peroxide immediately prior to use.
6. Stop solution: 0.5 *M* sulfuric acid (28 mL of 95–97% acid in 1000 mL of distilled water).

3. Methods

3.1. Biotin Labeling of Mucins in the Protein Moiety

1. Dissolve BNHS in DMF to give a final concentration of 20 mg/mL.
2. Dissolve 1 mg of mucin in 0.9 mL of PBS (larger batches can be prepared at the same mucin buffer ratio).
3. Add 0.1 mL of BNHS solution and incubate at 4°C overnight or at room temperature for 4 h.
4. Load the biotin/mucin incubation (1 mL) onto a Sephadex G25 column and run in PBS buffer, collecting 1-mL fractions up to 30 mL total volume. The Sephadex column is discarded (see **Note 2**).
5. Test fractions from the column for mucin using the slot blot assay with the periodic acid Schiff's stain (*see* Chapter 4).
6. Test fractions for their biotin labeling by adherence to 96-well microtiter plate assay (*see* **Subheading 3.4**).
7. Pool the labeled fractions (approx 5 × 1 mL) to give a concentration of 20 µg/mL mucin, and aliquot as 0.2-mL samples. Store at 4°C until used.

3.2. Biotin Labeling of Mucins in the Carbohydrate Moiety

Carbohydrate labeling requires the periodate oxidation of the carbohydrate moieties of the mucin before biotinylation of these oxidized residues. In the case of colonic mucins, the presence of *O*-acetyl esters will block this oxidation and a saponification step is needed first.

1. Dissolve 1 mg of mucin in 0.5 mL of 0.1 *M* sodium hydroxide and incubate for 45 min at room temperature. Neutralize to approx pH 7.0 with 0.05 mL of 1 *M* HCl. Check the pH.
2. Adjust to pH 5.5 with 0.1 *M* acetate buffer. Add sodium periodate so that the final concentration is 1 m*M*.
3. Incubate for 20–60 min at room temperature
4. Apply the oxidised mucin to a Sephadex G25 column as for the biotinylation of protein-labeled mucin (*see* **Note 6**) and collect fractions.
5. Detect mucin containing fractions mucin using the slot-blot assay with the periodic acid Schiff's stain (*see* Chapter 4), and pool these fractions (approx 5 × 1 mL).
6. Add BACH in DMSO to a final concentration of 1 m*M*.
7. Incubate at room temperature for 2 hours or overnight at 4°C.
8. Separate the biotinylated mucin on a Sephadex G25 column as for protein-labeled mucin.

9. Collect fractions and detect mucin using the slot blot assay with the periodic acid Schiff's stain and with the 96-well microtiter plate assay (*see* **Subheading 3.4.**).
10. Pool the labeled fractions (approx 8 × 1 mL) to give a concentration of 12.5 µg/mL mucin and aliquot as 0.2-mL samples. Store at 4°C until used.

3.3. Mucinase Assay with Radioactive Substrates (see Notes 7 and 8)

1. Mix 20–500 µL of enzyme extract or commercially available enzyme (*see* **Subheading 2.1., item 1**) with 5000–50,000 cpm of radiolabeled mucin in incubation buffer in a final volume of 1 mL, maximum (*see* **Notes 7** and **8**).
2. Incubate at 37°C for periods up to 24 h (usually 6, 12, or 24 h, but start with 24 h if the activity is unknown).
3. Prepare a blank and incubate under the same conditions as **step 2** above (*see* **Note 9**).
4. After incubation, either load onto Sepharose CL 2B column immediately, or freeze at –20°C until ready to start gel filtration.
5. Load as a 1 mL sample onto a column 30 × 1cm of Sepharose CL 2B equilibrated in 10 m*M* Tris-HCl, pH 8.0, and elute with the same buffer.
6. Collect 30 1-mL fractions
7. Mix the entire fraction (or 0.5 mL where >50,000 cpm counts are present) with scintillation fluid and measure the radioactivity.
8. Compare profiles of the test with the blank incubations. Identify the region of low molecular weight product that represents degraded mucin. **Figure 1** gives an example profile.
9. Subtract the blank incubation (background) from the test incubation, and assess the proportion of low molecular weight product formed from the high molecular weight substrate.

3.4. Mucinase Assay with Biotinylated Substrates

3.4.1. Coating Microtiter Plates

1. Dilute the biotinylated mucin in coating buffer (*see* **Notes 6** and **7**).
2. Carefully place the chosen volume in the bottom of the microtitre plate well (*see* **Note 5**). Typically about 50 µL is used.
3. Incubate the plates overnight at 4°C for the mucin to adsorb onto the plate.
4. Empty the plates carefully (*see* **Note 5**).
5. Wash once with incubation buffer 1 × 50 µL followed by 3 × 200 µl. The buffer used is the buffer that is to be used in the assay (see **Notes 7** and **8**).

3.4.2. Digestion of Coated Plates

1. Prepare a suitable dilution of the enzyme sample in incubation buffer (*see* **Notes 7, 8,** and **10**).
2. Place 60–100 µL of enzyme preparation in each well (i.e., more than the volume of mucin solution which was used to coat the plate).
3. Incubate at 37°C typically for 1–2 h.
4. Carefully remove the digestion media.
5. Wash four times with 200 µL of PBS.

3.4.3. Detection of Labeled Mucins

1. Block nonspecific binding with 150–300 µL of blocking buffer for 1 h at room temperature or overnight at 4°C.
2. Empty the plates and wash twice with 200 µL PBS per well.
3. Incubate with streptavidin-HRP solution (75 µL) for 60 min at room temperature.

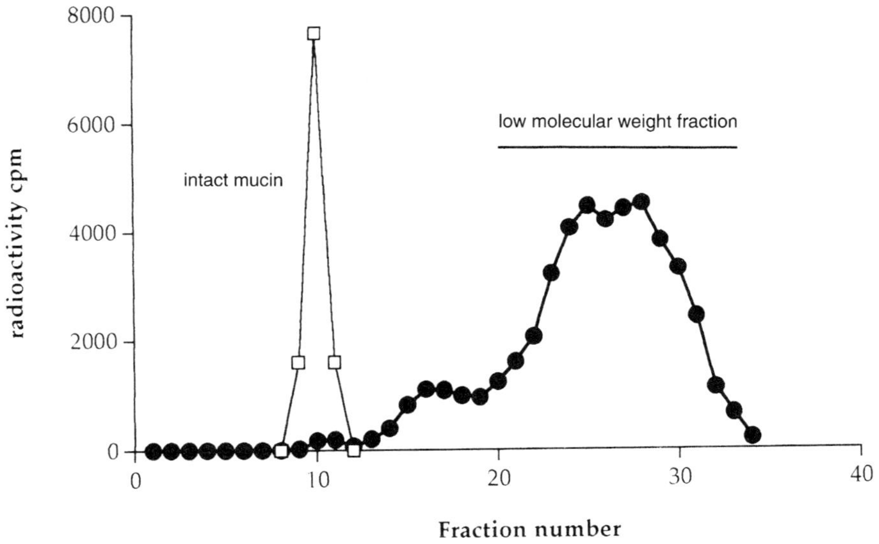

Fig. 1. Identification of the products of mucinase activity by Sepharose CL-2B chromatography. Intact colonic mucin labeled with [^{3}H]-D-glucosamine (□). The products of incubation with human fecal extract are shown (●) and the fractions pooled as low molecular weight.

4. Empty the plates and wash with PBS 1 × 100 µL and 3 × 200 µL.
5. Place 100 µL of OPD solution in wells and incubate in the dark for up to 60 min.
6. Stop the reaction and develop color with the addition of 100 µL of stop solution (0.5 *M* H_2SO_4).
7. Measure the absorbance at 490 nm (*see* **Note 11**).

4. Notes

1. The choice of the source mucin for mucinase assays is of great importance. If possible it is always best to prepare a mucin substrate from the mucosa that is the target for mucinase study. In some cases it is not possible to obtain suitable samples, e.g., owing to ethical considerations, or to the low abundance of material from minor mucosal surfaces. It may be possible to obtain equivalent material from normal animals. If "nontypical" mucins are chosen and used to probe the presence of mucinase activity in normal and/or disease situations, the interpretation of the physiological significance of the results needs careful consideration. Such mucins can of course be used simply to detect the presence of any mucinase activity.
2. The choice of plate type and buffer is critical. The amount of mucin that binds to the plate is dependent on both the type of plastic that the plate is composed of and any treatment it has undergone (such treatments affect the charge of the plastic). The pH affects the charge of the mucin and therefore the adherence to the plate. Preliminary experiments are crucial to assess the effect of both the plate type and the pH of the buffer used.
3. When preparing the plates, it is essential to dilute the mucin in the desired coating buffer before placing in the plate. When placing in the plate great care must be taken to not to touch the sides of the well where the mucin will not coat Also when transporting the plate, it is essential not to distribute the mucin outside the normal coating area. The mucins stick rapidly and disturbance will increase the coated surface area.

4. The chemical composition and pH of the coating buffer are critical. Avoid using amino containing buffers (Tris, HEPES, glycine), particularly with protein-labeled mucins. Instead, use buffers such as acetate or citrate.
5. The Sephadex G25 step removes free biotin. It is necessary to discard the column because irreversible adsorption of biotin occurs.
6. It is necessary to remove any unreacted periodate from the mucin before the biotinylation reaction is started because it interferes with the reaction. The use of a Sephadex G25 column is an easy method to do this and provides a high recovery.
7. Mucinase activity usually shows a broad pH range as the complete activity is a composite of enzymes acting together. Suitable incubation buffers should be chosen to cover the potential range of activity expected. Typically, acetate buffers at lower pHs and Tris at higher pHs. Highly purified enzymes available for commercial sources can be used to validate plate assays, and to generate specimen fragmentation profiles.
8. The use of phosphate buffers should be avoided if possible because phosphate may inhibit sulfatase activities (*see* Chapter 34). and thus affect overall mucinase activity.
9. Blank incubations can be prepared using enzyme extract that has been heat treated for 5 min at 95°C and centrifuged. Alternatively, the enzyme can be substituted by the same volume of incubation buffer. The suitability of either of these blanks should be tested for interference in each assay system.
10. The dilution of enzyme samples should be made in incubation buffer suitable for the enzyme itself and for the mucinase plate assay. This is easiest with commercial enzymes in which the amount of enzyme is known. In cases in which crude preparations are being used, such as fecal extracts, urine, serum, bacterial, or cell cultures as cell suspensions, or cell supernatants, the samples should be tested directly and as dilutions in incubation buffer. It is wise to test at high, low, and intermediate pH ranges.
11. If a 490-nm filter is not available, use the nearest available wavelength. If the plate reader uses a reference filter for standardization, wavelengths between 630 and 650 nm (or nearest available) should be used.

Acknowledgments

This work was supported by grants 046530/Z/96 and 051586/Z/97 from the Wellcome Trust.

References

1. Roussel, P. and Lamblin, G. (1996) Human mucosal mucins in diseases, in Glycoproteins and Disease. (Montreuil, J., Vliegenthart, J. F. G., and Schachter, H. eds.), Elsevier Science BV, Amsterdam, pp. 351–393.
2. Gendler, S. J. and Spicer, A. P. (1995) Epithelial mucin genes. *Annu. Rev. Physiol.* **57,** 607–634.
3. Dwarakanath, A. D., Campbell, B. J., Tsai, H. H., Sunderland, D., Hart, C. A., and Rhodes, J. M. (1995) Faecal mucinase activity assessed in inflammatory bowel disease using ^{14}C threonine labelled mucin substrate. *Gut* **37,** 58–62.
4. Corfield, A. P., Wagner, S. A., O'Donnell, L. J. D., Durdey, P., Mountford, R. A., and Clamp, J. R. (1993) The roles of enteric bacterial sialidase, sialate O-acetyl esterase and glycosulfatase in the degradation of human colonic mucin. *Glycoconj. J.* **10,** 72–81.
5. Roberton, A. M. and Corfield, A. P. (1999) Mucin degradation and its significance in inflammatory conditions of the gastrointestinal tract, in Medical Importance of the Normal Microflora. (Tannock, G. W. ed.), Kluwer Academic Publishers, Dordrecht, pp. 222–261.

6. Colina, A.-R., Aumont, F., Deslauriers, N., Belhumeur, P., and de Repentigny, L. (1996) Evidence for degradation of gastrointestinal mucin by *Candida albicans* secretory aspartyl proteinase. *Infect. Immun.* **64,** 4514–4519.
7. Colina, A.-R., Aumont, F., Deslauriers, N., Belhumeur, P., and de Repentigny, L. (1996) Development of a method to detect secretory mucinolytic activity from *Candida albicans*. *J. Med. Vet. Mycol.* **34,** 401–406.
8. Koritsas, V. M. and Atkinson, H. J. (1995) An assay for detecting nanogram levels of proteolytic enzymes. *Anal. Biochem.* **227,** 22–26.

32

Proteinase Activity

David A. Hutton, Adrian Allen, and Jeffrey P. Pearson

1. Introduction

Early studies concerning proteolytic degradation of mucins demonstrated that the protein core of mucins consisted of two distinct regions, glycosylated regions: protected from degradation by the densely packed carbohydrate side chains and nonglycosylated regions susceptible to proteases *(1,2)*. Since the late 1980s, sequencing of mucin genes has underlined these studies and provided a firm molecular basis for these concepts *(3)*. Gene-cloning studies have shown that the protein backbone of the subunits of secreted polymeric mucins can be up to 5000 amino acids in length (approx 20% by weight of the molecule) and consists of two major types of domain that alternate throughout the sequence *(3)*. One type of domain, situated centrally accounts for about 50% of the protein core and is characterized by tandem repeat (TR) sequences, rich in threonine, serine, and proline, and the hydroxyl amino acids form the sites of attachment of the oligosaccharide chains (approx 80% by weight of the molecule). The other major type of domain, situated at the N- and C-terminals and between regions of TR sequences, is relatively poor in these three amino acids and relatively rich in cysteine *(3)*. Some of these cysteine residues can form disulphide bridges with other mucin monomers (M_r 2–3 × 10^6) to form large polymeric mucins linked end to end (M_r ~ 10^7) and capable of forming gels *(4–6)*. The tandem repeat domains are protected from proteolysis by the carbohydrate side chains that sterically inhibit proteinases from gaining access to the protein core, however, proteinases can hydrolyze the cysteine-rich regions of accessible nonglycosylated protein, thereby fragmenting the polymeric mucins *(6)*. The soluble glycopeptides resistant to proteolysis remain of relatively high *Mr* (200–700 kD) and recent studies have suggested that individual mucin gene products may contain different types and lengths of glycosylated domains. For instance, analysis of high *Mr* glycopeptides produced by trypsin digestion of the MUC5B subunit indicated that it contained different types and lengths of glycosylated domains; one domain of *Mr* 7.3 × 10^5, two domains of 5.2 × 10^5 and a third domain of 2 × 10^5 *(7)*. Similarly rat small intestinal Muc2 mucin subunit contains two glycopep-

From: *Methods in Molecular Biology, Vol. 125: Glycoprotein Methods and Protocols: The Mucins*
Edited by: A. Corfield © Humana Press Inc., Totowa, NJ

tides with an estimated mass of 650 and 335 kDa *(8)*. The significance of the differences in size of these domains is unclear.

Protective essentially insoluble mucus gels adherent to mucosal surfaces are solubilised by proteinases, e.g., in the lumen of the gastrointestinal tract **(6)**. The mechanism of solubilisation is cleavage of the susceptible regions of the mucin protein core which are responsible for the gel-forming properties of mucus gels. This mucolysis results in degradation of the polymeric mucins (high viscosity, gel-forming) into relatively low *Mr* glycopeptides (low viscosity, soluble). A dynamic balance exists between secretion of gel-forming mucins and their degradation by proteinases, if this relationship is tilted in favor of proteolytic degradation the protective properties of the mucus gel will be compromized and disease may result. The proportion of polymeric mucins present in mucus gels has been shown to be an indicator of gel strength *(9)* and evidence exists for enhanced mucolysis by proteinases and/or inferior mucin polymerisation and consequent impairment of the mucus barrier in peptic ulceration and inflammatory bowel disease *(10,11)*. Techniques are therefore required for measurement of the integrity of mucin polymeric structure in large numbers of samples isolated from mucus gels and the mucosa secreting them and methods are needed to assess the mucolytic potential of host and pathogen secretions.

The digestion of mucin polymers and their reduced subunits by proteases has been followed by measuring molecular size using analytical ultracentrifugation or light scattering *(2,5)*. These techniques, whilst providing the most empirically quantitative descriptions of size, rely on highly purified samples, expensive equipment, a high level of expertise and can require relatively large quantities of sample (e.g., measurement of sedimentation coefficient). Therefore, they are generally unsuitable for the assessment of the polymeric integrity of large numbers of mucin samples or for the screening of the mucolytic potential of protease containing samples.

For most laboratories gel-filtration chromatography, commonly on Sepharose CL-2B or Sephacryl S-500, has provided the most convenient method of assaying mucin degradation in terms of both integrity of mucins and mucolytic activity *(5,10,12,13)*. Polymeric mucins are excluded in the void volume of the column, and proteolytically degraded subunits are partially included ($K_{av} \sim 0.5$), eluted mucins being assayed in solution or after blotting onto nitrocellulose membranes (*see* **Note 1**). However, this technique also requires relatively large quantities of sample and is time consuming in processing large numbers of samples. The use of radiolabelled mucin samples improves the sensitivity of the technique, however it remains laborious *(14)*.

Mucin polymers and their reduced subunits do not penetrate polyacrylamide gels (>3%), however, proteolytically digested subunits will enter gels of >4%. The advent of readily commercially available precast polyacrylamide gradient gels (4–15%) requiring small amounts of sample and with rapid running and staining times has vastly improved the analysis of the in vivo polymeric integrity of mucins isolated from adherent mucus gels and mucosa and is described below.

Assessment of the mucolytic potential of endogenous secretions or extracts from invasive pathogens is initially underpinned by measurement of nonspecific proteinase activity. General protease activity has often been assayed by methods distinguishing hydrolyzed protein from nonhydrolyzed protein by precipitation of the latter, e.g., with trichloroacetic acid. These methods are reliant on qualitative size and conformational size changes preventing protein precipitation *(15)*. A sensitive and more accurate method for estimating proteolytic activity involves measuring trinitrophenylated derivatives of new N-terminal groups which form on peptide bond hydrolysis *(16)*. The sensitivity of the assay is improved by blocking existing amino groups on the protein substrate, e.g., by succinylation. The preparation of succinyl albumin is described subsequently. The assay can also be used with mucin substrates *(12)* and the large size of mucin protein cores means that fewer N-terminals are present on a weight for weight basis and therefore blocking is unnecessary to achieve low background absorbances. Large quantities of highly purified mucin are, however, required. The assay has also been adapted for use after separation of proteinase isoenzymes on agarose gels *(16)*.

Attempts to measure specific mucolytic activity have been dogged by similar problems to those of general proteinase activity. Cetyl trimethylammoniumbromide (CTAB) or protamine sulfate precipitation has been used to precipitate only undigested mucins, but again depends upon qualitative assessment of precipitability of undegraded macromolecules *(17,18)*. Other methods that depend on the appearance of hydrolysis zones on petri dishes containing mucin/agarose gels often rely on crude mucin preparations (contaminated with hydrolyzable protein) to provide sufficient material or require large amounts of purified mucin *(19,20)*.

Dilute solution viscosity studies measuring mucolysis by following the fall in viscosity of mucin solutions *(10,12)* have a number of advantages over other methods. They provide information on the kinetics of digestion and the size (molecular space occupancy) of the degraded species, and crude mucus samples can be studied as well as purified mucins. The technique is readily compatible with other techniques, e.g., if mucolysis is inhibited at particular time points, aliquots of sample can be further studied by gel filtration or N-terminal analysis *(12)*.

The measurement of mucolysis by quantitating digoxigenin-labeled mucin bound to and released from microtitre plates requires little mucin substrate and may therefore be useful for screening large numbers of potentially mucolytic samples *(20)*. Differences in binding to the plates between undegraded and degraded mucins may, however, make interpretation of the results difficult (*see* **Note 1**).

1.1. Proteinases in the Gastrointestinal Tract

Proteinases from all regions of the gastrointestinal tract degrade mucus gel and mucins in vivo and in vitro to produce soluble glycopeptide fragments which are further degraded by glycosidases in the colon *(1)*.

In the stomach, the secreted mucins in the adherent mucus gel layer are faced with pepsins (maximum concentration ~0.7 mg/ml in humans). Of the seven pepsin isoenzymes in human gastric juice, pepsin 3 is the most abundant with some pepsin 5. Pepsin 1 is normally a minor component (<5%) but can account for up to 25% of the proteolytic activity in peptic ulcer patients. Pepsin 1 has increased collagenolytic and mucolytic activities compared with other pepsins *(15)*. In peptic ulcer patients, there is disruption of the mucus layer and an increase in the percentage of nongel forming lower molecular mass mucin in the adherent gel *(21)* and this is associated with increased mucolytic activity of the gastric juice owing to raised levels of pepsin 1 *(10)*.

The human pancreas produces 1–3 g of chymotrypsin, 1–3 g of trypsin and approx 0.5 g of elastase per day, however total fecal levels of pancreatic enzymes are ~ 1 mg/d *(22)*. This fall in levels is due partly to bacterial degradation and partly to autodegradation *(23)*. Human fecal extracts contain proteinases of pancreatic *(23)* and bacterial *(24)* origin whereas enzymes associated with inflammation, e.g., white cell elastase are present only in trace amounts *(25)*. The bacterial microflora can produce copious amounts of luminal proteinases *(24)*. Fecal proteinases have been demonstrated to have mucus degrading activity *(12,14)*. Human fecal proteinase activity is raised in ulcerative colitis patients compared to nonsymptomatic control subjects *(26,27)* and this raised luminal proteinase activity probably equates with elevated mucolytic activity. These observations could explain in part the thinner colonic mucus in ulcerative colitis *(28)* by allowing mucin degradation to outweigh mucus secretion.

1.2. Specific Proteolysis of Mucins

Specific sites for proteolytic cleavage are also known to exist in the amino acid sequences of mucin protein cores. cDNA studies on human and rat MUC2 have shown that these mucins share a proteolytic cleavage site sequence approx 700 amino acids from their C-terminus (TGWGD PH(Y/F*)VTFDGLYY) and cleavage of the aspartyl proline bond generates a C-terminal glycopeptide fragment of approximately 120 kDa *(29)*. This site is homologous with a site cleaved during the synthesis of rat mammary sialomucin at an early stage of intracellular transport. The timing of cleavage (during or after biosynthesis) of intestinal mucins has not been established. A heparin binding site sequence, SRRARRSPRHLGSG, is also cleaved from the protein core of MUC2 at an early stage of biosynthesis *(30)*.

2. Materials

1. Bovine serum albumin (BSA), Fraction V, (Sigma, Poole, Dorset, UK).
2. Succinic anhydride (Sigma).
3. Trinitrobenzenesulphonic acid (TNBS; 5% [w/v] aqueous solution) (Sigma).
4. Enzymes: porcine pepsin (pepsin A; EC 3.4.23.1), porcine pancreatic trypsin (E.C 2.4.21.4) and porcine pancreatic elastase (EC 3.4.21.36) (Sigma).
5. Proteinase inhibitor cocktail. 1.0 m*M* phenylmethylsulfonyl fluoride (PMSF), 50 m*M* iodoacetamide, 100 m*M* aminohexanoic acid, 5 m*M* benzamidine HCl, 1 m*M* *N*-ethylmaleimide, 1 mg/L soybean trypsin inhibitor (all from Sigma Chemical Company), 10 m*M* EDTA (B.D.H. Poole, Dorset, UK) in 0.67 *M* phosphate buffer, pH 7.5.
6. Sodium dodecyl sulfate (SDS)-polyacrylamide gels for electrophoresis: 4–15% Phast gels were purchased from Pharmacia, L.K.B. Biotechnology, Uppsala, Sweden.
7. Schiffs reagent, commercial solution (Sigma).
8. Sepharose CL-2B (Pharmacia) use in 1.5 ID × 150 cm glass columns equilibrated with 0.2 *M* NaCl/0.02%(w/v) NaN_3 (B.D.H., Poole, UK).
9. Pharmacia Phast-gel system (Pharmacia).
10. Contraves low shear 30 viscometer (Contraves A.G. Zurich, Switzerland).

3. Methods

3.1. Preparation of Substrate, Blocking of N-Terminals

1. Dissolve 10g of BSA in 100ml 0.1M phosphate buffer pH 7.5.
2. Add 1.4 g of succinic anhydride , stir to dissolve and maintain the pH at 7.5 with 2 *M* NaOH using a pH stat as the succinic anhydride dissolves (*see* **Note 2**).
3. Dialyze the solution exhaustively against distilled water at 4°C, freeze-dry and store the protein substrate with blocked *N*-terminals at –20°C (*see* **Note 3**).

3.1.1. N-Terminal Assay

1. Mix proteinase (*see* **Note 4**) or extract containing proteinase activity (200 μL, 0.1–0.5 μg) with 0.5 mL buffer of choice (*see* **Note 5**). Prepare control samples by heating the proteinase preparation at 100°C for 10 min to destroy activity. Alternatively add the substrate immediately prior to **step 4** below. (This will give a background value for any free *N*-terminals in the proteinase sample). Prepare a reagent blank by replacing enzyme or extract with buffer.
2. Add substrate (0.5 mL 8mg/mL succinyl albumin) to start the reaction, mix and incubate samples for 30 min at 37°C in a waterbath.
3. Add 0.5 mL of sodium bicarbonate (to increase the pH of the solution to pH 8.0) and add 0.5 mL 0.05% (w/v) TNBS in water (to trinitrophenylate any free amino groups formed). Incubate at 50°C for 10 min in a water bath to develop color.
4. Add 0.5 mL 10% (w/v) SDS to prevent protein precipitation. Then add 0.25 mL 1 *M* HCl to complete the reaction.
5. Read the absorbance at 340 nm in a spectrophotometer.
6. Calculations: Proteinase activity in millimoles new *N*-terminals/min/g extract is calculated using the following equation:

$$\frac{Vt \times \text{dilution} \times A_{340} \times 10^3}{E \times t \times Vs \times g}$$

where Vt = final tube volume (mL); E = molar extinction coefficient of trinitrophenyl amino acids (1.3×10^4 cm^2/mole); t = time (min); Vs = volume of enzyme; A_{340} = absorbance at 340 nm; g = wet weight of extract/mL (*see* **Note 6**).
The example for g is for proteinase activity in fecal extracts, and g refers to the weight of material present in 1 mL of a fecal homogenate.

3.1.2. Collection, Biopsies, and Brushings: Purification of Mucin

Samples of human mucin can be obtained from samples obtained at surgery, e.g., during routine endoscopy or colonoscopy. Adherent mucus gel is obtained by brushing the mucosa with a cytology brush. Mucosal biopsies provide both intracellular mucin and adherent mucus gel.

1. Place samples immediately into a cocktail of proteinase inhibitors, i.e., 1.0 m*M* phenylmethylsulphonyl fluoride (PMSF), 50 m*M* iodoacetamide, 100 m*M* aminohexanoic acid, 5 m*M* benzamidine HCl, 1 m*M* *N*-ethylmaleimide, 1 mg/L soybean trypsin inhibitor, 10 m*M* EDTA in 0.67 *M* phosphate buffer, pH 7.5, at 4°C to minimize proteolytic degradation (*see* **Note 5**) and store at –20°C until processed.
2. Solubilize mucins by brief homogenization (30 s, low-speed, hand-held homogenizer) and centrifuge (8000*g*, 1 h, 4°C) to remove cell debris.

3. Purify mucins by equilibrium density gradient centrifugation in CsCl (100,000*g*, 48 h, 4°C, starting density 1.42 g/mL) (*see* **Note 7**).
4. Dialyze the pooled mucins exhaustively against distilled water, freeze-dry and store at –20°C.

3.2. SDS Polyacrylamide Gel Electrophoresis

3.2.1. Phast-Gel Electrophoresis

This method refers to SDS polyacrylamide gel electrophoresis using 4–15% gels on the Pharmacia Phast-gel system (Pharmacia).

1. Solubilize freeze-dried mucin purified from human gastrointestinal mucosal brushings or biopsies in, e.g., 100 µL of 0.0125 *M* Tris-HCl, 0.4% (w/v) SDS, 2% (v/v) glycerol, and 0.0002% bromophenol blue, pH 6.8, at a concentration of 3 mg/mL.
2. Heat the samples for 2 min at 100°C.
3. Load 1-µL aliquots onto the gel (3 µg).
4. Electrophorese for 45 min, 15°C, 250 V, 10 mA.

3.2.2. Staining of SDS 4–15% Gels

1. Following electrophoresis incubate gels in 25 mL 7% (v/v) acetic acid for 1h. at room temperature.
2. Incubate in 25 mL of periodic acid solution (7% acetic acid containing 0.04% [v/v] periodic acid) for 1 h at room temperature.
3. At the same time as **step 2** above incubate a 24-mL solution of Schiffs reagent containing 0.4 g sodium metabisulphite at 37°C for 1 h.
4. Pour the periodic acid solution from the gel and incubate for 30 min at 37°C in Schiff's reagent.
5. Wash the gel several times (approx 15 min each wash) with 25 mL of 7.5% (v/v) acetic acid until the washing solution becomes colorless.
6. Store the gel overnight at 4°C in 7% acetic acid and scan at 555 nm with a gel scanner.

3.3. Solution Viscosity as an Assay for Mucolytic Activity

Mucolytic activity by proteinases can be monitored by measuring the fall in viscosity of mucin solutions on incubation with enzyme or extract. Viscosity can be measured using a Couette rotating cup viscometer (e.g., Contraves' Low Shear 30). An electronic speed programmer (Rheoscan 20) allows the speed of rotation of the cup to be increased under controlled conditions. The cup is filled with the solution under investigation (~2 mL), and a bob is gently lowered into the solution without introducing any air bubbles. The bob is attached to a torsion wire, and when the cup is rotated, the resistance offered by the solution is transferred to the bob and the torsion wire. This resistance is converted into an electrical signal recorded on a chart recorder. The temperature of the cup and bob can be maintained at 37°C using a water bath that circulates water through the viscometer block. The shear rate is increased linearly from 1.7 to 128.5/s.

1. Solubilize freeze-dried mucin (either crude or purified) at 2–5 mg/mL by stirring overnight at 4°C in buffer (e.g., 67 m*M* sodium phosphate buffer, pH 7.5) containing 0.08 *M* NaCl (*see* **Note 8**).
2. Equilibrate the mucin, the buffer and the enzyme preparation or extract at 37°C.

3. Add the enzyme to the mucin and measure the viscosity at various time intervals over a 24-h period. Prepare a control solution containing mucin and buffer and measure its viscosity over the same time period (*see* **Note 9**). Measure the viscosity of the buffer being used.
4. Calculations: The solution viscosity of the sample is directly proportional to the gradient of the curve plotted on the chart recorder, the *Y*-axis being the percentage deflection and the *X*-axis speed. The ratio of this gradient and the gradient produced for the solvent (buffer) alone gives the relative viscosity (η_{rel}).

$$\eta_{rel} = \frac{\text{gradient of test sample}}{\text{gradient of solvent}}$$

Specific viscosity may then be calculated as:

$$\eta_{sp} = \eta_{rel} - 1$$

Specific viscosity can then be plotted against time to follow the course of digestion. Reduced specific viscosity η_{red} can be calculated as η_{sp}/C, (where C = concentration of mucin), and intrinsic viscosity calculated by plotting η_{red} against mucin concentration and extrapolating to zero concentration (Huggins Plot). Alternatively, plotting ln η_{rel} as a function of concentration and extrapolating to zero also gives values of intrinsic viscosity (Kraemer Plot).

4. Notes

1. The high M_r glycopeptides produced by proteinase digestion bind poorly to nitrocellulose membranes, e.g., in the PAS slot blot assay ***(31)*** after gel filtration on Sepharose CL-2B. A method to improve binding by previously applying wheat germ agglutinin to the membranes has been described ***(32)***.
2. It is essential to maintain the pH at 7.5 to prevent denaturation of the albumin.
3. Allow expansion space in the dialysis tubing to avoid bursting since it will swell during dialysis.
4. Porcine pepsin (pepsin A; EC 3.4.23.1), porcine pancreatic trypsin (EC 2.4.21.4), and porcine pancreatic elastase (EC 3.4.21.36) (Sigma) have all been successfully used as standards in the *N*-terminal proteinase assay ***(12,16)***.
5. *Inhibition of proteinases.* It is vitally important to inhibit endogenous proteinases when attempting to isolate intact, native polymeric mucins. Unless rigorous precautions are taken during isolation by including a mixture of wide ranging proteinase inhibitors then "nicking" of the protein core occurs to give units which, while still retaining much of the intact protein core, are of smaller size. 1 m*M* phenylmethylsulfonyl fluoride has been found to be particularly important in inhibiting proteolysis (D. A. Hutton and A. Allen, unpublished observations). PMSF should be dissolved in a small quantity (~5 mL) of propan-2-ol and added to the rest of the cocktail immediately before use. An alternative inhibitor to PMSF is diisopropylphosphorofluoridate (DIFP) which while very effective ***(5)*** does require extremely careful handling
6. A protein assay carried out on the proteinase sample to determine the mg/mL concentration of protein will allow the activity to be expressed per milligram of protein.
7. For human mucin from biopsies and brushings the mucin containing fractions are best estimated using density, the concentration of mucin in each fraction being insufficient to measure even with the slot-blot method ***(31,32)***. Fractions having a density equivalent to mucins under these conditions (1.45–1.51 g/mL) should be pooled.

8. It is essential to have isotonic conditions, i.e., 300 mosmolar to shield the negative charge on the mucins otherwise anomalously high viscosities will result.
9. It is vital to include controls without enzyme in the mucolysis experiments. Crude mucus samples may contain endogenous protease activity and some interactions in pure mucin samples may be disrupted by repeated shearings causing an apparent loss of viscosity.

References

1. Allen, A. and Pearson, J. P. (1993) Mucus glycoproteins of the normal gastrointestinal tract. *Eur. J. Gastroenterol. & Hepatol* **5,** 193–199.
2. Scawen, M. and Allen, A. (1977) The action of proteolytic enzymes on the glycoprotein from pig gastric mucus. *Biochem. J.* **163,** 363–368.
3. Gendler, S. J. and Spicer, A. P. (1995) Epithelial mucin genes. *Annu. Rev. Physiol.* **57,** 607–634.
4. Snary, D., Allen, A., and Pain, R. H. (1970) Structural studies on gastric mucoproteins. Lowering of molecular weight after reduction with mercaptoethanol. *Biochem Biophys Res Commun.* **40,** 844–851.
5. Carlstedt, I. Sheehan, J. K., Corfield, A., and Gallagher, J. T. (1985) Mucus glycoproteins: a gel of a problem. *Essays Biochem.* **20,** 40–76.
6. Allen, A. (1981) Structure and Function of gastrointestinal mucus, in *Physiology of the Gastrointestinal Tract.* (Johnson, L. R., ed.), Raven Press, New York, pp. 617–640.
7. Thornton, D. J., Howard, M., Khan, N., and Sheehan, J. K. (1997) Identification of two glycoforms of the MUC5B mucin in human respiratory mucus. *J. Biol. Chem.* **272,** 9561–9566.
8. Carlstedt, I., Herrmann, A., Karlsson, H., Sheehan, J., Fransson, L. A., and Hansson, G. C. (1993) Characterisation of two different glycosylated domains from the insoluble mucin complex of rat small intestine. *J. Biol. Chem.*, **25,** 18,771–18,781.
9. Sellers, L. A., Allen, A., Morris, E. R., and Ross Murphy, S. B. (1988) Mucus glycoprotein gels: role of polymeric structure and carbohydrate side chains in gel formation. *Carb Res.* **178,** 93–110.
10. Pearson, J. P., Ward, R., Allen, A., Roberts, N. B., and Taylor, W. H. (1986) Mucus degradation by pepsin: comparison of mucolytic activity of human pepsin 1 and pepsin 3: implications in peptic ulceration. *Gut* **27,** 243–248.
11. Rankin, B. J., Sirvastava, E. D., Record, C. O., Pearson, J. P., and Allen, A. (1995) Patients with ulcerative colitis have reduced mucin polymer content in the adherent colonic mucus gel. *Biochem. Soc. Trans.* **23**, 1045.
12. Hutton, D. A., Pearson, J. P., Allen, A., and Foster, S. N. E. (1990) Mucolysis of the colonic mucus barrier by faecal proteineses: inhibition by interacting polyacrylate. *Clin. Sci.* **78,** 265–271.
13. Pearson, J. P., Allen, A., and Venables, C. W. (1980) Gastric mucus: Isolation and polymeric structure of the undergraded glycoprotein: Its breakdown by pepsin. *Gastroenterol.* **78,** 709–715.
14. Dwarkanath, A. D., Campbell, A. B., Tsai, H. H., Sunderland, D., Hart, C. A., and Rhodes, J. M. (1995) Faecal mucinase activity in inflammatory bowel disease using ^{14}C labelled mucin substrate. *Gut* **37,** 58–62.
15. Allen, A., Hutton, D. A., Leonard, A. J., Pearson, J. P., and Sellers, L. A. (1989) Pepsins, in: *Endogenous Mediators of Gastrointestinal Disease.* (Wallace, J. L., ed.), CRC, Boca Raton, Florida, pp. 53–69.
16. Hutton, D.A., Allen, A., Pearson, J. P., Ward, R., and Venables, C. W. (1986) Separations of pepsins in human gastric juice: analysis of proteolytic and mucolytic activity. *Biochem. Soc. Trans.* **14,** 735–736.

17. Young, D. B. and Broadbent, D. A. (1982) Biochemical characterisation of extracellular proteases from Vibrio cholerae. *Infection and Immunity*. **37,** 875–883.
18. Haider, K., Hossain, A., Wanke, C., et al. (1993) Production of mucinase and neuraminidase and binding of shigella to intestinal mucin. *J. Diarrhoeal. Dis. Res.* **11,** 88–92.
19. Macfarlane, G. T., Cummings, J. H., Macfarlane, S., and Gibson, G. R. (1989). Influence of mucin on glycosidase, protease and arylamidase activities of human gut bacteria grown in three stage continuous culture system. *J. Appl. Bacteriol.* **67,** 521–527
20. Colina, A. R., Aumont, F., Belhumeur, P., and De Repentigny, L. (1996) Development of a method to detect secretory mucinolytic activity from Candida albicans. *J. of Appl. Med. and Vet. Mycology*. **34,** 401-406.
21. Younan, F., Pearson, J. P., Allen, A., and Venables, C. W. (1982) Changes in the structure of the mucus gel on the mucosal surface of the stomach in association with peptic ulcer disease. *Gastroenterol.* **82,** 827–831.
22. Kukral, J. C., Adama, A. P., and Preston, F. W. (1965) Protein producing capacity of the human exocrine pancreas. *Annals Surg*. **162,** 67–73.
23. Bohe, M., Borgstrom, A., Gennel, S., and Ohlsson, K. (1983). Determination of immunoreactive trypsin, pancreatic elastase and chymotrypsin in extracts of human feces and ileostomy. *Digestion* **27**, 8–15.
24. Macfarlane, G. T., Allison, G. T., Gibson, S. A. W., and Cummings, J. H. (1988). Contribution of the microflora to proteolysis in the human large intestine. *J. Appl. Bacteriol.* **64,** 37–46.
25. Andus, T., Gross, V., Caesar, L., Krumm, D., Gerok, W., and Scholmerick, J. (1993) PMN-elastase in assessment of patients with inflammatory bowel disease. *Dig. Dis. Sci.* **38,** 1638-1644.
26. Corfield, A. P., Williams, A. J. K., Clamp, J. R., Wagner, S. A., and Mountford, R. A. (1988) Degradation by bacterial enzymes of colonic mucus from normal subjects and patients with inflammatory bowel disease: the role of sialic acid metabolism and the detection of a novel O-acetylsialic acid esterase. *Clin Sci*. **74,** 71–78.
27. Samson, H. J., Allen, A., Pearson, J. P., Cunliffe, W. J., Rhodes, M., and Rhodes, J. (1991) Faecal proteinase activity; raised levels in patients with ulcerative colitis. *Gut* **32,** A1235–A1236.
28. Pullan, R. D., Thomas, G. A. O., Rhodes, M., Newcombe, R. G., Willimas, G. T., Allen, A., and Rhodes, J. M. (1994). Thickness of adherent mucus gel on colonic mucosa in humans and its relevance to colitis. *Gut*. **35,** 353–359.
29. Forstner, J.F. and Forstner, G.G. (1994) Gastrointestinal Mucus, on: *Physiology of the Gastrointestinal Tract*. 3rd ed. (Johnson L. R., ed.), Raven Press, New York, pp. 1255–1283.
30. Xu, G., Forstner G. G., and Forstner, J. F. (1996) Interaction of heparin with synthetic peptides corresponding to the C-terminal domain of intestinal mucins. *Glycoconjugate J.* **13,** 81–90.
31. Thornton, D. J. Holmes, D. F., Sheehan, J. K., and Carlstedt, I. (1989) Quantitation of mucous glycoproteins blotted onto nitrocellulose membranes. *Anal. Biochem.* **182**, 160–164.
32. Ayre, D., Hutton, D. A., and Pearson, J. P. (1994) The use of wheat germ agglutinin to improve binding of heterogeneous mucin species to nitrocellulose membranes. *Anal. Biochem.* **219,** 373–375.

33

Glycosidase Activity

Anthony P. Corfield and Neil Myerscough

1. Introduction

The glycosidases and associated hydrolytic enzymes acting on glycoconjugate oligosaccharides form part of the total mucinase activity. This chapter describes some assay methods for the determination of these enzymes. Our knowledge of the number of enzymes required for mucin degradation and their regulation in physiological situations is scanty *(1)*. The degradation of both protein and carbohydrate domains require specific enzymes which are able to degrade mucin structure. Carbohydrate degradation may be dependent on prior or concomitant peptide cleavage in mucins. The issues that need to be addressed include the following:

1. Being able to demonstrate individual substrate specificity in relation to the known mucin structure (both protein and carbohydrate).
2. The pathogenic or nutritional role of bacterial degradation leading to the loss (degradation to create receptor binding sites as part of bacterial colonization or infection of the host) or recycling (utilization of the released products for energy production)
3. The need to consider the mode of growth of bacteria at the mucosal surface, in particular the role of biofilms (*see* Chapter 36).
4. The role of "additional" enzymes such as sulfatases, phosphatases, and lipases, which are important for rarer posttranslational modifications to mucin peptide and oligosaccharide structure *(1)*.

Many studies with glycosidases have been carried out using synthetic substrates, such as 4-nitrophenyl- and 4-methyl umbelliferyl-glycosides. These substrates only give information with regard to the anomeric configuration of the glycoside, but not with respect to the nature of linkage to the next sugar in an oligosaccharide. Thus, the results may have limited physiologic relevance. When the degradation of specific glycoconjugates such as mucins is to be assessed, alternative, novel substrates have to be prepared. Only in this way can the manner in which mucins are degraded in vivo be studied. The design of substrates has followed two directions. The first, the use of intact mucin/glycoprotein or glycopeptide substrates. These assays have relied on the

From: *Methods in Molecular Biology, Vol. 125: Glycoprotein Methods and Protocols: The Mucins*
Edited by: A. Corfield © Humana Press Inc., Totowa, NJ

detection of the individual monosaccharide being released. The sensitivity of these assays depends on the sensitivity of the monosaccharide product detection. Colorimetric, fluorimetric, radioactive, ultraviolet (UV) and pulsed amperometric detection define the limits of these assays. Often further separation techniques (gel filtration, high-performance liquid chromatography [HPLC], ion-exchange, etc.) are necessary to isolate and quantify the products. As a result, many of the relevant substrates for studies of degradation of mucin are not available commercially and must be prepared, and in some cases, the design of suitable glycoconjugate substrates has not proved possible. The second direction is the chemical synthesis of oligosaccharides with relevant structure and the identification of degradation by chromatographic methods.

This chapter addresses the question of glycosidase activity. Description of the use of mucin-related substrates together with some of the widely available synthetic substrates gives an approach to the identity of the general range of mucin-degrading glycosidase activities present in enzymatic preparations.

2. Materials

2.1. Glycosidase Substrates

2.1.1. Natural Glycosidase Substrates

Many naturally occurring glycoconjugates can be utilized as physiologic substrates for glycosidases in the assays described in this chapter. In addition to those listed here, other purified glycoconjugates can be used in the same way.

1. Fetuin type III (Sigma, Poole, UK).
2. Salivary gland glycoproteins from bovine *(2)* and ovine *(3)* sources.
3. α_1-Acid glycoprotein (Sigma).
4. Antifreeze glycoprotein *(4)*.
5. Porcine seminal gel glycoprotein *(5)*.
6. Asialoglycoproteins: Prepare in all cases by incubation of the sialoglycoprotein at 1 mg/mL in 0.1 *M* HCl at 80°C for 60 min. The sialic acid content is measured before and after the hydrolysis and additional hydrolysis carried out under the same conditions if significant sialic acid remains *(6)*.
7. Saponification of O-acetyl esters: Carry out saponification on sialic acids in bovine salivary gland mucin at 1 mg/mL in 0.1 *M* NaOH at room temperature for 45 min, and neutralize the solution to approx pH 7.0 with 1 *M* HCl.
8. α2-3 and α2-6 Sialyllactose (Sigma).

2.1.2. Synthetic Glycosidase Substrates

Synthetic substrates form the basis of rapid and sensitive colorimetric (4-nitro–phenyl-glycosides) and fluorimetric (4-methyl umbelliferyl-glycosides) assays. These substrates are available through several suppliers, including Sigma; Oxford Glycosciences, Abingdon, UK; Dextra, Reading, UK; Boehringer Mannheim, Lewes, UK; and Chemica Alta, Edmonton, Canada.

1. 4-Nitrophenyl β-galactose: Dissolve in 1/10 volume of methanol and make up to a final concentration of 4 m*M* in 100 m*M* sodium acetate, pH 6.0.
2. 4-Nitrophenyl α-galactose: Dissolve in 1/10 volume of methanol and make up to a final concentration of 4 m*M* in 100 m*M* sodium acetate, pH 6.0.

3. 4-Nitrophenyl β-N-acetylglucosamine: Dissolve in 1/10 volume of methanol and make up to a final concentration of 4 m*M* in 100 m*M* sodium acetate, pH 6.0.
4. 4-Nitrophenyl α-N-acetylgalactosamine: Dissolve in 1/10 volume of methanol and make up to a final concentration of 4 m*M* in 100 m*M* sodium acetate, pH 6.0.
5. 4-Nitrophenyl Galβ1-3GalNAc: Dissolve the substrate directly in 100 m*M* citrate-phosphate, pH 6.0, to give a final concentration of 2 m*M*.
6. 4-Nitrophenyl α-fucose: Dissolve in 1/10 volume of methanol and make up to a final concentration of 4 m*M* in 100 m*M* sodium acetate, pH 6.0.
7. 4-Methyl umbelliferyl sialic acid: Dissolve the substrate directly in 0.4 *M* sodium acetate, pH 4.2, to give a final concentration of 2 m*M*.

*2.1.3. Galactose/N-acetyl-D-Galactosamine-Labeled Glycoproteins Galactose Oxidase and Tritiated Borohydride (see **Notes 1** and **2**)*

1. Dissolve 5 mg of the glycoprotein, e.g., antifreeze glycoprotein, asialofetuin, asialo-ovine submandibular gland mucin, or α_1-acid glycoprotein, in 500 µL of 50 m*M* sodium phosphate and 5 m*M* NaCl, pH 7.0.
2. Add 5 U of galactose oxidase (Sigma) to each glycoprotein solution, and incubate for 24 h at 37°C.
3. Dilute the products five times with 50 m*M* sodium phosphate and 50 m*M* NaCl, pH 7.8.
4. Add 12.5 MBq of sodium boro-[^{3}H]-hydride (NaB[^{3}H]$_4$), typically 10 GBq/mmol (Amersham, UK) to each, and stir the solutions for 60 min at room temperature (**Hazard:** radioactive; *see* **Note 1**).
5. Add 1.5 mg of solid $NaBH_4$ (Sigma) to each incubation and stir for a further 30 min (**Hazard:** attacks respiratory tract mucous membranes).
6. Add 10-µL aliquots of glacial acetic acid (analytical grade; BDH/Merck, Poole, UK), to destroy borohydride, until no more bubbles are formed.
7. Add 2 mL of analytical grade methanol, and evaporate to dryness under reduced pressure below 30°C in a rotary evaporator. Repeat this extraction five times to remove borate.
8. Adjust the pH of the solution to approx 6.5.
9. Desalt the products by repeated runs on a column of Sephadex G25 (30 × 1 cm; Pharmacia, Milton Keynes, UK) in 0.2 *M* NaCl with a final run in distilled water.
10. Store the products as 100- to 500-µL aliquots in 0.02% sodium azide (BDH/Merck) at 4°C.

*1.4. Sialyl GalNAc [^{3}H]-ol from Boar Seminal Gel Mucin (see **Note 1**)*

1. Dissolve 25 mg of porcine seminal gel glycoprotein in 4 mL of 0.05 *M* NaOH.
2. Add 190 mg of solid $NaBH_4$ (**Hazard:** attacks respiratory tract mucous membranes).
3. Add 925 MBq of NaB[^{3}H]$_4$ (typically 10 GBq/mmol) (Amersham) in 1 mL of 0.05 *M* NaOH, and incubate at 45°C with stirring, for 16 h (*see* **Note 1**).
4. Add glacial acetic acid (analytical reagent grade; BDH/Merck) dropwise until no more bubbles are formed, to destroy excess borohydride (*see* **Note 1**).
5. Pass the solution through a column (20 mL) of Dowex 50 H^+ (200–400 mesh) (Bio-Rad, Hemel Hempstead, UK), and wash with 100 mL of distilled water.
6. Evaporate to dryness under reduced pressure below 30°C using a rotary evaporator. Add 2 mL of methanol and repeat this evaporation five times to remove borate.
7. Dissolve the sample in 5 mL of 0.1 *M* pyridine acetate, pH 5.0, apply to a column of Bio-Gel P4 (200–400 mesh, 150 × 2 cm) (Bio-Rad), and elute in the same buffer.
8. Collect 5-mL of fractions and measure radioactivity. Pool the major radioactive peak (Neu5Acα2-6GalNAc-[^{3}H]-ol) (*see* **Note 3**).

9. Evaporate to dryness under reduced pressure below 30°C using a rotary evaporator. Redissolve in 5 m*M* pyridine acetate, pH 5.0, and apply to a column (18 × 1 cm) of Dowex 1 × 2 (–400 mesh, acetate form) (Bio-Rad). Elute with a gradient of 2–350 m*M* pyridine acetate pH 5.0 (2 × 250 mL) and collect 5-mL fractions.
10. Pool the radioactive peak and remove the pyridine acetate by rotary evaporation.
11. Convert the oligosaccharide to the sodium salt by titration with Dowex 50 Na^+ (50–100 mesh) (Bio-Rad).
12. Store at 4°C in 2% aqueous ethanol.

2.1.5. Sialyl-[^3H]-Labeled Sialoglycoproteins

Any sialoglycoprotein can be labeled with tritium after periodate oxidation (*see* **Note 4**). Typically, α_1-acid glycoprotein and bovine submandibular gland mucin have been used. The sialic acid content of the glycoprotein must be determined first in order to make up the correct ratio of sialic acid to periodate in the mild oxidation step. In the case of some mucin substrates, e.g., bovine submandibular gland mucin, in which sialic acid *O*-acetyl esters are expected or suspected, a mild saponification step should be included before the periodate oxidation (*see* **Subheading 2.1.1., item 7**; **Note 5**).

1. Dissolve 10–50 mg sialoglycoprotein in 100 m*M* sodium acetate, 150 m*M* sodium chloride buffer, pH 5.5, to give a final concentration of 1 m*M* with respect to sialic acid and equilibrate at 4°C.
2. Add ice cold 10 m*M* sodium metaperiodate (Sigma) in water to give a final concentration of 1 m*M* periodate (approx 1/10 of the volume of the sialoglycoprotein solution) and stir.
3. Incubate for 10 min at 4°C with stirring.
4. Add glycerol (0.2 mL for each 10 mL of sialoglycoprotein solution) and stir for a further 10 min.
5. Dialyze the solution against three changes of 2.5 L of 0.05 *M* sodium phosphate, 0.15 *M* sodium chloride, pH 7.4, for 24 h at 4°C.
6. Add 500–1000 Mbq of $NaB[^3H]_4$ (typically 10 GBq/mmol) (Amersham) in 0.05 *M* NaOH adjusted to a final concentration of approx 0.1 *M* borohydride with unlabeled $NaBH_4$ in a volume of 1 mL and stir for 30 min at room temperature.
7. Add 1 mL of 0.1 *M* $NaBH_4$ and incubate for a further 30 min.
8. Dialyze the product against two changes of 3 L of 0.1 *M* sodium acetate, pH 5.5, and then against two changes of 3 L of distilled water.
9. Concentrate the sialoglycoprotein solution if necessary (*see* **Note 6**).

2.2. Buffers and Reagents for Enzyme Assay

1. General 4-nitrophenyl-glycoside assay buffer: 100 m*M* sodium acetate, pH 6.0.
2. General 4-nitrophenyl-glycoside stop solution: 10% trichloroacetic acid (TCA) in distilled water.
3. 0.5 *M* Sodium carbonate.
4. Ovalbumin (Sigma), 80 mg/mL in 0.1 *M* sodium phosphate, pH 7.0, buffer.
5. 5% phosphotungstic acid (PTA)/15% TCA: 5% (w/v) PTA (BDH/Merck) and 15% (w/v) TCA (BDH/Merck) in distilled water.
6. *O*-glycanase incubation buffer: 100 m*M* citrate-phosphate, pH 6.0. Prepare 100 m*M* citric acid and 200 m*M* disodium phosphate. Take 1 vol of citric acid and titrate to pH 6.0 with phosphate, and make up to a final volume equivalent to 1:1 of starting volume.
7. Sialidase incubation buffer (colorimetric assay): 100 m*M* sodium acetate, 20 m*M* $CaCl_2$, 150 m*M* NaCl, pH 5.5.

8. Reagents for the Warren assay of sialic acids:
   a. 0.25 *M* Periodic acid: 5.7 g of periodic acid (Sigma) in 75 mL phosphoric acid (BDH/Merck) make up to 100 mL with distilled water. Stock solutions will keep for approx 12 mo at room temp.
   b. 0.38 *M* sodium arsenite in 0.5 *M* sodium sulfate: 5 g of sodium arsenite (BDH/Merck), 7.1 g of anhydrous sodium sulphate (BDH/Merck) make up to 100 mL with distilled water. Solution is stable for several months at room temperature.
   c. Thiobarbituric acid in 0.5 *M* sodium sulfate: 0.9 g of thiobarbituric acid (Aldrich, Gillingham, UK), 7.1 g of anhydrous sodium sulfate (BDH/Merck) made up to 100 mL with distilled water. Solution is stable for approx 1 wk only (*see* **Note 7**).
   d. Cyclohexanone analytical reagent (AR) (Aldrich, Gillingham, UK).
   e. *N*-Acetyl neuraminic acid (sialic acid; Sigma): Standard solution for calibration of the assay is 0.31 mg/mL, use 10 or 20 μL of this solution in a final volume of 100 μL for the assay (*see* **Subheading 3.7.1.**).
9. Sialidase incubation buffer (fluorimetric assay): 400 m*M* sodium acetate pH 4.2.
10. Sialidase stop buffer (fluorimetric assay): 85 m*M* glycine/sodium carbonate buffer, pH 10.0.
11. Sialate *O*-acetyl esterase buffer: 100 m*M* triethanolamine, pH 7.8.
12. Acetic acid detection assay kit (Boehringer) (*see* **Note 8**).
13. Acylneuraminate pyruvate lyase incubation buffer: 200 m*M* potassium dihydrogen phosphate adjusted to pH 7.2 with KOH.
14. Acylneuraminate pyruvate lyase substrate: 10 m*M* sialic acid (Sigma) in potassium phosphate buffer, pH 7.2, containing 0.5–1.0 kBq/mL of [^{14}C]-*N*-acetylneuraminic acid (Amersham).
15. Acylneuraminate pyruvate lyase from *Clostridium perfringens* (Sigma).

3. Methods

3.1. β-Galactosidase (see Notes 9 and 10)

Several different assays are possible for β-galactosidase; synthetic substrates are widely available for either colorimetric or fluorimetric assay. Radioactive assays with glycoproteins give data on physiologically significant molecules (*see* **Notes 9–11**).

3.1.1. Synthetic Substrate

1. Mix 25 μL of 4 m*M* *p*-nitrophenyl β-galactose in 100 m*M* sodium acetate, pH 6.0 (*see* **Subheading 2.1.2., item 1**), with 25 μL of extract/enzyme.
2. Incubate for 20 min at 37°C.
3. Add 50 μL of 10% trichloracetic acid (*see* **Subheading 2.2., item 2**) to stop the reaction
4. Add 1 mL of 0.5 *M* sodium carbonate.
5. Centrifuge to remove any solid, and read the supernatant at 400 nm.
6. Prepare blank incubations by adding TCA to enzyme extract incubated alone for 20 min, and then add substrate incubated alone for 20 min. Add 1 mL of 0.5 *M* sodium carbonate, centrifuge and read at 400 nm as above (*see* **Note 12**).

3.1.2. General Mucin Galactosidase Assay Radioactive Substrates (see Note 13)

Radioactive asialoglycoproteins labeled in their terminal galactosyl residues (*see* **Subheading 2.1.3.**) can be used as substrates in this precipitation assay. Depending on the nature of the radioactive substrate, this assay can be specific for a particular glycosidic linkage (*see* **Note 14**).

1. Mix 50 µL of 0.85 kBq/mL asialo-glycoprotein substrate in 100 m*M* sodium acetate, pH 6.0, with 50 µL of extract/enzyme.
2. Incubate for 1 h at 37°C.
3. Stop the reaction by the addition of 100 µL of ice-cold ovalbumin (*see* **Subheading 2.2., item 4**).
4. Add 500 µL of 5% PTA/15% TCA (*see* **Subheading 2.2., item 5**) and mix.
5. Stand for 15 min at room temperature.
6. Centrifuge for 5 min at 15,000*g* (benchtop microcentrifuge).
7. Take 500 µL of the clear supernatant and count the radioactivity.

3.1.3. Glycoprotein Assay with β1-3 Linked- and β1-4 Linked-Galactose Radioactive Substrates (see **Notes 9, 10, and 14**)

The detection of β1-3- and β1-4-specific galactosidase activity can be achieved using antifreeze glycoprotein and asialo-α_1-acid glycoprotein, respectively (*see* **Subheading 2.1.3.**). In these assays, identification of the galactose product is made by gel filtration.

1. Mix 50 µL of 0.42–0.83 kBq/mL asialo-glycoprotein substrate in 100 m*M* sodium acetate, pH 6.0, with 50 µL of extract/enzyme (*see* **Note 10**).
2. Incubate for 1–24 h at 37°C.
3. Stop the reaction by adding of 1mL of 0.1 *M* pyridinium acetate, pH 5.0.
4. Apply the total incubation to a column of Bio-Gel P2 (200–400) (Bio-Rad), and elute with 0.1 *M* pyridinium acetate, pH 5.0, collecting 2-mL fractions.
5. Measure the radioactivity in the collected fractions and determine the proportion of tritium label migrating as free galactose (**Fig. 1**).

3.2. α-Galactosidase (see Notes 9–12, and 15)

Assay for α-galactosidase can be carried out rapidly using synthetic substrates. Commercial enzyme is available to use as a positive control. However, several different α-galactose linkages are known to occur, and further examination may be necessary to identify the specificity (*see* **Notes 9, 10,** and **15**).

1. Mix 25 µL of 4 m*M* 4-nitrophenyl α-galactose (*see* **Subheading 2.1.2., item 2**) in 100 m*M* sodium acetate, pH 6.0, with 25 µL of extract/enzyme.
2. Incubate for 20 min at 37°C.
3. Add 50 µL of 10% TCA to stop the reaction.
4. Add 1 mL of 0.5 *M* sodium carbonate.
5. Centrifuge to remove any solid, and read the supernatant at 400 nm.
6. Prepare blank incubations by adding TCA to enzyme extract incubated alone for 20 min, and then add substrate incubated alone for 20 min. Add 1 mL of 0.5 *M* sodium carbonate, centrifugem and read at 400 nm as in **step 5** (*see* **Note 12**).

3.3. β-N-acetylglucosaminidase (see Note 16)

Assay for β-*N*-acetylglucosaminidase can be carried out easily using synthetic substrates. Commercial enzyme is available to use as a positive control (*see* **Notes 9** and **10**).

1. Mix 25 µL of 4 m*M* 4-nitrophenyl β-*N*-acetylglucosamine in 100 m*M* sodium acetate, pH 6.0 (**Subheading 2.1.2., item 3**), with 25 µL of extract/enzyme.

2. Incubate for 20 min at 37°C.
3. Add 50 µL of 10% TCA to stop the reaction.
4. Add 1 mL of 0.5 *M* sodium carbonate.
5. Centrifuge to remove any solid, and read the supernatant at 400 nm.
6. Prepare blank incubations by adding trichloracetic acid to enzyme extract incubated alone for 20 min, and then add substrate incubated alone for 20 min. Add 1 mL of 0.5 *M* sodium carbonate, centrifuge, and read at 400 nm as in **step 5** (*see* **Note 12**).

3.4. α-N-acetylgalactosaminidase (see Note 17)

Assay for α-*N*-acetylgalactosamine can be carried out easily using synthetic substrates. Commercial enzyme is available to use as a positive control (*see* **Notes 9** and **10**).

3.4.1. Synthetic Substrate

1. Mix 25 µL of 4 m*M* 4-nitrophenyl α-*N*-acetylgalactosamine in 100 m*M* sodium acetate, pH 6.0, residues (*see* **Subheading 2.1.2., item 4**) with 25 µL of enzyme extract.
2. Incubate for 20 min at 37°C.
3. Add 50 µL of 10% TCA to stop the reaction.
4. Add 1 mL of 0.5 *M* sodium carbonate.
5. Centrifuge to remove any solid, and read the supernatant at 400 nm.
6. Prepare blank incubations by adding trichloracetic acid to enzyme extract incubated alone for 20 min, and then add substrate incubated alone for 20 min. Add 1mL of 0.5 *M* sodium carbonate, centrifuge, and read at 400 nm as in **step 5** (*see* **Note 12**).

*3.4.2. Mucin α-N-Acetylgalactosaminidase Assay with Radioactive Substrate (see **Notes 13** and **16**)*

Radioactive asialo-ovine salivary gland mucin is labeled in the terminal *N*-acetylgalactosaminyl residues (*see* **Subheading 2.1.3.**) and only contains this single sugar attached to the peptide backbone (*see* **Notes 13** and **14**).

1. Mix 50 µL of 0.85 kBq/mL asialo-ovine salivary gland mucin substrate in 100 m*M* sodium acetate, pH 6.0, with 50 µL of extract/enzyme.
2. Incubate for 1 h at 37°C.
3. Stop the reaction by adding 100 µL of ice-cold ovalbumin (*see* **Subheading 2.2., item 4**).
4. Add 500 µL of 5% PTA/15% TCA (*see* **Subheading 2.2., item 5**) and mix.
5. Stand for 15 min at room temperature
6. Centrifuge for 5 min at 15,000*g* (benchtop microcentrifuge)
7. Take 500 µL of the clear supernatant and count the radioactivity.

3.5. α-Fucosidase (see Note 18)

Assay for α-fucosidase can be carried out easily using synthetic substrates. Commercial enzyme is available to use as a positive control (*see* **Notes 9** and **10**)

1. Mix 25 µL of 4 m*M* 4-nitrophenyl α-fucose in 100 m*M* sodium acetate, pH 6.0, (*see* **Subheading 2.1.2., item 6**).with 25 µL of extract/enzyme.
2. Incubate for 30 min at 37°C.
3. Add 50 µL of 10% TCA to stop the reaction.
4. Add 1 mL of 0.5 *M* sodium carbonate.

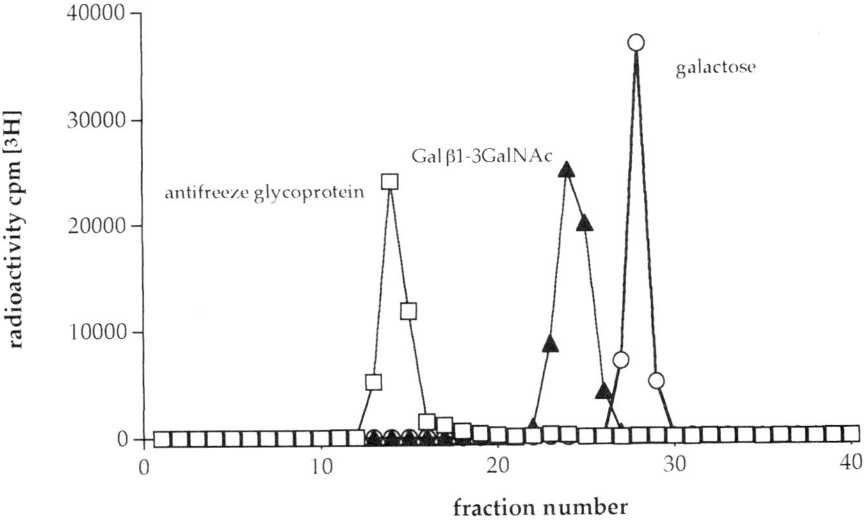

Fig. 1. Identification of the products of *O*-glycanase and β1-3-galactosidase activity on antifreeze glycoprotein substrate by Bio-Gel P2 chromatography. The positions of elution of intact [^{3}H]-labeled antifreeze glycoprotein (□), Galβ1-3GalNAc (▲) and galactose (○) are shown.

5. Centrifuge to remove any solid, and read the supernatant at 400 nm.
6. Prepare blank incubations by adding trichloracetic acid to enzyme extract incubated alone for 20 min, and then add substrate incubated alone for 20 min. Add 1 mL of 0.5 *M* sodium carbonate, centrifuge, and read at 400 nm as in **step 5** (*see* **Note 12**).

3.6. O-Glycanase (see Note 19)

Determination of *O*-glycanase activity can be made with synthetic or glycoprotein substrates.

3.6.1. Synthetic Substrate

1. Mix 25 µL of 2 m*M* 4-nitrophenyl Galβ1-3GalNAc in 100 m*M* citrate-phosphate, pH 6.0 (*see* **Subheading 2.1.2., item 5**), with 25 µL of extract/enzyme.
2. Incubate for 20 min at 37°C.
3. Add 50 µL of 10% TCA to stop the reaction.
4. Add 1 mL of 0.5 *M* sodium carbonate.
5. Centrifuge to remove any solid, and read the supernatant at 400 nm.
6. Prepare blank incubations by adding trichloracetic acid to enzyme extract incubated alone for 20 min, and then add substrate incubated alone for 20 min. Add 1 mL of 0.5 *M* sodium carbonate, centrifuge, and read at 400 nm as in **step 5** (*see* **Note 12**).

3.6.2. Tritiated Glycoprotein Substrate

1. Mix 50 µL of 0.42–0.83 kBq/mL antifreeze glycoprotein substrate (*see* **Subheading 2.1.3.**) in 100 m*M* citrate phosphate buffer, pH 6.0, with 50 µL of extract/enzyme (*see* **Note 9**).

2. Incubate for 1–24 h at 37°C.
3. Stop the reaction by adding 1 mL of 0.1 *M* pyridinium acetate, pH 5.0.
4. Apply the total incubation to a column of Bio-Gel P2 (200–400) (Bio-Rad), and elute with 0.1 *M* pyridinium acetate, pH 5.0, collecting 2-mL fractions.
5. Measure the radioactivity in the collected fractions and determine the proportion of tritium label migrating as free Galβ1-3GalNAc (*see* **Fig 1**).

3.7. Sialidase (see Notes 4, 5, 9, 13, and 20).

Sialidase activity can be measured using several types of assay. The variation in sialic acid substitution in glycoconjugates indicates the varied range of sialidase activities that may be expected and underlines the need for examination of different substrates and assay conditions (*see* **Notes 10** and **20**).

3.7.1. Colorimetric Sialidase Assay

Preliminary determination of the sialic acid content of the substrates is necessary to prepare the substrates for the assays given next.

1. Prepare substrates (bovine salivary gland mucin, saponified; α_1-acid glycoprotein; α2-3 sialyllactose) to give a final concentration of 1 m*M* sialic acid in 100 m*M* sodium acetate, 20 m*M* $CaCl_2$, and 150 m*M* NaCl, pH 5.5, and keep on ice.
2. Mix 50 µL of substrate with 50 µL of enzyme extract and incubate for 60 min at 37°C.
3. Remove from incubation block and mix with 20 µL of sodium periodate solution (*see* **Subheading 2.2., item 8a**).
4. Leave at room temperature for 30 min.
5. Add 200 µL of sodium arsenite solution (*see* **Subheading 2.2., item 8b**), mix until yellow-brown iodate color appears, and then the solution finally becomes colorless (may take several minutes, if unsure wait for 5 min).
6. Add 200 µL of thiobarbituric acid solution (*see* **Subheading 2.2., item 8c**) and incubate for 15 min at 95°C.
7. Cool on ice (10 min) and add 700 µL of cyclohexanone (*see* **Subheading 2.2., item 8d**).
8. Shake to mix the two layers.
9. Centrifuge at 12,000*g* for 2 min in a benchtop microcentrifuge to separate the two layers.
10. Read the organic (pink) layer at both 532 and 549 nm.
11. Calculate the ΔOD value for the released sialic acid, allowing for the interference by compound-forming chromophores at 532 nm using the formula 0.9 OD549–0.3 OD532.
12. Process a standard of 3 µg of Neu5Ac (Sigma) (*see* **Subheading 2.2., item 8e**) in 100 µL through the same assay, and correct with the same formula (0.9 OD549–0.3 OD532) (*see* **Note 21**).
13. Convert the ΔOD value into moles of Neu5Ac to give the activity of the enzyme.

3.7.2. Radioactive Sialidase Assay (see Note 13)

1. Mix 50 µL of radioactive substrate (0.8–1.5 kBq/mL; *see* **Subheading 2.1.5.**, e.g., bovine salivary gland mucin and α_1-acid glycoprotein) in incubation buffer (*see* **Subheading 2.2., item 7**) with 50 µL of extract/enzyme.
2. Incubate for 60 min at 37°C.
3. Stop the reaction by adding 100 µL of ice-cold ovalbumin (*see* **Subheading 2.2., item 4**).
4. Add 500 µL of 5% PTA/15% TCA (*see* **Subheading 2.2., item 5**) and mix.
5. Stand for 15 min at room temperature.

6. Centrifuge for 5 min at 15,000*g* (benchtop microcentrifuge).
7. Take 500 µL of the clear supernatant and count the radioactivity.

3.7.3. Fluorimetric Sialidase Assay (see **Note 11**)

1. Mix 50 µL of 2 m*M* 4-methyl umbelliferyl sialic acid (*see* **Subheading 2.1.2., item 7**) in acetate incubation buffer (*see* **Subheading 2.2., item 9**) with 50 µL of enzyme extract.
2. Prepare blanks using distilled water in place of enzyme extract (*see* **Note 12**).
3. Incubate for 60 min at 37°C.
4. Stop the reaction with 1 mL of stop buffer (*see* **Subheading 2.2., item 10**) and mix.
5. Centrifuge (microcentrifuge) for approx 40 s at 14,000*g*. Keep in the dark before reading in a fluorimeter.
6. Read in a fluorimeter at excitation 365 nm and emission 448 nm.
7. Calculate fluorescence after subtraction of the blanks from the test values.
8. Calibrate the results using standard curves of 4-methylumbelliferone.

3.8. Sialate O-acetyl Esterase (see Note 22)

The sialate *O*-acetyl esterase is assayed using bovine salivary gland mucin (see **Subheading 2.1.1., items 2** and **7**) and the released acetic acid is detected using a quantitative kit assay.

1. Mix 100 µL of bovine salivary mucin substrate in incubation buffer (*see* **Subheading 2.2., item 11**) with 100 µL of enzyme extract (*see* **Note 23**), and incubate for 60 min at 37°C.
2. Stop the reaction by heating at 95°C for 3 min.
3. Cool the incubation mixture on ice or freeze at –20°C until use.
4. Measure acetic acid in the incubation using the NAD/NADH kit assay (*see* **Subheading 2.2., item 12**).
5. Calculate the amount of acetic acid released during the initial incubation with the salivary mucin substrate.

3.9. Acylneuraminate Pyruvate Lyase (see Note 24)

1. Mix 50 µL of sialic acid substrate in phosphate buffer, pH 7.2 (*see* **Subheading 2.2., items 13** and **14**), with 50 µL of enzyme extract.
2. Incubate for 60 min at 37°C.
3. Stop the reaction with 1 mL of ice-cold water.
4. Centrifuge for 2 min in a bench microcentrifuge at 12–15,000*g*.
5. Apply the supernatant to a column of Dowex 1 × 8, 200–400 mesh, Cl^- form (1 mL wet weight), and wash with 5 mL of water directly into a 20-mL scintillation vial.
6. Place in an oven at 80°C until water has evaporated off.
7. Redissolve the residue in 0.5 mL of water and add scintillation fluid (1 mL).
8. Include a positive control with purified acylneuraminate pyruvate lyase from *Clostridium perfringens* (*see* **Subheading 2.2., item 15**).

4. Notes

1. It is advisable to carry out labeling experiments in a recognized radioisotope laboratory. The use of radioactive borotritide ($NaB[^3H]_4$) requires large amounts of [^{3}H] and results in the emission of tritium gas. It is important to check the registration limits for [^{3}H] in the laboratory to be used. Since a hood is required it is necessary that the tritium gas be properly trapped and that the necessary decontamination procedures be applicable. If regu-

lar (multiple) tritium labeling is envisaged, it is especially important to check the registration limits and decontamination procedures with the radioactive officer responsible for the laboratory.

2. Galactose oxidase will act on terminal galactose and *N*-acetyl-D-galactosamine residues to give oxidation at the primary alcohol group at carbon 6. Glycoproteins with both of these sugars in terminal nonreducing locations, e.g., asialoseminal gel mucin, will be labeled in both residues.
3. The use of porcine seminal gel has the advantage that the sialic acid is found as Neu5Ac (no Neu5Gc is present). In addition to Neu5Acα2-6GalNAc, the tetrasaccharide Neu5Acα2-3Galβ1-3[Neu5Acα2-6]GalNAc is also found. The β-eliminated, reduced oligosaccharides are well separated on Bio-Gel P4 and Dowex 1 × 2 acetate.
4. Lower concentrations of periodate are used to limit the oxidation to the C7–C9 end of the sialic acid molecule. Higher concentrations of periodate result in the oxidation of other sugars in glycoconjugate oligosaccharides *(7)*.
5. Mucins in the human colorectum typically have *O*-acetyl esters located on C7-C9 hydroxyl groups of sialic acids. In order to oxidize these hydroxyl groups with periodic acid, the esters must be removed. This can be done rapidly and under mild conditions by saponification. The conditions described (*see* **Subheading 2.1.1., item 7**) do not cause degradation of the glycoproteins used in the protocols given here. In the case of other glycoproteins, it is best to check the preparation after saponification for the loss of carbohydrate and peptide cleavage.
6. When the final volume containing the glycoconjugate is large and needs to be reduced, several methods are available for concentration. These include rotary evaporation under reduced pressure, freeze drying, and concentration cells. Because drying of samples may affect solubility, especially with mucins, the recovery and integrity of the products should be routinely checked.
7. The preparation of the thiobarbituric acid stock solution requires that the fresh solution be heated under hot tap water with shaking to completely dissolve all components. On storage a precipitate appears that may need to be filtered before use. The stock solution should be stored in a stoppered bottle at room temperature. Fresh stock solution should be prepared on a weekly basis.
8. The amount of acetic acid in the esterase incubation is determined using a commercial UV test kit (Boehringer). This is an NAD-linked assay system converting acetic acid to acetyl-CoA (acetyl CoA synthetase), acetyl-CoA to citrate (citrate synthase), and finally using the conversion of NAD to NADH with malate dehydrogenase. The change in absorbance is measured at 340 nm. The kit contains all enzymes and buffers and gives details of the calculation necessary to determine the amount of acetic acid in each sample.
9. It is recommended that each assay be validated using commercially available enzyme sources with known substrates and to run such incubations routinely as positive controls. This is especially important when separation techniques such as gel filtration and HPLC are utilized to identify the released monosaccharide product. Commercial enzymes and substrates can be obtained from a variety of suppliers including, Oxford Glycosciences; Sigma; Dextra; ICN Biomedicals, High Wycombe, UK; Boehringer Mannheim; Calbiochem-Novabiochem (UK), Nottingham, UK; and Seikagaku Kogyku, Tokyo, Japan. Remember that when an unknown substrate or enzyme is being tested, differences in substrate specificities exist for glycosidases. A negative result indicates that other possible substrates should be tested before the absence of enzymatic activity can be concluded. In the case of novel substrates that are resistant to commercial enzymes, a detailed structural analysis is necessary to identify the oligosaccharide type and monosaccharide linkage(s).

10. Individual galactosidases show a range of pH optima using 4-nitrophenyl and glycoprotein substrates. The examination of unknown samples for this activity should include a suitable range of pHs to cover the known ranges described. The commercial sources as mentioned in **Note 9** should be included to provide positive controls for these activities. Examples are *Escherichia coli* (200 m*M* Tris-HCl, pH 7.3); *Aspergillus niger* (200 m*M* sodium acetate, pH 4.0); coffee bean α-galactosidase (200 m*M* Tris-HCl, pH 6.5); bovine testes (100 m*M* citrate phosphate, pH 4.0).
11. In addition to the colorimetric assays given in this chapter, additional sensitivity can be attained by the use of fluorimetric assays with 4-methyl umbelliferyl-glycosides. Further information may be derived from the use of individual reducing oligosaccharide substrates that can be tagged radioactively or with fluorescent labels. Products can be separated from substrates by HPLC techniques. Remember that the synthetic substrate assays may not give physiologically relevant information and that oligosaccharides also represent partial physiologic substrates. Thus, the use of intact natural substrates should be considered. However, when impure enzyme mixtures are being analyzed, the use of natural substrates must include direct product (monosaccharide) identification. Depending on the type of assay used additional enzymatic activities may lead to fragmentation of the substrate, giving false-positive results (*see also* **Note 13** and Chapter 31 on total mucinase activity and glycoprotein fragmentation during degradation).
12. Blank incubations can be prepared using enzyme extract incubated under normal conditions but without substrate, stopping the reaction with TCA and combining the stopped enzyme with substrate also incubated alone. Alternatively, heat-treated (5 min at 95°C) enzyme extract can be used in a normal incubation. In many cases, the heat treatment of crude extracts results in interference with the colorimetric or fluorescent reaction and should be avoided.
13. Assays using precipitation of the substrate must be used with enzyme sources which generate no alternative soluble product. Crude extracts are likely to contain proteinases and peptidases that may generate soluble glycopeptide products, and these will give a "false-positive" result in the precipitation assay. If radioactive substrates are to be used with crude enzyme preparations separation of the monosaccharide product e.g. galactose, must be part of the assay procedure. This type of assay is described under **Subheading 3.1.3.**
14. The preparation of galactose oxidase/borotritide-labeled substrates, as detailed in **Subheading 2.1.3.** can be carried out with glycoproteins that have an exclusive glycosidic linkage type. Antifreeze glycoprotein contains only β1-3 linked-galactose and asialo-α_1-acid glycoprotein only β1-4 linked-galactose. The use of these substrates automatically gives a linkage-specific enzymatic assay. Glycoproteins that have terminal GalNAc residues are also labeled (*see* **Note 2**); hence, asialo-ovine salivary gland mucin contains only GalNAc linked to the peptide backbone and is a good substrate for the enzyme removing GalNAc from the peptide. Glycoproteins with terminal GalNAc as blood group A, (GalNAc α1-3Gal-) or Sda or Cad antigen (GalNAc β1-4Gal-) are also potential substrates. However, such glycoproteins may also have terminal galactose residues and need to be carefully analyzed before use in specific assays.
15. Blood group B activity is associated with the expression of terminal α1-3 galactose and is commonly associated with mucins. If mucin degradation is being studied, the blood group of the source material should be determined. In addition, the secretor status of individuals is relevant since only secretor-positive individuals will express blood group activity in mucins. There are also examples of galactose in other α-linkages, and examination of additional oligosaccharide substrates (*see* **Note 9**) by HPLC is an alternative way to identify the presence of α1-4 or α1-6 galactosidase activities.

16. The detection of *N*-acetyl-D-glucosaminidase activity is generally carried out with the 4-nitrophenyl- or 4-methyl umbelliferyl-synthetic substrates. Preparation of glycoprotein substrates labeled specifically in GlcNAc residues is difficult to achieve. The availability of GlcNAc-terminated oligosaccharides provides a source of substrates for HPLC based assays after radioactive or fluorescent tagging.
17. Glycoproteins with terminal GalNAc as blood group A (GalNAc α1-3Gal-) or as Tn antigen (GalNAc α-*O*-ser/thr) are substrates for different enzymes. The use of synthetic substrates must be compared with substrates containing GalNAc in either of these two arrangements. Oligosaccharide substrates are only possible with the blood group A structure. The presence of β-linked GalNAc has also been found in the Sda or Cad antigens (GalNAc β1-4Gal-). These are substrates for different enzymes and have not been examined in detail.
18. Fucose is present in glycoproteins in a number of different linkages, including α1-2, α1-3, α1-4, and α1-6, and as with other glycosidases, there are a number of different fucosidases that have varying specificity for these linkages in different oligosaccharides. Many of these oligosaccharides are available commercially (*see* **Note 8**) and can be used to examine the detailed specificity of fucosidase activity detected with synthetic substrates.
19. *O*-glycanase (endo-α-*N*-acetylgalactosaminidase) is an enzyme that removes the disaccharide Galβ1-3GalNAc from peptide-carbohydrate substrates by cleavage of the *O*-glycosidic linkage. Several of these enzymes have been identified and purified from bacteria. Commercial sources of *O*-glycanase enzyme are available and can be used to validate the assay. Almost all *O*-glycanases described to date will not act on Galβ1-3GalNAc substituted by any other sugars. It is possible to "generate" substrate by prior treatment of a glycoprotein with sialidase and fucosidase to strip the *O*-linked oligosaccharides to the basic Galβ1-3GalNAc disaccharide. In the enzyme assay it is important to identify the product as a disaccharide and in the radioactive assay described this is achieved by gel filtration. A clear discrimination between Galβ1-3GalNAc and Gal products should be made. In the case of the synthetic substrate Galβ1-3GalNAc-4-nitrophenyl it is possible that the joint action of β1-3-galactosidase and *O*-acetylgalactosaminidase could generate a false-positive result. Where such doubt may arise an alternative assay with direct disaccharide product identification is needed.
20. Sialidases represent a group of enzymes which act on the wide range of sialoglycoconjugates with a variety of linkages of sialic acids to different sugars *(8)*. Although the sialidases from microorganisms often show a broad substrate specificity there are also examples of limited substrate activity and new enzyme sources need to be evaluated using several substrates to determine its properties (*see* **ref.** *8* for examples). In the case of sialidases acting on mucin, sialic acids on *O*-linked oligosaccharides are the main targets. However, each mucin contains a characteristic mixture of sialo-oligosaccharides. Examination of individual oligosaccharide substrates analogous to those present in mucins, e.g., α2-3- or α2-6-sialyl lactose isomers may give further information concerning sialidase specificity. However, it will not predict the activity of enzymes on the intact mucin molecules where oligosaccharide structure and their arrangement on the polypeptide will play a role. It is thus important to use intact mucin substrates to examine sialidase activity.
21. The use of the correction formula for the colorimetric assay is valid for values of OD532 up to 0.80 × OD549. Above this value the correction becomes inaccurate.
22. Acylneuraminate *O*-acetyl esterase acts on ester groups in sialic acids. As these *O*-acetyl ester groups reduce the activity of sialidase activity the esterase is proposed to have a regulatory role in the rate of colonic mucin degradation where these ester forms of sialic acids are abundant *(9)*. Mucins in the human colon contain sialic acids with *O*-acetyl esters on the hydroxyl groups at carbons C7, C8, and C9 and these may be present as

mono-, di-, or tri-*O*-acetyl esters for any sialic acid residue. The esterase removing these esters can be assayed using bovine salivary gland mucin as a substrate *(5)*. This mucin substrate contains on average two moles of *O*-acetyl ester per mole of sialic acid. The sialic acid esters in this substrate are released by the enzyme and the acetic acid is detected using a commercial detection kit. Blanks are prepared using saponified mucin.

23. Care must be taken to calibrate the amount of acetic acid present in the enzyme extracts used to assay the acylneuraminate *O*-acetyl esterase. In some cases steps may be necessary to remove high or unpredictable levels of acetic acid in enzyme preparations before assay. Studies with human faecal extracts showed that prior precipitation with ammonium sulphate and subsequent gel filtration were needed to obtain an enzyme preparation that could be assayed reliably ***(10)***.
24. Acylneuraminate pyruvate lyase is an enzyme responsible for the degradation of sialic acids to *N*-acylmannosamine. This is part of the bacterial metabolic route to convert the sialic acids into glucose for utilization in energy metabolism. The enzyme is frequently found in bacteria which produce sialidase activity and sialic acid transporter proteins in order to recover the sialic acids released from extracellular substrates such as mucins *(8)*.

Acknowledgments

This work was supported by grant 051586/Z/97 from the Wellcome Trust.

References

1. Roberton, A. M. and Corfield, A. P. (1998) Mucin degradation and its significance in inflammatory conditions of the gastrointestinal tract, in *Medical Importance of the Normal Microflora.* (Tannock, G. W., ed.), Kluwer Academic Publishers, Dordrecht, pp. 222–261.
2. Tettamanti, G. and Pigman, W. (1968) Purification and characterization of bovine and ovine submaxillary mucins. *Arch. Biochem. Biophys.* **124,** 41-50.
3. Hill, H. D., Reynolds, J. A., and Hill, R. A. (1977) Purification, composition, molecular weight and subunit structure of ovine submaxillary mucin. *J. Biol. Chem.* **252,** 3791–3798.
4. Corfield, A. P., Higa, H., Paulson, J. C., and Schauer, R. (1983) The specificity of viral and bacterial sialidases for α(2-3) and α(2-6)-linked sialic acids in glycoproteins. *Biochim. Biophys Acta* **744,** 121–126.
5. Corfield, A. P., Williams, A. J. K., Clamp, J. R., Wagner, S. A., and Mountford, R. A. (1988) Degradation by bacterial enzymes of colonic mucus from normal subjects and patients with inflammatory bowel disease:the role of sialic acid metabolism and the detection of a novel O-acetylsialic acid esterase. *Clin. Sci.* **74,** 71–78.
6. Corfield, A. P., do Amaral Corfield, C., Veh, R. W., Wagner, S. A., Clamp, J. R., and Schauer, R. (1991) Characterization of the major and minor mucus glycoproteins from bovine submandibular gland. *Glycoconj. J.* **8,** 330–339.
7. Veh, R. W., Corfield, A. P., Sander, M., and Schauer, R. (1977) Neuraminic acid specific modification and tritium labelling of gangliosides. *Biochim. Biophys. Acta* **486,** 145–160.
8. Corfield, T. (1992) Bacterial sialidases - roles in pathogenicity and nutrition. *Glycobiology* **2,** 509–521.
9. Corfield, A. P., Wagner, S. A., Clamp, J. R., Kriaris, M. S., and Hoskins, L. C. (1992) Mucin degradation in the human colon: Production of sialidase, sialate O-acetyl esterase, N-acetylneuraminate lyase, arylesterase and glycosulfatase activities by strains of fecal bacteria. *Infect. Immun.* **60,** 3971–3978.
10. Corfield, A. P., Wagner, S. A., O'Donnell, L. J. D., Durdey, P., Mountford, R. A., and Clamp, J. R. (1993) The roles of enteric bacterial sialidase, sialate O-acetyl esterase and glycosulfatase in the degradation of human colonic mucin. *Glycoconj. J.* **10,** 72–81.

34

Assays for Bacterial Mucin-Desulfating Sulfatases

Anthony M. Roberton, Douglas I. Rosendale, and Damian P. Wright

1. Introduction

The regions of the gastrointestinal tract that are densely colonized by bacteria secrete mucus that stains as sulfomucus. A body of evidence suggests that the sulfation of mucins is protective against degradation by bacteria, and desulfation is one of the important rate-limiting steps in mucin degradation *(1)*. Bacterial sulfatases that carry out mucin desulfation have been described, but undoubtedly a group of such enzymes will be discovered with distinctive specificities for the differently sulfated sugars found in mucins, and a combination of such enzymes will be required to completely desulfate mucins.

This chapter describes assays for mucin-desulfating enzymes, based on model substrates or [^{35}S]mucins. The disadvantages of using model substrates are that the sulfatase may not recognize them, or that the sulfatase being measured may have a physiologic sulfated substrate other than mucin. It is therefore necessary to confirm the specificity for mucin at some point. The disadvantage of using [^{35}S]mucin is that a number of different sulfated sugar structures occur in mucins, and one must be certain that the mucin chosen as substrate contains the sulfated sugar of interest. Also, the free sulfate released during the assay may be the sum of several enzymic activities with different specificities.

1.1. Characterization of Sulfomucins

The mucus secreted in the mouth *(2)* and the colon *(3)* of humans contains a large portion of sulfomucins. Historically this characterization is based on the staining properties. Individual mucin oligosaccharide chains may contain no acidic groups, one or more sialic acid groups, one or more sulfate groups, or mixtures of sialic acid and sulfate groups *(4,5)*. The histochemical description of mucin as neutral, sialo-, and sulfomucin depends on whether the mixture of oligosaccharides in the molecule, and indeed the mixture of mucin molecules present, has low levels of acidic oligosaccharides, a preponderance of sialic acid-staining chains, or a preponderance of sulfate-staining chains.

From: *Methods in Molecular Biology, Vol. 125: Glycoprotein Methods and Protocols: The Mucins*
Edited by: A. Corfield © Humana Press Inc., Totowa, NJ

However, the terms *sialomucin* or *sulfomucin* do not imply an absence of sulfate or sialic acid, respectively, from the molecule or mixture of molecules. Indeed, human small intestinal mucins, which are characterized mainly as neutral and sialomucins *(6)*, contain significant quantities of sulfate *(7,8)*. A definition of the minimum quantity of sulfate that an isolated mucin must contain to qualify as a sulfomucin is not possible at this time. The intensity of staining appears to be qualitatively related to the sulfate content but may not be quantitatively proportional *(9)*.

1.2. Sulfated Sugars Found in Mucins

Sulfation of mucin oligosaccharides occurs by a variety of position linkages to *N*-acetylglucosamine and galactose. These structures need emphasizing in studies on mucin desulfation by bacteria, because distinct sulfatases will probably be required to hydrolyze the different sulfate esters (*see* **Note 1**). The predominant sulfated sugars in human mucins include *N*-acetylglucosamine-6-sulfate, galactose-6-sulfate and galactose-3-sulfate. The best documented structural studies on human sulfomucins have been on tracheobronchial mucins, where all three of these sulfated sugars have been found *(4,5,10–12)*. Specific sulfotransferases have been described that sulfate mucin oligosaccharide structures containing galactose or *N*-acetylglucosamine, giving rise to galactose-3-sulfate *(13)* and *N*-acetylglucosamine-6-sulfate *(14)*. The galactose-6-sulfate structure has been found in oligosaccharides from mucins secreted by the human colon cancer cell line CL.16E *(15)*, and the galactose-3-sulfate structure has been found in oligosaccharides from another human colon cancer cell line, LS174T-HM7 *(16)* and from human meconium mucin *(17)*. The presence of *N*-acetylglucosamine-6-sulfate in pig stomach mucin has been documented *(18)*, and *N*-acetylglucosamine-6-sulfate and a minor amount of galactose monosulfate have been found in rat stomach mucin *(19)*. A specific sulfating system has been described for galactose-3-sulfate in rat colon *(20)*. Structural studies on rat small and large intestinal mucin oligosaccharides *(21)* showed that sulfation is mainly present on C-6 of *N*-acetylglucosamine. The lack of structural studies on human sulfomucins in the normal colon leaves a gap in our knowledge of the type(s) of sulfated mucin sugars to which colonic bacteria will be exposed.

1.3. Bacteria Known to Desulfate Mucins

To date, only a few bacteria have been shown to remove sulfate from sulfomucin. The list includes colonic anaerobes *Bacteroides* strain RS13, *Clostridium* strain RS42, *Prevotella* strain RS2 *(22,23)*, *B. thetaiotaomicron*, *B. fragilis* *(24,25)*, *Ruminococcus torques* strain IX-70, *Bifidobacterium* strain VIII-210 *(26)*, the stomach microaerophilic pathogen *Helicobacter pylori* *(27)*, and a selection of oral *Streptococcus* species *(28)*.

Identification of further bacteria that possess mucin-desulfating enzymes will be aided by refining the techniques and strategies for detecting the enzymes. If mucin desulfation is used as the criterion for determining the presence of mucin-desulfating enzyme(s), then selection of a source of mucin that has the sulfated sugar(s) corresponding to the specificity of the bacterial sulfatase(s) is essential. It is important to look for sulfatase activity not only in the culture medium supernatant but also on the surface and in the cell extracts of bacteria *(29)* (*see* **Note 2**). If the sulfatase is an inducible enzyme, then sulfomucin should be present in the growth medium and the presence of other sulfated contaminants such as glycosaminoglycans should be minimized. The use of model substrates for assays during sulfatase purification must be consistent with assays on sulfomucin desulfation (*see* **Note 3**), because many other sulfatases besides mucin-specific sulfatases are known to be present in enteric bacteria, and many sulfated molecules other than mucin are also present in the digestive tract.

1.4. Relevance of Mucin-Desulfating Enzymes

Evidence that desulfation of sulfomucins is one of the rate-limiting steps in mucin degradation by bacteria has been summarized recently ***(1,9)***. The evidence suggests that a high level of sulfate in mucin decreases the removal of mucin sugars by glycosidases, and conversely, that mucin-specific sulfatases increase the susceptibility of a mucin to glycosidases. The subtlety and specificity that might be associated with the variety of sugar sulfation in different mucins remains to be explored.

2. Materials

2.1. Glucose-6-Sulfate Desulfation Reagents

1. Buffer/β-mercaptoethanol: imidazole-HCl buffer (20 m*M*, pH 7.4) to which β-mercaptoethanol (0.7 μL/mL) (10 m*M* final concentration) has been added.
2. Substrate: potassium glucose-6-sulfate (Sigma, St. Louis, MO, cat. no. G3899) (33 m*M*, 9.8 mg/mL) dissolved in buffer/β-mercaptoethanol.
3. Sulfatase from a bacterial source in buffer/β-mercaptoethanol to which 0.002% (w/v) diisopropyl fluorophosphate has been added to inhibit serine proteases if the enzyme is in the crude state. Dilute as necessary in buffer/β-mercaptoethanol.
4. Phenol reagent: Phenol (3% [w/v]) in water. Store in a dark bottle at 4°C.
5. Dye/buffer reagent: Dissolve 10 g of Na_2HPO_4 (anhydrous), 1 g of sodium azide, 0.3 g of 4-aminophenazone (4-aminoantipyrine) (Sigma, cat. no. A4382) in water to 1 L final volume. Store in a dark bottle at 4°C. Adjust the pH to 7.4 before adding the 4-aminophenazone.
6. Iodoacetamide reagent: iodoacetamide (0.5 *M*, 92 mg/mL of water). Make up fresh daily.
7. Glucose oxidase/peroxidase reagent: For 20 assays, a standard curve, and controls (2.5 mL per reaction), mix 1 mL of phenol (3% w/v) reagent, 100 mL of dye/buffer reagent, 2150 U of glucose oxidase (Sigma, cat. no. G9010), and 150 U of peroxidase (Sigma, cat. no. P8250). Add the glucose oxidase and peroxidase just before use.
8. Glucose standard: Dissolve 18 mg of glucose in 10 mL buffer/β-mercaptoethanol (10 m*M* concentrated glucose stock), and store at –20°C in 0.5-mL aliquots. Each day dilute an aliquot of the concentrated stock to 0.5 m*M* glucose standard by adding buffer/β-mercaptoethanol.

2.2. [^{35}S]Mucin Desulfation Reagents

2.2.1. [^{35}S]Mucin Preparation

The choice of mucin-secreting tissue will determine the mixture of mucin sugars which become labeled with [^{35}S]sulfate. The following protocol describes a method used to label mucin from the rat stomach corpus region ***(23,30)***. Methods for preparing [^{35}S]human colonic mucin have been described ***(25,31,32,33)***.

1. Modified minimal essential medium (S-MEM). Add the following materials to S-MEM (Joklil modified, free from sulfate): 200 mg/L anhydrous $CaCl_2$, 0.1 mg/L $FeCl_2$, 100 mg/L sodium pyruvate, 350 mg/L sodium bicarbonate, 8.9 mg/L L-alanine, 15 mg/L L-asparagine, 13.3 mg/L L-aspartate, 15 mg/L L-glutamate, 50 mg/L L-glycine, 50 mg/L L-proline, 50 mg/L L-serine, 36 mg/L L-tyrosine, 140 mg/L L-methionine, 100 mg/L L-fucose, 200 mg/L D-glucosamine, 200 mg/L D-galactose. Also add carbachol (1 m*M*), prostaglandin E_1 (1 μ*M*) and A23187 (10 μ*M*) to stimulate mucin production. Filter-sterilize the medium before use.
2. Excise the corpus region of the stomachs of three male Wistar rats (250 g). Wash in Dulbecco's buffer ***(34)*** and place in modified S-MEM (*see* **item 1**).
3. Cut the corpus tissue into cubes of <2 mm. Wash several times in modified S-MEM, and finally suspend in 10 mL in a glass Petri dish. Add 25 MBq of [^{35}S]sulfate. Incubate for 5 h at 37°C under 5% CO_2 in air, with gentle rocking.
4. At the end of the incubation, cool to 0°C, add EDTA (5 m*M*), phenylmethanesulfonyl fluoride (1 m*M*), and one grain of sodium azide. Homogenise the tissue and medium in a glass homogenizer. Centrifuge at 10,000*g* for 20 min. Chromatograph the mucin on Sepharose CL-4B column (100 mL) and collect the void volume peak. Further purify the mucin by centrifugation on a CsCl gradient (initial density 1.45 g/mL, 100,000*g* for 48 h), and collect the mucin peak. Desalt the mucin on a Sephadex G-25 column and freeze-dry. Suspend in 1.5 mL of buffer, and store aliquots at –70°C. The incorporation of [^{35}S] into mucin is about 8,000 dpm/0.1 mL.

2.2.2. Other Reagents

1. Sulfatase from a bacterial source in buffer containing diisopropyl fluorophosphate (0.002% w/v) to inhibit serine proteases, as described in **Subheading 2.1.**
2. Strips of Whatman No. 3 filter paper.
3. Electrophoresis buffer: ammonium acetate (0.5 *M*, pH 5.6).
4. Equipment for paper electrophoresis and scintillation counting.

2.3. [1-^{3}H]Lactitol-6'-Sulfate Desulfation Reagents

1. Lactose-6-sulfate can readily be prepared and purified from lactose and pyridine-sulfur trioxide complex [35]. β-Galactosyl-6-sulfate-(1–4)-glucit-[^{3}H]ol or [1-^{3}H]lactitol-6'-sulfate is then made by reduction with excess [^{3}H]borohydride. After destruction of excess borohydride, the product can be purified by ion-exchange chromatography *(36)*. A suitable radioactive product contains 0.5–1.0 GBq/mmol.
2. Sulfatase prepared from a bacterial source ***(26)*** or from human faecal extracts ***(36)***.
3. Buffer: sodium acetate (400 m*M*, pH 5.0)/$CaCl_2$ (40 m*M*).
4. Dowex AG1 × 8 formate anion-exchange resin in 1-mL columns with taps.

2.4. p-Nitrophenyl N-Acetyl-β-D-Glucosaminide-6-Sulfate Desulfation Reagents

1. Substrate: p-nitrophenyl *N*-acetyl-β-D-glucosaminide-6-sulfate (Industrial Research, Lower Hutt, NZ) (3.2 m*M*) dissolved in Tris chloride buffer (50 m*M*, pH 7.4).
2. *N*-acetyl-β-D-glucosaminidase. An *Aspergillis oryzae* extract (sold as a β-D-galactosidase by Sigma, cat. no. G7138) contains both β-D-galactosidase activity (76 nmol/min/mg) and *N*-acetyl-β-D-glucosaminidase activity (6 nmol/min/mg), as measured using the corresponding *p*-nitrophenyl glycosides as substrates. The powder is dissolved in Tris chloride buffer (1 mg/mL).

3. Sulfatase from a bacterial source in buffer/β-mercaptoethanol, containing diisopropyl fluorophosphate (0.002% w/v) to inhibit serine proteases, as described in **Subheading 2.1.**
4. Stopping reagent: glycine buffer (0.5 *M*, pH 9.6).
5. Product standard: *p*-nitrophenol (1.25 m*M*).

2.5. p-Nitrophenyl-β-D-Galactoside-6-Sulfate Desulfation Reagents

1. Substrate: *p*-nitrophenyl β-D-galactoside-6-sulfate (Industrial Research) (3.2 m*M*) dissolved in Tris chloride buffer (50 m*M*, pH 7.4).
2. β-Galactosidase (Sigma, cat. no. G5635) (0.1 mg protein/mL containing 720 U/mg protein) in Tris chloride buffer (50 m*M*, pH 7.4).
3. *See* **Subheading 2.4., item 3.**
4. *See* **Subheading 2.4., item 4.**
5. *See* **Subheading 2.4., item 5.**

2.6. p-Nitrophenyl β-D-Galactoside-3-Sulfate Desulfation Reagents

1. Substrate: *p*-nitrophenyl β-D-galactoside-3-sulfate (Industrial Research) (3.2 m*M*) dissolved in Tris chloride buffer (50 m*M*, pH 7.4).
2. *See* **Subheading 2.5., item 2.**
3. *See* **Subheading 2.5., item 3.**
4. *See* **Subheading 2.5., item 4.**
5. *See* **Subheading 2.5., item 5.**

3. Methods

3.1. Glucose-6-Sulfate Desulfation Assay

This assay uses glucose-6-sulfate as a model substrate for the *N*-acetylglucosamine-6-sulfate present in oligosaccharide chains of mucins. The bacterial sulfatase releases glucose from glucose-6-sulfate (incubation 1). β-Mercaptoethanol is necessary to stabilize some sulfatases and is removed after incubation 1 by adding iodoacetamide. Otherwise, β-mercaptoethanol will remove the H_2O_2 formed by glucose oxidase during incubation 2. The H_2O_2 reacts with peroxidase to form a red product in the presence of suitable substrates:

$$\text{Glucose-6-sulfate} \xrightarrow{\text{Sulfatase}} \text{glucose} + SO_4^{2-} \text{ (incubation 1)}$$

$$\text{Glucose} + O_2 \xrightarrow{\text{Glucose oxidase}} \text{D-gluconic acid} + H_2O_2 \text{ (incubation 2)}$$

$$H_2O_2 + \text{phenol} + \text{4-aminophenazone} \xrightarrow{\text{Peroxidase}} \text{red chromagen} + H_2O \text{ (incubation 2)}$$

An alternative strategy is to measure glucose formation using hexokinase/glucose-6-phosphate dehydrogenase as the auxiliary enzyme system ***(24)***.

1. Add 0.1 mL of prewarmed glucose-6-sulfate (33 m*M*) in imidazole buffer/β-mercaptoethanol to a disposable tube. Prewarm at 37°C for 5 min. Then add 0.1 mL sulfatase, suitably diluted. Also prepare control tubes in which buffer/β-mercaptoethanol replaces

sulfatase, and buffer/β-mercaptoethanol replaces substrate. Incubate for 30 min.
2. Place tubes on ice for 5 min. Then add 0.3 mL of 0.5 *M* iodoacetamide.
3. Using the 0.5 m*M* glucose standard make up tubes containing 0–100 nmol glucose in 0.2 mL buffer/β-mercaptoethanol. Place at 0°C, and add 0.3 mL of 0.5 *M* iodoacetamide.
3. Carry out a second incubation as follows. To experimental, control and standard tubes add 2.5 mL of glucose oxidase/peroxidase reagent containing the phenol and 4-aminophenazone at minute intervals. Incubate 15 min at 30°C. Read absorbances at 515 nm at minute intervals.
4. Construct a standard curve. Subtract the no enzyme controls from each experimental reading. Determine the glucose produced per 30 min/0.1 mL of sulfatase. Values greater than absorbance 0.2 are beyond the linear range and should be repeated with more dilute sulfatase. If the enzyme contains significant brown colour, carry out additional zero time controls in which iodoacetamide is added at zero time, and the tubes are held at 0°C instead of incubating at 37°C *(23)*.

3.2. [^{35}S]Mucin Desulfation Assay

1. Incubate 0.1 mL of [^{35}S]mucin with 0.3 mL sulfatase in imidazole-HCl buffer (20 m*M*, pH 7.4) containing β-mercaptoethanol (10 m*M*) at 37°C. Remove 0.66-mL aliquots at intervals into an equal volume of ethanol, and keep at –20°C.
2. Spot the sample on Whatman No. 3 filter paper at the origin, rinsing the tube with small amounts of ethanol, which are also spotted.
3. Wet the Whatman paper with ammonium acetate buffer (0.5 *M*, pH 5.6) and electrophorese for 1 h at 12.5 V per cm. Free sulfate separates well from sulfated sugars/oligosaccharides. Mucin stays almost at the origin. Dry the paper, cut into 1-cm strips for scintillation counting. Determine the counts that represent the inorganic sulfate released from the mucin.
4. Calculate the rate of [^{35}S]sulfate formation from the linear part of the reaction *(23)*.

3.3. [1-^{3}H]Lactitol-6'-Sulfate Desulfation Assay

1. Mix radioactive substrate, (8 kBq in 25 µL of [1-^{3}H]lactitol-6'-sulfate) with 25 µL of sodium acetate buffer (400 m*M*, pH 5.0) containing $CaCl_2$ (40 m*M*) and 50 µL of sulfatase. Carry out the incubation at 37°C for 60 min.
2. Terminate the incubation by adding 1 mL ethanol, and centrifuge (12,000*g*, 2 min) to remove protein. Pass the supernatant through a 1 mL Dowex AG1 × formate ion form column, and wash the column with 2 × 5 mL water. The desulfated sugar passes through, while sulfated sugar is retained.
3. Dry the unabsorbed fraction at 80°C. Dissolve the residue in 200 µL of water, mix with scintillation fluid and count. Calculate the specific activity from the known radioactivity of the substrate. The specificity of this assay is for galactose-6-sulfate groups within mucin *(36)*.

3.4. p-Nitrophenyl N-Acetyl-β-D-Glucosaminide-6-Sulfate Desulfation Assay

1. Mix p-nitrophenyl *N*-acetyl-β-D-glucosaminide-6-sulfate (0.05 mL of 3.2 m*M* stock) with 0.01 mL of *Aspergillus oryzae N*-acetylglucosaminidase solution, and prewarm for 3 min at 37°C. Add prewarmed bacterial sulfatase (0.1 mL). The substrate concentration is 1 m*M*. Continue the incubation at 37°C for 20 min. Use suitable enzyme dilutions to check that the time course is linear. The *N*-acetylglucosaminidase will act on *p*-nitrophenyl

N-acetyl-β-D-glucosaminide but not *p*-nitrophenyl *N*-acetyl-β-D-glucosaminide-6-sulfate. Carry out controls containing no sulfatase and no substrate.
2. Add glycine buffer (1.84 mL, 0.5 *M*, pH 9.6) to stop the reaction and ionise the p-nitrophenol formed. Measure the yellow p-nitrophenol absorbance at 410 nm.
3. Construct a standard curve by mixing standard p-nitrophenol (0–200 nmol) with glycine stopping buffer, and measure absorbances. Calculate the activities of enzyme using this standard curve.

3.5. p-Nitrophenyl β-D-Galactoside-6-Sulfate Desulfation Assay

1. Mix p-nitrophenyl β-D-galactoside-6-sulfate (0.05 mL of 3.2 m*M* stock) with 0.01 mL of β-galactosidase solution, and preheat for 3 min at 37°C. Add bacterial sulfatase (0.1 mL) and continue incubating at 37°C for 20 min. The β-galactosidase will deglycosylate *p*-nitrophenyl β-D-galactoside but not *p*-nitrophenyl β-D-galactoside-6-sulfate. Carry out controls containing no sulfatase and no substrate.
2. *See* **Subheading 3.4., step 2.**
3. *See* **Subheading 3.4., step 3.**

3.6. p-Nitrophenyl β-D-Galactoside-3-Sulfate Desulfation Assay

1. Mix *p*-nitrophenyl β-D-galactoside-3-sulfate (0.05 mL of 3.2 m*M* stock) with 0.01 mL of β-galactosidase solution, and preheat for 3 min at 37°C. Add bacterial sulfatase (0.1 mL), and continue incubating at 37°C for 20 min. The β-galactosidase will act on *p*-nitrophenyl β-D-galactoside but not *p*-nitrophenyl β-D-galactoside-3-sulfate. Carry out controls containing no sulfatase and no substrate.
2. *See* **Subheading 3.4., step 2.**
3. *See* **Subheading 3.4., step 3.**

4. Notes

1. Purification to homogeneity has been achieved for two mucin-desulfating enzymes. *Prevotella strain RS2* produces a *N*-acetylglucosamine-6-sulfate specific mucin-desulfating sulfatase. After purification this removes about one-third of the sulfate from [^{35}S]mucin purified from the body region of rat stomach ***(23)***. Another mucin-desulfating enzyme with different molecular size and pH optimum has been purified from faecal supernatant ***(25)***.
2. Use of [^{35}S]mucin as the assay substrate for crude bacterial desulfating enzymes gives the sum of the activities of all the desulfating enzymes with different sugar and position specificities present. The activity measured depends on whether appropriate sulfated sugars are present in the chosen source of mucin.
3. Mucin-desulfating enzymes have been located in bacterial growth medium (extracellular), in bacterial periplasm, and in bacterial cytoplasm. In early studies researchers have sometimes assumed enzymes that degrade macromolecules would be extracellular, but this is by no means correct ***(29,37)***.
4. It is not possible to compare specific activities of assays using different substrates. Assays involving low concentrations of radioactive substrate are likely to contain substrate concentrations well below the K_m. Assays involving model substrates will be affected by substrate specificity and the kinetic parameters K_m and V_{max}.
5. When a model sulfated substrate has been used to assay a sulfatase during purification, the pure enzyme must finally be tested on [^{35}S]mucin to show whether the latter is a physiologic substrate. There are many other sulfated molecules besides mucins present in the digestive tract which could be the physiologic substrate. It is also important to con-

firm the nature of the products formed from model substrates, by means such as chromatography, to eliminate the possibility of interference by glycosidase that remove sulfated sugars or sulfated oligossaccharides.

6. It is important to determine whether growth of the bacterium on mucin induces increased production of sulfatase. In studies on the *Prevotella* sulfatase, we have found that the degree of enzyme induction is very dependant on the quality and quantity of mucin present.
7. Some commercially-available sources of mucin are very crude, and contain only small amounts of mucin. When bacteria are grown on this "mucin," the results of sulfatase-induction experiments may be uninformative unless the mucin quantity and quality is checked and purified if necessary.
8. The 3-dimensional structures of two members of the sulfatase family of enzymes have been published ***(38,39)***. Unexpected features of these enzymes include the post-translational modification of a cysteine or serine group to give a formylglycine at the active centre ***(40,41)***, and the involvement of Ca^{2+} or Mg^{2+} in substrate binding. These features will have relevance to future molecular biology studies on sulfatase expression by genetic engineering. Corfield et al. ***(36)*** noted that EDTA inhibited sulfatase activity, and this could be reversed by adding an appropriate divalent cation.
9. Some sulfatases can be partially inhibited by phosphate, so phosphate buffers should be avoided in assays. Imidazole-HCl, Tris chloride, succinate and acetate buffers work well over their pH ranges in our experience.

References

1. Roberton, A. M. and Corfield, A. P. (1999) Mucin degradation and its significance in inflammatory conditions of the gastrointestinal tract, in: *Medical Importance of the Normal Microflora.* (Tannock, G. W., ed.), Kluwer Acad. Publ., Norwell, MA, pp. 222–261.
2. Levine, M. J., Reddy, M. S., Tabak, L. A., Loomis, R. E., Bergey, E. J., Jones, P. C., Cohen, R. E., Stinson, M. W., and Al-Hashim, I. (1987) Structural aspects of salivary glycoproteins. *J. Dent. Res.* **66,** 436–441.
3. Irimura, T., Wynn, D. M., Hager, L. G., Cleary, K. R., and Ota, D. M. (1991) Human colonic sulfomucin identified by a specific monoclonal antibody. *Cancer Res.* **51,** 5728–5735.
4. Mawhinney, T. P., Landrum, D. C., Gayer, D. A., and Barbero, G. J. (1992) Sulfated sialo-oligosaccharides derived from tracheobronchial mucous glycoproteins of a patient suffering from cystic fibrosis. *Carbohyd. Res.* **235,** 179–197.
5. Lo-Guidice, J-M., Wieruszeski, J-M., Lemoine, J., Verbert, A., Roussel, P., and Lamblin, G. (1994) Sialylation and sulfation of the carbohydrate chains in respiratory mucins from a patient with cystic fibrosis. *J. Biol. Chem.* **269,** 18,794–18,813.
6. Filipe, M. I. (1979) Mucins in the human gastrointestinal epithelium: a review. *Invest. Cell Pathol.* **2,** 195–216.
7. Wesley, A. W., Forstner, J. F., and Forstner, G. G. (1983) Structure of intestinal-mucus glycoprotein from human post-mortem or surgical tissue: inferences from correlation analysis of sugar and sulfate composition of individual mucins. *Carbohyd. Res.* **115,** 151–163.
8. Roberton, A. M., Rabel, B., Harding, C. A., Tasman-Jones, C., Harris, P. J., and Lee, S. P. (1991) Use of the ileal conduit as a model for studying human small intestinal mucus glycoprotein secretion. *Am. J. Physiol.* **261,** (Gastrointest. Liver Physiol. 24) G728–734.
9. Roberton, A. M. and Wright, D. P. (1997) Bacterial glycosulfatases and sulfomucin degradation. *Can. J. Gastroenterol.* **11,** 361–366.
10. Mawhinney, T. P., Adelstein, E., Gayer, D. A., Landrum, D. C., and Barbero G. J. (1992) Structural analysis of monosulfated side-chain oligosaccharides isolated from human tracheobronchial mucous glycoproteins. *Carbohyd. Res.* **223,** 187–207.

11. Lo-Guidice, J-M., Herz, H., Lamblin, G., Plancke, Y., Roussel, P., and Lhermitte, M. (1997) Structures of sulfated oligosaccharides isolated from the respiratory mucins of a non-secretor (O, Le(A+B-)) patient suffering from chronic bronchitis. *Glycoconjugate J.* **14,** 113–125.
12. Sangadala, S., Bhat, U. R., and Mendicino, J. (1993) Structures of sulfated oligosaccharides in human trachea mucin glycoproteins. *Mol. Cell. Biochem.* **126,** 37–47.
13. Lo-Guidice, J-M., Perini, J. M., Lafitte, J. J., Ducourouble, M. P., Roussel, P., and Lamblin, G. (1995) Characterization of a sulfotransferase from human airways responsible for the 3-O-sulfation of terminal galactose in N-acetyllactosamine-containing mucin carbohydrate chains. *J. Biol. Chem.* **270,** 27,544–27,550.
14. Degroote, S., Lo-Guidice, J-M., Strecker, G., Ducourouble, M-P., Roussel, P., and Lamblin, G. (1997) Characterization of an N-acetylglucosamine-6-O-sulfotransferase from human respiratory mucosa active on mucin carbohydrate chains. *J. Biol. Chem.* **272,** 29,493–29,501.
15. Capon, C., Laboisse, C. L., Wieruszeski, J.-M., Maoret, J-J., Augeron, C., and Fournet, B. (1992) Oligosaccharide structures of mucins secreted by the human colonic cancer cell line CL.16E. *J. Biol. Chem.* **267,** 19,248–19,257.
16. Capon, C., Wieruszeski, J-M., Lemoine, J., Byrd, J.C., Leffler, H., and Kim, Y. S. (1997) Sulfated Lewis X determinants as a major structural motif in glycans from LS174T-HM7 human colon carcinoma mucin. *J. Biol. Chem.* **272,** 31,957–31,968.
17. Capon, C., Leroy, Y., Wieruszeski, J. M., Ricart, G., Strecker, G., Montreuil, J., and Fournet, B. (1989) Structures of O-glycosidically linked oligosaccharides isolated from human meconium glycoproteins. *Eur. J. Biochem.* **182,** 139–182.
18. Slomiany, B. L. and Meyer, K. (1972) Isolation and structural studies of sulfated glyoproteins of hog gastric mucosa. *J. Biol. Chem.* **247,** 5062–5070.
19. Liau, Y. H. and Horowitz, M. I. (1982) Incorporation in vitro of [^{3}H]glucosamine or [^{3}H] glucose and [^{35}S]SO_4^{2-} into rat gastric mucosa. *J. Biol. Chem.* **257,** 4709–4718.
20. Kuhns, W., Jain, R. K., Matta, K. L., Paulsen, H., Baker, M. A., Geyer, R., and Brockhausen, I. (1995) Characterization of a novel mucin sulphotransferase activity synthesizing sulphate O-glycan core 1,3-sulphate-Gal-β-1-3-GalNAc-α-R. *Glycobiology* **5,** 689–697.
21. Karlsson, N. G., Herrmann, A., Karlsson, H., Johansson, M. E. V., Carlstedt, I., and Hansson, G. C. (1997) The glycosylation of rat intestinal Muc2 mucin varies between rat strains and the small and large intestine. A study of *O*-linked oligosaccharides by a mass spectrometric approach. *J. Biol. Chem.* **272,** 27,025–27,034.
22. Stanley, R. A., Ram, S. P., Wilkinson, R. K., and Roberton, A. M. (1986) Degradation of pig gastric and colonic mucins by bacteria isolated from the pig colon. *Appl. Environ. Microbiol.* **51,** 1104–1109.
23. Roberton, A. M., McKenzie, C., Scharfe, N., and Stubbs, L. (1993) A glycosulphatase that removes sulphate from mucus glycoprotein. *Biochem. J.* **293,** 683–689.
24. Tsai, H. H., Hart, C. A., and Rhodes, J. M. (1991) Production of mucin degrading sulphatase and glycosidases by *Bacteroides thetaiotaomicron. Lett. Appl. Microbiol.* **13,** 97–101.
25. Tsai, H. H., Sunderland, D., Gibson, G. R., Hart, C. A., and Rhodes, J. M. (1992) A novel mucin sulphatase from human faeces: its identification, purification and characterisation. *Clin. Sci.* **82,** 447–454.
26. Corfield, A. P., Wagner, S. A., Clamp, J. R., Kriaris, M. S., and Hoskins, L. C. (1992) Mucin degradation in the human colon: production of sialidase, sialate O-acetylase, N-

acetylneuraminate lyase, arylesterase, and glycosulfatase activities by strains of faecal bacteria. *Infect. Immun.* **60,** 3971–3978.

27. Murty, V. L. N., Piotrowski, J., Morita, M., Slomiany, A., and Slomiany, B. L. (1992) Inhibition of *Helicobacter pylori* glycosulfatase activity toward gastric sulfomucin by nitecapone. *Biochem. Int.* **26,** 1091–1099.
28. Smalley, J. W., Dwarakanath, A. D., Rhodes, J. M., and Hart, C. A. (1994) Mucin-sulphatase activity of some oral Streptococci. *Caries Res.* **28,** 416–420.
29. Salyers, A. A., Valentine, P., and Hwa, V. (1993) Genetics of polysaccharide utilisation pathways of colonic *Bacteroides* species, in: *Genetics and Molecular Biology of Anaerobic Bacteria*. Brock/Springer series in Contemporary Bioscience. (Sebald, M., ed.), New York, Springer-Verlag, pp. 505–516.
30. Goso, Y. and Hotta, K. (1989) Types of oligosaccharide sulphation, depending on mucus glycoprotein source, corpus or antral, in rat stomach. *Biochem. J.* **264,** 805–812.
31. Corfield, A. P. and Paraskeva, C. (1993) Secreted mucus glycoproteins in cell and organ culture, in: *Glycoprotein Analysis in Biomedicine*. vol. 14 (Hounsell, E. F., ed.), Humana, Totowa, NJ, pp. 211–232.
32. Corfield, A. P., Wagner, S. A., O'Donnell, L. J. D., Durdey, P., Mountford, R. A., and Clamp, J. R. (1993) The roles of enteric bacterial sialidase, sialate O-acetyl esterase, and glycosulfatase in the degradation of human colonic mucin. *Glycoconjugate. J.* **10,** 72–81.
33. Corfield, A. P., Myerscough, N., Einerhand, A. W. C., van Klinken, B. J. W., Decker, J. and Paraskeva, C. (1999) Biosynthesis of mucin cell and organ culture methods for biosynthetic study, in: *Glycoprotein Methods and Protocols: The Mucins* (Corfield, A. P., ed.), Humana, Totowa, NJ, pp. 219–226.
34. Bashor, M. M. (1979) Dispersion and disruption of tissues. *Methods Enzymol.* **58,** 119–131.
35. Mian, N., Anderson, C. E., and Kent, P. W. (1979) Effect of O-sulphated groups in lactose and N-acetylneuraminyl-lactose on their enzymic hydrolysis. *Biochem. J.* **181,** 387–399.
36. Corfield A. P., Wagner, S. A., and Clamp, J. R. (1987) Detection of a carbohydrate sulphatase in human faecal extracts. *Biochem. Soc. Trans.* **15,** 1089.
37. Salyers , A. A., Reeves, A., and D'Elia, J. (1996) Solving the problem of how to eat something as big as yourself: diverse bacterial strategies for degrading polysaccharides. *J. Indust. Microbiol.* **17,** 470–476.
38. Bond, C. S., Clements, P. R., Ashby, S. J., Collyer, C. A., Harrop, S. J., Hopwood, J. J., and Guss, J. M. (1997) Structure of a human lysosomal sulfatase. *Structure* **5,** 277–289.
39. Lukatela, G., Krauss, N., Theis, K., Selmer, T., Gieselmann, V., von Figura, K., and Saenger, W. (1998) Crystal structure of human arylsulfatase A: the aldehyde function and the metal ion at the active site suggest a novel mechanism for sulfate ester hydrolysis. *Biochemistry* **37,** 3654–3664.
40. Meich, C., Dierks, T., Selmer, T., von Figura, K., and Schmidt, B. (1998) Arylsulfatase from *Klebsiella pneumoniae* carries a formylglycine generated from a serine. *J. Biol. Chem.* **273,** 4835–4837.
41. Recksiek, M., Selmer, T., Dierks, T., Schmidt, B., and von Figura, K. (1998) Sulfatases, trapping of the sulfated enzyme intermediate by substituting the active site formylglycine. *J. Biol. Chem.* **273,** 6096–6103.

XI

Mucin-Bacterial Interactions

35

Mucin-Bacterial Binding Assays

Nancy A. McNamara, Robert A. Sack, and Suzanne M. J. Fleiszig

1. Introduction

Surface epithelia throughout the body are covered by mucus, a protective secretion that serves as a selective physical barrier between the epithelial cell plasma membrane and the extracellular environment. Mucin, the glycoprotein constituent of mucus, has been shown to bind bacteria at mucosal surfaces that line the lung, gut, bladder, oral cavity, and eye *(1–7)*. Since bacterial binding to an epithelial cell surface is generally thought to be an important prerequisite for infection *(8)*, the interaction between bacteria and mucin, together with normal mucosal clearance mechanisms, is believed to act as a defense against infection by inhibiting bacterial adherence to the underlying epithelial cell surface.

In support of mucin's role as a nonspecific defense mechanism, malfunctions in the production and/or clearance of mucin have been implicated in the etiology of many diseases. This has led to the development of methods that can be used to study the effects of disease and various interventions (e.g., drugs and medical devices) on the ability of mucin to protect the underlying tissue. In this chapter, we present methods that can be used to examine the interaction between bacteria and mucin, as well as the extent to which this interaction serves to protect the epithelial cell surface from bacterial invasion. We have used these methods to study the interactions of *Pseudomonas aeruginosa* with the ocular surface in both human and animal models; however, they can also be used to test bacterial/mucin interactions in other tissues with other organisms.

2. Materials

2.1. Preparation and Analysis of Human Precorneal Tear Film Components

1. Noninvasive corneal irrigation chamber *(9,10)*.
2. 0.9% NaCl (autoclaved).
3. Spectra/Por® cellulose ester membranes, 500 Da molecular weight (MW) cut-off (Fisher Scientific, Fair Lawn, NJ).
4. BCA Protein assay kit (Pierce, Rockford, IL).

From: *Methods in Molecular Biology, Vol. 125: Glycoprotein Methods and Protocols: The Mucins*
Edited by: A. Corfield © Humana Press Inc., Totowa, NJ

5. Precast, 4–20% Tris-HCl Ready Gel (Bio-Rad, Hercules, CA) for sodium dodecyl sulfate-polyacrylamide gel electrophoresis (SDS-PAGE).
6. Silver stain kit (Bio-Rad).
7. Coomasie blue R-250 (Sigma, St. Louis, MO).

2.2. Isolation and Analysis of Human Ocular Sialoglycoprotein

1. Glass microcapillary tubes (calibrated, fire-polished, disposable, 10 μL capacity).
2. SW 4000 size-exclusion column.
3. 0.5 *M* NaCl, 0.1 *M* phosphate buffer, pH 5.0.
4. 20% Methanol (v/v).
5. 3-kDa Cutoff centrifugal ultrafilters (Filtron, Northborough, MA).
6. Precast, 4% and 4-20% Tris-glycine gels for SDS-PAGE (Novex, San Diego, CA).
7. Kaleidoscope™ (Bio-Rad) and See-Blue ™ (Novex) molecular weight ladders, 6-250 kDa.
8. Human lysozyme, IgG, lactoferrin, albumin and sIgA (Sigma).
9. Pre-stained molecular weight markers IgM (990 kDa) and thyroglobulin (669 kDa) (Calbiochem®, La Jolla, CA).
10. Coomasie brilliant blue R-250 (Sigma).
11. Periodate silver and Alcian blue (AB) stains (Sigma).
12. Immobilon P membranes (Millipore, Bedford, MA).
13. Sialyl-Lewis epitope (clone 258-11413, O.E.M. Concepts, Toms River, NJ).

2.3. Preparation of Bacteria

1. *P. aeruginosa* (strain 6294, serogroup O6).
2. Trypticase soy agar (TSA) plates (PML Microbiologicals, Wilsonville, OR).
3. Phosphate-buffered saline (PBS), pH 7.4 (Sigma).
4. MacConkey agar plates (PML Microbiologicals).
5. Spectrophotometer.

2.4. Microtiter Plate Assay of Bacterial Adherence

1. Linbro/Titertek 96-well microtiter plate (ICN Biomedicals, Aurora, OH).
2. PBS, pH 7.4 (Sigma).
3. 0.50% Triton X-100 (LabChem, Pittsburgh, PA).
4. MacConkey agar plates (PML Microbiologicals).
5. Bovine submaxillary gland (BSG) mucin (Sigma).

2.5. Assay to Confirm Adherence of Mucin to Microtiter Wells

1. Same as **Subheading 2.4., step 1** (Linbro/Titertek 96-well microtiter plate [ICN Biomedicals]).
2. PBS, pH 7.4 (Sigma).
3. 1% Bovine serum albumin (BSA) in PBS (w/v).
4. Biotinylated wheat germ agglutinin (WGA) (Vector, Burlington, CA).
5. PBS/Tween: 0.25 mL Tween-20 (Sigma) in 500 mL PBS.
6. PBS/Tween:Streptavidin conjugated alkaline phosphatase (Jackson Immuno Research, West Grove, PN), 500:1 (v/v).
7. Development solution: 5 m*M* *p*-nitrophenyl phosphate (NPP) in 0.1 *M* alkaline buffer solution (Sigma).
8. Stop solution: 2 *M* Na_2CO_3 (2.12 g Na_2CO_3/10 mL distilled water).

2.6. Assay to Measure Bacterial Invasion of Corneal Epithelial Cells

2.6.1. Preparation of Rabbit Corneal Epithelial Cell Cultures

1. Cell culture-treated 96-well plates (Fischer).
2. Modified supplemental hormone essential medium (SHEM) containing 10 µg/mL of bovine pituitary extract *(11)*.

2.6.2. Gentamicin Survival Assay

1. Buffered minimal essential medium (BMEM): 9.53 g of MEM (Cellgro™) plus 2.2 g of sodium bicarbonate per liter of distilled water (pH to 7.4).
2. Gentamicin sulfate (BioWhittaker, Walkersville, MD).
3. 0.25% Triton X-100 (LabChem).
4. BSG mucin (Sigma).

3. Methods

3.1. Preparation and Analysis of Human Precorneal Tear Film Components

1. Collect precorneal tear film components (TFCs) from the ocular surface of human eyes using a noninvasive corneal irrigation chamber by irrigating each cornea for 30 s with 10 mL of sterile saline using a metered pump as previously described *(9,10)*.
2. Remove cells and debris by centrifuging the eyewash samples three times at 6000 rpm for 15 min.
3. Dialyze the final supernatant against several changes of distilled water at 4°C and concentrate to 200 µL using vacuum centrifugation.
4. Determine the protein content of each 200 µL eyewash sample using the BCA protein assay kit *(12)*.
5. Separate tear film proteins using a precast 4–20% Tris-HCl gel, and visualize using a silver staining procedure that is ideal for staining polysaccharides and highly glycosylated proteins as recommended by the manufacturer (Bio-Rad) *(13)*.
6. Counter-stain the silver-stained gels with 200 mL of 0.1% Coomassie brilliant blue R-250 in 25% methanol/7.5% acetic acid for 1 h, destain overnight, and then photograph in color as previously described *(14)* (*see* **Note 1**).

3.2. Isolation and Analysis of Human Ocular Sialoglycoprotein

1. Collect closed-eye tear samples (which are rich in high molecular weight sialoglycoprotein) *(15)* from four human subjects over a period of several weeks as previously described *(16)*.
2. Pool samples and centrifuge at 11,000 rpm in a refrigerated Eppendorf microfuge for 30 min, then repeat. Store the resultant supernatants at –70°C until needed.
3. Separate the supernatant isocratically in 15-µL aliquots on a SW 4000 size exclusion column in 0.5 *M* NaCl, 0.1 *M* phosphate buffer (pH 5.0), at a flow rate of 0.25 mL/min, while monitoring the eluent at 254 nm.
4. Concentrate each fraction using a 3-kDa centrifugal ultrafilter. To establish the elution profile for all the major tear proteins, run fractions under both reducing and non-reducing conditions at 125 V for one hour on a precast, 4–20% Tris-glycine gel for SDS-PAGE. Use Kaleidoscope™ and See-Blue™ MW ladders, as well as human lysozyme, IgG, lactoferrin, albumin, and sIgA as standards.
5. Using this method, the high molecular weight glycoprotein fractions are recovered slightly after the void volume in the first peak, which elutes off the HPLC column at approxi-

mately 24 min. To prepare the high molecular weight glycoprotein, collect the initial HPLC fraction, concentrate to 100 µL by centrifugal ultrafiltration, dilute 1:1 with HPLC solvent, and separate into 15-µL aliquots.

6. Run high molecular weight glycoprotein fraction under both reducing and nonreducing conditions on a precast 4% Tris-glycine gel with pre-stained molecular weight markers IgM (990 kDa) and thyroglobulin (669 kDa).
7. To detect sialoglycoprotein (SG), periodate treat gel and stain with alcian blue in 3% acetic acid as previously described ***(17)***.
8. For further characterization, transfer glycoprotein overnight onto Immobilon P in 20% methanol (v/v) at 30 V, followed by 80 V for 1 h.
9. Probe with a panel of antibodies to specific known mucins, their core proteins, and common sugar epitopes (*see* **Note 2**).

3.3. Preparation of Bacteria for Binding Assay

1. Grow bacteria overnight at 37°C on TSA plates.
2. Wash bacteria three times in PBS by centrifugation at 7000 rpm for 5 min ***(9)***.
3. Prepare the inoculum by resuspending the washed bacteria into PBS until the optical density at 650 nm reaches 0.1 (equivalent to 1×10^8 cfu/mL).
4. Quantify the starting inoculum used in each experiment (typically 1×10^6 cfu/mL) by serially diluting the sample and plating 10 µL (in duplicate) on MacConkey agar.

3.4. Microtiter Plate Assay of Bacterial Adherence

3.4.1. To Determine Whether or Not Bacteria Bind to Human Tear Film or Mucin (see **Note 3**)

1. Coat microtiter wells overnight at 37°C with 100 µL of tear film or mucin sample. Pretreat control wells with PBS which does not promote bacterial adherence to these wells ***(5,7)***.
2. Wash wells four times with PBS to remove non-adherent material.
3. Prepare bacterial inoculum containing 1×10^6 cfu/mL in PBS (10 µL of 1×10^8 cfu/mL + 990 µL PBS).
4. Add inoculum containing 30 µL of 1×10^6 cfu/ml *P. aeruginosa* 6294 to all wells.
5. Incubate plate at 37°C for 30 min.
6. Aspirate bacteria with a sterile pipette and wash wells 20 times with PBS to remove nonadherent bacteria.
7. Dislodge adherent bacteria from the well surface by adding 300 µL of 0.5% Triton X.
8. Incubate plate at 37°C for 30 min.
9. Vigorously stir each well with a sterile pipet and perform a viable count by plating 10 µL (in duplicate) on MacConkey agar.

3.4.2. To Determine Whether or Not Mucin Blocks Bacterial Adherence to Known Bacterial Binding Factors (see **Note 4**)

1. Coat microtiter wells overnight at 37°C with 100 µL of known bacterial binding factor (e.g., TFCs or BSG mucin).
2. Wash wells four times with PBS to remove nonadherent material.
3. Coat microtiter wells with 100 µL of SG for 18 h at 37°C. Treat control wells with 100 µL of PBS instead of SG since PBS does not affect bacterial binding to either TFCs or BSG mucin.
4. Prepare bacterial inoculum containing 1×10^5 cfu/mL in PBS (1 µL of 1×10^8 cfu/mL + 999 µL PBS).
5. Add inoculum containing 30 µL of 1×10^5 cfu/mL *P. aeruginosa* 6294 to all wells.

6. Incubate plate at 37°C for 30 min.
7. Wash wells 20 times with PBS to remove nonadherent bacteria.
8. Dislodge adherent bacteria from the well surface by adding 300 µL of 0.5% Triton X.
9. Incubate plate at 37°C for 30 min.
10. Vigorously stir each well with a sterile pipet and perform a viable count by plating 10 µL (in duplicate) on MacConkey agar.

*3.4.3. To Determine Whether or Not Treating Bacteria with Mucin Blocks Their Ability to Adhere to Known Bacterial Binding Factors (see **Note 4**)*

1. Coat microtiter wells overnight at 37°C with 100 µL of known bacterial binding factor (e.g., TFCs or BSG mucin).
2. Wash wells four times with PBS to remove nonadherent material.
3. Prepare bacterial inoculum containing 2×10^5 cfu/mL in PBS (2 µL of 1×10^8 cfu/mL + 998 µLl PBS).
4. Prepare starting inoculum by mixing 100 µL of 2×10^5 cfu/mL*P. aeruginosa* (prepared above) with 100 µL of SG and incubate at 37°C for 1 h. Prepare inoculum for controls by mixing 100 µL of 2×10^5 cfu/mL bacteria with 100 µL of PBS (*see* **Note 5**).
5. Add 30 µL of the starting inoculum containing *P. aeruginosa* 6294 (1×10^5 cfu/mL) and either SG or PBS (control) to wells.
6. Incubate plate at 37°C for 30 min.
7. Wash wells 20 times with PBS to remove nonadherent bacteria.
8. Dislodge adherent bacteria from the well surface by adding 300 µL of 0.5% Triton X.
9. Incubate plate at 37°C for 30 minu.
10. Vigorously stir each well with a sterile pipet and perform a viable count by plating 10 µL (in duplicate) on MacConkey agar.

3.5. Assay to Confirm Adherence of Mucin to Microtiter Wells

1. Coat 96-well microtiter plate overnight at 37°C with several dilutions of SG, TFC, and BSG mucin.
2. Wash wells 24 times with PBS.
3. Block for 2 h at room temperature (or overnight at 4°C) by adding 200 µL of 1% BSA/PBS to each well.
4. After blocking, wash wells twice with PBS/Tween and incubate with 100 µL of 5 µg/mL biotinylated wheat germ agglutinin (WGA) for 45 min at room temperature (wrap plate in plastic). WGA is a plant lectin that binds specifically to sialic acid residues, and thus, serves as a probe for quantifying the amount of sialylated glycoprotein that is bound to the microtiter well.
5. Wash wells six times with PBS/Tween.
6. Detect WGA-bound biotin by adding 100 µL of PBS/Tween:streptavidin-peroxidase solution to each well and incubating the wrapped plate at 37°C for 45 min.
7. Wash wells six times with PBS/Tween.
8. Add 100 µL of development solution (NPP in alkaline buffer) and incubate at room temperature. Detect the sialoglycoprotein-lectin-enzyme complexes by adding the enzyme substrate (NPP) which is converted to a colored product in the presence of enzyme. Immediately begin monitoring the intensity at 405 nm using a standard ELISA reader. Readings should be taken every 2 min since development occurs quickly. Controls consist of ovalbumin treated wells incubated with the WGA and/or streptavidin system alone ***(18,19)***.
9. Stop the reaction by adding 10 µL of stop solution. This step can be omitted if the absorbance is monitored continuously following the addition of development solution (*see* **Note 6**).

3.6. Assay to Measure Bacterial Invasion of Corneal Epithelial Cells (see Note 7)

3.6.1. Preparation of Rabbit Corneal Epithelial Cell Cultures

1. Place 100 µL of SHEM suspension containing rabbit corneal epithelial cells in cell culture-treated 96-well plates.
2. Maintain cells at 37°C while changing SHEM every 2 d until confluent (typically 4–7 d).

3.6.2. Preparation of Bacteria for Gentamicin Survival Assay (see Subheading 2.3. for Materials)

1. Grow bacteria overnight at 37°C on TSA.
2. Prepare the inoculum by resuspending bacteria in 5 mL of buffered Eagle's minimal essential medium (BMEM) using a sterile cotton swab until the optical density reaches 0.1 (equivalent to 1×10^8 cfu/mL at 650 nm).
3. The starting inoculum of 1×10^7 cfu/mL is prepared by placing 100 µL of the bacterial suspension into 900 µL of BMEM. Quantify the starting inoculum used in each experiment by serially diluting the sample and plating 10 µL (in duplicate) on MacConkey agar.

3.6.3. Gentamicin Survival Assay

1. Wash confluent cells by adding 200 µL of BMEM to each well and aspirating with a sterile pipet.
2. Coat cells with 70 µL of BMEM containing several concentrations of BSG mucin, TFC, or SG. Pretreat control cells with 70 µL of BMEM alone.
3. Incubate plate at 37°C for 2 h.
4. Add 8 µL of an inoculum containing 1×10^7 cfu/mL of *P. aeruginosa* 6294 to all wells.
5. Incubate plate at 37°C for 3 h.
6. Perform viable counts on each well after 3 h to estimate the number of bacteria in the extracellular medium (*see* **Note 8**).
7. Aspirate remaining medium and kill extracellular bacteria by adding 200 µL gentamicin solution (200 µg/mL) to each well.
8. Incubate plate at 37°C for 2 h.
9. Wash cells with 200 µL BMEM.
10. Lyse cells with 100 µL 0.25% Triton-X solution.
11. After 15 min, use a sterile pipet to vigorously stir each well. Perform a viable count to quantify intracellular bacteria by plating 10 µL (in duplicate) on MacConkey agar. Compare the susceptibility of cells to invasion by *P. aeruginosa* in the presence of mucin with that of control cells incubated with medium alone (*see* **Note 9**).

4. Notes

1. Using the silver/coomassie blue double-staining method described in **Subheading 3.1.**, all of the silver-stained glycoproteins and lipids stain by silver only and appear dark gray, whereas all of the unglycosylated proteins stain with coomassie blue and appear blue. Thus, polypeptides stained by coomassie brilliant blue and glycosylated proteins stained by silver can be differentially visualized in the same gel. With this method, whole human tear film samples are shown to contain a prominent high molecular weight silver-stained band of glycosylated protein (>200 kDa) that does not stain with coomassie brilliant blue. Others have identified a glycoprotein of similar size in human tear samples collected with a microcapillary tube ***(20–22)***.

2. SDS-PAGE in combination with periodate silver and AB staining (described in **Subheading 3.2.**) reveal that high molecuar weight glycoproteins isolated from the ocular surface consist primarily of three SG bands, one in the stacking gel and two in the running gel with apparent molecuar weight ranges of 500–450 kDa. In addition, all of the SGs isolated from human tear film exhibit a common reactivity with a mAb raised against a sialyl-Lewis epitope specific to salivary mucin (clone 258–11413, O.E.M. Concepts). Immunofluorescence microscopy reveals intense specific staining of the conjunctival epithelial plasma membrane but not the goblet cells on probing with the mAb to salivary mucin, suggesting that these SGs arise specifically from the epithelium ***(15)***.
3. In a previous study we showed that *P. aeruginosa* binds to factors present in whole human tears ***(23)***. To examine whether the bacterial binding factors in precorneal human tear film may include mucin, we used the bacterial adherence assay, described in **Subheading 3.4.**, and compared bacterial adherence to the SG isolated from the human ocular surface to binding elicited with whole human TFCs and BSG mucin.
4. Using the bacterial adherence assay, we found that *P. aeruginosa* adhered to wells coated with both whole human tear film and BSG mucin, but did not bind to wells coated with either ocular SG or PBS alone. This suggested that human ocular SG was not responsible for bacterial binding to whole tear film and that the bacterial binding factor(s) in human tears are either other glycoproteins or nonglycoprotein fractions. Having ruled out bacterial binding to SG, we developed two new methods, which were variations of the original bacterial adherence assay, to determining whether SG could inhibit bacterial interaction with known bacterial binding factors. The first variation of the bacterial adherence assay, described in **Subheading 3.4.2.**, sought to determine whether or not ocular SG could block bacterial binding to whole human tear film or to BSG mucin when microtiter plates were precoated with TFCs or mucin and then subsequently coated with SG. The second variation, described in **Subheading 3.4.3.**, was used to determined whether or not bacterial binding to mucin-coated microtiter plates could be blocked by adding SG to the bacterial inoculum.
5. When bacteria are pretreated with any factor prior to performing the bacterial adherence assay, one must demonstrate that preincubation with that factor does not reduce the viability or the replication rate of the bacteria. This can be accomplished by performing viable counts after 2 and 8 h to determine whether inocula containing treated and untreated bacteria are equivalent.
6. Using the biotinylated WGA assay described in **Subheading 3.6.**, we found that the absorbance reading for a 10^{-2} dilution of SG was similar to that observed for a 10^{-2} dilution of TFCs and 10 μg/mL of BSG mucin; thus, we used these concentrations in the bacterial binding assay to ensure that an equivalent number of sialic acid residues were bound to the bottom of the well for each preparation.
7. Bacterial association with the corneal surface is a crucial step towards bacterial colonization and invasion of corneal epithelial cells. Since soluble factors in whole human tears and BSG mucin bind *P. aeruginosa*, while human ocular SG does not, we used the gentamicin survival assay to examine whether these factors could block *P. aeruginosa* invasion of corneal epithelial cells. As was expected, treating cells with several different dilutions of BSG mucin or TFCs caused a two- to threefold decrease in susceptibility to bacterial invasion by *P. aeruginosa* strain 6294, whereas ocular SG did not significantly enhance or inhibit bacterial invasion of corneal epithelial cells compared to controls.
8. When studying bacterial invasion of epithelial cells using the gentamicin survival assay, it is important to confirm that an equivalent number of extracellular bacteria remain in the wells following each treatment (i.e., the treatment does not alter the viability or replica-

tion rate of the bacteria). This is best done by performing a viable count on one well in each different treatment group at the end of the 3-h incubation of cells with bacteria (just prior to adding the gentamicin).

9. For all the assays described in this chapter, we recommend that at least six wells are used for each sample group in each experiment and that experiments are repeated a minimum of three times. Nonparametric statistical techniques are typically used to make comparisons since the presence of outlying values may cause distributions to deviate from normality. Two group comparisons can be made with the Wilcoxan 2-group test for unpaired data and the Kruskal-Wallis test can be used for the comparison of three or more independent groups of data.

References

1. Florey, H. (1955) Mucin and the protection of the body. *Proc. R. Soc. Lond.* **143,** 147–158.
2. Liljemark, W. and Gibbons, R. (1972) Proportional distribution and relative adherence of *Streptococcus miteor (mitis)* on various surfaces in the human oral cavity. *Infect. Immun.* **6,** 852–859.
3. Parsons, C. and Mulholland, S. (1978) Bladder surface mucin. Its antibacterial effect against various bacterial species. *Am. J. Pathol.* **93,** 423–432.
4. Cohen, R., Rossoll, R., Cabelli, V., Yang, S., and Laux, D. (1983) Relationship between the mouse colonizing ability of a human fecal *Escherichia coli* strain and its ability to bind a specific mouse colonic mucous gel protein. *Infect. Immun.* **40,** 62–69.
5. Vishwanath, S. and Ramphal, R. (1984) Adherence of *Pseudomonas aeruginosa* to human tracheobronchial mucin. *Infect. Immunity.* **45,** 197–202.
6. Chick, S., Harber, M., Mackenzie, R., and Asscher, A. (1981) Modified method for studying bacterial adhesion to isolated uroepithelial cells and uromucoid. *Infect. Immun.* **34,** 256–261.
7. Fleiszig, S., Zaidi, T., Ramphal, R., and Pier, G. (1994) Modulation of *Pseudomonas aeruginosa* adherence to the corneal surface by mucus. *Infect. Immun.* **62,** 1799–1804.
8. Ramphal, R., McNiece, M., and Polack, F. (1981) Adherence of Pseudomonas aeruginosa to the injured cornea. A step in the pathogenesis of corneal infections. *Ann. Ophthalmol.* **100,** 1956–1958.
9. Fleiszig, S., Efron, N., and Pier, G. (1992) Extended contact lens wear enhances *Pseudomonas aeruginosa* adherence to human corneal epithelium. *Invest. Ophthalmol. Vis. Sci.* **33,** 2908–2916.
10. Fullard, R. and Wilson, G. (1986) Investigation of sloughed corneal epithelial cells collected by non-invasive irrigation of the corneal surface. *Curr. Eye Res.* **5,** 847–856.
11. Jumblatt, M. and Neugeld, A. (1983) Serotonergic responsiveness of rabbit corneal epithelial cells in culture. *Invest. Ophthalmol. Vis.Sci.* **24,** 1139–1143.
12. Lowry, O., Rosebrough, N., Farr, A., and Randall, R. (1951) Protein measurement with the folin phenol reagent. *J. Biol. Chem.* **193,** 265–275.
13. Merril, C., Goldman, D., Sedman, S., and Ebert, M. (1981) Ultrasensitive stain for proteins in polyacrylamide gels show regional variations in cerebrospinal fluid proteins. *Science* **211,** 1437.
14. Dzandu, J., Deh, M., Varratt, D., and Wise. G. (1984) Detection of erythrocyte membrane proteins, sialoglycoproteins and lipids in the same polyacrylamide gel using a double staining technique. *Proc. Natl. Acad. Sci. USA* **81,** 1733–1737.
15. Sack, R., Bogart, B., Beaton, A., Sathe, S., and Lew, G. (1997) Diurnal variations in tear glycoproteins. Evidence for an epithelial origin for the major non-reducible ≥ 450 kDa sialoglycoprotein(s). *Curr. Eye. Res.* **16,** 577–588.

16. Sack, R., Tan, K., and Tan, A. (1992) Diurnal tear cycle: evidence for a nocturnal inflammatory constitutive tear fluid. *Invest. Ophthalmol. Vis. Sci.* **33,** 626–640.
17. Wardi, A. and Michos, G. (1972) Alcian blue staining of glycoproteins in acrylamide disc electrophoresis. *Anal.. Biochem.* **49,** 607–609.
18. Kaku, H., Mori, Y., Goldstein, I., and Shibuya, N. (1993) Monomeric, monovalent derivative of *Maackia amurensis* leukoagglutinin. *J. Biol. Chem.* **268,** 13,237–13,241.
19. Rogerieux, R., Belaise, M., Terzidis-Trabelsi, H., Greffard, A., Pilatte, Y., and Lambre, C. (1993) Determination of the sialic acid linkage specificity of sialidases using lectins in a solid phase assay. *Anal. Biochem.* **211,** 200–204.
20. Moore, J. and Tiffany, J. (1979) Human ocular mucus. Origins and preliminary characterization. *Exp. Eye. Res.* **29,** 291–301.
21. Chao, C., Vergnes, J., and Brown, S. (1983) *O*-glycosidic linkage in glycoprotein isolates from human ocular mucus. *Exp. Eye. Res.* **37,** 533–541.
22. Jay, G. and Hong, B. (1992) Characterization of a bovine synovial fluid lubricating factor. II. Comparison with purified ocular and salivary mucin. *Conn. Tissue Res.* **28,** 89–98.

36

Growth of Mucin Degrading Bacteria in Biofilms

George T. Macfarlane and Sandra Macfarlane

1. Introduction

Mucins are important sources of carbohydrate for bacteria growing in the human large intestine. As well as being produced by goblet cells in the colonic mucosa, salivary, gastric, biliary, bronchial, and small intestinal mucins also enter the colon in effluent from the small bowel. Particulate matter, such as partly digested plant cell materials, are entrapped in this viscoelastic gel, which must be broken down to facilitate access of intestinal microorganisms to the food residues. It is estimated that between 2 to 3 g of mucin enter the large bowel each day from the upper digestive tract *(1)*, however, the rate of colonic mucus formation is unknown. Complex polymers, such as mucin must be degraded by a wide range of hydrolytic enzymes to smaller oligomers and their component sugars and amino acids before they can be assimilated by intestinal bacteria.

Pure and mixed culture studies have established that in many intestinal bacteria, synthesis of these enzymes, particularly β-galactosidase, *N*-acetyl β-glucosaminidase, and neuraminidase *(2–4)*, is catabolite regulated, and is therefore dependent on local concentrations of mucin and other carbohydrates. Although some colonic microorganisms can produce several different glycosidases, which allows them to completely digest heterogeneous polymers *(5–8)*, the majority of experimental data points to the fact that the breakdown of mucin and other complex organic molecules is a cooperative activity.

In the large bowel, bacteria occur in a multiplicity of different microhabitats and metabolic niches, on the mucosa, in the mucous layer, and in the colonic lumen, where they exist in microcolonies, as free-living organisms, or on the surfaces of particulate materials *(9,10)*.

Wherever there are surfaces, bacteria form biofilms. They are usually complex microbial assemblages that develop in response to the chemical composition of the substratum and other environmental constraints. The available evidence shows that in the colon, these microbiotas are heterogeneous entities that form rapidly on the sur-

From: *Methods in Molecular Biology, Vol. 125: Glycoprotein Methods and Protocols: The Mucins*
Edited by: A. Corfield © Humana Press Inc., Totowa, NJ

faces of partly digested foodstuffs in the intestinal lumen, and in the mucous layer covering the mucosa ***(9)***. Sessile microorganisms in biofilms often behave very differently from their nonadherent counterparts, and, in particular, the nature and efficiency of their metabolism may be changed ***(10–12)***. Close spatial relationships between bacterial cells in biofilms are ecologically important in that they minimize potential growth limiting effects on crossfeeding populations ***(13)***.

Although the large gut is often likened to a continuous culture system, this is an oversimplification. Consideration of colonic motility and the way in which intestinal material is processed suggests that only the cecum and ascending colon exhibit characteristics of a continuous culture. The inaccessibility of the large bowel for experiments on the digestion of mucin by colonic microorganisms inevitably means that the majority of studies are made in vitro. A variety of models are available that enable pure and mixed populations of intestinal bacteria to be grown under anaerobic conditions, ranging from small screw-capped or serum bottles to more complex batch and continuous fermentation systems (chemostats).

The effectiveness of in vitro model systems varies depending on the problem to be investigated, and each method has advantages and disadvantages. For example, fermentation experiments made using serum bottles are inexpensive, allow screening of a number of substrates and/or fecal samples from different individuals, and require small amounts of substrate and test sample ***(14)***. Depending on bacterial cell numbers, the experiments can be designed to be of short duration, thereby minimising potential distortions in the data resulting from the selection of nonrepresentative populations of microorganisms. Longer-term experiments effectively become enrichment cultures, selecting bacteria that are most efficient at utilizing the test substrates.

However, these fermentations are uncontrolled, and yield little information on bacterial metabolism, the organisms involved, or how the processes are regulated. The limited data provided by such experiments relate simply to the input of substrates and the output of products. Other problems may also be encountered; for example, if high-substrate concentrations are used, strong nonphysiological buffers are needed to control pH, which may have unpredictable effects on bacterial metabolism. If culture pH is not regulated, the environment in the vessel will change rapidly such that the fermentation conditions become physiologically irrelevant.

Another factor to be considered with batch fermentations is that they are closed systems, in which bacterial metabolic activities and environmental conditions in the cultures are constantly changing. Thus, at the beginning of bacterial growth, substrate concentrations are high, and become depleted as the cells grow, whereas bacterial fermentation products and other autoinhibitory metabolites progressively accumulate in the culture.

Many of these problems are avoided in continuous cultures, since they are open systems that work efficiently at high bacterial population densities. Cell growth is strictly controlled by the concentrations of limiting nutrients in the feed medium. For this reason, the organisms grow suboptimally at specific growth rates (μ) set by the experimenter, through alterations in dilution rate (D), which is regulated by varying the rate at which culture medium is fed to the fermentation vessel.

The principal advantage of the chemostat in physiologic and ecologic studies on microorganisms is that it enables long-term detailed investigations to be made under a multitude of externally imposed steady-state conditions that are not possible with closed batch-type cultures. A two-stage continuous culture model can be used for studying the formation of mucin-degrading biofilms under nutrient-rich and nutrient-depleted conditions. The system comprises paired glass fermentation vessels fitted with modified lids (**Fig. 1**), and several extra sampling ports fitted to the vessel sides to which removable mucin baits or mucin gel cassette holders (**Fig. 2**) are attached. The fermenters are connected in series, with fresh culture medium being fed to vessel 1 (V1), and spent culture from this vessel being pumped into vessel 2 (V2). This facilitates the study of mucin colonization under relatively carbohydrate-rich V1 and extremely carbon-limited (V2) environmental conditions, comparable to the proximal and distal colons.

Three main protocols are outlined in this chapter for studying (1) mucous-degrading bacterial consortia occurring in biofilms on the rectal mucosa, (2) mucinolytic species growing in artificial mucin biofilms in continuous culture models of the colon in the laboratory, and (3) mucinolytic microorganisms colonizing the surfaces of food particles in fecal material.

Methods outlined for the isolation of bacteria from the rectal mucosa are essentially destructive, and primarily provide details of the types and numbers of different species that take part in this process. They do not contribute information concerning the multicellular organization of biofilm communities. However, the chemostat modeling protocols afford useful comparative data on the enzymology and physiology of the breakdown of mucin by adherent (mucin baits) and planktonic bacterial communities, under varying environmental conditions, while the use of mucin-coated glass slides attached to cassettes facilitates microscopic examination of biofilm development.

2. Materials

1. Samples for scanning electron microscopy (SEM) are placed in 3% (v/v) glutaraldehyde in 1 M PIPES buffer, pH 7.0. Then fix the samples with 4% (w/v) aqueous OsO_4, dehydrated stepwise in ethanol, which involves three changes (10 min) in each of 50, 75, 95, and, finally, 100% ethanol. Then dry the samples on a Poleron E 5000 critical-point drier, place on stubs, and gold-coat to a depth of 30 nm.
2. Glass tubes, mucin gel cassettes, and fermentation vessels for baiting studies are manufactured by Soham Glass, Ely, Cambs, UK.
3. All formulated bacteriologic culture media and growth supplements are supplied by Oxoid. Use as per manufacturer's instructions. Unless stated otherwise, all chemicals are obtained from Sigma Aldrich Ltd. (Poole, UK). Agars for isolating specific bacterial groups in mucin-degrading consortia are as follows:
   a. Nutrient agar (total facultative anaerobes).
   b. MaConkey agar no. 2 (lactose-fermenting and nonlactose-fermenting enterobacteria, enterococci).
   c. Azide blood agar base (facultative anaerobic cocci, some Gram-positive anaerobic cocci).
   d. Wilkins-Chalgren agar (total anaerobe counts).

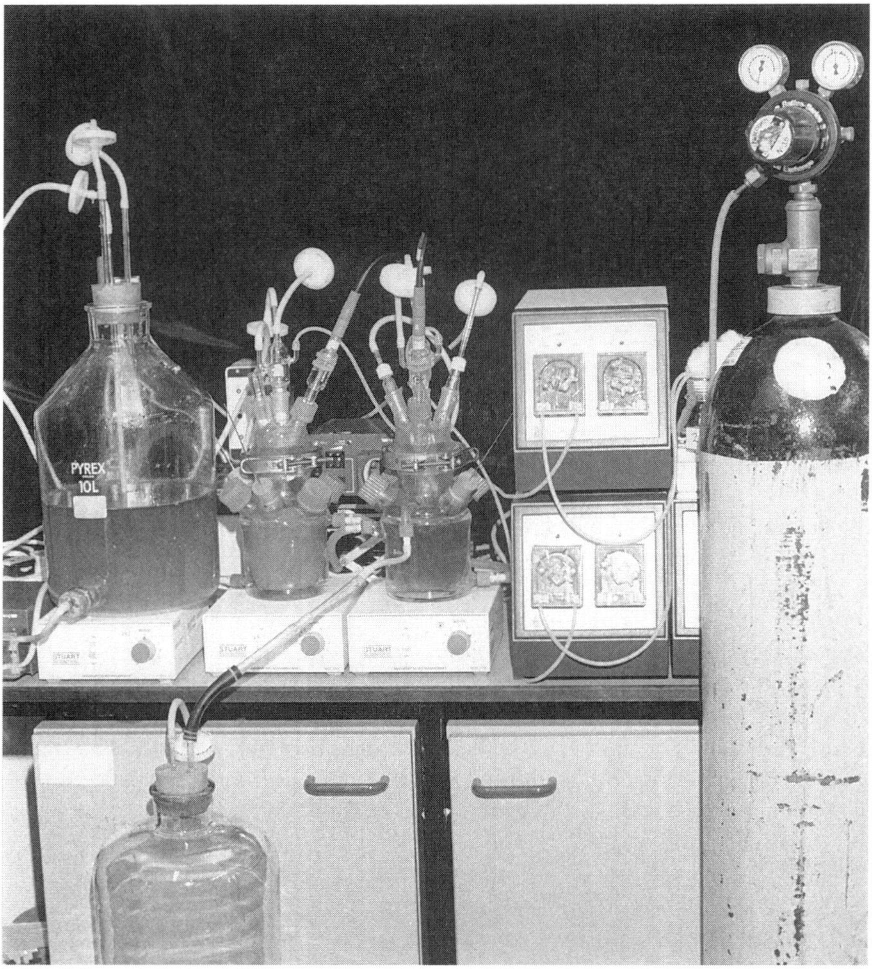

Fig. 1. Two-stage continuous culture model used to study mucin-degrading biofilms under carbon-excess (vessel 1, left) and carbon-limited (vessel 2, right) conditions.

e. Wilkins-Chalgren agar plus nonsporing supplements (nonsporing anaerobes). The supplements contain hemin, menadione, sodium pyruvate, and nalidixic acid.
f. Wilkins-Chalgren plus Gram-negative supplements (Gram-negative anaerobes). The selective agents in this culture medium are hemin, menadione, sodium succinate, nalidixic acid, and vancomycin.
g. MRS agar (lactobacilli).
h. Perfringens agar and supplements (*Clostridium perfringens* and certain other clostridia). The selective supplements (A and B) contain sulfadiazine, oleandomycin phosphate, and polymyxin B. All antibiotic additions are added at 50°C after autoclaving for 121°C at 15 min.

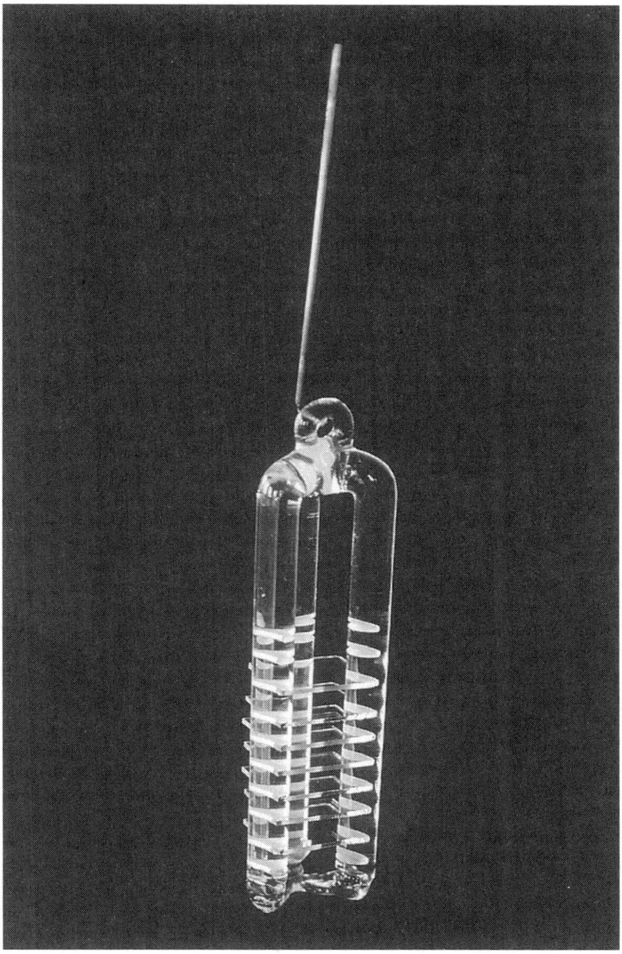

Fig. 2. Mucin gel cassette used for investigating bacterial colonization of mucous surfaces in continuous culture experiments. The glass frame contains several slots into which are fitted removable mucin-coated glass plates or microscope cover slips.

i. Fusobacterium agar (fusobacteria). This comprises: 37.0 g/L Brucella agar base, 5.0 g/L Na_2HPO_4, 1.0 g/L NaH_2PO_4, 1.0 g/L $MgSO_4 \cdot 7H_2O$, 0.005 g/L hemin, at pH 7.6. This is autoclaved, cooled to 50°C and the following antibiotics are then added after filter sterilisation in 5 mL distilled water: 20 mg/L neomycin, 10 mg/L vancomycin, 6.0 mg/L josamycin (ICN Biomedicals, Aurora, Ohio).
j. Beerens Agar for selective isolation of bifidobacteria is made as follows: 42.5 g/L Columbia agar, 5.0 g/L glucose, 0.5 g/L cysteine HCl, 1.5 g/L purified bacteriologic agar. Five milliliters of propionic acid is added to these constituents, after they have been boiled and cooled to 70°C. The pH of the medium is then adjusted to 5.0 before pouring the agar into Petri plates.

   k. Bacteroides mineral salts medium for selective isolation of members of the *B. fragilis* group consists of: 1.5 g/L KH_2PO_4, 1.0 g/L K_2HPO_4, 9.0 g/L NaCl, 1.2 g/L cysteine HCl, 1.2 g/L $NaHCO_3$, 0.1 g/L $CaCl_2{\cdot}2H_2O$, 0.15 g/L $MgCl_2{\cdot}6H_2O$, 0.05 g/L $MnCl_2{\cdot}4H_2O$, 0.05 g/L $CoCl_2{\cdot}6H_2O$, 0.001 g/L $FeSO_4.7H_2O$, 0.005 g/L hemin, 0.005 g/L vitamin B_{12}, 1.0 g/L NH_4SO_4, 5.0 g/L glucose, 20 g/L purified bacteriologic agar. After autoclaving and cooling to 50°C, 5 mL of a filter sterilized antibiotic solution is added, containing: 3.0 mg/L vancomycin, 10.0 mg/L nalidixic acid.
4. Freezer vials containing Wilkins-Chalgren broth supplemented with 10% glycerol and 2% porcine gastric mucin (Sigma Type III, partially purified), with pH adjusted to 6.5.
5. Neuraminidase substrate (1 mg/mL *N*-acetylneuraminlactose). The neuraminidase standard is 1 mg/mL *N*-acetylneuraminic acid (NANA). Make both in 0.1 *M* acetate buffer (pH 5.5).
   a. Solution A: 0.2 *M* sodium periodate (meta) in 9 *M* phosphoric acid.
   b. Solution B: 10% sodium arsenite in 0.5 *M* sodium sulfate/0.2 *M* H_2SO_4.
   c. Solution C: 0.6% thiobarbituric acid in 0.5 *M* sodium sulfate.

   Store solutions A and B at room temperature, and make solution C fresh daily.
6. Make glycosidase assays using the following *p*-nitrophenyl substrates: *N*-acetyl α-D-galactosaminide, α-L-fucopyranoside, *N*-acetyl β-D-glucosaminide, and β-D-galactopyranoside, all prepared as 15 m*M* solutions in 0.01 *M* Tris buffer, pH 6.5. The stop solution is a mixture of 0.5 *M* Na_2CO_3 and 0.5 *M* $NaHCO_3$. A standard curve using varying dilutions of *p*-nitrophenol is used to calculate enzyme activities.
7. PYG broth: 20 g/L glucose, 10.0 g/L Yeast extract, 5.0 g/L Tryptone Soya broth, 5.0 g/L Peptone water, 0.5 g/L cysteine HCl, 0.005 g/L hemin (*see* **item 8**). Add 40 mL of PYG salt solution, 0.2 mL of vitamin K_1 solution (*see* **item 8**), and 10 mL of Tween-80 to 950 mL of distilled water.
   a. To make PYG salt solution, add 0.2 g of $CaCl_2{\cdot}2H_20$ and 0.2 g of $MgSO_4$ to 300 mL of distilled H_2O and dissolve by mixing. Then add a further 500 mL of H_2O, together with 1.0 g K_2HPO_4, 1.0 g of $KH_2\ PO_4$, 10.0 g of $NaHCO_3$, and 2.0 g of NaCl. Finally, make up the volume to 1 L with distilled water.
8. MIDI PYG broth is made as follows: 5.0 g/L peptone water, 5.0 g/L Pepticase (Quest International, Norwich, New York), 10.0 g/L Yeast extract, 0.5 g/L cysteine HCl, 10.0 g/L glucose. In addition, the following solutions are added: 40.0 mL salts solution, 10 mL hemin solution, 0.2 mL vitamin K_1. Add the hemin solution, vitamin K_1, and cysteine after the medium is boiled, but before it is dispensed into metal capped glass Universal bottles at 100°C and autoclaved. The salt solution is made in the same way as for normal PYG medium, but the NaCl concentration is increased to 50 g. The haemin solution is made as follows: Dissolve 50 mg of hemin in 1 mL of 1 *M* NaOH; make to 100 mL with distilled water, then autoclave at 121°C for 15 min. Store at 4°C. Vitamin K_1: Dissolve 0.15 mg in 30 mL 95% of ethanol. Store at 4°C in a brown bottle. Discard after 1 mo. For identification of Gram-positive organisms, add 2.5 mL of 1:10 Tween-80 in distilled water at the same time as cysteine to the medium.
9. To make Balch trace elements solution ***(15)***, add the following constituents to 600 mL of distilled water: 3.0 g $MgSO_4{\cdot}7H_2O$, 0.45 g $MnCl_2{\cdot}4H_2O$, 1.0 g NaCl, 0.10 g $FeSO_4{\cdot}7H_2O$, 0.18 g $CoSO_4{\cdot}7H_2O$, 0.10 g $CaCl_2{\cdot}2H_2O$, 0.18 g $ZnSO_4{\cdot}7H_2O$, 0.01 g $CuSO_4{\cdot}5H_2O$, 0.018 g $Al(SO_4)_2{\cdot}12H_2O$, 0.01 g H_3BO_4, 0.01 g $NaMoO_4{\cdot}2H_2O$, 0.19 g $Na\ _2SeO_4$, 0.092 g $NiCl_2{\cdot}6H_2O$. Adjust the solution to pH 7.0 with 1*M* KOH, then make up to 1 L. Store at 4°C until use.

3. Methods

3.1. Enumeration and Identification of Mucinolytic Bacteria in Rectal Biopsies

1. Rectal biopsy material is obtained from hospital out-patients. Tissue samples should be immediately placed in preweighed sterile Bijoux bottles containing 4 mL of a suitable anaerobic transport medium, such as Wilkins-Chalgren broth (*see* **Note 1**).
2. Weights and sizes of the samples are measured before placing them in an anaerobic cabinet (atmosphere 10% H_2, 10% CO_2, 80% N_2) at 37°C. Speed is important during this step (*see* **Note 2**).
3. Mascerate the biopsy material using a sterile glass tissue homogeniser. One mL of this sample is serially diluted (10-fold dilutions to 10^{-5}) in test-tubes containing 9 mL half-strength sterile anaerobic Peptone water (*see* **Note 1**).
4. Plate out 50 µL of the original sample and 100 µL of all dilutions to 10^{-5} in triplicate, onto a range of selective and nonselective culture media, using sterile tips and glass spreaders (see **Subheading 2., item 3**). Plates for aerobic incubation are removed from the anaerobic cabinet and incubated at 37°C.
5. Aerobic plates are incubated for 2 d, and anaerobic plates for up to 5 d, with periodic examination, before counting of colonies.
6. The bacteria are then characterized on the basis of their Gram staining characteristics, cellular morphology, fermentation products *(16)*, and cellular fatty acid methyl ester (FAME) profiles (*see* **Note 3**).
7. Fermentation products (short chain fatty acids, lactate, succinate) are analysed by growing the organisms as pure cultures in PYG broth (*see* **Subheading 2., item 7**) for 24 h, then centrifuging (13,000*g*, 10 min) to obtain a clear supernatant for GC or HPLC analysis.
8. Bacterial cellular fatty acids are extracted from overnight cultures of the organisms in MIDI PYG broth (see **Subheading 2., item 8**). After centrifugation to obtain a cell pellet, FAMEs are produced by saponification, methylation, and finally, solvent extraction. FAMEs are then separated using a 5898 A Microbial Identification System. (Microbial ID, Newark, DE).
9. FAMEs are automatically integrated and numerical analysis done using standard MIS Library Generation Software which identifies the organisms.
10. Colonies for further study are grown on agar plates and removed with sterile swabs into 2-mL freezer vials which are then stored at –80°C (*see* **Subheading 2., item 4**).

3.2. Mucin-Degrading Enzymes in Mucosal Bacteria

1. Grow individual isolates at 37°C in Wilkins-Chalgren Broth, supplemented with 5 g/L partially purified porcine gastric mucin, in anaerobic Universal bottles (prepare by boiling and dispensing the media into the bottles at 100°C, and then autoclaving).
2. After the cultures have grown, keep a portion of the whole culture, and harvest some of the bacteria by centrifugation (13,000*g*, 30 min). Retain the cell-free supernatants and the whole-cell cultures for comparative determinations of cell-bound and extracellular mucin-degrading enzymes.
3. Calculate culture dry weights (*see* **Note 4**) by spinning down 1 mL of the culture in a microcentrifuge at 13,000*g* for 5 min. Discard the supernatant and add a further 1 mL, repeating the process until a total of 5 mL of culture have been collected. Finally, wash the bacterial pellets with distilled water. Place the microcentrifuge tubes containing the bacteria in a drying oven at 90°C for 3 d, or until dry. Determine the culture weights by weighing the sample and calculating the dry weight per milliliter of original culture.

4. Neuraminidase assay: Test solution (0.05 mL) and boiled controls are incubated with 0.1 mL *N*-acetylneuraminlactose for 1–2 h at 37·C. Stop the reaction by boiling at 100·C for 2 min. Add solution A (0.1 mL), mix, and allow to stand for 20 min at room temperature. Next add solution B (0.4 mL) and mix until the yellow colour disappears. Then add solution C (1.0 mL), and heat the mixture in a boiling water bath for 15 min, before cooling in cold water for 5 min. After centrifuging at 13,000*g* for 5 min to remove the precipitate, read the absorbance of the supernatant at 549 nm. Prepare a standard curve by using known amounts of NANA and developing these with the test after the boiling stage (*see* **Subheading 2., item 5**).
5. In glycosidase assays, incubate 0.5 mL of test solution at 37°C with 0.25 mL of substrate, until a yellow color begins to appear, or for 1 h. Terminate the reaction by adding 0.75 mL of stop solution and then centrifuge at 13,000*g* for 5 min before reading the absorbance at 420 nm (*see* **Subheading 2., item 6**).

3.3. In Vitro Modeling System Using Mucin Baits and Mucin Gel Cassettes

1. Use glass fermentation vessels (560 mL working volume) with modified lids, containing several extra sampling ports in these experiments (*see* **Fig. 1**).
2. Use fresh feces to prepare 20% (w/v) inocula in 100 m*M* anaerobic sodium phosphate buffer (pH 6.0), by macerating the stool in a stomacher for 5 min and then sequentially filtering through 500- and a 250-μm metal sieves to remove particulate material. Add 200 mL of this inocula to 200 mL of double-strength culture medium (*see* **step 4**) in the fermenter.
3. Constantly stir the fermentation vessels and set the dilution rates at 0.1/h (*see* **Note 5**) operating pH at 6.0, and temperatures at 37°C. Maintain anaerobic conditions by sparging cultures with O_2-free N_2 at a low gas flow rate (2.4 L/h).
4. A suitable culture medium (*see* **Note 6**) comprises: 2.0 g/L starch (soluble), 0.5 g/L pectin (citrus), 0.5 g/L inulin, 0.5 g/L xylan (oatspelt) 0.5 g/L arabinogalactan (larchwood), 0.5 g/L guar gum, 2.0 g/L mucin, 3.0 g/L Tryptone, 3.0 g/L Peptone water, 4.5 g/L Yeast extract, 0.015 g/L hemin, 4.5 g/L NaCl, 2.5 g/L KCl, 0.45 g/L $MgCl_2 \cdot 6H_2O$, 0.2 g/L $CaCl_2 \cdot 6H_2O$, 0.4 g/L KH_2PO_4, 0.8 g/L cysteine, 0.4 g/L Bile salts No.3, 20 mL Balch trace elements (*see* **Subheading 2., item 9**); 0.5 mL Tween-80.
5. When the chemostats reach steady state, at least 10 turnovers in culture volume, as indicated by analysis of short chain fatty acid (SCFA) profiles (*see* **Note 7**), extracellular and cell-associated samples of the lumenal populations, for comparative purposes, are taken for both chemical and enzymic measurements, and for viable counts of bacteria, using a range of selective and nonselective agars (*see* **Subheading 2., item 3**). The range of dilutions of samples for plating should be increased to 10^{-3}–10^{-8}, to take into account the greater numbers of bacteria in these samples.
6. Place sterile mucin gels in glass tubes (17 × 12 mm, 2 mL/vol), or mucin gel cassettes in the fermenters, in each of the five side sample ports. Prepare the baits by placing the glass tubes in a covered glass beaker, autoclave them, and when cool, pour over 2% (w/v) porcine gastric mucin with the addition of 0.2% (w/v) purified bacteriologic agar, after autoclaving and cooling. Then place the gels in an anaerobic cabinet to set. Make the gel cassettes by autoclaving a solution containing 0.8% (w/v) purified bacteriologic agar and 2% (w/v) mucin in distilled water, then aseptically coating sterile glass microscope coverslips, or custom made glass plates, with this solution at 60°C, before fitting them with sterile forceps to the cassette holder.
7. Remove gels periodically over 48 h for analysis. Wash the surfaces gently with 100 m*M* anaerobic sodium phosphate buffer, pH 6.0 to remove loosely adherent planktonic micro-

organisms. Resuspend the gel material in 10 mL anaerobic glycosidase buffer at pH 6.5 for enzymic analysis, and bacterial enumeration (methods as for biopsy samples). Use the mucin-coated glass coverslips directly for microscopic analysis. Take the samples and freeze at –20°C for carbohydrate analysis (see below).

8. Keep samples of the planktonic populations and fermenter media for measurements of mucin carbohydrate uptake. Rates are calculated as follows: $q_s = D\,(S_o – S)/x$, where D = dilution rate, S_o = substrate entering fermenter, S = residual substrate in fermenter, and x = community dry weight (q_s = substrate utilized/[min·mg dry weight bacteria]).
9. Mucin oligosaccharides are determined by hydrolysing samples in 2 *M* H_2SO_4 for 2 h at 100°C. A standard sugar mix containing 1 mg/mL of fucose, galactosamine, glucosamine, galactose, glucose and mannose is also hydrolyzed in 2 *M* of H_2SO_4. To 100 mL of hydrolysate, or standard sugar mix (in 2 *M* of H_2SO_4), add 5 mL of internal standard solution (0.02 mg/mL deoxygalactose in high purity water), mix, and then run on a Dionex DX 500/ED 40 analytical system.
10. Neutral and amino sugars are separated by high-pressure anion exchange chromatography with pulsed amperometric detection (HPAEC-PAD) on a Dionex CarboPac PA 10 (4 × 250 mm) column equipped with a Dionex PA 10 guard column (4 × 50 mm) and a Dionex ED 40 detector using the Dionex DX500 system (*see* **Note 8**). High-purity deionized water (18 MΩ cm) should be employed in these tests, after being filtered through 2-mm filters. Sodium hydroxide (50%, low in carbonate) is purchased from BDH, Poole, Dorest, UK. Solution 1 is 0.2 *M* NaOH, and solution 2 is distilled water. During preparation of these solutions the water is sparged with helium for 15 min before and during the addition of NaOH. Carry out monosaccharide detection using a gold cell and preset carbohydrate waveforms. Achieve isocratic separation of neutral and amino sugars at 1.0 mL/min with 30 m*M* NaOH. After 20 min, the column is purged with 100 m*M* NaOH for 10 min, then re-equilibrated with the starting conditions for 10 min before the next sample is injected. Use a PC 10 Pneumatic controller to introduce 0.3 *M* NaOH at a flow rate of 0.5 mL/min to the column effluent, before the PAD cell, which minimizes baseline drift and increases the analytical signal. Use a Dionex Eluant De-gas Module to saturate the eluants with helium gas to minimize CO_2 absorption. Transfer the samples to polyvials with 20.mm filters and inject with a Dionex AS40 Automatic sampler via a Dionex high pressure valve. Use a Dionex Peaknet Software data handling system to plot and integrate results.
11. Determine NANA by hydrolyzing samples in 0.05 *M* of H_2SO_4 for 1 h at 80°C. Then visualize released NANA colorimetrically as in the neuraminidase assay (*see* **Subheading 3.2., item 4**), after the boiling stage.

3.4. Short-Term Fermentation Studies on Biofilm and Lumenal Populations in Chemostats

1. Take culture from both culture vessels, together with material from mucin baits or gel cassetes. At this time, also remove the biofilms that form on the vessel walls. After washing and resuspension in 0.1 *M* sodium phosphate buffer, pH 6.0, the samples are centrifuged at 13,000*g* for 20 min. Resuspend each of the resulting pellets in 20 mL anaerobic 0.4 *M* phosphate buffer, pH 6.0. Add 10 mL of each suspension to 40 mL of chemostat medium in 70 mL Wheaton serum, bottles under N_2, at 37°C. Take samples hourly for 6 h (*see* **Note 9**), centrifuge at 13,000*g* for 10 min then freeze the supernantants at –20°C for subsequent measurement of SCFA and other organic acids. Also freeze samples for analysis of residual mucin. Make dry weight determinations on the samples as in **Subheading 3.2., item 3** for calculations of specific rates of substrate uptake and utilization.

3.5. Desorption of Mucinolytic Bacteria from Food Materials in Feces

1. Fresh fecal samples are homogenized in anaerobic 0.1 mol/L sodium phosphate buffer (pH 6.5) to give 10% (w/v) slurries. Pass fecal slurries sequentially through 500- and 250-mm diameter sieves. Retain filtrates containing nonadherent bacteria under anaerobic conditions for enzymic analysis, fermentation studies, and bacterial counts.
2. Material retained on the filters is washed twice with 500 mL of the anaerobic buffer to remove loosely adherent organisms. Washed food particles are subsequently incubated at 37°C under anaerobic conditions (O_2-free N_2 atmosphere) in phosphate buffer in the presence of a surfactant such as 0.001% (w/v) cetyltrimethylammonium bromide (CTAB) (BDH) for 30 min, with mixing. Samples are then refiltered to remove food materials. Retain filtrates containing adherent bacterial populations and residual food materials under anaerobic conditions (*see* **Note 10**).
3. Place samples of particulate material, washed particulate material and CTAB treated particles in 3% (v/v) glutaraldehyde in PIPES buffer (0.1 *M*, pH 7.4) at 4°C for SEM (*see* **Subheading 2., item 1**).
4. Perform enzymic analysis on bacteria extracted directly from feces and organisms removed from particulate materials with CTAB resuspended in 0.1 *M* sodium phosphate buffer (pH 6.5), as described in **step 2**.
5. Serially dilute unattached fecal bacteria and organisms desorbed from particulate material with 0.001% CTAB on a variety of selective and nonselective agars for enumeration (*see* **Subheading 2., item 3**).

3.6. Mucin Fermentation Experiments with Biofilm and Nonadherent Fecal Bacteria

1. Incubate biofilm and nonadherent faecal bacteria from fecal material at 37°C under O_2-free N_2 in 0.1 *M* sodium phosphate buffer (pH 6.5), in sealed 70-mL serum bottles (Wheaton) with mucin. Take samples (2 mL) periodically over a period of 6 h (*see* **Subheading 3.4.**) and freeze for analysis of fermentation products and residual mucin carbohydrate. Determine culture dry weights (*see* **Subheading 3.2., item 3**) to calculate specific rates of substrate utilization and fermentation product formation.

4. Notes

1. The benefits of using rectal biopsies to study mucosal bacterial populations are that, for most of the time, the rectum is empty and the mucosa is clean, and uncontaminated with lumenal material, and samples are relatively easy to obtain since the patients/volunteers do not need to be cleaned, or otherwise specially prepared, as would be required when removing tissue from the proximal or distal bowel during colonoscopy. Wilkins-Chalgren broth is sterilized by autoclaving (121°C, 15 min). The bottles are prereduced by being placed in an anaerobic chamber or gas jar (Don Whitley Scientific, Shipley, Yorks) with loose lids, and allowed to cool. This is also used to prepare anaerobic peptone water for the dilution series.
2. Weights and physical dimensions of the biopsy samples are needed to estimate bacterial cell densities, either as per unit area or as per unit tissue weight. Rapid handling of samples is essential to prevent growth of facultative anaerobes, and inactivation of strict anaerobes during transport.
3. Bacterial CFAs are highly stable and reproducible taxonomic markers. This allows phenotypic analysis of pure and mixed populations of intestinal microorganisms to be undertaken by extracting their CFAs and comparing patterns of the methyl esters by GC, using

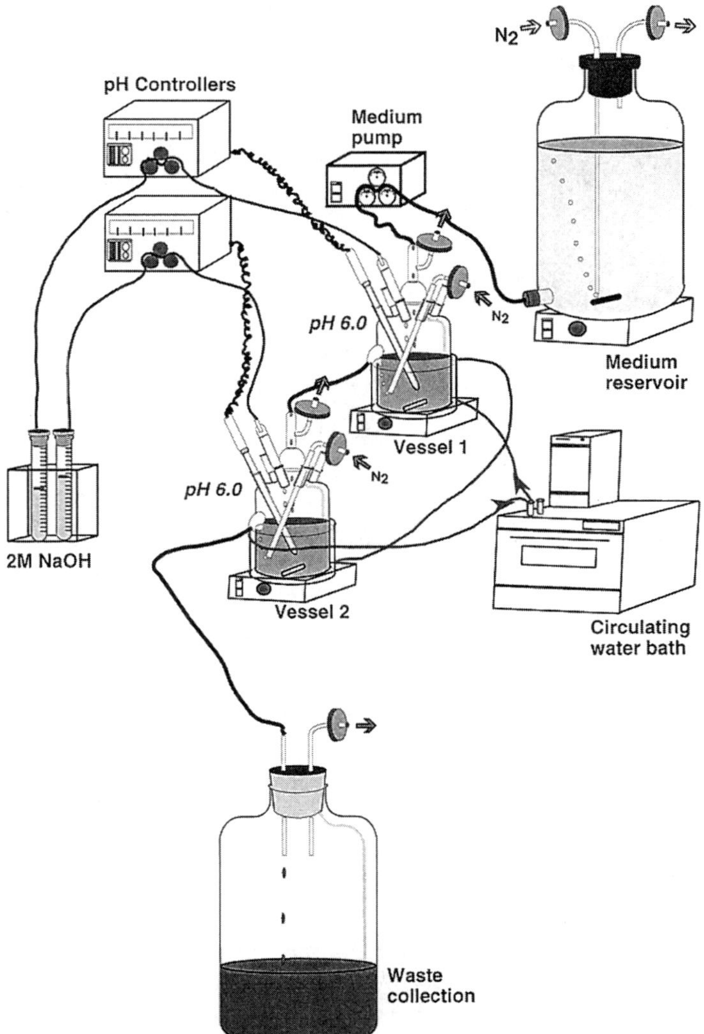

Fig. 3. Schematic diagram showing a simple single-stage continuous culture system.

the MIDI system. This highly automated procedure is controlled by a microprocessor, and contains several computer-generated libraries comprising data on over several hundred different aerobic and anaerobic species.

4. Culture dry weights or, alternatively, the protein content of the sample (e.g., Lowry method) are used for calculating specific enzyme activities.
5. The basic experimental setup of a semicontinuous or continuous culture type fermentation system, fitted with automatic pH control is shown in **Fig. 3**. The fermentation vessel can be operated as either a batch or a continuous culture. The reactor is stirred magneti-

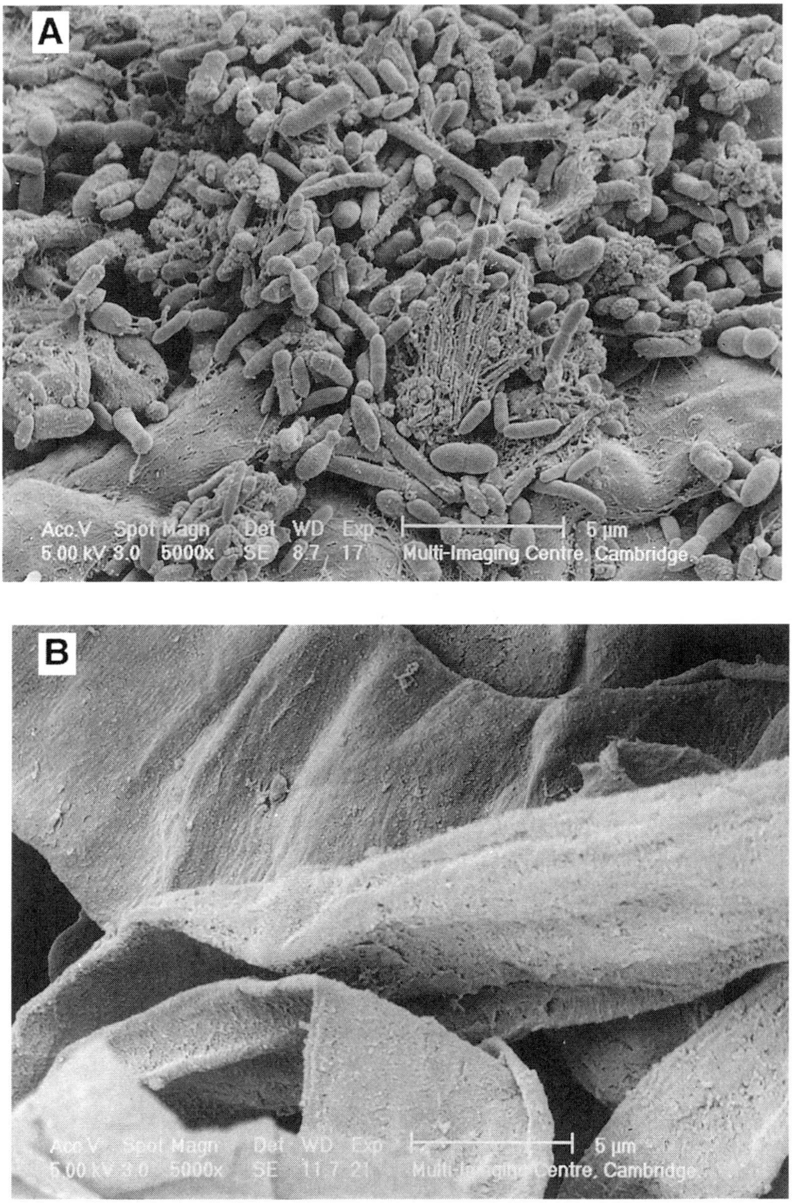

Fig. 4. Scanning electron micrographs of **(A)** bacterial biofilm colonising the surface of digestive materials in human cecal contents, and **(B)** particulate material in which the bacteria have been removed by surfactant treatment.

cally, and in the continuous culture configuration, where sterile growth medium is continuously supplied to the fermenter, spent culture medium leaves the fermentation vessel by gravity, via an overflow syphon. More sophisticated systems stir the cultures using a direct-drive impeller, with spent culture medium and bacterial cells being removed from the reactor using a dedicated pump. Dilution rate (*D*) is calculated as the working volume of the fermentation vessel divided by the flow rate of growth medium into the system. Specific growth rates (μ) of bacteria in a chemostat are equal to *D*, under steady-state conditions. The mean residence time of organisms in the fermenter can be calculated as the reciprocal of *D*, and mean doubling times of the cells (T_d) are derived from *In2/D*.

6. The growth medium contains a complex mixture of proteins, carbohydrates and mucin, to maximize species diversity in the ecosystems. When the chemostats reach steady-state, sterile mucin gels in baits, or glass coverslips, are fitted to the fermenters and removed periodically for analysis. Partially purified porcine gastric mucin is used in these experiments because of its compositional and structural similarity to human mucins.
7. Bacteriologic analysis is time-consuming and labor-intensive, especially when working with strict anaerobes. Daily measurements of SCFA (acetate, propionate, butyrate) in the cultures provides a rapid and simple test of metabolic steady state.
8. Sulfate ions in the hydrolysate decrease retention times and result in coelution of peaks. Incorporation of a AG5 guard column, with column switching after 2.4 min prevents sulfate ions from reaching the analytical column.
9. Incubations should be kept as short as possible, to prevent major changes occurring in the species composition of the samples.
10. **Figure 4** shows the effectiveness of CTAB in removing strongly adherent bacteria from digestive materials. Do not use higher concentrations of the surfactant, since it is inhibitory to many intestinal bacteria.

References

1. Stephen, A. M., Haddad, A. C., and Phillips, S. F. (1983) Passage of carbohydrate into the colon. Direct measurements of humans. *Gastroenterology* **85**, 589–595.
2. Macfarlane, G. T., Macfarlane, S., and Sharp, R. (1997) Differential expression of virulence determinants in *Clostridium septicum* in relation to growth on mucin and the swarm cell cycle. *Biosci. Micro.* **16**, 28.
3. Macfarlane, G. T. and Gibson, G. R. (1991) Formation of glycoprotein degrading enzymes by *Bacteroides fragilis*. *FEMS Microbiol. Lett.* **77**, 289–294.
4. Macfarlane, G. T., Hay, S., and Gibson, G. R. (1989) Influence of mucin on glycosidase, protease and arylamidase activities of human gut bacteria grown in a 3-stage continuous culture system. *J. Appl. Bacteriol.* **66**, 407–417.
5. Macfarlane, G. T., Hay, S., Macfarlane, S., and Gibson, G. R. (1990) Effect of different carbohydrates on growth, polysaccharidase and glycosidase production by *Bacteroides ovatus*, in batch and continuous culture. *J. Appl. Bacteriol.* **68**, 179–187.
6. Pettipher, G. L. and Latham, M. (1979) Production of enzymes degrading plant cell walls and fermentation of cellobiose by *Ruminococcus flavifaciens*. *J. Gen. Microbiol.* **110**, 29–38.
7. Berg, J. O. (1981) Cellular location of glycoside hydrolases in *Bacteroides fragilis*. *Curr. Microbiol.* **5**, 13–17.
8. Degnan, B. A. (1993) *Transport and Metabolism of Carbohydrate by Anaerobic Gut Bacteria*. Ph.D Thesis, Cambridge University, Cambridge.
9. Englyst, H. N., Hay, S., and Macfarlane, G. T. (1987) Polysaccharide breakdown by mixed populations of human faecal bacteria. *FEMS Microbiol. Ecol.* **95**, 163–171.
10. Macfarlane, S., McBain, A. J., and Macfarlane, G. T. (1997) Consequences of biofilm and sessile growth in the large intestine. *Adv. Dent. Res.* **11**, 59–68.

11. Mozes, N. and Rouxhet, P. G. (1992) Influence of surfaces on microbial activity, in *Biofilms-Science and Technology* (Melo, L. F., Bott, T. R. and Capdeville, B., eds.), Kluwer Academic Publishers, Doorderecht, Germany, pp. 125–136.
12. Van Loosdrecht, M. C. M., Lyklema, J., Norde, W., and Zehnder, A. J. B. (1990) Influence of interfaces on microbial activity. *Microbiol. Rev.* **54**, 75–87.
13. Conrad, R., Phelps, T. J., and Zeikus, J. G. (1985) Gas metabolism evidence in support of the juxtaposition of hydrogen-producing and methanogenic bacteria in sewage sludge and lake sediments. *Appl. Environ. Microbiol.* **50**, 595–601.
14. Barry, J-L., Hoebler, C., Macfarlane, G. T., Macfarlane, S., Mathers, J. C., Reed, K. A., Mortensen, P. B., Norgaard, I., Rowland, I. R., and Rumney, C. J. (1995) Estimation of the fermentability of dietary fibre in vitro: A European interlaboratory study. *Brit. J. Nutr.* **74**, 303–332.
15. Balch, W. E., Fox, G. E., Magrum, L. J., Woese, C. R., and Wolfe, R. S. (1979) Methanogens: reevaluation of a unique biological group. *Microbiol. Rev.* **43**, 260–296.
16. Macfarlane, G. T., Cummings, J. H., and Allison, C. (1986) Protein degradation by human intestinal bacteria. *J. Gen. Microbiol.* **132**, 1647–1656.

XII

Cellular and Humoral Responses to Mucins

37

Generation of MUC1 Cytotoxic T-Cells in Mice and Epitope Mapping

Vasso Apostolopoulos, Ian F. C. McKenzie, and Geoffrey A. Pietersz

1. Introduction

A successful vaccine for cancer immunotherapy, particularly for solid tumors, would require a suitable target antigen and the production of a cytotoxic T-cell response *(1)*. In the mid- to late-1980s, there was a focus on monoclonal antibodies (MAbs) for the treatment of common cancers, such as those of the colon, breast, and lung. However, with the difficulties of using such agents, there is now a clear focus on cellular immunity for several reasons. First, using genetic engineering techniques, peptide epitopes have been identified and can be produced in large amounts, particularly as synthetic peptides and as recombinant molecules. Second, the description of many cytokines, combined with the knowledge of antigen processing and presentation by class I and class II pathways, has led to a degree of sophistication and knowledge in how to immunize to produce the desired response. These developmets are proving useful in the generation of new and improved vaccines and the future holds much promise for the production of effective vaccines to prevent, control, and possibly eradicate many diseases, including cancer. It is now theoretically possible to induce either antibodies or CTLs to defined polypeptides. However, it remains to be determined which will be the most effective.

As it will be apparent, MUC1 peptides bind to Class I molecules with variable affinity, some peptides containing appropriate anchors that bind with high affinity whereas others contain no known anchors but still bind and induce high-avidity CTLs *(2,3)*. Furthermore, MUC1 is an interesting molecule and there is good evidence, to be presented herein, that it is presented by Class I molecules in an unusual manner so that the peptides are exposed to anti-MUC1 antibodies—being the only peptide to be described thus far that has this property *(3a)*.

In breast and other cancers in which mucin is expressed, MUC1 appears to be a useful target for immunotherapy. As has been described extensively, there is an

From: *Methods in Molecular Biology, Vol. 125: Glycoprotein Methods and Protocols: The Mucins*
Edited by: A. Corfield © Humana Press Inc., Totowa, NJ

upregulation (~100-fold) in MUC1 production in cancer. Such material can clearly be demonstrated by antibody on the surface of cells, but of greater relevance for this chapter is that large amounts of material produced intracellularly could lead to presentation of peptides by the class I pathway, as has been described for the endogenous pathway of antigen presentation. Thus, MUC1 could be a suitable target for CTLs.

The first descriptions of cellular immunity to MUC1 were unusual, in that non-major histocompatibility complex (MHC) restricted cytotoxic cells were described ***(4,5)***. These studies were important because they showed that patients with cancer had CTL precursors in their lymph nodes, and it was an enticing suggestion to consider that stimulation of patients by peptides could activate these cells for successful immunotherapy. These findings were provocative and stimulating and led us and others to commence trials with immunotherapy with MUC1 peptides. Our initial observations using MUC1 peptides from the variable numbers of tandem repeats (VNTR) (*see* Chapter 30) led to the production of antibodies in mice and to little cellular immunity ***(6)***. Nonetheless, a clinical trial resulted, which was the first to use MUC1 peptides in a phase I study, and demonstrated that immune responses were weak or nonexistent; there were no tumor responses and small MUC1 peptides used alone were not considered to be satisfactory ***(7)***. However, other studies have been used with synthetic MUC1 peptides in patients in which DTH responses were noted but CTL responses were poor. We then revised our immunisation procedure to find a strategy where MUC1 was coupled to mannan to target the mannose receptor (under oxidizing conditions), leading to the satisfactory production of CTLs to MUC1. The strategy was based on a firm foundation: we had the knowledge that mice that had rejected tumors expressing human MUC1 could produce CTLs. The aim of the study was therefore to use synthetic MUC1 peptides or fusion proteins to induce CTLs. Our mode of immunization with mannan was the first to describe the production of CTLs in any strain of mice to human MUC1. Subsequently, other modes of immunization, particularly with DNA or delivered by vaccinia, have also led to the production of CTLs. Here, we describe the induction of CTLs in mice to human MUC1, the definition of epitopes detected, and how the studies were extended to use HLA-A*0201 transgenic mice in which further epitopes were also defined.

2. Materials

1. Chemicals: Mannan, ethylene glycol, glutathione (reduced form), glutathione-agarose, isopropyl-β-D-thiogalactopyranoside (IPTG), phytohemagglutinin (PHA), and ethanediol were from Sigma, St. Louis, MO). Sodium periodate, Triton X-100, and glutaraldehyde were from BDH, Poole, Dorset, UK. PD-10 columns were from Pharmacia, Uppsala, Sweden. ^{51}Cr-sodium chromate was from Amersham, Amersham, CA. Ampicillin was were from Boehringer Mannheim, CA. DME, RPMI, penicillin, streptomycin, and fetal calf serum (FCS) were from Commonwealth, Melbourne, Australia. Bacto-tryptone and bacto-yeast were from Difco, Detroit, MI.
2. Culture media:
   a. Growth medium: DME or RPMI supplemented with 10% FCS, 2m*M* glutamine, 100 IU/mL penicillin, 100 μg/mL streptomycin, 0.05 m*M* 2-mercaptoethanol.
   b. Bacterial growth medium: Luria broth (LB) was prepared by mixing bacto-tryptone (10 g), bacto–yeast extract (5 g), and sodium chloride (10 g) in 950 mL of deionized

water and adjusting the pH to 7.0 with 5 *M* sodium hydroxide. The volume was adjusted to 1 L and sterilized by autoclaving for 20 min at 15 lb/in.^{2} on liquid cycle.

3. Cell lines: Tumor target cells used were: the mastocytoma cell line P815 (DBA/2 strain origin, H-2^d); MUC1 transfected P815 cells (containing the cDNA of the membrane-anchored form of MUC1) ***(8)***, RMA (C57BL/6 strain origin, H-2^b thymoma) cells, and the C57BL/6 TAP-deficient cell line RMA-S pulsed with peptide. L-cells transfected with K^b, D^b, D^d, or L^d with or without peptide pulsing can also be used as target cells. Human Epstein-Barr virus (EBV)-transformed B-cells, human breast cancer cell lines (MCF-7 [HLA-A*0201$^+$, MUC1$^+$], BT-20 [HLA-A*0201$^-$, MUC1$^+$]), or melanoma cell line (MF272 [HLA-A*0201$^+$, MUC1$^-$]). All cell lines and lymphoblasts are grown at 37°C, 10% CO_2 in RPMI, supplemented with 10% FCS, 2 m*M* glutamine, 100 IU/mL of penicillin, and 100 µg/mL of streptomycin.
4. Buffers and solutions:
   a. Phosphate buffered saline (PBS): 2.85 g of disodium hydrogen phosphate dihydrate, 0.624 g of sodium dihydrogen phosphate dihydrate, and 8.766 g of sodium chloride were dissolved in distilled water and the pH adjusted to 7.2 and volume to 1 L.
   b. Bicarbonate buffer: 20 mL of a 0.2 *M* solution of sodium carbonate and 230 mL of sodium bicarbonate were mixed and the volume adjusted to 1 L.
   c. Phosphate buffer: 0.1 *M* sodium dihydrogen phosphate titrated with concentrated sodium hydroxide to pH 6.0.
5. Synthetic peptides: Peptides containing two VNTRs (Cp13-32) were synthesized using an Applied Biosystems Model 430A automated peptide synthesizer. Overlapping 9-mer peptides spanning the MUC1 VNTR with single amino acid sequence changes were synthesized by Chiron Mimotopes, Victoria, Australia.
6. Inbred mice: Balb/c, C57BL/6, CBA; recombinant mice: H-$2K^kD^b$B10.A(2R), H-$2K^bD^d$, B10.A(5R). HLA-A*0201/K^b ***(9)*** mice were obtained from the Scripps Clinic and Research Foundation, California. All mice were 8 to 10 wk old.

3. Methods

3.1. Production of Soluble MUC1 Fusion Protein

1. Grow *Escherichia coli* transformed with pGEX-3X plasmids containing coding region of five repeats of the MUC1 VNTR ***(10–12)*** overnight in LB containing ampicillin (100 µg/mL) .
2. Dilute the culture 1:25 with fresh (LB) medium and grow bacteria for a further 1 h at 37°C.
3. Add 0.1 m*M* IPTG to induce the production of recombinant protein and incubate a further 3 h at 37°C.
4. Centrifuge at 2,500*g* for 15 min at 4°C and resuspend the pellet in 1:10 culture volume of PBS. Perform all subsequent steps at 4°C (*see* **Note 1**).
5. Lyse cells by sonication (3 × 30 s) and add 1% Triton X-100.
6. Centrifuge cells at 10,000*g* for 15 min at 4°C.
7. Mix the supernatant containing the soluble fusion protein with a 50% solution of glutathione-agarose beads (supernatant:agarose beads, 5:1) on a rotating platform.
8. Collect the beads by centrifugation at 500*g* for 5 min, and wash three times with PBS.
9. Elute the fusion protein with free glutathione using three 5-min washes with 1.5 bead volume of 50 m*M* Tris-HCl (pH 8.0) containing 5 m*M* reduced glutathione (*see* **Note 2**).
10. Dialyze the supernatants into PBS.
11. Measure the optical density at 280 nm, and calculate the concentration using the following formula:concentration (mg/mL) = (OD_{280} × 10)/14.3.
12. Store at –20°C in aliquots.

3.2. Conjugation of MUC1 Fusion Protein or Peptides to Mannan

1. Disssolve 14 mg of mannan in 1 mL of 0.1 *M* phosphate buffer and leave on ice *(13)*.
2. Make a 0.1 *M* solution of sodium periodate and add 100 µL to the mannan and leave on ice for 1 h.
3. Stop the oxidation by adding 10 mL of ethylene glycol and leave on ice for a further 30 min.
4. Pass the reaction mixture through a PD-10 column (void volume = 2.5 mL) and collect 2 mL of oxidized mannan fraction.
5. Add 0.7 mg of MUC1 FP to the oxidized mannan and react overnight at room temperature to form the conjugate (MFP) (*see* **Note 3**).
6. Store the MFP in aliquots at -20°C.

3.3. Mice and Immunization Schedule (14)

1. Immunize mice intraperitoneally with 5 µg of MFP (5 µg = to the amount of FP) or the same amount of peptides (linked to keyhole limpet hemocyanin [KLH]) weekly for 3 wk (*see* **Note 4**).

3.4. Preparation of Target Cells

1. Prepare lymphoblast target cells by placing 2×10^6 spleen cells in wells of a 24-well plate with 1 µg/mL of PHA-L, and incubate for 48 h at 37°C in 10% CO_2 to form blasts cells.
2. Incubate blast cells overnight with 20 µ*M* peptide (either 20 mer or short 9 mers for epitope mapping).
3. Use the tumor target cells, P815, P815 transfected with the MUC1 cDNA, RMA, single H-2 alleles expressed after L-cell transfection, and TAP-deficient RMA-S cells (used as CTL targets and in stabilization assays). For HLA-A*0201 mice, use the following target cells: PHA blasts from autologous mice, human EBV transformed B cells, PHA pulsed human peripheral blood mononuclear cells (PBMC), human breast cancer cell lines, (MCF-7 [HLA-A*0201$^+$, MUC1$^+$]), BT-20 (HLA-A*0201$^-$, MUC1$^+$) or melanoma cell line (MF272 [HLA-A*0201$^+$, MUC1$^-$]).
4. Radiolabel 10^6 peptide-pulsed blast cells or 10^6 tumor target cells with 100 µCi of $Na_2{}^{51}CrO_4$ (^{51}Cr) for 60 min at 37°C.

3.5. Cytotoxic T-Lymphocyte Assay

1. Immunize mice with MFP.
2. Sacrifice mice 7–10 d after the final injection, collect their spleen cells, remove red cells, and wash with 2% FCS/PBS.
3. Radiolabel 10^6 target cells with 100 µCi of $Na_2{}^{51}CrO4$ (Amersham) for 60 min at 37°C.
4. Resuspend spleen cells and target cells, in culture medium, and then combine in various effector-to-target ratios (100:1 to 5:1) in triplicate, in 96-well U-bottomed plates.
5. To ascertain spontaneous release (medium alone) and maximum release (after treatment with 10% sodium dodecyl sulfate [SDS]) set up six wells with labeled target cells alone.
6. Centrifuge the plates at 100*g* for 3 min to initiate cell contact, and incubate for 4 h at 37°C in 10% CO_2.
7. After incubation, centrifuge again, collect supernatants, and quantitate radioactivity in a gamma counter.
8. Determine specific ^{51}Cr release as follows: [(experimental – spontaneous)/(maximum – spontaneous)] × 100%.

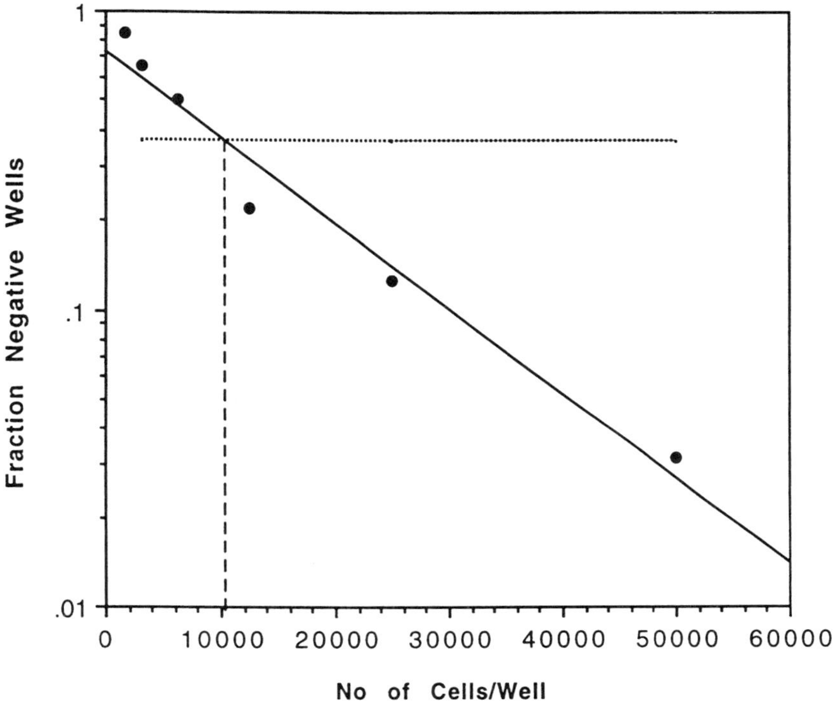

Fig. 1. Typical example of a graph depicting the CTLp frequency (10,000) for MFP immunized mice. (············), 0.37% negative wells; (———), CTLp frequency.

3.6. Cytotoxic T-Lymphocyte Precursor (CTLp) Assay

1. Ten to 14 d after the last immunization, sacrifice mice and collect effector spleen cells.
2. Set up 32 replicates for at least six effector cell numbers (1×10^3–1.28×10^5) in U-bottomed microtiter trays together with 5×10^5 mitomycin C-treated spleen cells (or spleen cells irradiated with 50,000 rad) in RPMI supplemented with 10% FCS, antibiotics, 5 μM VNTR peptide, and 10 U/mL of rhIL-2.
3. Set up control wells containing stimulator cells alone, with peptide, and with IL-2 only.
4. Seven days later replace 100 µL of supernatant with 100 µL of target cell suspension containing 10^4 ^{51}Cr-labeled targets.
5. After 4 h quantitate the radioactivity in a gamma counter.
6. Calculate the mean ± standard deviation (SD) of specific ^{51}Cr release from the controls.
7. Calculate the fraction of wells with cytolytic activity by counting the number of wells from 32 of each effector cell number with radioactivity greater than mean + 3 SDs.
8. Graph the fraction of negative wells on a logarithmic scale (*y*-axis) vs the effector cell number (*x*-axis).
9. Determine the CTLp frequency from the graph as the responder cell dose required to generate 37% negative wells (*see* **Note 5** and **Fig. 1**).

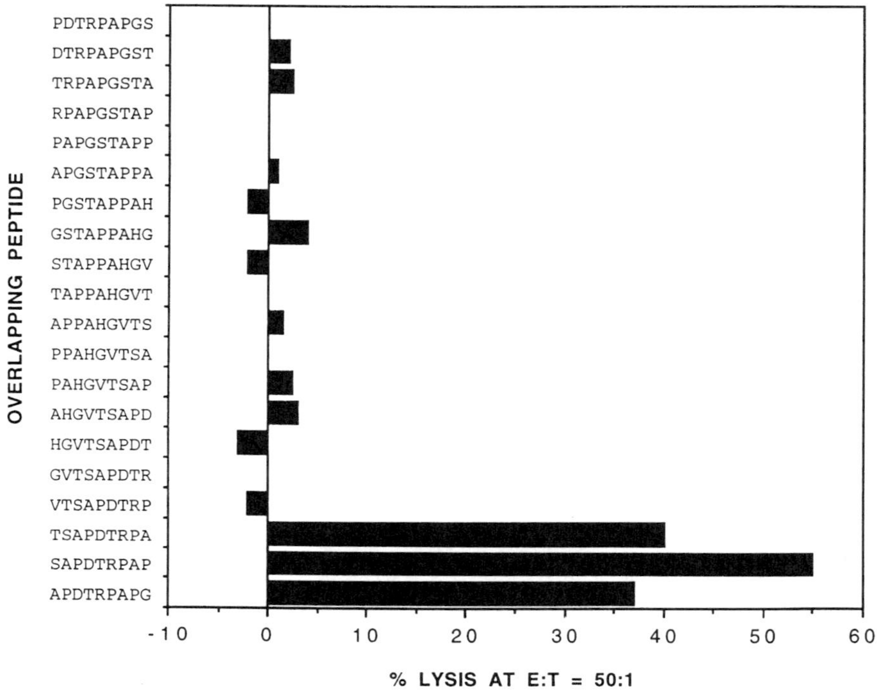

Fig. 2. Typical example of a graph showing the mapping of the K^b epitope (SAPDTRPAP) presented by MUC1. CTL assay using C57BL/6 ($H2^b$) effectors and B10.A(5R) (K^bD^d) PHA blasts.

3.7. Epitope Mapping (3,15,16) *(see Note 6)*

1. For generation of CTL, see Subheading 3.3.1.
2. For generation of targets, **Subheading 3.4.**
3. Peptide pulsing: Incubate target cells overnight with 20 μ*M* peptide and label with ^{51}Cr as described under **Subheading 3.4.** (*see* **Note 7**).
4. Set up CTL assays as under **Subheading 3.5., item 4** at a constant effector-to-target ratio.
5. Graph the 9-mer sequence (*y*-axis) and specific lysis (*x*-axis) as a horizontal bar graph to determine the epitope (**Fig. 2**)
6. Examples of epitope mapping: Effector spleen cells from C57BL/6 ($H2K^bD^b$) mice immunized with M-FP are mixed with B10.A(5R) ($H2K^bD^d$) peptide pulsed lymphoblast target cells so that only the reaction with $H2K^b$ is determined. Similarly, $H2D^b$ is mapped using recombinant mice, with L-cells transfected with D^b cDNA, and specificity is confirmed in RMA-S cells. H2d is mapped similarly. For HLA-A*0201 epitope mapping, EBV-immortalized B-cells (HLA-A*0201, HLA-A1, HLA-A11, and so forth) and PHA lymphoblast target cells derived from HLA-A*0201/K^b mouse spleen cells or from human PBMC are used.

4. Notes

1. The pellet may be stored at –70°C for a couple of months before resuspending in PBS.
2. As an alternative to batch processing, a chromatography column packed with GST-agarose beads may be used. The supernatant can be recirculated through the column, washed

and eluted with GSH using a fast-performance liquid chromatography (FPLC) or high-performance liquid chromatography (HPLC).

3. Peptides were linked to mannan via KLH. Peptides were linked to KLH using glutaraldehyde. Briefly, dissolve the peptide (2 mg) in 1.75 mL of PBS and mix with 0.25 mL of KLH (2 mg/mL). Add 0.25% glutaraldehyde (1 mL) and mix overnight at room temperature in the dark. Dialyze the mixture overnight into PBS with several changes. Add the dialyzed mixture to 1 mL of oxidized mannan.
4. CTL can readily be generated in any strain of mice and measured directly. We have immunized 15 strains and without fail obtain 60% lysis at an effector-to-target ratio of 20:1. Three immunizations are optimal—more or less gives less lysis. As an alternative to three immunizations by the IP route, one injection of in vitro sensitized macrophages/dendritic cells will suffice. For this, adherent cells are exposed to MFP overnight, washed, and then injected. The CTLp frequency (*see* **Note 5**) is ~1/10,000—the same as obtained with three IP injections.
5. The CTLp frequency obtained depends on the number of immunizations performed: three injections gives ~1/10,000 or greater, one injection gives ~1/80,000, and two injections gives ~1/30,000. It appears that a frequency of >1/20,000 is required to cause tumor rejection.
6. Mucin 1 peptides are processed by different cells and act as antigen-presenting cells; we have used adherent cells, PHA blasts, or EBV-infected cells. Peptides 30–105 mer are processed and presented as 8 to 9 mers. Alternatively, small peptides (8 to 9 mers) can be surface loaded on to cells.
7. Epitopes are mapped by preparing direct CTL on peptide-loaded/pulsed target cells and determining which peptides lead to lysis.

References

1. Singer, A. (1994) News: Time and truth for cancer vaccines - a new generation. *J. Natl. Cancer Inst.* **86,** 330–330.
2. Dahl, A. M., Beverley, P. C., and Stauss, H. J. (1996) A synthetic peptide derived from the tumor-associated protein mdm2 can stimulate autoreactive, high avidity cytotoxic T lymphocytes that recognize naturally processed protein. *J. Immunol.* **157,** 239–246.
3. Apostolopoulos, V., Haurum, J. S., and McKenzie, I. F. C. (1997) MUC1 peptide epitopes associated with 5 different H2 Class I molecules. *Eur. J. Immunol.* **27,** 2579–2587.

3a. Apostolopoulos, V., Chelvanayagam, G., Xing, P. X., and McKenzie, I. F. L. (1998) Anti-MUC1 antibodies react directly with MUC1 peptides presented by Class1 H2 and HLA molecules. *J. Immunol.* **161,** 767–775.

4. Barnd, D. L., Lan, M. S., Metzgar, R. S., and Finn, O. J. (1989) Specific major histocompatibility complex-unrestricted recognition of tumor associated mucins by human cytotoxic T cells. *Proc. Natl. Acad. Sci. USA* **86,** 7159–7163.
5. Ioannides, C. G., Fisk, B., Jerome, K. R., Irimura, T., Wharton, J. T., and Finn, O. J. (1993) Cytotoxic T cells from ovarian malignant tumors can recognize polymorphic epithelial mucin core peptides. *J. Immunol.* **151,** 3693–3703.
6. Apostolopoulos, V., Xing, P. X., and McKenzie, I. F. C. (1994) Murine immune response to cells transfected with human MUC1: immunisation with cellular and synthetic antigens. *Cancer Res.* **54,** 5186–5193.
7. Xing, P. X., Apostolopoulos, V., Michaels, M., Prenzoska, J., Bishop, J., and McKenzie, I. F. C. (1995) Phase I study of synthetic MUC1 peptides in cancer. *Int. J. Oncology* **6,** 1283–1289.

8. Acres, B., Hareuveni, M., Balloul, J. M., and Kieny, M. P. (1993) VV-MUC1 immunisation of mice-immune response and protection against the growth of murine tumours bearing the MUC1 antigen. *J. Immunother.* **14,** 136–143.
9. Vitiello, A., Marchesini, D., Furze, J., Sherman, L. A., and Chestnut, R. W. (1991) Analysis of the HLA-restricted influenza-specific CTL response in transgenic mice carrying a chimeric human-mouse class I MHC molecule. *J. Exp. Med.* **173,** 1007.
10. Siddiqui, J., Abe, M., Hayes, D., Shani, E., Yunis, E., and Kufe, D. (1988) Isolation and sequencing of a cDNA coding for the human DF3 breast carcinoma-associated antigen. *Proc. Natl. Acad. Sci.* **85,** 2320–2323.
11. Smith, D. B. and Johnson, K. S. (1988) Single-step purification of polypeptides expressed in E. coli as fusions with glutathione S-transferase. *Gene* **67,** 31–40.
12. Apostolopoulos, V., Xing, P. X., Trapani, J. A., and McKenzie, I. F. C. (1993) Production of anti-breast cancer monoclonal antibodies using a glutathione-S-transferase-MUC1 bacterial fusion protein. *Br. J. Cancer* **67,** 713–720.
13. Apostolopoulos, V., Pietersz, G. A., and McKenzie, I. F. C. (1996) Cell-mediated immune responses to MUC1 fusion protein coupled to mannan. *Vaccine* **14,** 930–938.
14. Pietersz, G. A., Wenjun, L., Popovski, V., Caruana, J.-A., Apostolopoulos, V., and McKenzie, I. F. C. (1998) Parameters in using mannan-fusion protein (MFP) to induce cellular immunity. *Cancer Immunol. Immunother.,* **45,** 321–326.
15. Apostolopoulos, V., Loveland, B. E., Pietersz, G. A., and McKenzie, I. F. C. (1995) CTL in mice immunised with human Mucin 1 are MHC-restricted. *J. Immunol.* **155,** 5089–5094.
16. Apostolopoulos, V., Karanikas, V., Haurum, J., and McKenzie, I. F. C. (1997) Induction of HLA-A2 restricted cytotoxic T lymphocytes to the MUC1 human breast cancer antigen. *J. Immunol.* **159,** 5211–5218.

38

Analysis of the Frequency of MHC-Unrestricted MUC1-Specific Cytotoxic T-Cells in Peripheral Blood by Limiting Dilution Assay

John R. McKolanis and Olivera J. Finn

1. Introduction

Mucins are highly glycosylated proteins present on the lumenal side of ductal epithelial cells. MUC1 is the only mucin with a transmembrane region anchoring it to the cell surface. The extracellular domain of MUC1 is composed of numerous tandem repeats of a 20-amino acid sequence *(1)*. Normal cells produce a highly glycosylated and sialylated form of MUC1. The *O*-linked carbohydrate side chains of MUC1 on tumor cells are shorter and less abundant, exposing previously unrecognized antigenic sites on the polypeptide core of the molecule *(2)*. This underglycosylated MUC1 molecule on tumor cells can be recognized as a tumor-specific antigen by T-cells. The major cytotoxic T-lymphocyte (CTL) response to tumor-specific MUC1 is T-cell receptor (TCR) mediated but major histocompatibility complex (MHC)-unrestricted. Owing to the high density of repeating antigenic epitopes extending along each MUC1 molecule, it is postulated that a large number of TCRs can be triggered simultaneously to activate the CTLs to kill tumor cells or proliferate *(3–5)*. Antibodies specific for the TCR, or for the defined MUC1 epitope recognized by the TCR, inhibit CTL recognition. Antibodies against MHC molecules have no effect.

In addition to the MHC-unrestricted CTLs, some MUC1-specific MHC-restricted CTLs have also been identified. Patients who are HLA-A11 and HLA-A3 have CTLs that recognize a nine amino acid peptide from the tandem repeat region bound to these alleles *(6)*. Another peptide has been reported that binds HLA-A2.1 allele *(7)*. Although it is not yet clear whether these CTLs, MHC unrestricted or MHC restricted, have a therapeutic function, it has been considered important to measure their frequency pre- and post vaccination with MUC1-based immunogens used in various clinical trials *(7–9)* in order to be able to correlate this specific component of the antitumor response with the outcome of the trial.

From: *Methods in Molecular Biology, Vol. 125: Glycoprotein Methods and Protocols: The Mucins*
Edited by: A. Corfield © Humana Press Inc., Totowa, NJ

We have established a limiting dilution assay (LDA) to evaluate the frequency of MHC-unrestricted MUC1-specific CTL. This assay is convenient because the same conditions and reagents can be applied to all patients regardless of their HLA type. We believe that these CTLs will be an important corollary of a tumor-specific immune response because they recognize an epitope that is expressed only on tumor cells and not on normal epithelial cells. The same assay may be used to determine the frequency of MHC-restricted CTL. However, the assay must be custom tailored to each patient, or at least to several specific HLA alleles.

2. Materials

1. RPMI-1640 tissue culture medium supplemented with 10% fetal calf serum (FCS) or human AB serum, 2 m*M* L-glutamine, 100 Units/mL penicillin, 100 µg/mL streptomycin.
2. Human male AB serum purchased from Gemini Bio Products, Calabasas, CA.
3. FCS purchased from Life Technologies, Grand Island, NY.
4. Glutamine, purchased as 100X stock from Life Technologies, Grand Island, NY.
5. Penicillin-streptomycin, purchased as 100x stock from Life Technologies.
6. Hepes buffer, purchased as 100X stock from Life Technologies.
7. Recombinant interleukin-2 (rIL-2), obtained from several sources.
8. Phytohemagglutinin (PHA-P), purchased from Sigma, St. Louis, MO.
9. Trypan blue, 0.4% stock solution, purchased from Life Technologies.
10. Phosphate buffered physiologic saline (1.9 mM NaH_2PO_4, 8.1 m*M* $NaHPO_4$, 154 m*M* NaCl, pH 7.2); may also be purchased as a 10X stock solution.
11. ^{51}Sodium chromate and ^{3}H thymidine, purchased from Amersham, Arlington Heights, IL.
12. Ficoll density gradient, purchased as LSM from Organon Technica Corp, Durham, NC.
13. DNAase, purchased from Calbiochem, La Jolla, CA. Filter-sterilized stock solution at 10 mg/mL in RPMI-1640 containing HEPES buffer should be kept frozen at –80°C.
14. Dimethyl sulfoxide (DMSO) purchased from Sigma.
15. Tumor cells that express MUC1 can be obtained from American Type Culture Collection (ATCC), Rockville, MD, or as gifts from various investigators. Epstein-Barr virus (EBV) immortalized B-lymphoblastoid cell lines or other cells transfected with the MUC1 gene can be obtained from various investigators. Natural killer and lymphokine-activated killer (LAK) cell target K562 can be purchased from ATCC.
16. Styrofoam boxes for slow freezing of cells, purchased from Sarstadt, Newton, NC.
17. 96-well, U-bottomed microtiter plates, purchased from Becton Dickinson, Franklin Lakes, NJ.
18. Cytotoxicity assay harvesting materials, purchased from Scatron Instruments, Sterling, VA.

3. Methods

3.1. Isolation of Mononuclear Cells from Peripheral Blood

Blood samples are obtained from patients at specific time intervals, depending on the goals of the study. Peripheral blood mononuclear cells are separated from the rest of the blood using cenrifugation through Ficoll density gradient (LSM). This involves layering 10 mL of blood diluted 1:2 in RPMI medium without serum, over 3 mL of LSM in a 15-mL conical centrifuge tube. Following centrifugation at 400*g* for 20 min, mononuclear cells can be removed from the interface between the two layers. The cells are washed three times in phosphate-buffered saline (PBS) to remove contaminating LSM. From 1 mL of blood we obtain 1–2 × 10^6 cells.

3.2. Cryopreservation of Cells

Mononuclear cells are frozen in 90% FCS and 10% DMSO at a concentration of 5–10 × 10^6 cells/mL. The cells are first resuspended in cold FCS at one-half of the final volume. The remaining volume of cold FCS with 20% DMSO is then added dropwise while swirling the tube. A controlled rate of freezing can be simulated by simply placing the vials in a styrofoam box in a –80°C freezer overnight. The vials are then transferred to liquid nitrogen for long-term storage.

At the time of the assay, the vials are thawed rapidly in a 37°C water bath. The contents are transferred to 10 mL of RPMI-1640 containing 30 μg/mL of DNAase. Following a 30-min incubation at 37°C, the cells are washed three times in PBS. Viable cells are counted by diluting the sample 1:2 in trypan blue and counting in a hemocytometer the cells that have not taken up the color blue.

3.3. Setting up the LDA Cultures

In the LDA culture assay, hundreds of peripheral blood lymphocyte microcultures are stimulated once in vitro with the appropriate antigen. The cells are grown in 96-well U bottomed plates.

1. At each dilution, seed peripheral blood lymphocytes in a 100-μL volume into 24 wells of a round-bottomed 96-well microtiter plate (see **Fig. 1**). We routinely use twofold dilutions of cells starting with 2 × 10^5 cells/well and ending with 6.25 × 10^3 cells/well. (*see* **Note 1**). Culture medium is RPMI-1640 supplemented with 10% human AB serum, penicillin, and streptomycin.
2. Add irradiated (6000 RADS) stimulator cells expressing MUC-1 in a 50-μL volume at a concentration of 2 × 10^4/well.
3. Use irradiated (3000 rad) autologous peripheral blood lymphocytes (PBLs) as "feeder" cells, and add at a concentration of 2 × 10^4/well in a 50-μL volume. Suspend the "feeder" cells in culture medium containing one half of the optimal concentration of IL-2 (*see* **Notes 1–4**).
4. Following 7 d, in culture, remove 100 μL of culture medium from each well and replace with fresh medium containing IL-2. Perform the CTL assay on d 10.

3.4. Setting up the CTL Assay

1. If adherent, trypsinize target cells and wash 1 × 10^6 cells three times in PBS. After the last wash, carefully aspirate the supernatant so that the smallest possible volume remains.
2. Resuspend the cell pellet in 100 μL of fresh ^{51}Cr, specific activity 1 mCi/mL, and incubate at 37°C for 1–2 h. Shake the tube periodically to keep the cells in suspension.
3. Following three washes, place 1 × 10^3 cells in a gamma counter to determine the uptake of radioactivity and thus ensure proper labeling. If less than 500 counts are obtained with tumor cells or 300 counts with EBV-transformed B-cells, a new batch of targets should be prepared. The new targets should be labeled for a longer period of time or newer chromium should be used.
4. While the targets are being prepared, the rest of the assay can be set up. Each well of the original plates should be carefully mixed with a micropipettor and transferred into two wells of a 96-well microtiter V-bottomed plate ("split-well" assay). Thus, row A of the original 96-well plate becomes rows A and B of the assay plate (*see* **Fig. 1**). Each new well contains 100 μL of the original 200-μL cell cuture.

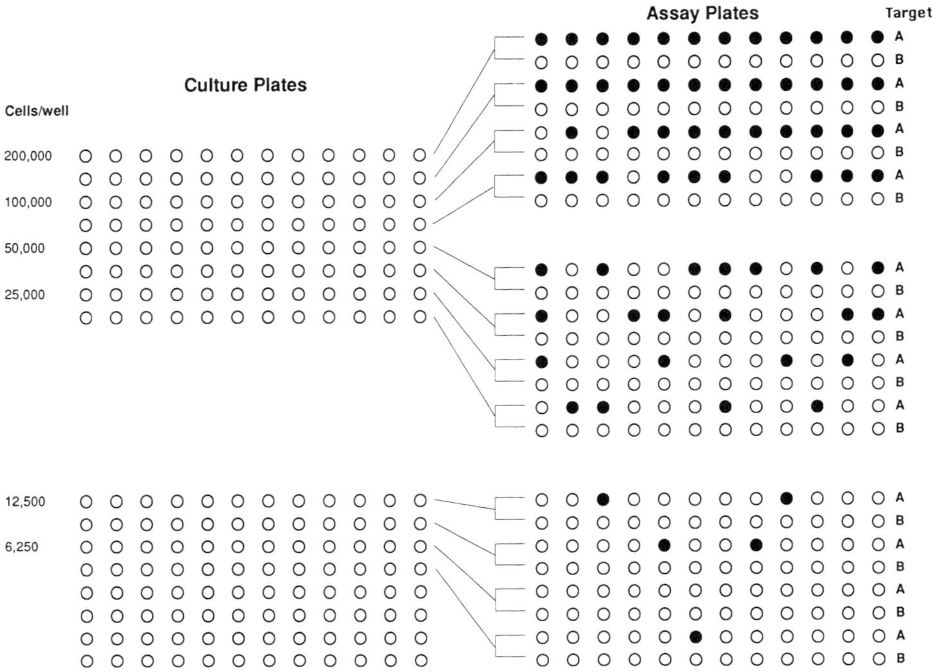

Fig. 1. An example of an experiment utilizing limiting dilution culture, showing the process of setting up the original culture plates and the final assay plates. In the culture plates, each cell concentration is represented by two rows (24 wells). Each culture well is split into two assay wells. The black wells in the assay plates were scored positive for CTL activity. Rows A contain the MUC1⁺ target, and rows B contain the negative control target.

5. To each new well, add 2.5×10^4 unlabeled LAK target cells in a volume of 50 µL. For CTL assays with tumor cells as labeled target cells, we use unlabeled K562 cells. When MUC1 transfected EBV B-cells are used as targets, we use unlabeled parental EBV cells to compete out the LAK lysis.
6. At each dilution, add 1×10^3 of the labeled specific targets to one of the split wells and to the other well 1×10^3 labeled control targets, in 50-µL volumes. Include several wells containing only target cells for spontaneous and maximal release values
7. Centrifuge the plates at 150*g* for 5 min and incubate at 37°C for 4 h.
8. Harvest supernatant from the wells and count the radioactive chromium released from lysed cells. Obtain maximal release values following lysis of the cells in designated wells with acid or hypertonic saline.
9. Calculate specific release using the following formula: 100 × (sample release – spontaneous release/maximal release – spontaneous release).
10. Calculate percentage of specific release for each well containing the specific target and compare this with the corresponding well containing the control target. Wells are scored positive if specific target cell lysis is 15% greater than the control (*see* **Note 7**).

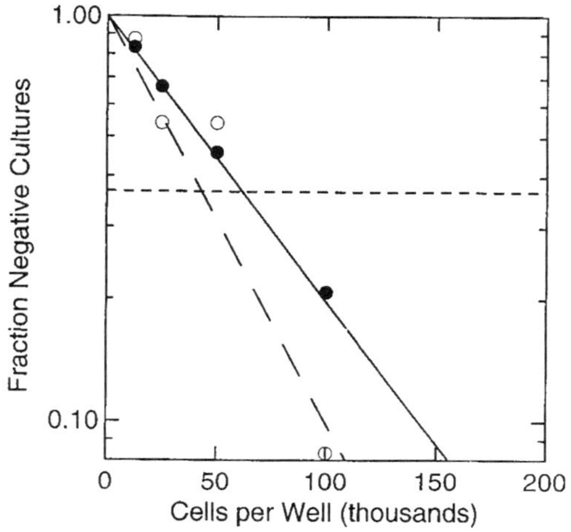

Fig. 2. MUC1-specific CTL frequency derived from the results depicted in **Fig. 1**. The solid line represents CTL frequency derived in the absence, and the broken line represents CTL frequency derived in the presence of anti-HLA class I antibody W632, illustrating the MHC-unrestricted nature of the CTL. The log of the fraction of negative cultures is plotted on a log scale against the number of cells per well. Linear regression analysis was used to calculate the slope. The wells containing 200,000 cells had no negative wells; thus, the value is 0 and is not plotted. A frequency of 1/65,949 for untreated cultures and 1/57,542 for antibody-treated cultures is determined when 37% of the cultures are negative. This is shown at the intersection of the two plotted lines with the horizontal dashed line.

3.5. Calculation of the CTL Frequency

Inasmuch as positive wells may contain more than one responding T-cell, the frequency of antigen-specific cells is determined by the number of negative wells in each dilution. The number of negative wells is divided by the total number of wells at each dilution. This fraction is converted to its base 10 log and is plotted on a log scale against the number of cells added to each well. The Maximal Likelihood method is one of the best statistical estimators used to find the best fitting straight line going through the origin *(10)*. Linear regression can also be utilized. It is assumed in these assays that all parameters for T-cell growth are optimal and that any well containing antigen-specific cells will be positive in the assay. The single hit model of Poisson distribution predicts that when 37% of the test cultures are negative, there is an average of one precursor cell per well *(11)*. Several statistical computer programs are available to simply evaluate the data. **Figure 2** shows graphic data from the assay represented by **Fig. 1** on an individual with slightly elevated frequency of MUC1-specific cells.

3.6. LDA Special Considerations

3.6.1. Time of Assay

LDA assays can be used to quantitate specific CTL precursors that need in vitro reactivation, or recently activated specific effector CTL. Precursor CTL are determined following 10 or 11 d in culture with antigen. Activated CTL could be assayed after 3–5 d of culture.

3.6.2. Number of Cells

The frequency of MUC1-specific precursor cells in most healthy individuals is about 1 in 10^6 cells. This frequency is increased to varying degrees in cancer patients and is in the range of 1 in 150,000 to 1 in 300,000. Following immunization of cancer patients in phase I trials, we have observed that the frequency in some patients may increase between two- to 11-fold *(8)*. These are, nevertheless, relatively low frequencies that require cell concentrations for the assays to start at 2×10^5 cells/well. We have found, however, that greater than 4×10^5 cells/well can result in overcrowding of the well and inconsistent results. Optimal statistical analysis requires at least 24 replicates of each dilution. Thus, the minimum number of cells required is 1.3×10^7 for each patient to be tested.

3.6.3. LAK vs Specific CTL Activity

Owing to the presence of IL-2 in the culture medium and additional IL-2 secreted following T-cell activation, some nonspecific cytokine-activated cell lysis can occur, which increases the background noise. We use unlabeled K562 cells as cold-target competitors with labeled targets to minimize the background. Other unlabeled (cold) targets could be used to compete out possible alloreactive responses or EBV-specific responses (*see* **Subheading 3.6.4.**).

3.6.4. Target Cells

EBV-transformed MUC1-transfected B-cells are excellent stimulator cells because they also express costimulation molecules necessary for optimal T-cell activation. However, unless these cell lines are derived from each patient and only autologous cells are used in the assay, it is possible to generate allospecific CTL during the culture period. If autologous cells are used, often EBV-specific CTL are activated together with the MUC1-specific CTLs. One way to overcome this problem is to use the parental untransfected B-cell line as the cold-target inhibitor. Appropriate control target cell in the split-well assays would be the same untransfected EBV-transformed cell, or a tumor cell target that expresses MUC1 but does not share MHC alleles with the stimulator cells.

The MHC-unrestricted nature of the MUC1-specific CTL allows us to use a more standardized and less labor-intensive approach. We stimulate the PBL in vitro with one $MUC1^+$ tumor cell line and test CTL function against another $MUC1^+$ tumor cell line that shares no MHC alleles with the stimulator tumor. Allospecific cytotoxicity, if any generated in the cultures, would be ineffective against the second target.

4. Notes

1. It is most convenient to prepare dilutions of cells directly in the microtiter plates. The first two rows receive 4×10^5 cells in 200 µL. Using a 12 -well multichannel pipettor, 12 serial dilutions can then be performed simultaneously by transferring 100-µL volumes into lower wells containing 100 µL of medium alone.
2. Sufficient IL-2 is added to the culture to initiate the growth of antigen stimulated T-cells but not the growth of nonspecific lymphokine activated T-cells. Thus, if 20 U/mL is the optimal concentration of IL-2, the stimulator cells are resuspended in 10 U/mL of IL-2. Thus, in this example, each well receives 0.5 U of the cytokine.

3. It is also possible to culture the cells for 3 d before adding IL-2. This may reduce nonspecific background from cytokine-activated T-cells.
4. Depending on the source, IL-2 is purchased in units/milliliter or micrograms/milliliter quantities. Each new batch should be evaluated to determine the concentration that promotes optimal growth. Dilutions of IL-2 from 0.1 to 100 U/mL or 0.1–100 ng/mL are prepared in RPMI-1640 containing 10% human AB serum, penicillin/streptomycin, glutamine, and HEPES. Using 96-well microtiter U-bottomed plates, triplicate 100-μL aliquots of each dilution are added to 100 μL of human PHA blasts at a concentration of 1 × 10^5/mL. PHA blasts are prepared by culturing 5 × 10^5 Ficoll-separated peripheral blood mononuclear cells in 5 mL of 10% human AB serum with 10 μg/mL PHA for 3 d at 37°C in a humidified CO_2 incubator. The cells can be frozen in 10% DMSO and 90% serum, and stored in small aliquots indefinitely in liquid nitrogen. Before use, the PHA blasts are thawed and rested by overnight incubation in 10% FCS at 37°C to obtain an accurate viable count. Following incubation in different IL-2 dilutions for 3 d at 37°, 1 μCi of ^{3}H thymidine is added. After 18–20 h at 37°C the cells are harvested and counted. The dilution of IL-2 that gives 50% of maximal growth is considered the optimal concentration.
5. Successful culture of T-cells requires enriched media such as RPMI-1640 or AIM-5, human male AB serum, and IL-2. Owing to lot-to-lot variation, AB serum should also be tested for supporting optimal cell growth. Testing of human serum is done with PHA blasts prepared as described above. Vendors will supply small samples of several serum lots on request. Cells (1 × 10^5) are thawed and cultured for 4 d in RPMI-1640 with 10% test serum, and the optimal concentration of IL-2 determined as described in **Note 4**. A viable cell count is a reliable way to determine proliferation. A second method is to assess incorporation of ^{3}H thymidine by mitotic cells. This is done in triplicate using 96-well microtiter plates. Viable cells (3 × 10^4) are cultured overnight with 1 μCi of ^{3}H thymidine. The cells are then harvested and counted for uptake of radionuclide by dividing cells.
6. It may be possible to assay simultaneously more than two target cells by splitting the original well into a number of test wells, depending on how vigorously the T-cells have proliferated during the 10-d culture period.
7. The arbitrary discrimination between positive and negative cultures is usually a 15% difference in specific release. Some patients have low levels of specific release and little or no observed background LAK activity. It may be possible in these patients to consider lower specific lysis as positive.

References

1. Hilkins, J. and Bujis, F. (1988) Biosynthesis of MAM-6, an epithelial sialomucin. Evidence for involvement of a rare proteolytic cleavage step in the endoplasmic reticulum. *J Biol. Chem.* **263,** 4214–4222.
2. Girling, A., Bartkova, J., Burchell, J., Gendler, S., Gillet, C., and Taylor-Papadimitriou, J. (1989) A core protein epitope of the polymorphic epithelial mucin detected by monoclonal antibody SM-3 is selectively exposed in a range of primary carcinomas. *Int. J. Cancer* **43,** 1072–1076.
3. Finn, O. J. (1992) Antigen-specific, MHC-unrestricted T cells. *Biotherapy* **4,** 239–249.
4. Magarian-Blander, J., Ciborowski, P., Hsia, S., Watkins, S. C., and Finn, O. J. (1998) Intercellular and intracellular events following the MHC-unrestricted TCR recognition of a tumor-specific peptide epitope on the epithelial antigen MUC-1. *J. Immunol.* **160,** 3111–3120.
5. Magarian-Blander, J., Hughey, R. P., Kinlough, C., Poland, P. A., and Finn, O. J. (1996) Differential expression of MUC1 on transfected cell lines influences its recognition by MUC1 specific T cells. *Glycoconjugate J.* **13,** 749–756.

6. Domenech, N., Henderson, R. A., and Finn, O. J.(1995) Identification of an HLA-A11-Restricted Epitope from the tandem repeat domain of the epithelial tumor antigen mucin. *J. Immunol.* **155,** 4766–4774.
7. Apostolopulos, V., Karanikas, V., Haurum, J. S., and McKenziee, I. F .C. (1997) Induction of HLA-A2-restricted CTLs to mucin 1 human breast cancer antigen *J. Immunol.* **159,** 5211–5218.
8. Goydos, J. S., Elder, E., Whiteside, T. L., Finn, O. J., and Lotze, M. T. (1996) A phase I trial of a synthetic mucin peptide vaccine. *J. of Surg. Res.* **63,** 298–304.
9. Karanikas,V., Hwang, L., Pearson, J., and Ong, V., (1998) Antibody and T cell responses of patients with adenocarcinoma immunized with mannan-MUC1 fusion protein. *J. Clin. Invest.* **100,** 2783–2793.
10. Taswell, C. (1981) Limiting dilution assays for the determination of immunocompetent cell frequencies. *J. Immunol.* **126,** 1614-1619.
11. Sharrock, C. E. M., Kaminski, E., and Man, S. (1990) Limiting dilution analysis of human T cells a useful clinical tool. *Immunol. Today* **11,** 281–285.

39

Expression of MUC1 in Insect Cells Using Recombinant Baculovirus

Pawel Ciborowski and Olivera J. Finn

1. Introduction

MUC1 mucin undergoes multistep posttranslational modifications before it is finally expressed on the apical surface of mammalian ductal epithelial cells. Two early precursor proteins are both *N*-glycosylated and differ in molecular weight owing to a proteolytic cleavage of a 20-kDa fragment. Proteolytically modified form is transported to the Golgi, where it undergoes extensive, although not complete, *O*-glycosylation on serine and threonine residues within the tandem repeat (TR) region. MUC1 is then transported to the cell surface. For additional glycosylation and sialylation, surface MUC1 is internalized and directed to *trans*-Golgi compartments. Mature form is again transported to the cell surface *(1)*.

MUC1 expressed by malignant epithelial cells such as breast and pancreatic adenocarcinomas is underglycosylated (aberrantly glycosylated), which makes it structurally and antigenically distinct from that expressed by normal cells *(2)*. As such, it may be an excellent target for immunotherapy. One of the ways to utilize tumor-specific forms of this molecule is as immunogens. Purifying these forms from tumor cells is not feasible because it is a labor-intensive process that gives low yields. A much more desirable approach is purification of a recombinant molecule from an appropriate expression system. Recombinant MUC1 expressed in a convenient prokaryotic system that does not glycosylate proteins, such as *Escherichia coli,* undergoes rapid and random proteolytic degradation. To obtain underglycosylated recombinant tumorlike forms of MUC1 in mammalian cells through expression vectors such as vaccinia virus, retroviral vectors, and plasmid vectors requires a prolonged treatment of infected or transfected cells with toxic and expensive inhibitors of *O*-linked glycosylation *(3,4)*. Furthermore, vaccinia and retroviral constructs spontaneously recombine out most TRs that characterize the major portion and the most immunogenic portion of MUC1 *(5)*.

We explored the baculovirus system that allows expression of MUC1 mucin in *Spodoptera frugiperda* Clone 9 (Sf-9) insect cells. We found that these cells, when

From: *Methods in Molecular Biology, Vol. 125: Glycoprotein Methods and Protocols: The Mucins*
Edited by: A. Corfield © Humana Press Inc., Totowa, NJ

infected with a MUC1 recombinant baculovirus, produce fully glycosylated, full-size (no deletion of TRs) molecules that are expressed on the cell surface ***(6)***. Moreover, under specific starvation growth conditions that we determined empirically, Sf-9 cells can also produce underglycosylated MUC1, similar to the MUC1 produced by tumor cells. The state of glycosylation of various forms can be evaluated by their migration in sodium dodecyl sulfate-polyacrylamide gel electrophoresis (SDS-PAGE) gels and reactivity with different anti-MUC1 antibodies in Western blot analysis ***(6,7)***. In this chapter, we present the techniques of expression of MUC1 mucin using three baculoviral vectors: pBlue*Bac*III, pFastBac, and pIE1-4. Additional vectors are commercially available and, as one can expect, more will emerge on the market in the future. In our opinion, they provide an ideal expression system to study different forms of MUC1 protein, their function, and utility.

2. Materials

2.1. Cloning Reagents

1. Vectors: pBlue*Bac*III was purchased from Invitrogen, San Diego, CA (*see* **Note 1**); pFastBac was purchased from Gibco, Life Technologies, Grand Island, NY; and pIE1-4 was purchased from Novagen, Madison, WI.
2. Competent *E. coli* cells such as MAX Efficiency DH5α™ Competent Cells and MAX Efficiency DH10Bac™ Competent Cells were obtained from Gibco-BRL.
3. Restriction enzymes, agarose, ligase, and other reagents for cloning may be obtained from any supplier of molecular biology reagents. Wizard™ Minipreps and Wizard™ Megapreps were obtained from Promega, Madison, WI. Cationic liposomes InsecticinPlus™ were obtained from Invitrogen, but can be also obtained from other commercial sources. BluoGal and isopropyl-β-D-thiogalactopyranoside (IPTG) were purchased from Sigma, St. Louis, MO; X-gal was purchased from Boehringer Mannheim, Indianapolis, IN; and SeaPlaque agarose was purchased from FMC BioProducts, Rockland, ME.

2.2. Cells, Media, and Antibodies

1. The insect cell line Sf-9 can be obtained from American Type Culture Collection (Rockville, MD) or from other suppliers such as Invitrogen, San Diego, CA.
2. Hink's TNM-FH Insect Medium can be obtained from several sources such as JRH Biosciences, Lenexa, KS. Penicillin, streptomycin, fungizone, and geneticin can be obtained from Gibco. Fetal bovine serum (FBS) was from Gibco-BRL.
3. Anti-MUC1 antibodies used in this study are not commercially available. Monoclonal antibodies (MAbs) used for Western blot and flow cytometry analysis are listed in **Table 1**. The TD-4 MUC1 Workshop (*see* **ref. 7**, pp. 1–152) provides the most up-to-date list of anti-MUC1 antibodies, their specific reactivities, and their sources.
4. Tissue culture flasks, plates, roller bottles, and disposable plastic tubes of various sizes can be obtained from various sources, e.g., Sarsted, Falcon, etc. Any 27°C incubator can be used, although one with a water jacket is recommended.

2.3. Western Blot

All reagents and equipment for PAGE and Western blot, except nitrocellulose, were purchased from Bio-Rad, Hercules, CA. Other suppliers can also be used. Nitrocellulose BioBlot-NC was purchased from Corning Costar, Corning, NY. Chemilumines-

Table 1
MUC1 Specific Antibodies[a]

Antibody	Isotype	Specificity
SM-3	IgG1	APDTRP[b], underglycosylated[c]
VU-4-H5	IgG1	PDTRPAP, underglycosylated[c]
VU-3-C6	IgG1	PDTRPAP, all forms
BC-3	IgM	APDTR, all forms
BC-2	IgG1	APDTR, all form
232A1	IgG	Proteolytic cleavage site

[a]For more details about antibodies see ISOBM TD-4 International Workshop on Monoclonal Antibodies against MUC1, *Tumor Biology*, 1998, **19 (Suppl. 1),** 1–152.

[b]Single letter code for amino acids. A, alanine; P, proline; T, threonine; D, glutamic acid; R, arginine.

[c]Tumor specific, recognizing underglycosylated but not fully glycosylated MUC1.

cence Western blotting detection kit was purchased from Amersham, Buckinghamshire, England.

3. Methods

3.1. Vector Construction

The cDNAs coding for MUC1 of various lengths owing to various numbers of TRs were obtained from previously made plasmid constructs. Plasmid expression vectors encoding MUC1 with 22 repeats (22TRMUC1) and two repeats (2TRMUC1) were made in our laboratory *(3)*. The cDNAs can be isolated from the plasmid vectors as *Hin*dIII cassettes (**Fig. 1**). Plasmid expression vectors containing MUC1 cDNA with 42 TRs (42TRMUC1) and MUC1 cDNA without TRs (TR^-MUC1), both *Bam*HI cassettes, were obtained from Dr. A. Hollingsworth, University of Nebraska, Omaha.

3.1.1. Cloning into pBlueBacIII Transfer Vector

An example we will use for cloning of the 3.2-kbp cDNA MUC1 with 22 TRs (22TRMUC1) and the 1.8-kbp cDNA MUC1 with 2 TRs (2TRMUC1). The resulting pBlue*Bac*III-22TR-MUC1 recombinant transfer plasmid is used for inserting MUC1 cDNA into the genome of the wild-type *Autographa californica* Multiple Nuclear Polyhedrosis Virus (wtAcMNPV), as described under **Subheading 3.3.1.**

1. Digest pBlue*Bac*III transfer vector with *Hin*dIII or *Bam*HI and treat with calf intestine phosphatase (CIP) to protect against self-ligation using standard methodology (*see* **Note 1** and **2**).
2. Prepare MUC1 cDNA cassette by *Hin*dIII digestion.
3. Purify fragments by electrophoresis in 0.7% agarose.
4. Ligate the cDNA cassette into the pBlue*Bac*III vector using T4 DNA ligase at 16°C overnight.

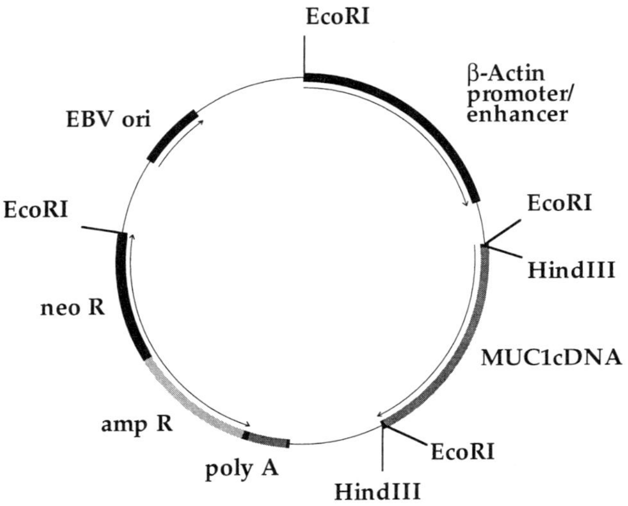

Fig. 1. MUC1 cDNA expression plasmid. The MUC1 cDNA is downstream from translational start codon. Constructs with 2 or 22 TRs that were made in our laboratory are contained in the *Hin*dIII.

5. Transform the ligated construct into *E. coli* MAX Efficiency DH5α Competent Cells following the protocol provided by the manufacturer.
6. Select recombinants using Luria agar with 10 µg/mL of ampicillin *(8)*.
7. Amplify ampicillin-resistant clones in 5-mL Luria broth/ampicillin *(8)* cultures. Use 1.5–3.0 mL of the culture to isolate plasmid DNA using Wizard Minipreps.
8. Analyze recombinant DNA by restriction enzyme digestion for orientation of the insert.

*3.1.2. Cloning into pFast*Bac *Transfer Vector*

As an example, we will use cloning of 4.6-kbp cDNA with 42 TRs (42TRMUC1). Fragment of MUC1 cDNA coding for transmembrane and cytoplasmic domains was replaced with a sequence linking the outer membrane portion of MUC1 with glycosylphosphatidylinositol (GPI) anchor of human decay accelerating factor. This new construct (42TRMUC1-GPI) was made in our laboratory and remains as a *Bam*HI cassette (Alter, M., unpublished data). The resulting pFast*Bac*-42TRMUC1-GPI recombinant transfer plasmid is used for inserting MUC1 cDNA into the genome of the wtAcMNPV, as described under **Subheading 3.3.2.**

1. Linearize pFast*Bac* transfer vector with *Bam*HI digestion and protect it with CIP against self-ligation using standard methodology.
2. Cut out the 42TRMUC1-GPI cDNA cassette by *Bam*HI digestion.
3. Purify a fragment of the correct size by electrophoresis in 0.7% agarose.
4. Ligate the cDNA cassette into the vector using T4 DNA ligase at 16°C for overnight.
5. Transform *E. coli* MAX Efficiency DH5α Competent Cells with the ligated construct.
6. Select recombinants using Luria agar with 10 µg/mL of ampicillin.
7. Select ampicillin-resistant clones, and amplify and purify plasmid DNA using Wizard Minipreps.
8. Analyze recombinant DNA by restriction enzyme digestion for orientation of the insert.

3.1.3. Cloning into the Episomal Transfer Vector pIE1-4

The vector pIEI-4 is used to provide stable expression of a cloned gene from the baculovirus *ie1* promoter. Cells are cotransfected with pIE1-neo providing neomycin selection marker expressed from *ie1* promoter. As an example, we will use cloning of 1.4-kbp MUC1 cDNA that lacks TRs (TR^-MUC1). The resulting pIE1-4TR^-MUC1-GPI recombinant transfer plasmid is used for cotransfection of Sf-9 cells with pIE-neo, as described under **Subheading 3.5.**

1. Linearize the pIE1-4 transfer vector with *Bam*HI and protect it with CIP against self-ligation using standard methodology (*see* **Note 3**).
2. Prepare TR^-MUC1 cDNA cassette by *Bam*HI digestion.
3. Purify the desired fragment by electrophoresis in 0.7% agarose.
4. Ligate the cDNA cassette into the vector using T4 DNA ligase at 16°C for overnight.
5. Transform *E. coli* MAX Efficiency DH5α Competent Cells with the ligated construct.
6. Select recombinants using Luria agar with 10 μg/mL of ampicillin. Amplify ampicillin-resistant clones and purify plasmid DNA using Wizard Minipreps.
7. Analyze recombinant DNA by restriction enzyme digestion for orientation of the insert.

3.2. Conditions for Culturing the Sf-9 Cells

Sf-9 insect cells are cultured in Hink's TNM-FH Insect Medium supplemented with 5 or 10% FBS and penicillin/streptomycin/fungizone at the concentrations of 100 U/mL, 100 μg/mL, and 2.5 μg/mL, respectively. Cells are grown as a monolayer at 27°C. For small scale growth, 75-cm^2 vented flasks are used (Costar, Cambridge, MA). Typically 5×10^5 cells and 20 mL of medium are used to start the culture of this size. For larger-scale growth, roller bottles are used. Cultures are usually started at the cell density of 10^6 cells/mL. During the logarithmic phase of growth, cells typically double every 24 h. Therefore, equal amounts of fresh medium are added each day to the roller bottle for up to 300 mL total volume. **Figure 2** shows the kinetics of growth in a typical roller bottle culture (*see* **Note 4**).

3.3. Production of Recombinant Virus by Cotransfection with Viral and Recombinant Transfer Vector DNAs

The wtAcMNPV viral DNA and the recombinant transfer vector DNA are shuttled into Sf-9 cells by cationic liposomes. Within the cells, transfer vector DNA and viral DNAs recombine, incorporating the gene of interest into the viral genome. Depending on the transfer vectors different protocols can be used to make recombinant virus. Two protocols are given next.

*3.3.1. Using pBlue*Bac*III Vector*

When using pBlue*Bac*III vectors, recombination leads to the replacement of the viral polyhedrin gene (phenotypically *occ*$^+$) with part of the transfer vector containing *lacZ* gene and gene of interest. Therefore, the selection is based on the phenotypic observation—lack of occlusion bodies (*occ*$^-$) and expression of β-galactosidase (*lacZ*$^+$).

1. Seed Sf-9 cells in a 6-well plate (10^6 cells/well) prior to the cotransfection, and rock them gently side-to-side for 1 h at room temperature to evenly distribute and attach the cells.

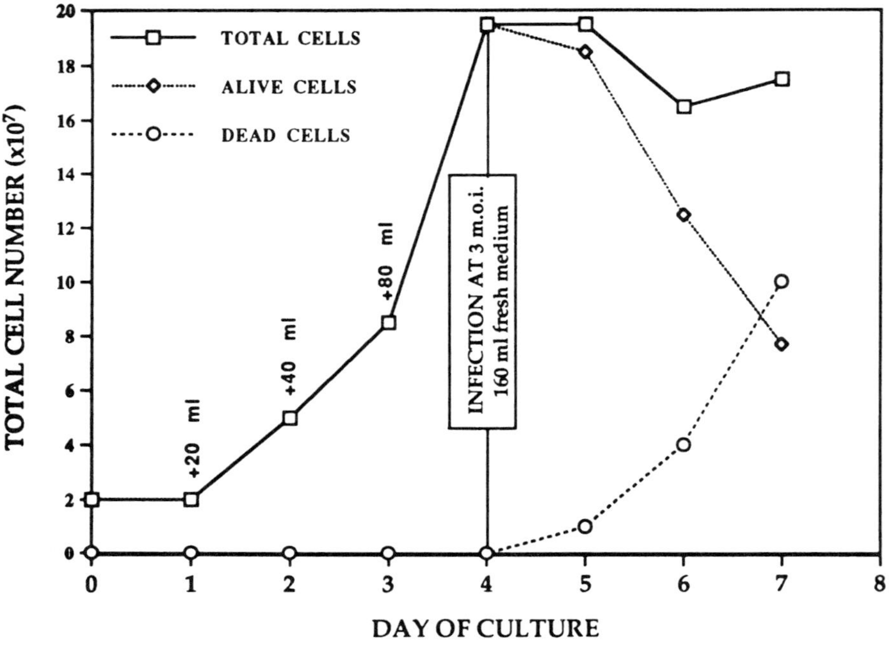

Fig. 2. Growth of SF-9 cells in a typical roller bottle culture. Infection was on d 4 and no new medium was added afterward. On d 5 all cells were expressing β-galactosidase (see **Subheading 3.6.1.** for details). Cells were usually harvested after 72 h.

2. Remove nonattached cells and medium, gently wash the adherent monolayer once with serum-free medium, cover with 2 mL of serum free medium, and incubate for 30 min at room temperature.
3. Prepare five independent transfection mixtures. Mix 100, 200, 500, and 750 ng, or 1 μg of the recombinant pBlue*Bac*III transfer plasmid, respectively, with 500 ng of linearized AcMNPV DNA, 40 mL Insectin-Plus Liposomes, and 1 ml of Hink's TNM-FH Insect Medium.
4. Vortex transfection mixtures vigorously for 10 s and incubate at room temperature for 30 min.
5. Remove serum-free medium from the cells, cover cell monolayer with one of the transfection mixtures, swirl to mix, and incubate for 4 h at room temperature with slow rocking.
6. Add 2 mL of complete Hink's TNM-FH Insect Medium (containing 10% FBS) to each well, wrap plates with clear plastic wrap, and incubate at 27°C for 48 h.
7. Take 100 μL of the culture supernatant that contains viruses produced by the transfected cells from each well and screen by plaque assay for the presence of double recombinants (*occ*$^-$, *lacZ*$^+$). Transfer the remaining medium to sterile microcentrifuge tubes and store at 4°C.

*3.3.2. Using pFast*Bac *Transfer Vector*

pFast*Bac* transfer vector is a part of the Bac-To-Bac™ Baculovirus Expression System developed by Gibco. In the first step, competent MAX Efficiency DH10Bac *E. coli* cells are transformed with pFast*Bac* donor plasmid with a gene of interest. The competent DH10Bac *E. coli* cells contain baculovirus shuttle vector (bacmid) and a helper plasmid. Bacmid propagates in *E. coli* and, besides viral DNA, contains several other elements such as attachment sites for transposon Tn7 and an open reading frame of *lacZα* peptide. Recombination between pFast*Bac* transfer vector and bacmid occurs within bacterial cells by transposition, with the aid of helper plasmid. Another feature of this system is that insertion of the gene of interest into the viral genome causes disruption of *lacZ* gene and *E. coli* containing recombinant bacmid grow as white colonies in the presence of Bluo-gal and IPTG. This feature makes the system easy to use and eliminates posttransfection isolation of recombinant viruses.

1. Transform *E. coli* MAX Efficiency DH10Bac Competent Cells that contain Bacmid DNA and helper plasmid. After transformation with recombinant pFastBac-*MUC1-42TR-GPI*, the transposition occurs inside *E. coli* cells disrupting *LacZ* gene within Bacmid DNA.
2. Select white growing colonies from Luria agar supplemented with kanamycin (50 μg/mL), tetracycline (10 μg/mL), gentamicin (7 μg/mL), BluoGal (100 μg/mL), and IPTG (40 μg/mL) to make a larger amount of the recombinant Bacmid DNA.
3. Amplify selected clones and purify Bacmid DNA using Wizard Minipreps.
4. Seed 10^6 Sf-9 cells/well in a 6-well plate immediately prior to transfection, and rock them gently side-to-side for 1 h at room temperature to evenly distribute and attach the cells. Alternatively, seed 2.5×10^5/well and grow them until desired density, usually 1 to 2 d.
5. Remove nonattached cells and medium, wash gently the cell monolayer once with serum free medium, cover with 2 mL of serum-free medium, and incubate for 30 min at room temperature.
6. Prepare three independent transfection mixtures: 100 to 200 ng of Bacmid DNA, mixed with 20, 40, or 60 μL Insectin-Plus Liposomes, and 1 mL of Hink's TNM-FH Insect Medium.
7. Vortex transfection mixtures vigorously for 10 s and incubate at room temperature for 30 min.
8. Remove serum-free medium from the wells, cover cell monolayers with the transfection mixture, swirl to mix, and incubate for 4 h at room temperature with slow rocking.
9. Add 2 mL of complete Hink's TNM-FH Insect Medium (containing 10% of FBS) to each well, wrap with clear plastic wrap, and incubated at 27°C for 48 h.
10. Transfer the culture supernatants that contain viruses produced by the transfected cells to sterile microtubes and store at 4°C.

3.4. Isolation of Recombinant Baculovirus

The culture supernatants from cells cotransfected with the pBlue*Bac*III as a transfer vector contain a mixture of recombinant and wild-type viruses. To isolate recombinant viruses, a plaque assay is performed followed by an end-point dilution round of purification.

3.4.1. Plaque Assay

1. Seed Sf-9 cells in 6-cm dishes with 2×10^6 cells/dish, and rock them gently side-to-side for 1 h at room temperature to evenly distribute and attach the cells.
2. Grow cells at 27°C to approx 80% confluency. Alternatively, seed more cells, and after 2 to 3 h during which the cells attach, the plates are ready for plaque assay.
3. Make a serial 10-fold dilution of the medium harvested from transfected cells, in full Hink's TNM-FH Insect Medium. Dilutions should range from 10^{-1} to 10^{-5}.
4. Remove the medium from the wells and add 1 mL of the culture supernatant containing a mixture of wild-type and recombinant viruses (viral inoculum) to the side of the dish, and

tilt the dish slowly to cover evenly all cells. It is important to do this gently in order not to disturb the attached cells. Incubate at room temperature for 1 h.

5. Melt the required volume of 3% SeaPlaque agarose, cool down to 45°C, and keep in a water bath. Prewarm to 37°C an equal volume of Hink's TNM-FH Insect Medium to which was added 120 µg/mL of X-gal.
6. Remove the viral inoculum by tilting the plate and aspirating from the edge.
7. Mix warm agar with medium and overlay dishes with 5 ml of this medium. Leave leveled until agarose sets.
8. Incubate at 27°C until blue plaques develop, usually 5–7 d.
9. Using a sterile Pasteur pipet, pick isolated plaques with the recombinant virus (blue plaques without occlusion bodies), transfer to tubes containing 2 to 3 mL of Hink's TNM-FH Insect Medium, and vortex for 30 s. Allow viral particles to diffuse from agar for another hour at room temperature. This could be used for screening and further purification.

3.5. Production of an Episomal Recombinant Vector for Stable Expression

3.5.1. Using pIE1-4 Vector

1. Seed Sf-9 cells in a 6-well plate (10^6 cells/well) prior to the cotransfection, and rock them gently side-to-side for 1 h at room temperature to evenly distribute and attach the cells.
2. Remove nonattached cells and medium, gently wash the adherent monolayer once with serum-free medium, cover with 2 mL of serum free medium, and incubate for 30 min at room temperature.
3. Prepare five independent transfection mixtures. Mix 3 µg of recombinant transfer plasmid pIE1-4-*TR⁻-MUC1*; 400 ng of pIE-neo plasmid DNA, and 20, 40, 60, 80, or 100 µL of Insectin-Plus liposomes. For mocktransfection, use 20 µL of Insectin-Plus liposomes.
4. Vortex transfection mixtures vigorously for 10 s and incubate at room temperature for 30 min.
5. Remove serum-free medium from the wells, cover cell monolayer with the transfection mixture, swirl to mix, and incubate for 4 h at room temperature with slow rocking.
6. Add 2 mL of complete Hink's TNM-FH Insect Medium (containing 10% FBS) to each well, wrap the plates with clear plastic wrap, and incubate at 27°C for 48 h.
7. Replace medium after 48 h with new medium containing 600 µg/mL of neomycin and grow cells for another 7 d.
8. After 7 d cells can be tested for MUC1 expression by flow cytometry and Western blot (*see* **Notes 5** and **6**).

3.6. Infection of Sf-9 Cells and MUC1 Production

3.6.1. Flask Cultures

1. Grow cells to approx 100% confluency.
2. Aspirate all medium and cover the cell monolayer with the minimal volume of a viral stock at 3 multiplicity of infection (*see* **Note 7**). This is usually 4 to 5 mL/75-cm² flask.
3. Rock the flask for 1 h at room temperature.
4. Add 20 mL of serum-free medium and incubate at 27°C for a desired time. To control yield of infection with recombinant virus carrying *lacZ* gene, aliquot a small sample of the culture (cells and supernatant) into a 1.5-mL microtube containing 1 µL of X-Gal at a concentration of 40 mg/mL, and incubate for 30 to 60 min at room temperature. After that time, infected cells should exhibit blue color when observed microscopically owing to β-galactosidase expression, and the culture supernatant should turn blue. If recombinant virus does not carry and express *lacZ* gene, other signs of infection such as swollen nuclei

can be used to assess the efficiency of infection. Typically more than 80% of cells are infected within the first day.

3.6.2. Roller Bottle Cultures

1. Start a roller bottle culture with Sf-9 cells in 20 mL of in Hink's TNM-FH Insect Medium, at a density of 10^6 cells/mL. The cell number usually doubles once every 24 h. Expand cells by adding every day an equal amount of fresh medium: 20 mL on d 2, 40 mL on d 3, 80 mL on d 4, and so forth, up to half the desired final volume of the culture (*see* **Note 8**).
2. Infect cells with the viral stock. There are two procedures for infection of a roller bottle culture. In the first, the culture is infected by adding the viral stock directly to the bottle and supplementing the culture with an equal amount of fresh medium 2 to 3 h after infection. The second method is to collect cells by centrifugation, resuspend in minimal volume of medium, usually one-tenth of the original, add the viral stock, and rock for 1 h at room temperature. Transfer infected cells back to the bottle, and add fresh medium in the amount equal to that prior to infection. We found both methods worked equally well. The only difference is that in the first method, in order to accomplish complete infection of all insect cells within 24 h, cultures should be infected at 5 m.o.i. or higher. In the second method, although more laborious, less viral stock is used and infection at 3 m.o.i. is sufficient.

3.7. Starving Sf-9 Cells in Culture to Obtain Underglycosylayted Forms of MUC1

3.7.1. Flask Cultures

1. Grow cells to 100% confluency, and prolong the time of culture for another day or two without changing the medium in order to deplete most of the nutrients. At this time, the number of dead cells slightly increases.
2. Aspirate all medium, clear by centrifugation and filtration through a 0.22-µm filter, and save.
3. Cover the cell monolayer with a minimal volume of the viral stock, usually 4 to 5 mL/75-cm^2 flask.
4. Rock the flask for 1 h at room temperature.
5. Add the saved nutrients-depleted medium and incubate the culture at 27°C for a desired time. To control efficiency of infection with recombinant virus carrying *lacZ* gene, follow the procedure described under **Subheading 3.6.2.**

3.7.2. Roller Bottle Cultures

1. Start and expand a roller bottle culture as described under **Subheading 3.6.**
2. Starve cells by growing at high density for 48 h without supplementing with fresh medium.
3. Infect cells with a viral stock as described under **Subheading 3.6.** In starved cultures, if cells are harvested for infection, nutrient-depleted medium should be collected, cleared by centrifugation and filtration through a 0.22-µm filter, and added back to the bottle.

3.8. Analysis of Recombinant MUC1 Produced by Sf-9 Cells

The quickest method to test the expression of MUC1 in insect cells is by immunostaining with specific MAbs. There is a large number of well characterized MAbs against different peptide and sugar epitopes on MUC1. Western blot is the method of choice for testing total expression, and flow cytometry is used for testing cell surface expression. The pattern of reactivity of baculovirus expressed MUC1 with

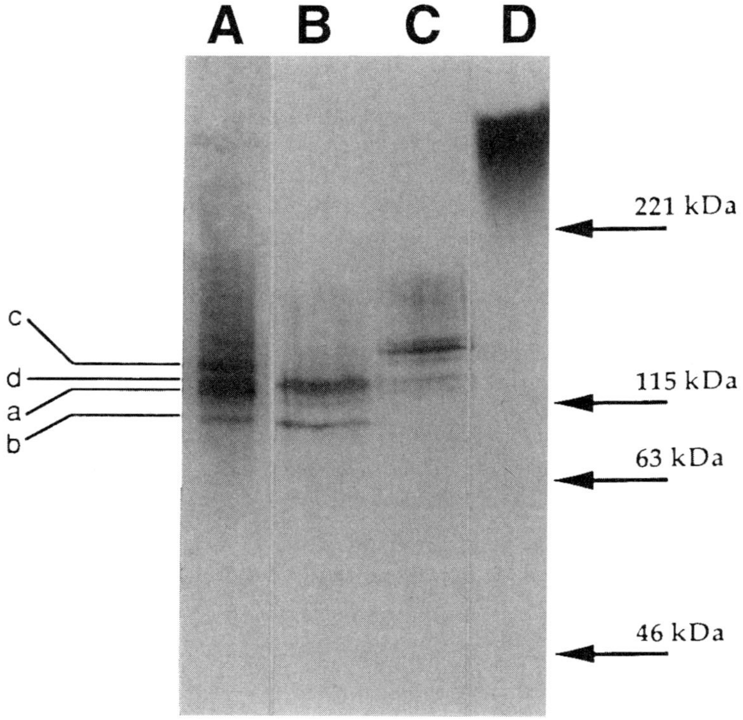

Fig. 3. Different forms of MUC1 expressed in Sf-9 cells using recombinant baculovirus. Form designated as (**A**) is a protein precursor which undergoes a proteolytic modifiction yeilding a 20 kDa smaller protein (**B**). Form (**C**) has not been characterized yet, and form (**D**) is MUC1 released from the cell surface.

various anti-MUC1 MAbs does not reflect the degree of glycosylation of the recombinant product. Although we do not provide a detailed protocol for Western blot and flow cytometry in this chapter, some technical details could be found in the figure legends (*see* **Note 9**).

MUC1 is the only transmembrane mucin known to date. It is transported to the apical surface of the ductal epithelial cells and anchored in the cell membrane via its transmembrane domain. MUC1 is removed from the cell surface by proteolytic cleavage in the membrane proximal domain, or, to lesser extent, it is internalized and degraded in the phagolysosomes. In insect cells, MUC1 also undergoes complex process of posttranslational modification as in mammalian cells. It is also transported to the surface of Sf-9 cells *(6)*.

A single, large band with an apparent molecular weight above the 221 kDa, represents the only form of MUC1 expressed by insect cells when they are grown in fully supported medium containing 10% FBS. However, when cells are starved for at least two days prior to infection and then grown in nutrient-depleated medium, the majority of the MUC1 product is of low molecular weight, as shown in **Fig. 3**, lane B (72 h post infection), and lane C (96 h postinfection). Another important fact can be derived from **Fig. 3**: The baculovirus expression system is capable of producing the full-length mucin polypeptide core (approx 106 kDa corresponding to 3.2 kbp cDNA) without the tandem repeat region undergoing deletions due to homologous recombination, a problem previously encountered with other expression systems ***(5)***.

Based on previous studies of MUC1 and processing ***(9,10)*** three out of four major immunoblotted products shown in **lane A** of **Fig. 3** can be identified. Protein precursor form (a) undergoes a proteolytic modification yielding a 20 kDa smaller protein designated as form (b). Form (c) has not been characterized yet, and form (d) is MUC1-released from the cell surface (**Fig. 4**, lane 72 h/s). It is important to note that MUC1 is mostly cell associated and only a very small fraction of the low-molecular-weight form is found in the supernatant. It must be noted, however, that in this experiment the culture supernatant was not cleared by short ultracentrifugation (120,000*g*/1 h), and therefore, the form of MUC1 found in the supernatant may have also been associated with fragments of cell membrane, rather than being a *bona fide* secreated form.

We have determined that MUC1 is also inserted into the insect cell membrane (**Fig. 5**). Surface expression was tested by flow cytometry analysis of immunostained cells. Panels A, B, and C represent cells stained respectively with IgG-control, BC-3, or VU-4H5 primary antibody followed with fluorescein isothiocyanate (FITC)-labeled secondary antibody. Antibody BC-2, like antibody BC-3, recognizes MUC1 independent of its state of glycosylation, while antibody VU-4H5 recognizes underglycosylated, tumor-specific MUC1 similarly to SM-3. It has been recently reported by Wright ***(12)*** and us (submitted for publication) that neither of these forms is glycosylated. It is not clear why insect cells do not *O*-glycosylate this recombinant product. GalNAc transferases purified from insect cells are able to glycosylate in vitro a synthetic peptide corresponding to one tandem repeat but not the recombinant product.

3.9. Large-Scale Purification of Recombinant MUC1

The most commonly used method for purification of a protein from complex mixtures is affinity chromatography using specific antibodies as ligands. This method is not easy to use for MUC1 purification because most MAbs are against epitopes within the TR region. High density of repeating epitopes leads to high-avidity binding to the affinity columns, which makes it almost impossible to elute bound MUC1 from IgG-Sepharose without boiling in sodium dodecyl sulfate (SDS). Insect cells appear to retain most of the MUC1 on the cell surface, and, thus, the cells rather than the supernatant are the primary source for MUC1 purification.

Purification of recombinant MUC1 containing highly hydrophobic transmembrane domain is a challenge. Fractionation of whole insect cell lysate by ultracentrifugation in cesium chloride gradient does not separate recombinant mucin effectively. Recombinant MUC1 is found in the top layer containing lipids. The form of MUC1 that is shed from the cell surface and does not have transmembrane and cytoplasmic domains migrates to the middle of the CsCl gradient (approx 1.4 g/mL). MUC1 binds nonspecifically to various resins such as glass (high-performance liquid chromatogra-

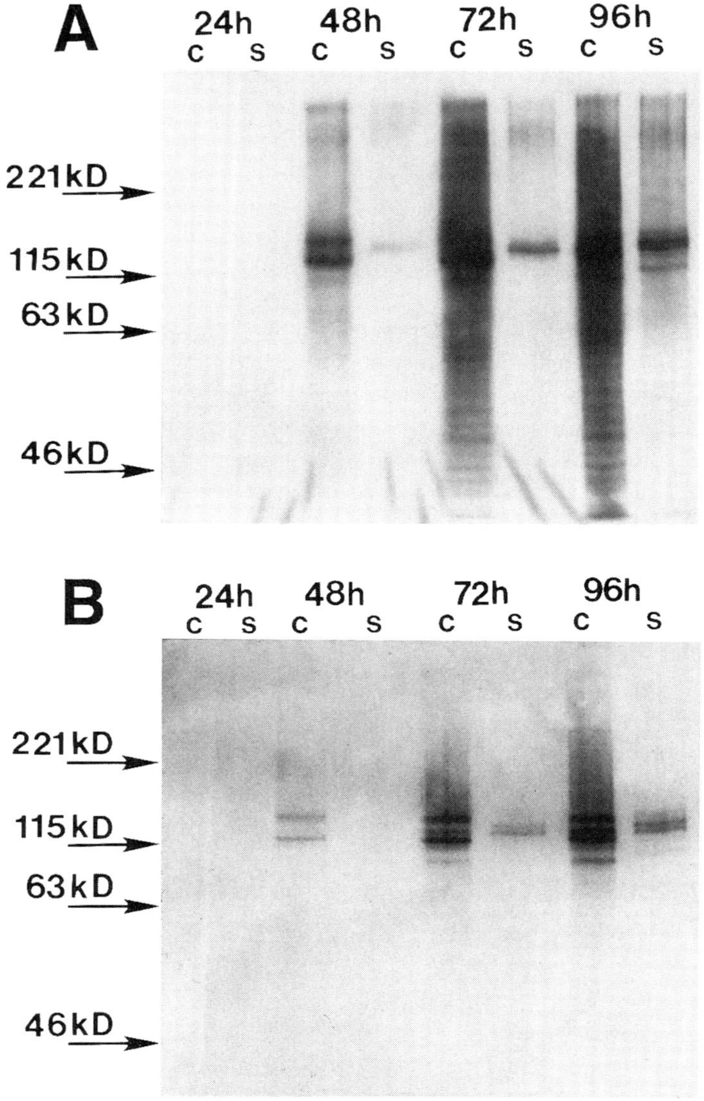

Fig. 4. Time course of MUC1 expression. Cell lysates (C) and supernatants (S) were analyzed by SDS-PAGE in 7% gel under reducing conditions. After transblotting, one membrane was probed with BC-3 MAb (**A**), and the other was probed with SM-3 MAb (**B**). The arrows on the left indicate the migration distances of protein molecular weight standards. (Reprinted with permission from **ref. *8***).

phy precolumn from Toso-Haas, Stuttgart, Germany), silica (HPLC size-exclusion column from Shodex, Showa Denko K.K., Tokyo, Japan), and Sephadex (CM-Sephadex from Pharmacia, Uppsala, Sweden), significantly decreasing yield of sepa-

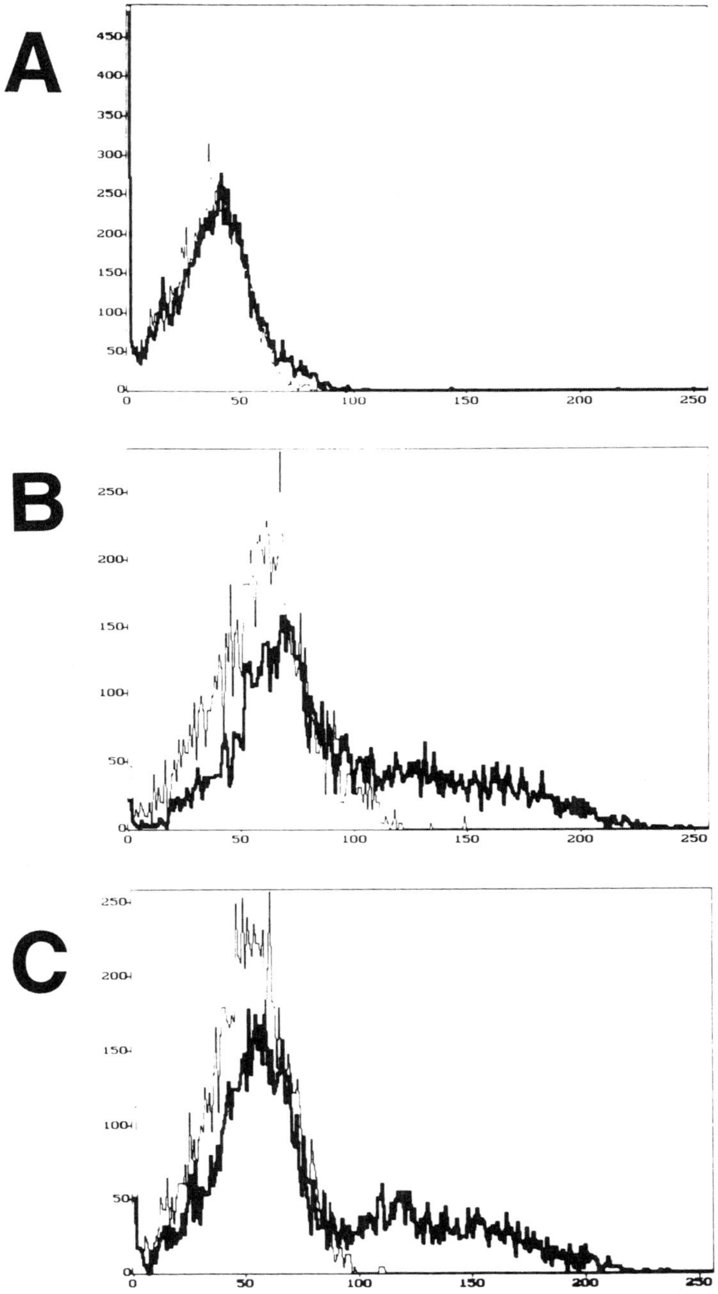

Fig. 5. Fluorescence-activated cell sorter analysis of surface expression of MUC1 on infected Sf-9 cells. Pairs of infected (bold line) and uninfected (faint line) cells were reacted with irrelevant γ1 control antibody (**A**), and antimucin antibodies BC-2 (**B**) and 4H5 (**C**), followed by immunostaining with FITC labeled secondary antibody (Reprinted with permission from **ref.** ***8***).

ration and shortening the lifetime of these resins. Recombinant MUC1 binds, although with low yield, to CM-Sephadex C-50 at pH 6.5 and can be eluted with a linear gradient of 0–1.0 *M* NaCl.

Recently, we developed a double-step affinity chromatography procedure to purify recombinant MUC1.

1. Anti-MUC1 antibody (3C6 MAb) on ProteinA-Sepharose (Pharmacia) column and washed with 10 bed volumes of PBS.
2. Insect cells expressing MUC1 are lysed in 50 m*M* Tris-HCl buffer, pH 6.5 containing 1% NP-40.
3. The lysate is precleared by two consecutive centrifugations of 15 min at 3000 rpm (Sorvall, Newtown, CT) and 1 h ultracentrifugation at 100,000*g* (Beckman, Palo Alto, CA).
4. Clear cell lysate is diluted with PBS at 1:1 ratio and loaded onto a Protein A–Sepharose-3C6 MAb affinity column. MUC1 binds to 3C6 antibody, creating MUC1-3C6 complexes that are eluted from the column with 50 m*M* glycine buffer, pH 2.8.
5. Mucin is released from the antibody by dissociation of the antibody chains with 1 m*M* dithiothreitol, followed by alkylation with 1 m*M* iodoacetamide.
6. After dialysis, heavy and light chains of 3C6 MAb are removed by affinity chromatography using antimouse IgG antibody covalently bound to CNBr-Sepharose (Sigma).

This method requires a ready supply of monoclonal anti-MUC1 antibody. Further separation of different MUC1 forms can be accomplished using a molecular sieving step.

4. Notes

1. pBlue*Bac*III is one of the early baculoviral vectors developed. It is large (10.2 kbp) but it can still handle very well MUC1 cDNA inserts as large as 3.1 kbp. We successfully inserted the *Hin*dIII cassettes into the polylinker, but were not successful in ligating the *Bam*HI cassettes cut out of the pcDNA3 vector. Recently, pBlue*Bac*III transfer vector has been replaced with pBlue*Bac*4.5 vector and linearized Bac-N-Blue™ AcMNPV DNA, which are now available from Invitrogen as an easy-to-use transfection kit. pBlue*Bac*4.5 vector is 4.9 kbp and offers more sites in the polylinker, including *Hin*dIII and *Bam*HI. We expect that it will also handle MUC1 cDNA properly.
2. By "standard methodology" we imply methodology described in several commonly used laboratory manuals such as *Current Protocols in Molecular Biology (8)* or similar.
3. Neither pIE1-4 nor pIE1-3 have *Hin*dIII restriction site in the polylinker. It makes it more complicated to subclone a *Hin*dIII cassette, such as 22TR- or 2TRMUC1 cDNAs.
4. We usually grow Sf-9 cells in 5% FBS and we have not noticed any difference in the expression of MUC1 when compared with cells cultured in 10% FBS. In some instances, e.g., in cultures started from thawed cells, we use 10% FBS to support better their recovery from the shock of freezing and thawing.
5. We recommend the use of CsCl purified plasmid DNA in this procedure. Moreover, various ratios between viral and vector DNAs help to obtain a good yield of recombinants in at least one of five transfections. We also recommend the use of more liposomes than the manufacturer suggests in order to increase further the yield of transfection.
6. Although we successfully expressed MUC1 mucin in the pIE-4/pIE-neo system, it was difficult to maintain long-term expression. The best level of expression was approx 30% of MUC1-positive cells in the Neo-resistant cell population. Expansion of such culture resulted in a progressive decrease in the percentage of MUC1/neo-positive cells. The

MUC1/neo-positive population of cells sorted by flow cytometry died within 2 wk without proliferation. Cloning of MUC1/neo-positive cells by limiting dilution in a 96-well plate resulted in progressive death of these cells, whereas neo-positive only cells expanded. We do not have a good explanation for this. We successfully expressed MUC1 of various molecular weights in insect cells using other vectors. We also successfully expressed MUC1 in human normal and malignant cells using vectors such as pcDNA3 or pRC/CMV, and in dendritic cells using MFG retroviral vector, without any signs of toxicity. A similar "toxic phenomenon" was observed by others who attempted expression of NMDA receptor in Sf-9 cells using pIE1-4 vector (Koslovsky, T. and Casio, M., personal communication). This suggests that the problem may be related to the pIE1 vector itself in which recombinant protein expression is under control of *ie1* promoter, whereas in other baculoviral vectors, *polh* promoter is used.

7. Viral stock is a cell-free, filtered (0.22 μm) culture supernatant with titrated viral particles. For details on how to prepare viral stock, *see* **ref. *11*** or a general laboratory manual. We routinely use the end-point silution method, which is easier than plaque assay. There is no need to concentrate or further purify viral stock. We recommend infecting cultures at 3 m.o.i. Infecting cultures at m.o.i. higher than 3 did not result in either a better protein expression or greater yield of infection. For virus amplification, we usually infect cultures at 0.1–0.5 m.o.i.
8. A 1-d lag phase might be observed in cell growth in roller bottle cultures. This is shown in **Fig. 1**.
9. To detect MUC1 without the TR region (TR^-MUC1), 232A1 MAb was used. This antibody is specific for the proteolytic cleavage site on the membrane proximal domain and is highly conformation dependent. It can be used for flow cytometry but not for Western blot. In the latter method, we used a polyclonal rabbit antibody against an epitope within the cytoplasmic domain (a gift from R. Hughey, University of Pittsburgh).

References

1. Litvinov, S. V., and Hilkens, J. (1993) The epithelial sialomucin, episialin, is sialylated during recycling. *J. Biol. Chem.* **268,** 21,364–21,371.
2. Barnd, D. L., Lan, M., Metzgar, R., and Finn, O. J. (1989) Specific, MHC-unrestricted recognition of tumor associated mucins by human cytotoxic T cells. *Proc. Natl. Acad. Sci. USA* **86,** 7159–7163.
3. Jerome, K. R., Bu, D., and Finn, O. J. (1992) Expression of tumor-associated epitopes on Epstein-Barr Virus-immortalized B-cells and Burkitt's Lymphomas transfected with epithelial mucin complimentary DNA. *Cancer Res.* **52,** 5985–5990.
4. Magarian-Blander, J., Domenech, N., and Finn, O. J. (1993) Specific and effective T-cell recognition of cells transfected with truncated human mucin cDNA. In: Specific Immunotherapy of Cancer with Vaccines. *Ann. NY Acad. Sci.* **690,** 231–243.
5. Bu, D., Domenech, N., Lewis, J., Taylor-Papadimitriou, J. and Finn, O. J. (1993) Recombinant vaccinia mucin vector: *in vitro* analysis of expression of tumor-associated epitopes for antibody and human cytotoxic T-cell recognition. *J. Immunother.* **14,** 127–135.
6. Ciborowski, P. and Finn, O. J. (1995) Recombinant epithelial cell mucin (MUC-1) expressed in baculovirus resembles antigenically tumor associated mucin, target for cancer immunotherapy. *Biomedi. Peptides, Proteins, Nucleic Acids* **1,** 193–198.
7. Ciborowski, P., Konitzki, W. M., Blander, J. M., and Finn, O. J. (1998) Screening of anti-MUC1 antibodies for reactivity with native (ascites) and recombinant (baculovirus) MUC1 and for blocking MUC1 specific cytotoxic T-lymphocytes. *Tumor Biol.* **19(Suppl. 1),** 147–151.

8. *Current Protocols in Molecular Biology* (Ausubel, F. M., Brent, R., Kingston, R. E., Moore, D. D., Seidman, J. G., Smith, J. A., and Struhl, K., eds.), Wiley, New York.
9. Hilkens, J. and Bujis, F. (1988) Biosynthesis of MAM-6, an epithelial sialomucin. Evidence for involvement of rare proteolytic cleavage step in the endoplasmic reticulum. *J. Biol. Chem.* **263,** 4215–4222.
10. Magarian-Blander, J., Hughey, R. P., Kinlough, C., Poland, P. A., and Finn, O. J. (1996) Differential expression of MUC1 on transfected cell lines influences its recognition by MUC1 specific T cells. *Glycoconjugate J.* **13,** 749–756.
11. O'Reilly, D. R., Miller, K., and Luckow, V. A. (1994) In: *Baculovirus Expression Vectors. A Laboratory Manual.* Oxford University Press, New York, pp. 124–138.
12. Hu, P. and Wright, S. E. (1998) Recombinant breast carcinoma-associated mucins expressed in a baculovirus system contain a tumor specific epitope. *Immunotechnology* **4,** 97–105.

40

Mucin Antigen Presentation Using Dendritic Cells

Joy Burchell, Rosalind Graham, and Joyce Taylor-Papadimitriou

1. Introduction

The study of humoral and cellular responses to mucins requires many of the standard immunologic techniques, although working with molecules as large as mucins sometimes leads to logistic problems. This chapter focuses on some of the techniques that may be used to analyze the immune response to mucins using dendritic cells to present mucin peptides.

Changes in the glycans, carried both on proteins and lipids, has long been associated with the change to malignancy *(1)*. In the case of proteins, many of these glycans are carried on mucins *(2–4)*. This aberrant glycosylation may make the mucin antigenically distinct to that expressed by normal cells, and it is relatively easy to envisage how this would affect the humoral response to the molecule *(5)*. However, there is an increasing amount of evidence that glycosylated peptides can be presented to T-cells via the major histocompatibility complex (MHC) molecules *(6,7)*. Galli-Stampino et al. have shown that peptides carrying simple *O*-linked core 1 sugars (Galβ1-3GalNAc) can bind to MHC class II molecules and be presented to T-cells, whereas the more complex core 2 sugars cannot. This is of particular relevance with regard to the immunogenicity of mucins because often the aberrant glycosylation of mucins that is observed in carcinomas results in the expression of the simple core 1 structure and Tn and STn *(8,9)*, and the revealing of normally cryptic peptide epitopes.

The mucin that has been most extensively studied with regard to its immunogenicity is MUC1, a membrane-bound epithelial mucin. Much attention has been focused on MUC1 as a potential target for active specific immunotherapy because this mucin is over- expressed and aberrantly glycosylated by many carcinomas but particularly those of the breast and ovary *(10)*. Like all mucins, MUC1 has a large domain of tandemly repeated amino acids allowing potential epitopes to be repeated many times. Humoral responses have been identified in breast cancer patients *(5)* and cytotoxic T-lymphnodes (CTLs) have been isolated from cancer patients that can kill MUC1-expressing target cells in a non-MHC-restricted manner (*see* Chapter 39). However,

From: *Methods in Molecular Biology, Vol. 125: Glycoprotein Methods and Protocols: The Mucins*
Edited by: A. Corfield © Humana Press Inc., Totowa, NJ

classic MHC class I epitopes have also been identified within the tandem repeat (TR) of MUC1 *(11)* and T-helper epitopes may also be present.

Dendritic cells are the most potent antigen-presenting cells (APCs) of the body and are involved in the presentation of antigens to naive T-cells. Human dendritic cells can be isolated from peripheral blood mononuclear cells (PBMCs) by culturing the adherent cells in interleukin-4 (IL-4) and GM-CSF *(12,13)*, whereas mouse dendritic cells can be obtained from bone marrow cells by culturing in the presence of granulocyte-macrophage colon-stimulating factor (GM-CSF) alone. These APCs can be used to present peptides or glycopeptides to T-cells. The use of the mouse system obviously has many advantages, but when studying human mucins the murine response to a foreign antigen is being analyzed. This can be overcome, to a certain extent, by the use of mice transgenic for human mucins, but only mice transgenic for MUC1 and MUC7 are currently available *(14,15)*. However, even when using transgenic mice, the murine response to the immunogen is still being evaluated. The use of transgenic mice crossed with MHC class I A2 molecules *(11)* may make the system more applicable to the human situation.

This chapter describes the isolation of dendritic cells from human and murine sources and their use in presenting mucin peptides or glycopeptides to autologous T-cells.

2. Materials

1. Buffers and cell culture medium:
   a. Blood collection buffer: 1,400 mL of RPMI-1640 HEPES-buffered medium plus 336 mL of 3.3% trisodium citrate in distilled water and 14 mL 5 μ*M* β-mercaptoethanol. Aliquot 25 mL of this into sterile 50-mL conical screw-capped tubes and store at 4°C until use.
   b. Hank's buffered salt solution (HBSS).
   c. AIM V medium (Gibco-BRL, Gaithersburg, MD) containing 2 m*M* glutamine, 50 μ*M* β-mercaptoethanol.
   d. Minimal essential medium (MEM).
   e. IMDM medium with glutamax (Gibco-BRL) containing 50 μ*M* β-mercaptoethanol, 5 μg/mL of transferrin (Sigma, St. Louis, MO), 100 IU/mL of penicillin, and 100 μg/ml of streptomycin.
   f. OPTIMEM reduced serum medium (Gibco-BRL).
2. Fetal calf serum (FCS).
3. Ficoll-Paque (Pharmacia Biotech).
4. Human cytokines: GM-CSF (Sandoz) and IL-4 (Genzyme), both made up in AIM V-medium and stored aliquoted at –20°C.
5. Phytohemagglutinin (PHA) (Sigma), made up at 2 mg/mL and stored in aliquots at –20°C.
6. ^{3}H thymidine stock at 1 mCi/mL (Amersham).
7. Automated cell harvester such as a Micro 96 manufactured by Skatron, but a simpler version will work just as well.
8. X63 cells secreting murine GM-CSF.
9. G418 (Gibco-BRL).
10. Peptides: NeoSystems has been found to be an excellent manufacturer, particularly of large peptides. The production of glycopeptides is a very specialized procedure especially if more than one sugar is required to be added. It is usually necessary to collaborate with a chemist who is familiar with the synthesis of glycopeptides.

11. Tissue culture plastics: 96-well flat-bottomed plates (Nunc, Nagle, UK), 33-mm dishes (Nunc), 15-mL conical sterile tubes (Falcon, Merck), and tissue culture flasks (Falcon).
12. 37°C, 5% CO_2 incubator.

3. Methods

3.1. Preparation and Culturing of Human Dendritic Cells (see Notes 1–4)

Our method of preparing and culturing human dendritic cells is an adaptation of the methods published in **refs. *12*** and ***13***.

1. Collect 20 mL of donor blood into 25 mL of blood collection buffer (*see* **Subheading 2., item 1a**).
2. Layer 22.5 mL of blood solution onto 18 mLof Ficoll (*see* **Subheading 2., item 3**). Spin at 400*g* for 20min.
3. Remove buffy coat taking up as little Ficoll as possible. Make up the volume of the buffy coat to 100 mL with HBSS (*see* **Subheading 2., item 1b**) and spin at 400*g* for 15–20 min. Wash the pelleted cells twice in HBSS.
4. Resuspend the cells in 10 mL AIM V medium (*see* **Subheading 2., item 1c**) and spin at 1200 rpm for 10 min.
5. Resuspend in 10 mL of AIM V medium (*see* **Note 1**), count and plate about 7×10^6 in 3 mL of medium onto 33-mm tissue culture dishes (*see* **Note 2**). Incubate at 37°C for 2 h in 5% CO_2 incubator.
6. Remove medium and nonadherent cells by pipetting up and then relatively gently washing the medium over the dish (*see* **Note 3**). To each dish, add 3 mL AIM V medium containing 800 U/mL of human GM-CSF and 500 U/mL of human IL-4 (*see* **Subheading 2., item 4**).
7. Culture for 7 d at 37°C under 5% CO_2. Cells require feeding (by removal of 1 mL of medium and replace with 1 mL of medium containing fresh IL-4 and GM-CSF) every 2–3 d. On d 7, the cells should be ready for use (*see* **Note 4**).

3.2. Proliferation Assay Using Autologous Human T-Cells (see Notes 5–8)

1. Harvest the dendritic cells on d 7 of culture by vigorously pipetting the medium up and down. Wash once in AIM V medium with ***no*** cytokines. Count (*see* **Note 5**) and incubate 2×10^5 cells in 1 mL of AIM V medium in 15-mL conical bottom tubes with or without 100 μg/mL of test peptides or glycopeptides (*see* **Note 6**) at 37°C for 2 h.
2. Count the nonadherent cells from **Subheading 3.1., step 6** (*see* **Note 7**) and dispense 2×10^5 cells per well of a 96-well flat-bottomed tissue culture dish. Include enough wells of PBLs to have six wells with no dendritic cells and six wells for a nonspecific stimuli like PHA (*see* **Subheading 2., item 5**).
3. Add 100 μL per well of the peptide pulsed dendritic cells from **Subheading 3.2., step 1**. If possible have at least six wells per sample. Add 10 μg of the appropriate peptide or glycopeptide to the wells. To the wells with PBLs alone and PBLs with PHA make the volume up to 200 μL with AIM V medium and where appropriate add 2 μg per well of PHA. Incubate for 6 d at 37°C in a 5% CO_2 incubator.
4. On d 6, add 1 μC/well of ^{3}H thymidine (*see* **Subheading 2., item 6**) and incubate at 37°C for 16–18 h.
5. Harvest the cells and count the ^{3}H thymidine incorporated (*see* **Note 8, Subheading 2., item 7**).

3.3. Isolation of Murine Dendritic Cells (see Notes 9–11)

1. Mouse femur and tibia are removed and placed in MEM (*see* **Subheading 2., item 1d**) plus 2% FCS in a 90-mm tissue culture dish. Using forceps and a scalpel, as much muscle and connective tissue as possible is removed from the bones.
2. Transfer the bones into a fresh dish containing MEM plus 2% FCS and snip off the ends of each bone creating a hollow tube. Wash out the marrow with a fine needle and 5-mL syringe containing MEM plus 2% FCS. Prepare a single cell suspension from the marrow using the syringe and needle or vigorous pipetting.
3. Transfer the marrow into a 20-mL universal. Allow any fragments of bone or muscle to fall to the bottom of the tube under gravity and the transfer the supernatant to a fresh universal.
4. Count cells and resuspend at 3.33×10^5/mL in IMDM (*see* **Subheading 2., item 1e**) plus 5%FCS and mouse GM-CSF (*see* **Subheading 3.4.** and **Note 9**). Cells are either cultured in T25 flasks (10 mL) or T75 flasks (30 mL) at 37°C, 5% CO_2.
5. After 2 d of culture, some clumps of adherent cells should be apparent. Nonadherent cells and media are removed and replaced with fresh media containing GM-CSF (*see* **Note 10**).
6. Replace one third of the media with fresh media on d 5 of culture.
7. Harvested the nonadherent dendritic cells on d 7, resuspend in a smaller volume of fresh media and cultured overnight in a tissue culture dish before use (*see* **Note 11**).

3.4. Production of Murine GM-CSF by X63 Cells (see Note 12)

1. X63 cells (*see* **Subheading 2., item 8**) are cultured in IMDM plus 5%FCS plus 1 mg/mL G418 (*see* **Subheading 2., item 9**).
2. Grow up to about 3×10^7 cells in the above medium.
3. Spin cells and wash twice in IMDM plus 5%FCS to remove the G418.
4. Resuspend the cells in media without G418 at a concentration of 3×10^5 cells/mL in one T75 tissue culture flask. Incubate the flask upright at 37°C 5% CO_2 for 48 h.
5. Pellet the cells and collect the GM-CSF containing supernatant (*see* **Note 12**).

3.5. Immunization of Mice with Peptide Pulse Dendritic Cells (see Notes 6 and 11)

1. Dendritic cells are pulsed with peptides (*see* **Note 6**) overnight d 7 to d 8 of the culture. On d 7 DCs are harvested and resuspended at 2×10^6 cells/mL in OPTIMEM medium (*see* **Subheading 2., item 1f**) with or without 100 μg/mL peptide. 3 mL are then dispensed into 33-mm tissue culture dishes and incubated overnight at 37°C in a 5% CO_2 incubator (*see* **Note 11**).
2. Harvest dendritic cells by vigorous pipetting, wash once in PBS, and suspend in PBS at between $1–5 \times 10^6$ cells/mL.
3. Immunize mice subcutaneously with peptide pulse dendritic cells or control dendritic cells at $1–5 \times 10^5$/mouse in 100 μL of PBS.

3.6. Murine Proliferation Assay Using Splenocytes (see Notes 8, 13, and 14)

1. Seven to 14 d after the mice received the autologous dendritic cells, sacrifice the mice and remove the spleens into IMDM culture medium.
2. Disperse the spleens by passing through a sterile grid and achieve a single cell suspension by vigorous pipetting. Then count the cells, ignore the red blood cells (*see* **Note 13**), and resuspend at 2.5×10^6 cells/mL.

3. Dispense 100 µL of the cells into the wells of a 96-well flat-bottomed tissue culture dish and add 5–50 µg of peptide per well (*see* **Note 14**). For each spleen, as a positive control PHA (*see* **Subheading 2., item 5**) is added to four wells and as a negative control medium alone is added to four wells. Incubate at 37°C in a 5% CO_2 incubator for 5 d.
4. On d 5, add 1 µCi/well of ^{3}H thymidine (*see* **Subheading 2., item 6**) and incubate at 37°C for 16–18 h.
5. Wash and harvest the cells and count the ^{3}H thymidine incorporated (*see* **Note 8**).

4. Notes

1. Human dendritic cells are cultured in serum-free medium, AIM V, as this reduces the background in the proliferation assay where the DCs are used as antigen presenting cells.
2. From 20 mL of blood there is normally enough PBMCs to put up three to five 33-mm dishes.
3. This dislodges the lymphocytes that can be collected and used as effector cells in the proliferation assay.
4. The monocytes differentiate into dendritic cells which by d 7 will form about 20–30% of the cultured cells. In the isolation of dendritic cells from PBMCs no proliferation occurs only differentiation in the presence of IL-4 and GM-CSF. DCs come off the tissue culture dish in clumps and in addition isolated dendritic cells can often be seen floating in the medium. The cells are relatively large (compared to lymphocytes) and processes or dendrites can clearly be seen and the dendritic cells are often described as having "veils." By d 7, the dendritic cells should express high levels of MHC class I and class II, high levels of CD40 and B7 and should be negative when stained for the monocyte marker CD14. The phenotype can be analysed by the use of FACscan. To show the DCs are functionally active they can be used as stimulators of an allogeneic mixed lymphocyte reaction.
5. Accurate counting of the dendritic cells is difficult as they only form 20–30% of the cell population. Where possible count the large cells and the cells that have processes.
6. Peptides corresponding to any part of a mucin molecule can theoretically be used but using peptides to a tandem repeat region will cover a large part of the molecule. In the case of MUC1 the peptides have been confined to the tandem repeat and have consisted of one or more repeats. Up to three TR (60 amino acids) can be synthesised successfully by commercial companies. Feeding large peptides, e.g., 60 mer, to the dendritic cells from the outside makes the assumption that the peptide is taken up by endocytosis and enters the MHC class II pathway.
7. The antigen presenting cell i.e. the dendritic cells, must be autologous to the nonadherent cells used as the responders.
8. The use of an automated cell harvested is essential.
9. The dendritic cells can be cultured in serum-free medium consisting of AIM V medium containing mercaptoethanol. However, in tumour challenge experiments better results have been obtained when the cells are cultured in serum-containing medium although some nonspecific protection may be observed.
10. Over the next week, the clumps of adherent cells release nonadherent dendritic cells into the medium. Unlike the human dendritic cells cultured from PBMCs some proliferation does occur.
11. During the overnight culture some cells will adhere to the tissue culture dish, these should not be used.
12. The supernatant from the X63 cells is used at a dilution of 1:10. Commercial murine GM-CSF is available but is rather expensive.
13. Red blood cells can be lysed before setting up the proliferation assay which makes the lymphocytes easier to count. The protocol is as follows:

a. Lysis buffer: 8.29 g NH_4Cl, 1 g $KHCO_3$, 37.2 mg Na_2EDTA. Add 800 mL H_2O and adjust pH to 7.2–7.4 with 1 *M* HCl. Adjust volume to 1 L with water and filter sterilize through 0.2-µm filter. Store at room temperature.
b. Method: Spin splenocytes to pellet the cells and resuspend the cells from one spleen in 1 mL lysis buffer. Incubate at 4°C for 30 s to 1 min. Spin down cells and wash 2 × 50 mL medium.

14. The peptide used in the proliferation assay can be the same as that used for pulsing the dendritic cells or it can be a smaller peptide. For example when using MUC1 peptides a 60 mer corresponding to three TR are used to pulse the dendritic cells whereas a 24 mer is used in the proliferation assay.

References

1. Hakomori, S.-I. (1989) Aberrant glycosylation in tumors and tumor-associated carbohydrate antigens. *Adv. Cancer Res.* **52,** 257–331.
2. Kim, Y. S., Gum, J., and Brockhausen, I. (1996) Mucin glycoproteins in neoplasia. *Glycocon. J.* **13,** 693–707.
3. Brockhausen, I., Yang, J. M., Burchell, J. M., Whitehouse, C., and Taylor-Papadimitriou, J. (1995) Mechanisms underlying aberrant glycosylation of MUC1 mucin in breast cancer cells. *Eur. J. Biochem.* **233,** 607–617.
4. Corfield, A. P., Myerscough, N., Gough, M., Brockhausen, I., Schauer, R., and Paraskeva, C. (1995) Glycosylation patterns of mucins in colonic disease. *Biochem. Soc. Trans.* **23,** 840–845.
5. von Mensdorff-Pouilly, S., Gourevitch, M. M., Kenemans, P., Verstraeten, A. A., Litvinov, S. V., van Kamp, G. J., Meijer, S., Vermorken, J., and Hilgers, J. (1996) Humoral immune responses to polymorphic epithelial mucin (MUC1) inpatients with benign and malignant breast tumours. *Eur. J. Cancer* **32,** 1325–1331.
6. Galli-Stampino, L., Meinjohanns, E., Frische, K., Meldal, M., Jensen, T., Werdeelin, O., and Mouritsen, S. (1997) T-cell recognition of tumour-associated carbohydrates: The nature of the glycan moiety plays a decisive role in determining glycopeptide Immunogenicity. *Cancer Res.* **57,** 3214–3222.
7. Haurum, J. S., Arsequell, G., Lellouch, A. C., Wong, S. Y. C., Dwek, R., McMichael, A. J., and elliott, T. (1994) Recognition of carbohydrate by major histocompatibility complex class I-restricted, glycopeptide-specific cytotoxic T lymphocytes. *J. Exp. Med.* **180,** 739–744.
8. Campbell, B. J., Finnie, I. A., Hounsell, E. F., and Rhodes, J. (1995) Direct demonstration of increased expression of Thomsen-Friedenreich (TF) antigen in colonic adenocarcinoma and ulcerative colitis mucin and its concealment in normal mucin. *J. Clin. Invest.* **95,** 571–576.
9. Lloyd, K. O., Burchell, J. M., Kudryashov V., Yin, B. W. T., and Taylor-Papadimitriou, J. (1996) Comparison of O-linked carbohydrate chains in MUC 1 mucin from Normal Breast Epithelial Cell Lines and Breast Carcinoma cell lines - Demonstration of a simpler and fewer glycan chains in tumor cells. *J. Biol. Chem.* **271,** 33,325–33,334.
10. Graham, R. A., Burchell, J. M., and Taylor-Papadimitriou, J. (1996) The Polymorphic Epithelial Mucin: Potential as an Immunogen for a Cancer Vaccine. *Cancer Immunol. Immunother.* **42,** 71–80.
11. Apostolopoulos, V., Karanikas, V., Haurum, J., and McKenzie, I. F. C. (1997) Induction of HLA-A2-restricted CTLs to the Mucin 1 human breast cancer antigen. *J. Immuno.* **159,** 5211–5218.
12. Romani, N., Gruner, S., Brang, D., Kampgen, E., Lenz, A., Trockenbacher, B., Konwalinka, G., Fritsch. P. O., Steinman, R. M., and Schuler, G. (1994) Proliferating dendritic cell progenitors in human blood. *J. Exp. Med.* **180,** 83–93.

13. Sallusto, F. and Lanzavecchia, A. (1994) Efficient presentation of soluble antigen by cultured human dendritic cells is mantained by granulocyte/macrophage colony-stimulating factor plus interleukin 4 and downregulated by tumour necrosis factor α. *J. Exp. Med.* **179,** 1109–1118.
14. Peat, N., Gendler, S. J., Lalani, E.-N., Duhig, T., and Taylor-Papadimitriou, J. (1992) Tissue-specific expression of a human polymorphic epithelial mucin (MUC1) in transgenic mice. *Cancer Res.* **52,** 1954–1960.
15. Bobek, L. A. Li, H., Rojstaczer, N., Jones, C., Gross, K. W., and Levine, M. J. (1998) Tissue-specific expression of human salivary mucin gene, MUC7 in transgenic mice. *Transgenic Res.* **7,** 195–204.

41

Detection of Humoral Immune Responses to Mucins

Silvia von Mensdorff-Pouilly, Claus Vennegoor, and Joseph Hilgers

1. Introduction

MUC1 is displaced and altered in malignancies originating in epithelial tissues. As a consequence, a molecule with short carbohydrate side chains and exposed repetitive epitopes on its peptide core is shed into the interstitial space surrounding a tumor and gains access to the circulation *(1–3)*, coming in contact with the immune system. Humoral immune responses to MUC1 have been described in healthy subjects and in patients with benign and malignant diseases *(4–6)*.

Breast and ovarian carcinoma patients have antibodies reactive with the MUC1 molecule, bound into circulating immune complexes *(7)*. Preliminary results suggest that the presence of such immune complexes in breast cancer patients at the time of diagnosis is related to a favorable disease outcome *(8)*.

Unbound MUC1 IgG and IgM antibodies are present in the circulation of both healthy subjects and breast cancer patients *(9)*. Screening of a population of healthy women with the bovine serum albumin (BSA)-conjugated peptide assay (described under **Subheading 2.1.**) gave median IgG and IgM MUC1 antibody levels of optical density units (OD) 0.580 (range 0.304–1.539) and OD 0.823 (range 0.227–1.589), respectively. In pretreatment serum samples from breast cancer patients, the results obtained were OD 0.651 (range 0.328–1.537) for IgG and OD 0.779 (range 0.178–1.655) for IgM MUC1 antibodies. A benefit in survival was found in early stage breast cancer patients with circulating MUC1 antibodies at first diagnosis (von Mensdorff-Pouilly, et al., *J. Clin. Oncol.*, in press), suggesting that active specific immunotherapy of breast cancer patients with MUC1-derived (glyco)peptides after primary surgery, in an adjuvant setting, might favorably influence the outcome of disease.

The enzyme-linked immunosorbent assays (ELISAs) described here are useful for pretreatment screening of carcinoma patients for the presence of humoral immune responses to MUC1 as well as for monitoring MUC1 antibody levels during MUC1 vaccine therapy. Other uses and applications include screening for suitable donors of

From: *Methods in Molecular Biology, Vol. 125: Glycoprotein Methods and Protocols: The Mucins*
Edited by: A. Corfield © Humana Press Inc., Totowa, NJ

MUC1-primed B-lymphocytes to obtain B-cell clones producing MUC1 antibodies ***(10)***, creating human phage (anti)body libraries that may hold antibodies that recognize MUC1 ***(11)***, monitoring antibody production in culture supernatant of MUC1-primed B-lymphocytes ***(10)***, and epitope mapping of MUC1 or other antibodies ***(12)***.

2. Materials

2.1. BSA-Conjugated Peptide Assay (PEM.CIg)

1. Serum or plasma samples, fresh or frozen. Repeated freezing and thawing of the samples (more than four times) should be avoided because it may alter results.
2. ELISA microtiter 96-well plates (Costar, Cambridge MA).
3. A synthetic 60mer peptide corresponding to three tandem repeats (TRs) of the MUC1 peptide core, i.e., NH_2-(HGVTSAPDTRPAPGSTAPPA)$_3$-COOH. The 60 mer peptide is conjugated to BSA (Sigma, St. Louis, MO, RIA grade, cat. no. A-7888) using a 1-ethyl-3-(3-dimethylaminopropyl) carbodiimide hydrochloride conjugation kit (Imject™ Immunogen EDC Conjugation Kit for KLH and BSA, Pierce, Rockford, IL). After conjugation, determine the total protein content of the solution and prepare the BSA-conjugated peptide solution for coating the plates on the basis of the measured value. Store stock solution at –20°C.
4. 10 m*M* Sodium phosphate/0.15 *M* sodium chloride buffer, pH 7.0. To prepare 2 L of 5X concentrated phosphate/salt buffer, dissolve 9.56 g of Na_2HPO_4, 4.5 g of $Na_2H_2PO_4 \cdot H_2O$ and 87.66 g of NaCl in demineralized water. Add demineralized water to 2 L. Experience shows that the pH of the concentrated buffer is 6.6–6.7. No pH adjustments are needed: after 5X dilution with demineralized water the pH of the phosphate/salt buffer should measure 7.0.
5. Phosphate/salt buffer containing 0.5% BSA, to dilute the conjugate.
6. Blocking buffer (also for serum/plasma dilutions): phosphate/salt buffer containing 1% BSA and 0.1% sodium azide.
7. Washing buffer-phosphate/salt buffer with 0.05% Tween-20. Approximately 30 mL of washing buffer are necessary per washing turn (480 mL/plate).
8. Horseradish peroxidase-labeled rabbit antihuman IgG and IgM, both from Dako A/S, Glostrup, Denmark. Dilute the conjugate 1:10,000 in phosphate/salt buffer containing 0.5% BSA.
9. Substrate solution containing 0.06 mg/mL of tetramethylbenzidine (TMB) diluted in 0.1 *M* sodium acetate/citric acid buffer solution, pH 4.0, and 0.03% hydrogen peroxide. Substrate solution should be prepared immediately before use as required.
   a. Stock solution TMB: 6 mg 3,3',5,5'-tetramethylbenzidine (Sigma, cat. no. T-2885) in 1 mL of dimethyl sulfoxide. Store in the dark at room temperature.
   b. 0.1 *M* sodium-acetate/citric acid buffer, pH 4.0. This pH is reached with a solution containing 0.0627 *M* sodium acetate and 0.0374 *M* citric acid.
   c. 30% Hydrogen peroxide.
10. 1.6 *N* sulfuric acid.

2.2. Biotinylated Peptide Assay

1. Streptavidin (Sigma-Aldrich, Zwijndrecht, Netherlands). Store stock solution (4 mg/mL) in 50-µL portions at –80°C. Streptavidin solution for coating the plates: 4 µg/mL of streptavidin in 10 m*M* sodium phosphate/0.15 *M* sodium chloride, pH 7.0, containing 0.02% sodium azide.
2. A 60mer synthetic peptide corresponding to three TR of the MUC1 peptide core biotinylated at the amino terminal end. Three alanines should be incorporated at the amino

terminal end to account for the streptavidin pocket. The final sequence is biotin-AAA-(HGVTSAPDTRPAPGSTAPPA)$_3$-COOH. Stock solution peptide: 1mg/ml in phosphate-buffered saline (PBS). Keep at –70°C.

3. All other materials as in **Subheading 2.1.**

3. Methods

3.1. BSA-Conjugated Peptide Assay (PEM.CIg)

1. Apply 250 ng of BSA-conjugated peptide (*see* **Note 1**) diluted in 75 µL of phosphate/salt buffer per well and 75 µL of 1% BSA solution per well to plates in duplicate alternate rows. Seal the wells with plastic plate sealers and incubate overnight at room temperature. Plates can be stored, well sealed, for several months at –70°C.
2. Rinse the plates twice with demineralized water, shake the plates empty, and fill the wells with demineralized water. Repeat once.
3. Fill the wells with 300 µL of blocking buffer, seal the plates, and incubate overnight at 4°C (or 3 h at 37°C).
4. Wash twice with washing buffer, shake the plates empty, and fill the wells with 300 µL of washing buffer per well. After the last wash leave the washing buffer in the wells until immediately before applying serum dilutions, because the wells should be left dry for only the shortest time possible.
5. Shake the plates empty, one by one, immediately before applying serum. Apply 75 µL of previously diluted human serum per well (*see* **Note 2**). Dilute human serum samples 1:100 (for IgG determinations) or 1:500 (for IgM determinations) in blocking buffer. Apply each serum sample in duplicate, i.e., fill two wells coated with peptide and two wells coated with BSA (*see* **Notes 3–6**). In each plate, fill one duplicate set of wells with blocking buffer only. Seal the plates and incubate overnight at 4°C.
6. Wash seven times with washing buffer: after the last wash leave the washing buffer in wells until ready to fill with conjugate. Shake the plates empty, one by one, and dry upper and underside of plates with clean tissue carefully before applying conjugate.
7. Drop into each well 75 µL of horseradish peroxidase-conjugated antihuman IgG (or antihuman IgM) diluted 1:10,000 in 0.5% BSA solution. Avoid touching the bottoms and sides of wells. Seal the plates. Incubate for 60 min at room temperature. Keep a small amount of conjugate to test the substrate solution before you apply it. At equal amounts, the mixture should immediately become an intense blue colour.
8. Wash seven times with washing buffer and rinse the plates once with demineralized water, proceeding as in **step 6**.
9. Drop 75 µL of substrate solution into each well, avoid touching the bottoms and sides of wells. Incubate for 60 min in the dark at room temperature. Positive wells will color blue. BSA-coated wells should color pale blue, very faintly; to see the color place the plate against a white background.
10. Add into each well 100 µL of 1.6 *N* sulfuric acid to stop the reaction. Blue will turn to yellow.
11. Measure the reaction at 450 nm in a microwell plate reader.

3.2. Biotinylated Peptide Assay

1. Apply 75 µL of streptavidin solution to each well, seal the plates, and incubate overnight at 4°C. Plates can be stored, well sealed, for several months at -70 C.
2. Wash the wells three times with washing buffer, shake the plates empty, and fill the wells with 300 µL of washing buffer per well. After the last wash leave the buffer in the wells.
3. Shake the wells until empty and fill with blocking buffer, 300 µL/well. Seal the plates and incubate at room temperature for 60 min (or overnight at 4°C).

Table 1
Examples of IgG and IgM Readings of Individual Wells

IgG Determinations					
Sample	60mer	60mer	BSA	BSA	OD
Blocking buffer	0.150	0.150	0.044	0.044	0.106
Standard 1:250	1.091	1.111	0.099	0.095	1.004
Standard 1:500	0.801	0.808	0.077	0.080	0.726
Standard 1:1000	0.539	0.534	0.065	0.070	0.469
Standard 1:2000	0.384	0.385	0.056	0.057	0.328
Low sample	0.322	0.333	0.062	0.059	0.267
Middle sample	0.913	0.949	0.060	0.087	0.858
High sample	1.735	1.783	0.098	0.110	1.655
IgM Determinations					
Sample	60mer	60mer	BSA	BSA	OD
Blocking buffer	0.216	0.215	0.046	0.047	0.169
Standard 1:250	1.417	1.422	0.073	0.073	1.347
Standard 1:500	1.244	1.244	0.064	0.062	1.181
Standard 1:1000	1.024	1.051	0.055	0.056	0.982
Standard 1:2000	0.785	0.811	0.053	0.053	0.745
Low sample	0.238	0.237	0.073	0.068	0.167
Middle sample	0.917	0.926	0.073	0.078	0.846
High sample	1.523	1.524	0.079	0.085	1.442

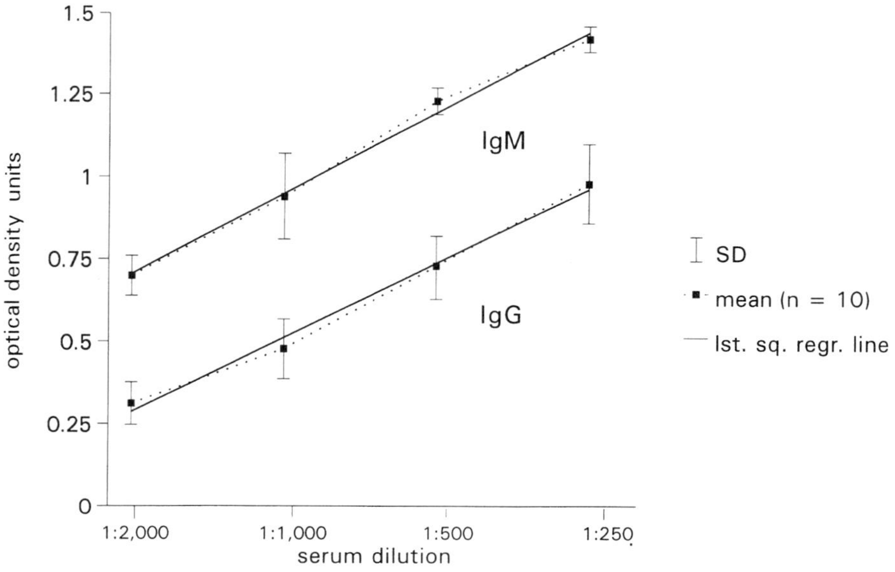

Figure 1.

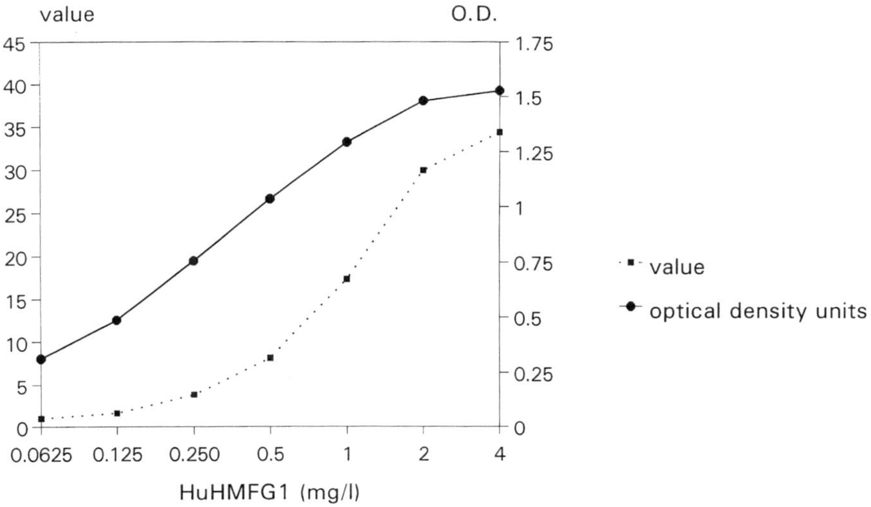

Figure 2.

4. Wash the wells twice with washing buffer. After the last wash, leave the buffer in wells.
5. Aspirate the buffer from the wells and fill with 75 µL of biotinylated peptide solution, diluted to a concentration of 1 µg/mL peptide in 1% BSA solution. Seal the plates and incubate for 60 min at room temperature.
6. Wash the wells three times as described in **step 2**. Proceed further as in **Subheading 2.2., steps 5–11**.

4. Notes

1. A comparison of the two assay methods showed the BSA-conjugated peptide assay to be more sensitive for the detection of human MUC1 antibodies in serum and plasma samples than the one with biotin-conjugated peptide. The reason could be the more random presentation of the peptide when conjugated to BSA, in comparison to its more orderly presentation, which could lead to steric hinderance, in the streptavidin-biotin system. The biotinylated peptide assay has been used with excellent results to detect and monitor the production of MUC1 antibodies by MUC1-primed B-lymphocytes in culture.
2. In our experience, prediluted serum and plasma samples (1:20 in phosphate/salt buffer/0.1% BSA/0.1% sodium azide) give reproducible results even after 2 to 3 mo of storage at 4°C.
3. Individual results are calculated as the mean difference between the readings in optical density units in the two peptide-coated and the two BSA-coated wells. **Table 1** illustrates some results.
4. To standardize the assay, construct a four-point standard curve for each individual plate by testing a positive serum sample in four consecutive dilutions. Ascribe an arbitrary value of 1 to the highest serum dilution, and calculate the value of the samples tested within each plate in relation to the standard curve by least-square regression analysis (**Fig. 1**). Measurements that fall outside the standard curve should be retested at higher dilutions.
5. The choice of standards should be a matter of careful consideration. Humanized monoclonal antibodies provide suitable standards for IgG determinations (**Fig. 2**) whereas

screening of a number of serum samples will provide suitable candidates for an IgM standard. Perform a dilution curve and choose four-point dilutions from the linear portion of the curve (OD 0.3–OD 1.3). Choose a sample from which a suitable amount is available (1 mL of serum will be enough for ~400 plates).

6. The intra- and interassay coefficients of variation of the PEM.CIg assay are, respectively, 2 and 12% for the IgG determinations and 1.2 and 3% for the IgM determinations.

References

1. Hilkens, J., Kroezen, V., Bonfrer, J. M. G., De Jong-Bakker, M., and Bruning, P. F. (1986) MAM-6, a new serum marker for breast cancer monitoring. *Cancer Res.* **46,** 2582–2587.
2. Sikut, R., Zhang, K., Baeckström, D., and Hansson, G. C. (1996) Distinct sub-populations of carcinoma-associated MUC1 mucins as detected by the monoclonal antibody 9H8 and antibodies against the sialyl-Lewis a and sialyl-Lewis x epitopes in the circulation of breast cancer patients. *Int. J. Cancer* **66,** 617–623.
3. Bon, G. G., Kenemans, P., Van Kamp, G. J., Yedema, C. A., and Hilgers, J. (1990) Review on the clinical value of polymorphic epithelial mucin tumor markers for the management of carcinoma patients. *J. Nucl. Med. Allied. Sci.* **34,** 151–62.
4. Hinoda, Y., Nakagawa, N., Nakamura, H., Makiguchi, Y., Itoh, F., Adachi, M., Yabana, T., Imai, K., and Yachi, A. (1993) Detection of a circulating antibody against a peptide epitope on a mucin core protein, MUC1, in ulcerative colitis. *Immunol. Lett.* **35,** 163–168.
5. Kotera, Y., Fontenot, D. J., Pecher, G., Metzgar, R. S., and Finn, O. J. (1994) Humoral immunity against a tandem repeat epitope of human mucin MUC-1 in sera from breast, pancreatic, and colon cancer patients. *Cancer Res.* **54,** 2856–2860.
6. Rughetti, A., Turchi, V., Ghetti, C. A., Scambia, G., Panici, P. B., Roncucci, G., Mancuso, S., Frati, L., and Nuti, M. (1993) Human B-cell immune response to the polymorphic epithelial mucin. *Cancer Res.* **53,** 2457–2459.
7. Gourevitch, M. M., von Mensdorff-Pouilly, S., Litvinov, S. V., Kenemans, P., Van Kamp, G. J., Verstraeten, A. A., and Hilgers, J. (1995) Polymorphic epithelial mucin (MUC-1)-containing circulating immune complexes in carcinoma patients. *Br. J. Cancer* **72,** 934–938.
8. von Mensdorff-Pouilly, . S, Gourevitch, M. M., Kenemans, P., Verstraeten, A. A., Litvinov, S. V., Van Kamp, G. J., Meijer, S., Vermorken, J., and Hilgers, J. (1996) Humoral immune response to polymorphic epithelial mucin (MUC-1) in patients with benign and malignant breast tumours. *Eur. J. Cancer* **32A,** 1325–1331.
9. von Mensdorff-Pouilly, S., Gourevitch, M. M., Kenemans, P., Verstraeten, A. A., Van Kamp, G. J., Kok, A., Van Uffelen. K., Snijdewint, F. G. M., Paul, M. A., Meijer, S. and Hilgers, J. (1998) An enzyme-linked immunosorbent assay for the measurement of circulating antibodies to polymorphic epithelial mucin (MUC1). *Tumor Biol.* **19,** 186–195.
10. Petrarca, C., Casalino, B., von Mensdorff-Pouilly, S., Rughetti, A., Rahimi, H., Scambia, G., Hilgers, J., Frati, L., and Nuti, M. Isolation of MUC1-primed B lymphocytes from tumor draining lymph nodes using immunomagnetic beads. *Cancer Immunol. Immunother.* **47,** 272–277.
11. Tureci, O., Sahin, U., and Pfreundschuh, M. (1997) Serological analysis of human tumor antigens: molecular definition and implications. *Mol. Med. Today* **3,** 342–349.
12. Vennegoor, C. J. G. M., Nijman, H. W., Drijfhout, J. W., Vernie, L., Verstraeten, A. A., von Mensdorff-Pouilly, S., Hilgers, J., Verheijen, R. H. M., Kast, W. M., Melief, C. J. M., and Kenemans, P. (1997) Autoantibodies to p53 in ovarian cancer patients and healthy women: a comparison between whole p53 protein and 18-mer peptides for screening purposes. *Cancer Lett.* **116,** 93–101.

Index

A

ABC immunoperoxidase technique, 37
α-N-Acetyl-D-galactosaminidase, 409
β-N-Acetyl-D-glucosaminidase, 408
N-Acetyl-D-hexosaminidase, 387, 408
α1-Acid glycoprotein, 404–406
Acylneuraminate pyruvate lyase, 412, 416
Adherent mucus gel barrier, 57–64
thickness, 57, 59, 60
Alkaline borohydride degradation, 182, *see also* β-elimination
Amino acid analysis, 79, 83, 113
[^{35}S]-Amino acids, 66, 69, 228, 233, 234
2-Aminobenzamide, 182, 184, 185, 187
Antibodies, 32, 37, 49, 71, 129, 241, 244, 245, 264, 353, 473
- affinity purification, 132, 137
- antigen source354
- replacement net (RNET) analysis, 132, 135, 139
- monoclonal, 353, 355
  - ascites culture, 361
  - clonal stability, 364
  - feeder layer cells, 358
- filter plaque technique, 361
- fusion, 359
- hybridoma culture, 360, 376
- liquid culture technique, 358
- myeloma cells, 359
- spleen cells, 359
- subcloning, 360
- VNTR peptide, 369, 375
- polyclonal, 353, 355
  - rabbits, 357
- screening, 355, 363, 377
- specificity, 365, 366

Antibody capture immunoassay, 48, 52
Antifreeze glycoprotein, 404, 405
Antigen retrieval, 37, 41
Ascites fluid, 21
ASGP1, ASGP2, 15, 17, 21, 22
Asialoglycoproteins, 404, 405

B

Bacterial adherence to microplates, 430, 432
Bacterial growth, 439
Bacterial invasion of corneal epithelial cells, 431, 434
Baculovirus, recombinant, 471
- episomal recombinant vector, 478
- infection of Sf-9 cells, 478
- isolation, 477

Benzyl 2-acetamido 2-deoxy-β-D-galactopyranoside, 46, 150, 261, 264
β-elimination, 182, 184, 191
Biofilms, 439
- short term fermentation, 447

Biosensor, testing of antibody affinity, 378
Blotting, 49, 52, 82, 146, 149, 231, 264
- Northern blot analysis, 305
- Southern blot analysis, 310, 316, 341
- Western blot analysis, 146, 149, 231, 264, 472
- vacuum blot analysis, 146, 149, 231, 264

Boar seminal gel glycoprotein, 404, 405
Bovine salivary gland mucin, 404–406
Brefeldin A, 252, 256, 2557
BSA-peptide conjugates, 134

C

Carbonyl cyanide M-chlorophenylhydrazone (CCCP), 251, 252, 256, 257

Cells, as antigens for antibody production, 354
Cell lines, 222, 242, 262, 298, 457, 472
  A431, 222, 234, 242
  adenoma lines, 222, 234
  apoptotic cells, 222
  Caco-2, 222, 234, 242
  dendritic, 489
  HT29 MTX, 262, 298
  LS174T, 222, 234, 242, 262, 298
  Sf-9, 475
  thymocyte, 223
Cesium chloride, 7, 8, 10, 103
Collagen coated culture flasks, 221, 223
Cone and plate mechanical spectroscopy, 102, 104
Continuous culture system, fecal bacteria, 449
Core 1 Disaccharide (Galb1-3GalNAcα –), 183–185, 262
Creep testing, 101

D

Deglycosylation, 38,354
Dendritic cells, 487
  human, 489, 491
  immunization of mice, 490
  murine, 490, 491
  proliferation assay, 489, 490, 492
Desialylation of glycopeptides, 123
Detection and quantitation of mucins, 29–44, 45–55, 57–64, 65–73
Disulfide bonds, 143, 253
Dolichos biflorus agglutinin, 251, 252
Domain constructs using PCR, 144, 147
  ligation, 145, 148
  transfection, 146, 148

E

Edman degradation, 121, 125
EGF domains, 17, 19
Electrophoresis, 68
  agarose gel, 79, 143, 231, 235, 242, 246, 306, 308, 313, 314, 316, 341,
  SDS-PAGE, 68, 71, 143, 146, 149, 235, 242, 249, 251, 254, 264, 398
Endoglycosidase H, 251, 253

F

Fetuin, 404
9-Fluorenylmethyl chloride, 113
Fmoc derivatized amino acids, 114–116
α-Fucosidase, 387, 409

G

Galactose oxidase, 405, 413
α-Galactosidase, 387, 408
β-Galactosidase, 387, 407, 408, 446
D-[^{3}H]-Galactose, 228, 233, 234, 240, 243, 267
D-[^{3}H]-Glucosamine, 228, 233, 234, 387
Gel chromatography, 5, 8, 231
Superose 12, 78
Gentamycin survival assay, 431, 434
Glucose-6-sulfate, 419, 421
Glycopeptides, 489
Glycosidase activity, 403, 446
  endoglycosidase H, 251, 253
  O-glycanase, 38, 184, 387
  N-glycosidase F, 150, 153, 251, 253, 403
  radioactive substrates, 405–409, 413
  synthetic substrates, 403–405, 413
Glycosylation, 147, 149, 261
  inhibition, 261
Glycosylation sites, 121,
Glycosyltransferases, 273
  assay, 282
    α6-sialyltransferase, 288
    α3-sialyltransferase, 288
    β4-Gal-transferase, 288
    C18 column, 282
    core 1 β3-Gal-transferase, 288
    core 2 β6-GlcNAc-transferase, 287
    core 3 β6-GlcNAc-transferase, 287
    core 4 β6-GlcNAc-transferase, 287
    Dionex, 283
    elongation β3-GlcNAc-transferase, 287
    high voltage electrophoresis, 283
    i β3-GlcNAc-transferase, 287
    I β6-GlcNAc-transferase, 287

Glycosyltransferases (*cont.*)
assay (*cont.*)
ion-exchange, 282
polypeptide GalNAc-transferase, 286
oligosaccharide substrates, 278, 279
Guanidine hydrochloride, 4, 9, 80

H

Histology, 29–44
Hydrazinolysis, 181, 183, 185
Humoral immune response to mucins, 495

I

Immunization, 358
cytotoxic T cells, 458
monoclonal antibody production, 358, 375, 379
polyclonal antibody production, 358
Immunoaffinity purification, 21, 22
Immunoassay, 47, 51, 130, 264, 376, 495
Immunoprecipitation, 239, 241, 251
In situ hybridization, 323
detection, 332, 333
digoxygenin labeling of oligonucleotides, 327, 330, 331, 333
fixation of tissue, 325,
fluorescent labeling, 329, 331, 333, 334
hybridization, 332
oligonucleotide probes, 324, 327
prehybridization, 330
posthybridization, 332, 334
ribonucleotide probes, 324, 333
[^{35}S]-labeled oligonucleotides, 327, 330, 331, 333
Ion-exchange high performance chromatography, 5
Mono Q, 5, 9, 78–80, 84
Isopycnic density gradient centrifugation, 5, 7, 231,

L

[1-^{3}H]-lactitol-6′-sulfate, 420, 421
Lectins, 31, 36, 49, 234, 264, 270
digoxygenin (DIG) labeled, 265, 270,

M

Magnetic rheogoniometry, 101, 103, 105
Mass spectrometry (MS), 191, 203–20
FAB-MS, 203
MALDI-MS, 204
ESI-MS, 205
Metabolic labelling, 65, 69–72, 227, 230, 239, 264
continuous labeling, 230,
pulse/chase labeling, 230, 243, 251
L-[^{35}S]-Methionine, 228, 233, 240
4-Methyl umbelliferyl-glycosides, 403
4-methyl umbelliferyl α-sialic acid, 405
Milk fat globule membrane, 16, 20, 21, 374
Monosaccharides, 159
composition, 159
electrophoresis, 161, 163
fluorophores assisted carbohydrate electrophoresis (FACE) 175
gas liquid chromatogrphy (GLC), 161, 162
methylglycosides, 167
alditol acetates, 161, 169
hexosamines, 161, 170–172
N-acetyl-D-galactosamine, 161, 191
N-acetyl-D-glucosamine, 161, 191
HPLC, 161, 162
amino bonded silica, 162, 169
Dionex, 447
HPAEC-PAD, 163, 170, 186
reversed phase, 162, 169
N-acetyl-D-neuraminic acid (sialic acid), 30, 159, 162, 164–166, 191
neutral, 161, 170
Mouse mammary carcinoma (TA3), 15, 20
MUC1 CA microsatellite PCR, 342
MUC1 cytotoxic T-lymphocytes, 455
assay, 458, 465
epitope mapping, 460
immunization, 458
peripheral blood levels, assay, 463, 465
precursor assay, 459
target cells, 458, 468
MUC1 expression in insect cells, 471, 478
analysis of MUC1, 479
large scale production, 481
underglycosylated MUC1, 479

MUC1 fusion protein, soluble, 457
conjugation to peptide and mannan, 458
MUC1 glutathione fusion protein, 374
MUC1 immunoassay, 495–500
BSA-peptide conjugate assay, 497
biotinylated-peptide assay, 497
MUC1 vector construction, 473
[^{35}S]-Mucin, 419, 422
Mucin
adherence to microtitre wells, 430, 433
antibodies, 32, 37, 49, 71, 129, 241, 244, 245, 264, 353–367, 369
antigen presentation in dendritic cells, 487
bacterial binding assays, 429
biosynthesis, 65, 70, 219, 227, 249, 273
concentration (enrichment), 232
degrading bacteria, 439
in vitro modeling system, 446
degrading enzymes, 385, 396, 403, 417, 445
desulfating bacteria, 418, 423
detection, 45, 50, 66, 249, 264
dimers, 145, 152
cDNA, 297
cloning, 297, 300
5′/3′ RACE PCR, 301
preparation, 315
pulsed field gel electrophoresis, contour-clamped homogenous electric field (CHEF), 313, 314, 316
radioactive probing of blots, 317
restriction enzymes, 315, 341–6
Southern blot analysis, 310, 316, 341
synthesis, 299
dot, slot, and Western blotting, 49, 52, 82
domains, 3, 143
elasticity, 106
extraction, 4, 6, 20–23, 80, 103, 397
gel forming, 3, 57, 99, 103
gene
MUC1, 3, 15, 16, 20, 21, 129, 331, 337, 343, 344, 369, 455, 463 471, 487, 495,
MUC2, 3, 65, 77, 143, 249, 254, 255, 313, 318, 319, 331, 337, 343, 344, 369
Mucin (*cont.*)
gene (*cont.*)
MUC3, 3, 15, 19, 254, 255, 331, 337, 343, 369
MUC4, 3, 254, 255, 298, 331, 337, 343, 344, 345, 369
MUC5AC, 3, 77, 81, 143, 254, 255, 298, 313, 318, 319, 331, 337, 343, 345
MUC5B, 3, 77, 81, 143, 254, 255, 298, 313, 318, 319, 331, 337, 346
MUC6, 3, 77, 143, 254, 255, 313, 318, 319, 331, 337
MUC7, 3, 331, 337, 346
MUC8, 3, 337–339
polymorphism, 337
glycoforms, 77, 80
histology, 29–44
Alcian Blue, 30, 35, 58, 60, 62, 267
cryostat sections, 39, 60, 61–63
DAB, 37, 40
fixation, 29, 39, 59, 60
high iron diamine (HID), 31, 35
immunohistochemistry, 32, 37, 377
interpretation and reporting, 32, 33
lectin, 31, 36
periodic acid Schiff, 30, 35, 58, 62
humoral immune response, 495
ion-exchange chromatography (Mono Q), 5, 9, 78–80, 84,
immunoassay, 47, 51, 130, 264, 376, 495
ELISA, 47, 130, 134, 136, 264, 376, 495, 497
radioassay, 47
intracellular processing, 249
mature mucin, 69, 241
membrane, 15
N-terminal peptide sequencing, 83
precursor, 68, 69, 72, 241, 245, 249, 254
high mannose glycan containing, 253, 257
proteolytic fragments, 5, 80, 393
quantitation, 45, 51, 58, 65, 68, 70
reduction, 5, 79, 80, 152
removal of proteoglycans, 232,
rheology, 99

Mucin (*cont.*)
mRNA, 68, 69, 299
electrophoresis, 306, 308
estimation of sizes, 309
filter hybridization, 307, 308
in situ hybridization, 323, 324
Northern blot analysis, 305
preparation, 306, 307
transfer and crosslinking, 306, 308
trypsin digestion, 79, 83, 393, 399
separation, 5–10, 77, 231
subunits, 3, 5, 79,
sulfate determination, 211
variable number tandem repeats (VNTR), 337, 338, 369, 456
viscosity, 99, 104, 106, 398
Mucinase, 385
assay radioactive substrates, 389
assay biotinylated substrates, 389
biotinylated substrate, peptide labeled, 388
biotinylated substrate, carbohydrate labeled, 388
radioactive substrate, 387
Mucinolytic bacteria, 445
desorption from food materials in feces, 448
ennumeration and identification, 445
rectal biopsies, 445, 448
Mucus, 59
thickness, 57, 59, 60

N

N-acetyl-neuraminlactose (sialyllactose), 446
Neuraminidase, 38, 150, 446
p-Nitrophenyl N-acetyl-β-D-glucosaminide-6-sulfate, 420, 421
p-Nitrophenyl β-D-galactoside-3-sulfate, 421, 423
p-Nitrophenyl β-D-galactoside-6-sulfate, 420, 423
4-Nitrophenyl-glycosides, 403–405,
4-nitrophenyl α-fucose, 405, 444
4-nitrophenyl Galb1-3GalNAc, 405
4-nitrophenyl α-galactose, 404
4-nitrophenyl β-galactose, 404, 444
4-nitrophenyl β-N-acetylglucosamine, 405, 444
4-nitrophenyl α-N-acetylgalactosamine, 405
Nuclear magnetic resonance (NMR) spectroscopy, 191, 193–202

O

O-Acetyl sialic acid, 30, 413
Ocular sialoglycoprotein, human, 430, 431
O-Glycanase, 38, 184, 387, 410, 415
Oligosaccharides, N-linked, 181, 253, 257
Oligosaccharides, O-linked, 181
core structures, 274
glycosyltransferases, 273
HPLC (GalNAc-O-benzyl derivatives), 268
HPLC glycosyltransferase substrates, 279, 283
poly N-acetyllactosamine, 274
release, 181
structural analysis, 191
Organ culture, 219, 223
submerged cultures, 223, 225

P

Peptide analysis, 79, 121
glycopeptide adsorption onto Hyperbond or PVDF, 125
protein coupling to arylamine membranes, 123
Peptide conjugation to BSA, for MUC1 immunoassay, 496
Periodic acid Schiff, 30, 35, 49, 234
Phenylisothiocyanate, 121
Phenylthiohydantoin-(PTH) amino acids, 121
Phosphatase, 281
3′-Phosphoadenosine 5′-phosphosulfate (PAPS), 275
PhosphorImager, 68, 71, 246
Porcine seminal gel glycoprotein, 404, 405, *see also* Boar seminal gel glycoprotein
Precorneal tear film components, human, 429, 431
Preimmune serum, 357
Protease inhibitors, 4, 10, 80, 103, 231, 233, 240, 251
Proteinase, 393-400
elastase, 396, 399,
gastrointestinal tract, 395
inhibition, 399

Proteinase (*cont.*)
mucin proteolysis, 396
N-terminal assay, 397
pepsin, 396, 399
substrate preparation, 397
trypsin, 396, 399
viscosity assay, 398
Pseudomonas aeruginosa, 429, 430, 433,435
Purification of mucin, 3–13, 15–26
N-2-Pyridyl glycamines, 173, 174
Pyrophosphatase, 281

R

Recombinant proteins, as antigens for antibody production, 355
Rough endoplasmic reticulum proteins, 252
Rheology, 99

S

Salivary gland glycoproteins, 404
Saponification of O-acetyl esters, 404
Scanning electron microscopy, 441, 450
Sialate O-acetyl esterase, 412, 415
Sialic acids, 30, 159, 162, 164–166, 407, 446
α-Sialidase, 387, 407, 411, 415
colorimetric assay, 411
fluorimetric assay, 412
radioactive assay, 412
Sialomucin complex (SMC), 15, 17, 22
Sialyl -GalNAc[^{3}H]-ol (Neu5Acα2-6GalNAc-[^{3}H]-ol, 405
Sialyl-[^{3}H]-labelled sialoglycoproteins, 406
Sialyllactose, 404, 415, 446
Sulfatases, mucin desulphating, 417
[^{35}S]-Sulfate, 66, 69, 228, 233, 234, 240, 243, 387
Sulfate determination, 211
Sulfated sugars, 418
Sulfomucin, 417
Sulfotransferase, 275
gal 3-sulfotransferase, 288
Swiss 3T3 fibroblast feeder cells, 221
Synthetic peptides, 129, 355, 373, 457
solid phase synthesis, 134, 370
multipin peptide synthesis, 135, 374
biotinylation for immunoassay of MUC1, 496

T

Tethered peptide libraries, 130
L-[^{3}H]-Threonine, 228, 233, 234, 240, 387
Tissue, as antigen for antibody production, 354
Tissue culture, 222
Trifluoromethanesulfonic acid (TFMSA), 182, 184, 185
Tritiated borohydride, 405, 412
Tunicamycin, 150, 152, 251, 253, 257

U

Ulex europaeus agglutinin I (UEA1), 36

V

Virus, recombinant, production, 475
Von Willebrand factor, 143

W

Wheat germ aglutinin, 234